The Transformed Phenotype

CANCER CELLS

COLD SPRING HARBOR LABORATORY
1984

1
The Transformed Phenotype

Edited by
Arnold J. Levine
State University of New York at Stony Brook

George F. Vande Woude
National Cancer Institute, National Institutes of Health

William C. Topp
Cold Spring Harbor Laboratory

James D. Watson
Cold Spring Harbor Laboratory

CANCER CELLS
1 / The Transformed Phenotype
2 / Oncogenes and Viral Genes

The Transformed Phenotype
© 1984 by Cold Spring Harbor Laboratory
Printed in the United States of America
Cover design by Emily Harste

Library of Congress Cataloging in Publication Data

Main entry under title:

The Transformed phenotype.

   (Cancer cells series, ISSN 0743-2194 ; v. 1)
   Bibliography: p.
   Includes index.
   1. Cancer—Genetic aspects. 2. Cancer cells.
3. Gene expression. 4. Carcinogenesis. I. Levine,
Arnold J. (Arnold Jay), 1939-   . II. Cold Spring
Harbor Laboratory. III. Series.
RC268.4.T73   1984   616.99'4071   83-26318
ISBN 0-87969-168-9

Cover: Immunoperoxidase staining of normal cellular pp60$^{c\text{-}src}$ (brown) in the neural retina of the chick embryo at stage 36 of development (see Sorge et al., this volume). (Photo courtesy of P.F. Maness, Department of Biochemistry and Cancer Research Center, University of North Carolina School of Medicine, Chapel Hill.)

All Cold Spring Harbor Laboratory publications are available through booksellers or may be ordered directly from Cold Spring Harbor Laboratory, Box 100, Cold Spring Harbor, New York 11724.   SAN 203-6185

Conference Participants

Aaronson, Stuart, NCI, National Institutes of Health, Bethesda, Maryland

Albrecht-Buehler, Guenter, Department of Cell Biology, Northwestern University School of Medicine, Chicago, Illinois

Andersen, Philip, Department of Molecular Biology, Abbott Laboratories, North Chicago, Illinois

Appert, Hubert, Department of Surgery, Medical College of Ohio, Toledo

Arai, Ken-Ichi, DNAX Research Institute, Palo Alto, California

Axelrod, David E., Department of Biological Sciences, Rutgers University, New Brunswick, New Jersey

Baltimore, David, Center for Cancer Research, Massachusetts Institute of Technology, Cambridge

Baluda, Marcel, Department of Pathology, University of California School of Medicine, Los Angeles

Barbacid, Mariano, NCI, National Institutes of Health, Bethesda, Maryland

Barnes, David W., Department of Biological Sciences, University of Pittsburgh, Pennsylvania

Barrett, J. Carl, National Institute of Environmental Health Sciences, Research Triangle Park, North Carolina

Baserga, Renato, Department of Pathology, Temple University School of Medicine, Philadelphia, Pennsylvania

Basilico, Claudio, Department of Pathology, New York University Medical Center, New York

Beemon, Karen, Department of Biology, Johns Hopkins University, Baltimore, Maryland

Belsham, Graham, Department of Biochemistry, National Institute for Medical Research, London, England

Benade, Leonard, Department of Virology, American Type Culture Collection, Rockville, Maryland

Bender, Timothy, NCI, National Navy Medical Center, Bethesda, Maryland

Benjamin, Thomas, Department of Pathology, Harvard Medical School, Boston, Massachusetts

Bennett, David, Department of Viral Oncology, Veteran's Administration Hospital, Salt Lake City, Utah

Bennett, Ellen, Department of Microbiology, McGill University, Montreal, Canada

Blair, Donald, NCI, Frederick Cancer Research Facility, Frederick, Maryland

Blumberg, Peter, NCI, National Institutes of Health, Bethesda, Maryland

Bos, J.L., Department of Medical Biochemistry, Silvius Laboratories, Leiden, The Netherlands

Boschek, C. Bruce, Department of Virology, Justus Liebig University, Giessen, Federal Republic of Germany

Bosselman, Robert, Amgen, Inc., Thousand Oaks, California

Brackenbury, Robert, Rockefeller University, New York, New York

Bravo, Rodrigo, European Molecular Biology Laboratory, Heidelberg, Federal Republic of Germany

Breitman, Martin, Department of Genetics, Hospital for Sick Children, Toronto, Canada

Brenner, Sidney, MRC, Cambridge, England

Brickell, Paul, Department of Biochemistry, Imperial College, London, England

Brugge, Joan, Department of Microbiology, State University of New York, Stony Brook

Butel, Janet S., Department of Virology, Baylor College of Medicine, Houston, Texas

Calberg-Bacq, C.M., Department of General Microbiology, University of Liege, Belgium

Campisi, Judith, Department of Cell Growth and Regulation, Dana-Farber Cancer Institute, Boston, Massachusetts

Canaani, Eli, Department of Immunology, Weizmann Institute of Science, Rehovot, Israel

Carley, William, Department of Cell Biology, Yale University, New Haven, Connecticut

Carroll, Robert B., Department of Pathology, New York University Medical Center, New York

Casnellie, John, Department of Pharmacology, University of Washington, Seattle

Cattoni, Sebastiano, Cancer Center, Columbia University, New York, New York

Celis, Julio, Biostructural Division, Aarhus University, Denmark

Chakrabarti, Sekhar, NCI, National Institutes of Health, Bethesda, Maryland

Chen, Lan-Bo, Dana-Farber Cancer Institute, Boston, Massachusetts

Chen, M.J., Department of Molecular Biology, Smith, Kline and Beckman, Swedeland, Pennsylvania

Chinnadurai, G., Department of Molecular Virology, St. Louis University Medical Center, Missouri

Chipperfield, Randall, Department of Biology, Massachusetts Institute of Technology, Cambridge

Clark, Robin, Cetus Corporation, Emoryville, California

Colby, Wendy W., Department of Molecular Biology, Genetech Inc., San Francisco, California

Cole, John S. III, NCI, National Institutes of Health, Bethesda, Maryland

Cole, Michael, Department of Biochemistry, St. Louis University School of Medicine, Missouri

Conrad, Susan, Department of Molecular Biology, University of California, Berkeley

Cook, Donald N., Department of Microbiology, McGill University, Montreal, Canada

Cooper, Jonathan, Department of Molecular Biology and Virology, Salk Institute, San Diego, California

v

Courtneidge, Sara, Department of Biochemistry, National Institute for Medical Research, London, England
Covey, Lori, Department of Biological Sciences, Columbia University, New York, New York
Crawford, Lionel, Imperial Cancer Research Fund, London, England
Croce, Carlo M., Wistar Institute, Philadelphia, Pennsylvania
Cuzin, Francois, Biochemistry Center, University of Nice, France
Danos, Olivier, Department of Molecular Biology, Pasteur Institute, Paris, France
Defendi, V., Department of Pathology, New York University Medical Center, New York
Dermer, Gerald, Department of Pathology, University of North Carolina, Chapel Hill
Deuel, Thomas F., Jewish Hospital at Washington University School of Medicine, St. Louis, Missouri
Diamond, Alan M., Dana-Farber Cancer Institute, Boston, Massachusetts
Dickson, Clive, Imperial Cancer Research Fund, London, England
Di Mayorca, G., University Medical and Dental School of New Jersey, Newark
Duesberg, Peter, Department of Molecular Biology, University of California, Berkeley
Dunn, Ashley F., Melbourne Tumor Biology Division, Ludwig Institute for Cancer Research, Victoria, Australia
Eisenmann, Robert, Fred Hutchinson Cancer Research Center, Seattle, Washington
Eva, A., NCI, National Institutes of Health, Bethesda, Maryland
Feramisco, James R., Cold Spring Harbor Laboratory, New York
Fink, Leslie, NCI, National Institutes of Health, Bethesda, Maryland
Fisher, Richard, Biogen Research Corporation, Cambridge, Massachusetts
Franke, Werner, Institute of Cell and Tumor Biology, German Cancer Research Center, Heidelberg, Federal Republic of Germany
Fried, Michael, Imperial Cancer Research Fund, London, England
Friedman, Beth Ann, Dept. of Toxicology, Massachusetts Institute of Technology, Cambridge
Fuchs, Elaine, Department of Biochemistry, University of Chicago, Illinois
Furth, Mark, Department of Molecular Oncogenesis, Memorial Sloan-Kettering Cancer Center, New York, New York
Gallimore, Phillip, Department of Cancer Studies, University of Birmingham, England
Garrels, James, Cold Spring Harbor Laboratory, New York
Gattoni-Cellis, S., Columbia University College of Physicians & Surgeons, New York, New York
Geiger, Benjamin, Department of Chemical Immunology, Weizmann Institute of Science, Rehovot, Israel
Gentry, Lawrence, Department of Viral Oncology, Fred Hutchinson Cancer Research Center, Seattle, Washington
Gernot, Walter, Immunobiology Institute, University of Freiberg, Federal Republic of Germany
Gilmer, Tona M., Department of Microbiology and Immunology, University of North Carolina, Chapel Hill
Goldberg, Allan, Rockefeller University, New York, New York
Gooding, Linda R., Department of Microbiology, Emory University School of Medicine, Atlanta, Georgia
Graesser, Friedrich, Department of Immunobiology, University of Freiburg, Federal Republic of Germany
Gray, Harry, Dana-Farber Cancer Institute, Boston, Massachusetts
Graziani, Y., Department of Pathology, University of Colorado Health Sciences Center, Denver
Green, Howard, Department of Physiology and Biophysics, Harvard University Medical School, Boston, Massachusetts
Greenberg, Michael, Department of Biochemistry, New York University Medical Center, New York
Gross, Ludwik, Veterans Administration Hospital, Bronx, New York
Grosveld, Gerard, Department of Cell Biology and Genetics, Erasmus University, Rotterdam, The Netherlands
Halligan, Brian, Department of Physiological Chemistry, Johns Hopkins University School of Medicine, Baltimore, Maryland
Hanafusa, Hidesaburo, Rockefeller University, New York, New York
Hann, Steve, Fred Hutchinson Cancer Center, Seattle, Washington
Hare, David, Amgen Development, Inc., Boulder, Colorado
Harlow, Edward, Cold Spring Harbor Laboratory, New York
Hassell, John, Department of Microbiology, McGill University, Montreal, Canada
Hawley, Robert, Biological Research Division, Ontario Cancer Institute, Toronto, Canada
Hayday, Adrian C., Center for Cancer Research, Massachusetts Institute of Technology, Cambridge
Hayman, M.J., Imperial Cancer Research Fund, London, England
Hayward, William, Memorial Sloan-Kettering Cancer Center, New York, New York
Herrlich, Peter, Department of Genetics, University of Karlsruhe, Federal Republic of Germany
Hill, M., Institute of Cancerology and Immunogenetics, Villejuif, France
Hillova, Jane, Institute of Cancerology and Immunogenetics, Villejuif, France
Hirschhorn, Ricky R., Department of Pathology, Temple University Health Sciences Center, Philadelphia, Pennsylvania
Hirt, Bernard, Swiss Cancer Institute, Lausanne, Switzerland

Hoffman, Robert, Department of Pediatrics, University of California, San Diego
Hsiao, Wendy, Douglaston, New York
Huang, Jung San, Jewish Hospital at Washington University School of Medicine, St. Louis, Missouri
Hunter, Tony, Salk Institute, San Diego, California
Hyland, Julia K., Department of Pathology, Temple University School of Medicine, Philadelphia, Pennsylvania
Ikawa, Yoji, Riken Institute, Saitama, Japan
Ikegaki, Naohiko, Department of Human Genetics, University of Pennsylvania School of Medicine, Philadelphia
Ip, Stephen, Cambridge Research Laboratory, Massachusetts
Ito, Yoshiaki, NCI, National Institutes of Health, Bethesda, Maryland
Janssen, J.W.G., Department of Experimental Cytology, Netherlands Cancer Institute, Amsterdam
Jenkins, J.R., Marie Curie Memorial Foundation Research Institution, Surrey, England
Jonak, Gerald, DuPont Experimental Station, Wilmington, Delaware
Kakunaga, Takeo, NCI, National Institutes of Health, Bethesda, Maryland
Kamen, Robert, Genetics Institute, Boston, Massachusetts
Kaplan, Paul, Salk Institute, San Diego, California
Khoury, George, NCI, National Institutes of Health, Bethesda, Maryland
Koprowski, Hilary, Department of Anatomy and Biology, Wistar Institute, Philadelphia, Pennsylvania
Kozma, Sara, Institute for Genetics and Toxology, Karlsruhe, Federal Republic of Germany
Kriegler, Michael, Department of Molecular Biology, University of California, Berkeley
Krueger, James, Rockefeller University, New York, New York
Kuehl, Mike, NCI, National Navy Medical Center, Bethesda, Maryland
Land, Hartmut, Massachusetts Institute of Technology, Cambridge
Leder, Philip, Department of Genetics, Harvard University Medical School, Boston, Massachusetts
Lee, William, Department of Medicine, University of California, San Francisco
Lehto, Veli-P., Department of Pathology, University of Helsinki, Finland
Lens, P.F., Department of Chemical Carcinogesis, Netherlands Cancer Institute, Amsterdam
Levine, Arnold, Department of Microbiology, State University of New York, Stony Brook
Levinson, Arthur, Genentech, Inc., South San Francisco, California
Lewin, Benjamin, *Cell*, Cambridge, Massachusetts
Lewis, C.M., Department of Immunobiology, Glaxo Group Research, Middlesex, England
Linzer, Daniel, Department of Molecular Biology and Genetics, Johns Hopkins University School of Medicine, Baltimore, Maryland
Lipp, Martin, Biochemistry Institute, Ludwig-Maximilians University, Munich, Federal Republic of Germany
Livingston, David M., Dana-Farber Cancer Institute, Boston, Massachusetts
Luciw, Paul, Chiron Corporation, Emeryville, California
Maltzman, Warren, Department of Biological Sciences, Rutgers University, New Brunswick, New Jersey
Maness, Patricia, Department of Biochemistry, University of North Carolina, Chapel Hill
Maniatis, Thomas, Department of Biochemistry and Molecular Biology, Harvard University, Cambridge, Massachusetts
Mannino, Raphael, Department of Microbiology and Immunology, Albany Medical College, New York
Mansfield, Brian, Department of Molecular Biology and Genetics, Johns Hopkins University School of Medicine, Baltimore, Maryland
Mao, Jen-I., Collaborative Research Inc., Lexington, Massachusetts
Marcu, Kenneth, Department of Biochemistry, State University of New York, Stony Brook
Marks, Paul A., Memorial Sloan-Kettering Cancer Center, New York, New York
Martin, G. Steven, Department of Zoology, University of California, Berkeley
Martin, M.A., National Institutes of Health, Bethesda, Maryland
Marx, Jean L., *Science*, Washington, DC
Matsumura, Fumio, Cold Spring Harbor Laboratory, New York
McClain, Kenneth, University of Minneapolis Hospital, Minnesota
McClure, Don, University of California, San Diego
Mendelson, Ella, NCI, National Institutes of Health, Bethesda, Maryland
Mercer, W. Edward, Department of Pathology, Temple University School of Medicine, Philadelphia, Pennsylvania
Mes-Masson, A., Department of Microbiology, McGill University, Montreal, Canada
Miles, Vincent, Amersham International, Amersham, England
Miyatake, Sho, DNAX Research Institute, Palo Alto, California
Moelling, Karin, Department of Molecular Genetics, Max-Planck Institute, Berlin, Federal Republic of Germany
Mroczkowski, B., Department of Biochemistry, Vanderbilt University, Nashville, Tennessee
Murphy, Cheryl, Dana-Farber Cancer Institute, Boston, Massachusetts
Murphy, David, Imperial College, London, England
Nathans, Daniel, Department of Molecular Biology and Genetics, Johns Hopkins University Medical School, Baltimore, Maryland
Nevins, Joseph, Rockefeller University, New York, New York
Newmark, Peter, *Nature*, London, England

Nishizuka, Yasutomi, Department of Biochemistry, Kobe University School of Medicine, Japan
Noda, Makoto, NCI, National Institutes of Health, Bethesda, Maryland
Nunn, Michael, University of California, Berkeley
Nusse, R., Netherlands Cancer Institute, Amsterdam
O'Farrell, Minnie, Department of Biology, University of Essex, England
Oren, Moshe, Department of Immunology, Weizmann Institute of Science, Rehovot, Israel
Oroszlan, Stephen, NCI, Frederick Cancer Research Facility, Frederick, Maryland
Osborn, Mary, Department of Biophysical Chemistry, Max-Planck Institute, Goettingen, Federal Republic of Germany
Ozanne, Brad, Department of Microbiology, University of Texas Health Science Center, Dallas
Papas, Takis S., NCI, Frederick Cancer Research Facility, Frederick, Maryland
Parada, Luis, Massachusetts Institute of Technology, Cambridge
Parker, Richard, Department of Microbiology, University of California, San Francisco
Parsons, J.T., Department of Microbiology, University of Virginia, Charlottesville
Patch, Cephas, NCI, National Institutes of Health, Bethesda, Maryland
Pearson, Mark L., NCI-Frederick Cancer Research Facility, Frederick, Maryland
Pellicer, Angel, Department of Pathology, New York University Medical Center, New York
Perucho, M., Department of Biochemistry, State University of New York, Stony Brook
Peters, Gordon, Imperial Cancer Research Fund, London, England
Pipas, James M., Department of Biological Sciences, University of Pittsburgh, Pennsylvania
Radke, Kathryn, Department of Veterinary Medicine, University of California, Davis
Rall, Leslie B., Chiron Corporation, Emeryville, California
Rapp, Ulf, NCI-Frederick Cancer Research Facility, Frederick, Maryland
Reid, Lola, Department of Molecular Pharmacology, Albert Einstein College of Medicine, Bronx, New York
Rein, A., NCI-Frederick Cancer Research Facility, Frederick, Maryland
Reynolds, Fred, NCI, Frederick Cancer Research Facility, Frederick, Maryland
Reynolds, Steven, Department of Tumor Virology, Fred Hutchinson Cancer Research Facility, Seattle, Washington
Rheinwald, James F., Dana-Farber Cancer Institute, Boston, Massachusetts
Rhim, J., NCI, National Institutes of Health, Bethesda, Maryland
Riemen, Mark W., Merck, Sharp and Dohme Research Laboratories, West Point, Pennsylvania
Rigby, P.W.J., Imperial College of Science and Technology, London, England
Rohrschneider, Larry, Fred Hutchinson Cancer Research Facility, Seattle, Washington
Rosner, Marsha, Toxicology, Massachusetts Institute of Technology, Cambridge
Rotter, Varda, Department of Cell Biology, Weizmann Institute of Science, Rehovot, Israel
Rowley, Janet D., Department of Medicine, University of Chicago, Illinois
Ruley, Earl, Cold Spring Harbor Laboratory, New York
Rusch, Harold, McArdle Laboratory, University of Wisconsin, Madison
Sabrin, Ira, Pall Corporation, Glen Cove, New York
Sager, Ruth, Dana-Farber Cancer Institute, Boston, Massachusetts
Saito, Haruo, Cancer Research Center, Massachusetts Institute of Technology, Cambridge
Schecter, Alan, Center for Cancer Research, Massachusetts Institute of Technology, Cambridge
Schlom, Jeffrey, NCI, National Institutes of Health, Bethesda, Maryland
Schneider, C., Imperial Cancer Research Fund, London, England
Schwab, Manfred, University of California, San Francisco
Schwartz, Stephen, Department of Pathology, University of Chicago, Illinois
Sefton, Bart, Salk Institute, San Diego, California
Senger, Donald, Department of Pathology, Beth Israel Hospital, Boston, Massachusetts
Shall, S., Department of Cellular and Molecular Biology, University of Sussex School of Biological Science, Brighton, England
Shalloway, David, Pennsylvania State University, University Park.
Sharp, Phillip, Center for Cancer Research, Massachusetts Institute of Technology, Cambridge
Shenk, Thomas, Department of Microbiology, State University of New York, Stony Brook
Sherr, C.J., Department of Human Tumor Cell Biology, St. Jude Children's Hospital, Memphis, Tennessee
Shoyab, Mohammed, NCI, National Institutes of Health, Bethesda, Maryland
Shulman, Marc J., Wellesley Hospital, Toronto, Canada
Siebert, Gary R., Department of Cell Biology, Becton Dickinson Research Center, Research Triangle Park, North Carolina
Smart, John E., Biogen, Inc., Cambridge, Massachusetts
Smith, Alan E., Department of Biochemistry, MRC National Institute of Medical Research, London, England
Snitman, David, Amgen Development, Boulder, Colorado
Spandidos, Demetrios, Beatson Institute for Cancer Research, Glasgow, Scotland
Sporn, Michael, NCI, National Institutes of Health, Bethesda, Maryland

Spurr, N.K., Imperial Cancer Research Fund, London, England
Srinivason, A., NCI, National Institutes of Health, Bethesda, Maryland
Stabinsky, Yizahak, Amgen Development, Inc., Boulder, Colorado
Stephenson, John R., NCI, National Institutes of Health, Bethesda, Maryland
Stern, David, F., Massachusetts Institute of Technology, Cambridge
Stiles, Charles, Dana-Farber Cancer Institute, Boston, Massachusetts
Sun, T.-T., Department of Dermatology and Pharmacology, New York University School of Medicine, New York
Tevethia, Satvir, Department of Microbiology, Pennsylvania State University College of Medicine, Hershey
Tjian, Robert, Department of Biochemistry, University of California, Berkeley
Todaro, George, Oncogen, Inc., Seattle, Washington
Tonegawa, Susumu, Massachusetts Institute of Technology, Cambridge
Tooze, John, EMBO, Heidelberg, Federal Republic of Germany
Tsichlis, Phillip, Fox Chase Cancer Institute, Philadelphia, Pennsylvania
Turek, Lubomir, Department of Pathology, University of Iowa, Iowa City
van der Eb, Alex, Sylvius Laboratories, Leiden, The Netherlands
Vande Woude, George, NCI, Frederick Cancer Research Facility, Frederick, Maryland
Verbeek, Joseph, Geert Girdote Plein Noores, Nymegen, The Netherlands
Verma, Inder M., Department of Molecular Biology and Virology, Salk Institute, San Diego, California
Vincenzo, S., Department of Cell Biology, Sezione Science Microbiologiche, Rome, Italy
Vogt, Peter, Department of Microbiology, University of California, Los Angeles
Wallner-Philipp, Barbara, Biogen, Inc., Cambridge, Massachusetts
Wang, Jean, Massachusetts Institute of Technology, Cambridge
Wang, Lu-Hai, Rockefeller University, New York, New York
Waterfield, M.D., Imperial Cancer Research Fund, London, England
Weber, Klaus, Department of Biophysical Chemistry, Max-Planck Institute, Goettingen, Federal Republic of Germany
Weber, Michael, Department of Microbiology, University of Illinois, Urbana
Weinstein, I.B., Columbia University, New York, New York
Weiss, Robin, Chester Beatty Laboratories, Institute for Cancer Research, London, England
Weissman, Sherman, Yale University School of Medicine, New Haven, Connecticut
Wigler, Michael, Cold Spring Harbor Laboratory, New York
Wilhelmsen, K.C., McArdle Laboratory, University of Wisconsin, Madison
Williams, Lewis T., Massachusetts General Hospital, Boston
Wilsnack, Roger E., Becton Dickinson Research Center, Research Triangle Park, North Carolina
Wirschubsky, Zvi, Department of Tumor Biology, Karolinska Institutet, Stockholm, Sweden
Witte, Owen, Molecular Biology Institute, University of California, Los Angeles
Wolf, David, Department of Cell Biology, Weizmann Institute of Science, Rehovot, Israel
Wong, Tai Wai, Rockefeller University, New York, New York
Wong-Staal, Flossie, NCI, National Institutes of Health, Bethesda, Maryland
Wu, Ying-Jye, Cambridge Research Laboratory, Massachusetts
Yang, Liu, Department of Physiological Chemistry, Johns Hopkins University School of Medicine
Zack, Jerry, Department of Microbiology, University of Texas Health Science Center, Dallas
Zimmer, S., Department of Pathology, University of Kentucky Medical Center, Lexington

First row: A. Levine; M. Mathews, A. Dunn; G. Khoury, M. Martin
Second row: G. Vande Woude; E. Harlow; B. Ozanne, R. Weiss; J. Garrels
Third row: M. Weber, S. Courtneidge; M. Pearson, K. Weber; L. Gooding, W. Topp
Fourth row: L. Gooding, G. Walter, B. Sefton; C. Dickson, J. Jenkins, N. Spurr

Preface

During the past few years we have seen an intensive application of the experimental tools developed in immunology and molecular biology to the problem of the cancer cell. The areas of cell biology and tumor immunology (concerned with the transformed cell phenotype) and of molecular biology and virology (concerned with oncogenes, viral genes, and their products) have necessarily become interdependent. In this inaugural volume of the *Cancer Cells* series and in the succeeding volume "Oncogenes and Viral Genes," there are numerous examples of the marriage between cell biology and molecular biology. The identification of activated or functional oncogenes in cells derived from tumors or tumor tissue depends entirely upon the experimental definition of the transformed phenotype. The relationships between oncogene products and growth factors and between chromosome translocations and the presence of oncogenes found near the sites of the breakpoints are additional examples of the interplay between these areas of research. The interesting observation that different oncogenes and different viral genes can alter distinct portions of the transformed cell phenotype has brought DNA and RNA tumor virologists together in proposing common mechanisms for the transformation of cells in culture. The clear necessity for biological assays to test cloned oncogenes, viral genes, purified growth factors, and tumor promotors dictated that cell biologists join molecular biologists on a common meeting ground. It was this realization that gave rise to The Cancer Cell meeting held at Cold Spring Harbor in September 1983.

As organizers of this conference, we sought the advice and counsel of many of our colleagues. We are particularly grateful to Earl Ruley, Mike Wigler, Mike Bishop, Howard Temin, Harold Varmus, and David Baltimore for helping to pinpoint critical research developments. Although we tried to cover the field as broadly as possible, some important contributions were no doubt overlooked. Nevertheless, the sessions were very full and stimulating, accurately reflecting the excitement that the research has generated over the past few years. It was clear that few scientists at the meeting doubted that oncogenes play a role in one or more steps on the road to the production of the cancer cell phenotype.

Embued with enthusiasm generated by the prospect of a highly exciting and productive meeting, we felt the occasion of this meeting would be an appropriate time to pay tribute to an individual who has made possible much of what has been accomplished. The meeting, and likewise this volume, was therefore dedicated to Mrs. Albert D. Lasker, who has contributed so much to the National Cancer Institute and the American Cancer Society, organizations that have funded much of the work presented at the conference.

We are most grateful to Gladys Kist and the staff of the Cold Spring Harbor Meetings Office. With cheerfulness and aplomb, they handled the complications of the registration, housing, arrivals, and departures of the more than 400 participants. Herb Parsons and his helpers deserve much thanks for their skill and care with the audiovisuals, so crucial to the success of any scientific conference. For providing financial support for this meeting, we wish to thank the National Cancer Institute, Fogarty International Center, and Becton-Dickinson and Co. Our special thanks go to the American Cancer Society for their rapid and generous response to our last-minute appeal for crucial funding.

This meeting, the first in a new series on the cancer cell, supersedes the Cold Spring Harbor Conferences on Cell Proliferation, which have been held over the past ten years. "The Transformed Phenotype" and "Oncogenes and Viral Genes" are volumes 1 and 2, respectively, of *Cancer Cells*, the new series of publications to emerge from these conferences. In content and format, these publications are a reflection of the changing nature of this field and the need of researchers in these diverse areas to have important results reported in a timely, readable fashion.

As editors of these volumes, we owe a great deal to the Cold Spring Harbor Laboratory Publications Department headed by Nancy Ford. Without the help, organization, prodding, and good humor of Douglas Owen, Judith Cuddihy, Karen Sundin, and Joan Ebert, these two volumes would not exist.

The Editors

This inaugural volume of *Cancer Cells* is dedicated to Mrs. Albert D. Lasker, without whose unfailing efforts over the last 40 years we would not now be so near to understanding the molecular basis of cancer. Through her advocacy the American Cancer Society was reorganized and began to actively support research into the causes of cancer. Through the Albert Lasker Medical Research Awards she helped raise public consciousness of important advances in scientific research. Through her tireless petitioning of Congress she helped expand the vital work of the National Institutes of Health and secure passage of the National Cancer Act. Her single-minded campaign against diseases that kill and cripple established forever a prominent place for medical research on the public agenda.

In 1969 Mrs. Lasker received the Medal of Freedom, the highest civilian award in the United States. The citation from President Johnson read:

Humanist, philanthropist, activist—Mary Lasker has inspired understanding and productive legislation which improved the lot of mankind. In medical research, in adding grace and beauty to the environment, and in exhorting her fellow citizens to rally to the cause of progress, she has made a lasting imprint on the quality of life in this country. She has led her president and the Congress to greater heights for justice for her people and beauty for her land.

Mrs. Albert D. Lasker

Contents

Conference Participants, v
Preface, xi

GROWTH FACTORS

Isolation and Characterization of Type β Transforming Growth Factors from Human, Bovine, and Murine Sources	M.B. Sporn, M.A. Anzano, R.K. Assoian, J.E. De Larco, C.A. Frolik, C.A. Meyers, and A.B. Roberts	1
The Amount of EGF Receptor Is Elevated on Squamous Cell Carcinomas	G. Cowley, J.A. Smith, B. Gusterson, F. Hendler, and B. Ozanne	5
Is the *ras* Oncogene Protein a Component of the Epidermal Growth Factor Receptor System?	T. Kamata and J.R. Feramisco	11
In Vitro Correlates of Tumorigenicity of REF52 Cells Transformed by Simian Virus 40	D.B. McClure, M. Dermody, and W.C. Topp	17
Relationship between the Transforming Protein of Simian Sarcoma Virus and Platelet-derived Growth Factor	M.D. Waterfield, G.T. Scrace, N. Whittle, P.A. Stockwell, P. Stroobant, A. Johnsson, A. Wasteson, B. Westermark, C.-H. Heldin, J.S. Huang, and T.F. Deuel	25
Close Similarities between the Transforming Gene Product of Simian Sarcoma Virus and Human Platelet-derived Growth Factor	K.C. Robbins, H.N. Antoniades, S.G. Devare, M.W. Hunkapiller, and S.A. Aaronson	35
The Platelet-derived Growth Factor Receptor Protein Is a Tyrosine-specific Protein Kinase	S.S. Huang, J.S. Huang, and T.F. Deuel	43
Cell-cycle Genes Regulated by Platelet-derived Growth Factor	B.H. Cochran, A.C. Reffel, M.A. Callahan, J.N. Zullo, and C.D. Stiles	51

CYTOSKELETON

Microfilaments in Normal and Transformed Cells: Changes in the Multiple Forms of Tropomyosin	J.J.-C. Lin, S. Yamashiro-Matsumura, and F. Matsumura	57
A Point Mutation and Other Changes in Cytoplasmic Actins Associated with the Expression of Transformed Phenotypes	T. Kakunaga, J. Leavitt, T. Hirawaka, and S. Taniguchi	67
Mitochondria in Tumor Cells: Effects of Cytoskeleton on Distribution and as Targets for Selective Killing	L.B. Chen, I.C. Summerhayes, K.K. Nadakavukaren, T.J. Lampidis, S.D. Bernal, and E.L. Shepherd	75
Movement of Nucleus and Centrosphere in 3T3 Cells	G. Albrecht-Buehler	87

GENE EXPRESSION

Onco-fetal Genes and the Major Histocompatibility Complex	P.W.J. Rigby, P.M. Brickell, D.S. Latchman, D. Murphy, M.R.D. Scott, K.-H. Westphal, V. Simanis, D.P. Lane, and K. Willison	97
Stimulation of Gene Expression by Viral Transforming Proteins	J. Brady, L.A. Laimins, and G. Khoury	105
Changes in Specific mRNAs Following Serum Stimulation of Cultured Mouse Cells: Increase in a Prolactin-related mRNA	D.I.H. Linzer and D. Nathans	111
pp60$^{c\text{-}src}$ Expression in Embryonic Nervous Tissue: Immunocytochemical Localization in the Developing Chick Retina	L.K. Sorge, B.T. Levy, T.M. Gilmer, and P.F. Maness	117
Gene Expression in Human Epidermal Basal Cells: Changes in Protein Synthesis Accompanying Differentiation and Transformation	J.E. Celis, S.J. Fey, P.M. Larsen, and A. Celis	123
Transformation-sensitive Proteins of REF52 Cells Detected by Computer-analyzed Two-dimensional Gel Electrophoresis	B.R. Franza, Jr., and J.I. Garrels	137
Changes Induced by Epidermal Growth Factor in the Polypeptide Synthesis of A431 Cells	R. Bravo	147

KERATINS

Intermediate Filaments—from Wool α-Keratins to Neurofilaments: A Structural Overview	K. Weber and N. Geisler	153
Differential Expression of Two Classes of Keratins in Normal and Malignant Epithelial Cells and Their Evolutionary Conservation	E. Fuchs, M.P. Grace, K.H. Kim, and D. Marchuk	161
Classification, Expression, and Possible Mechanisms of Evolution of Mammalian Epithelial Keratins: A Unifying Model	T.-T. Sun, R. Eichner, A. Schermer, D. Cooper, W.G. Nelson, and R.A. Weiss	169
Cytokeratins: Complex Formation, Biosynthesis, and Interactions with Desmosomes	W.W. Franke, D.L. Schiller, M. Hatzfeld, T.M. Magin, J.L. Jorcano, S. Mittnacht, E. Schmid, J.A. Cohlberg, and R.A. Quinlan	177
Conventional and Monoclonal Antibodies to Intermediate Filament Proteins in Human Tumor Diagnosis	M. Osborn, M. Altmannsberger, E. Debus, and K. Weber	191
Dynamic Rearrangements of Cytokeratins in Living Cells	B. Geiger, T.E. Kreis, O. Gigi, E. Schmid, S. Mittnacht, J.L. Jorcano, D.B. von Bassewitz, and W.W. Franke	201

Expression of Specific Keratin Subsets and Vimentin in Normal Human Epithelial Cells: A Function of Cell Type and Conditions of Growth during Serial Culture	T.G. Rheinwald, T.M. O'Connell, N.D. Connell, S.M. Rybak, B.L. Allen-Hoffmann, P.J. LaRocca, Y.-J. Wu, and S.M. Rehwoldt	217

PROMOTERS

Multistage Carcinogenesis Involves Multiple Genes and Multiple Mechanisms	I.B. Weinstein, S. Gattoni-Celli, P. Kirschmeier, M. Lambert, W. Hsiao, J. Backer, and A. Jeffrey	229
Protein Kinase C and the Mechanism of Action of Tumor Promoters	U. Kikkawa, R. Miyake, Y. Tanaka, Y. Takai, and Y. Nishizuka	239
Membrane and Cytosolic Receptors for the Phorbol Ester Tumor Promoters	P.M. Blumberg, N.A. Sharkey, B. König, S. Jaken, K.L. Leach, and A.Y. Jeng	245
Isolation and Characterization of a Specific Receptor for Biologically Active Phorbol and Ingenol Esters from Murine Brain: The Homogeneous Receptor Is a Ca^{++}-independent Protein Kinase	M. Shoyab	253

SURFACE ANTIGENS

Antibody and Cellular Detection of SV40 T-Antigenic Determinants on the Surfaces of Transformed Cells	L.R. Gooding, R.W. Geib, K.A. O'Connell, and E. Harlow	263
Localization of Antigenic Sites Reactive with Cytotoxic Lymphocytes on the Proximal Half of SV40 T Antigen	S.S. Tevethia and M.J. Tevethia	271
Unique Molecular Structure of the Leukemogenic Cell Membrane Glycoprotein (gp55) Encoded by the Friend Spleen Focus-forming Virus	Y. Ikawa, M. Obata, N. Sagata, and H. Amanuma	277
Detection and Enhancement (by Recombinant Interferon) of Carcinoma Cell Surface Antigens, Using Monoclonal Antibodies	J. Greiner, P. Horan Hand, S. Pestka, P. Noguchi, P. Fisher, D. Colcher, and J. Schlom	285
Human Tumor Antigens	H. Koprowski	293

Author Index, 299

Subject Index, 301

Isolation and Characterization of Type β Transforming Growth Factors from Human, Bovine, and Murine Sources

M.B. Sporn, M.A. Anzano, R.K. Assoian, J.E. De Larco, C.A. Frolik, C.A. Meyers, and A.B. Roberts

Laboratory of Chemoprevention, National Cancer Institute, Bethesda, Maryland 20205

The transforming growth factors (TGFs) have been defined to include the set of polypeptides that confer the transformed phenotype on untransformed indicator cells (Sporn et al. 1981; Marquardt et al. 1983; Roberts et al. 1983a). For fibroblastic cells, the transformed phenotype has been operationally defined by the loss of density-dependent inhibition of growth in monolayer, overgrowth in monolayer, characteristic changes in cellular morphology, and acquisition of anchorage independence, with the resulting ability to grow in soft agar. The first of the TGFs to be discovered was sarcoma growth factor (SGF), described by De Larco and Todaro (1978) in the conditioned medium of murine 3T3 cells transformed by Moloney sarcoma virus. Although it was originally believed that the transforming activity was the property of a single peptide, it has subsequently been shown by Anzano et al. (1983) that the biological activity in the conditioned medium that is responsible for the growth of cells in soft agar results from the combined action of two distinct polypeptides, which have been classified as type α and type β TGFs (Roberts et al. 1983a). Neither type α nor type β TGF alone has the ability to promote growth of normal rat kidney (NRK) fibroblasts in soft agar; this biological activity is the result of the concerted action of both peptides (Anzano et al. 1982). Type α TGFs all bind to a common receptor, generally known as the EGF (epidermal growth factor) receptor. Members of the TGF-α family that have been characterized thus far all consist of a single peptide chain, cross-linked by three disulfide bridges; their molecular weight is between 5000 and 7000 (Marquardt et al. 1983). Included in this group are human and murine EGF as well as several other TGFs isolated from a variety of neoplastic cell lines (Todaro et al. 1980; Marquardt and Todaro 1982; Twardzik et al. 1982; Marquardt et al. 1983). Type β TGFs do not bind to the EGF receptor; their receptor is presently being characterized (C.A. Frolik et al., unpubl.). Members of the TGF-β family that have been purified thus far all consist of two polypeptide chains, cross-linked by disulfide bridge(s); the dimer has a molecular weight of approximately 25,000. Type β TGFs have been isolated from a variety of neoplastic and nonneoplastic cells, including murine sarcoma cells, murine kidney, bovine kidney, human placenta, and human blood platelets. We will briefly summarize the procedures that have been developed in our laboratory for the isolation of TGF-β from bovine kidney, human placenta, and human platelets, and then review what is known about the chemical and biological properties of this family of molecules.

Isolation of TGF-β from bovine kidney, human placenta, and human platelets

We have recently published detailed procedures for the purification of TGF-β from bovine kidney (Roberts et al. 1983b), human placenta (Frolik et al. 1983), and human platelets (Assoian et al. 1983). Tables 1 and 2 summarize the purification from kidney and placenta, respectively. The basic procedure involves acid/ethanol extraction of 8–14 kg of tissue, followed by gel exclusion chromatography in 1 M acetic acid, ion exchange chromatography, and finally two steps of high-pressure liquid chromatography (HPLC) on C_{18} and CN columns, using acetonitrile-trifluoroacetic acid and propanol-trifluoroacetic acid gradients. Since the original acid/ethanol extraction rep-

Table 1 Purification of TGF-β from Bovine Kidney

Purification step	Protein recovered (mg)	Total activity (units × 10^{-3})	ED_{50} (ng/ml)	Specific activity (units/μg)	Degree of purification (-fold)	Recovery of activity (%)
Acid/ethanol extract	89,000	5,100	13,000	0.057	1.0	100
Bio-Gel P-30	5,600	2,400	1,700	0.43	7.5	47
Cation exchange	99	350	200	3.5	61	6.9
HPLC-C_{18}	3.7	460	5.8	120	2,100	9.0
HPLC-CN	0.04	510	0.056	13,000	230,000	10

Data from Roberts et al. (1983b).

Table 2 Purification of TGF-β from Human Placenta

Purification step	Protein recovered (mg)	Total activity (units × 10^{-3})	ED_{50} (ng/ml)	Specific activity (units/μg)	Degree of purification (-fold)	Recovery of activity (%)
Acid/ethanol extract	239,000	21,510	7,600	0.09	1.0	100
Bio-Gel P-30						
pool A	73,900	3,695	15,000	0.05	0.6	17
pool B	27,720	—	—	—	—	—
pool C	1,900	3,800	360	2.0	22	18
Bio-Gel P-6						
pool B	8,700	14,790	410	1.7	19	69
Ion exchange						
pool B	140	1,610	62	11.5	128	31
pool C	46.3	390	85	8.4	93	1.8
HPLC-C_{18}						
pool B	0.27	1,928	0.10	7,142	79,000	37
pool C	0.26	155	1.2	595	6,610	0.7
HPLC-CN						
pool B	0.025	248	0.072	9,920	110,000	4.8
pool C	0.022	245	0.064	11,160	124,000	1.1

Data from Frolik et al. (1983).

resents at least a 10-fold purification from the starting tissue, the overall purification is greater than a millionfold. Dose responsiveness of the assays used to monitor the purification is shown in Figures 1 and 2. From several kilograms of kidney or placenta, only a few parts per billion of pure TGF-β is obtained. For this reason, other sources have been investigated, and only one tissue, namely blood platelets, has been found to be superior; in this case the yield of pure TGF-β is approximately one part per million. The procedure for TGF-β purification from platelets, however, is considerably simpler since two steps of sequential chromatography on Bio-Gel P-60 columns, first in the absence and then in the presence of urea, suffice to yield pure material.

Chemical characterization of TGF-β from kidney, placenta, and platelets

All three TGF-β preparations, from kidney, placenta, and platelets, behave identically when subjected to electrophoresis in polyacrylamide gels in the presence of sodium dodecyl sulfate (SDS) (Fig. 3). Under nonreducing conditions, a single band of approximately 25,000 daltons is seen for each TGF-β. Biological activity (promotion of growth of NRK cells in soft agar) can be shown

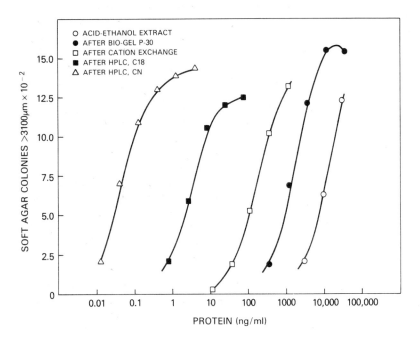

Figure 1 Dose-response curves of the ability of TGF-β from bovine kidney to promote growth of NRK cells in soft agar, determined at various stages of purification of TGF-β. Appropriate aliquots of samples were assayed in the standard soft-agar assay in the presence of 5 ng/ml of EGF. (Reprinted, with permission, from Roberts et al. 1983b.)

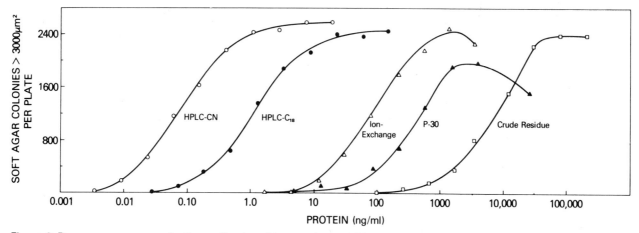

Figure 2 Dose-response curves for the purification of human placental TGF-β.

for each of the three TGF-β bands, when assayed in the presence of TGF-α (Fig. 3). No activity is seen from the gel in any region other than the location of the TGF-β bands. A single band, with a molecular weight of 12,500–13,000, is seen when SDS-polyacrylamide gel electrophoresis is performed under reducing conditions. The reduction of TGF-β causes total loss of biological activity.

The amino acid composition of all three TGF-β preparations is strikingly similar (Table 3) and markedly different from that of platelet-derived growth factor (PDGF) (Deuel et al. 1981; Heldin et al. 1981). In collaboration with S. Stein and his colleagues at the Department of Molecular Genetics of Hoffmann-La Roche, Inc. (Nutley, New Jersey), a partial aminoterminal amino acid sequence of TGF-β from bovine kidney and human placenta has been determined (Roberts et al. 1983b; C.A. Frolik et al., unpubl.). The sequence of the first 15 amino acids is identical for the peptides from both kidney and placenta and is as follows: H_2N-Ala-Leu-Asp-Thr-Asn-Tyr-Cys-Phe-Ser-Ser-Thr-Glu-Lys-Asn-Cys-. Initial and repetitive yields were found to be equal to the yields calculated for myoglobin as a standard protein. The data thus suggest that the two chains of TGF-β are similar and possibly identical, although further studies are obviously needed to determine this.

Biological properties of TGF-β

TGF-β from kidney, placenta, and platelets has an ED_{50} in the standard NRK cell soft-agar growth assay of 50–100 pg/ml ($2 \times 10^{-12} - 4 \times 10^{-12}$ M) in the presence of 2–5 ng/ml of EGF. By itself, TGF-β has essentially no mitogenic activity on NRK cells grown in monolayer. The intrinsic biological role of TGF-β in the tissues in which it has been found is unknown. Finding this peptide in relatively high concentrations in the platelet suggests that one of its roles is to participate in tissue repair and wound healing, and we have obtained some experimental evidence in support of this concept (Sporn et al. 1983). The role of TGF-β in oncogenesis is also unknown. It has been shown that the release of TGF-α into the medium of transformed cells can be correlated with the expression of the transformed phenotype (De Larco et al. 1981; Twardzik et al. 1982). Whether TGF-β expression is also involved in a coordinated manner with the expression of TGF-α is not known at present.

The hypothesis that malignant transformation of cells may result from inappropriate, later expression of autocrine factors that were required by cells during normal embryogenesis (Sporn and Todaro 1980) has attracted increasing attention with the recent discovery that the *sis* oncogene encodes a protein similar to PDGF (Doolittle et al. 1983; Waterfield et al. 1983). The entire question of the relationships of oncogenes to peptide growth factors has now become a central problem in the mechanism of carcinogenesis. It remains to be determined whether the TGF-β gene will be found to be yet another oncogene or whether the role of TGF-

Table 3 Amino Acid Composition of TGF-β from Bovine Kidney, Human Placenta, and Human Blood Platelets

Amino acid	Kidney	Placenta	Platelet
Aspartic acid	25 ± 1	24 ± 2	23 ± 2
Threonine	7 ± 1	8 ± 1	9 ± 1
Serine	16 ± 2	17 ± 1	17 ± 1
Glutamic acid	26 ± 1	25 ± 1	26 ± 1
Proline	n.d.	n.d.	n.d.
Glycine	15 ± 3	17 ± 4	21 ± 5
Alanine	18 ± 1	18 ± 1	17 ± 1
Half-cystine	16 ± 2	16 ± 2	16 ± 2
Valine	17 ± 1	15 ± 2	14 ± 1
Methionine	3 ± 0	2 ± 1	2 ± 0
Isoleucine	11 ± 1	11 ± 1	10 ± 1
Leucine	25 ± 1	24 ± 2	23 ± 2
Tyrosine	17 ± 2	17 ± 1	14 ± 1
Phenylalanine	8 ± 1	8 ± 1	6 ± 1
Histidine	8 ± 1	7 ± 1	6 ± 1
Lysine	22 ± 2	19 ± 2	18 ± 2
Tryptophan	n.d.	n.d.	n.d.
Arginine	12 ± 1	11 ± 1	10 ± 1

Values are expressed as residues/mole and are shown as mean ± range. For details, see Roberts et al. (1983b); Frolik et al. (1983); and Assoian et al. (1983).
(n.d.) Not done.

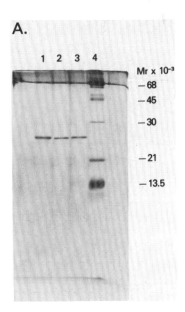
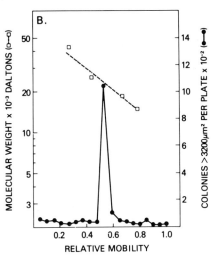

Figure 3 SDS-polyacrylamide gel electrophoresis of purified type β TGFs. (*A*) 50 ng each of purified TGF-β from bovine kidney (lane *1*), human placenta (lane *2*), and human platelets (lane *3*) were run on a polyacrylamide gel (1.5 mm, 12.5% acrylamide) under nonreducing conditions and stained with silver. Molecular-weight standards (lane *4*) were bovine serum albumin (68,000), ovalbumin (45,000), carbonic anhydrase (30,000), soybean trypsin inhibitor (21,000), and cytochrome *c* (13,500). (*B*) 240 ng of human placental TGF-β was applied to a gel (1.5 mm, 12.5% polyacrylamide). After electrophoresis, gel slices of 2 mm were extracted and assayed for colony-forming activity in the presence of 5 ng/ml of EGF. C^{14}-labeled protein standards run in an adjacent well consisted of ovalbumin (43,000), α-chymotrypsinogen (25,700), β-lactoglobulin (18,400), and lysozyme (14,300).

β in malignant transformation of cells will be found to be of a more indirect nature.

Acknowledgments

R.A. was supported by a postdoctoral fellowship from BASF Aktiengesellschaft, Ludwigshafen, Germany. We thank Margaret Konecni for help in preparation of the manuscript.

References

Anzano, M.A., A.B. Roberts, J.M. Smith, M.B. Sporn, and J.E. De Larco. 1983. Sarcoma growth factor from conditioned medium of virally transformed cells is composed of both type α and type β transforming growth factors. *Proc. Natl. Acad. Sci.* **80:** 6264.

Anzano, M.A., A.B. Roberts, C.A. Meyers, A. Komoriya, L.C. Lamb, J.M. Smith, and M.B. Sporn. 1982. Synergistic interaction of two classes of transforming growth factors from murine sarcoma cells. *Cancer Res.* **42:** 4776.

Assoian, R.K., A. Komoriya, C.A. Meyers, D.M. Miller, and M.B. Sporn. 1983. Transforming growth factor-β in human platelets. *J. Biol. Chem.* **258:** 7155.

De Larco, J.E. and G.J. Todaro. 1978. Growth factors from murine sarcoma virus–transformed cells. *Proc. Natl. Acad. Sci.* **75:** 4001.

De Larco, J.E., Y.A. Preston, and G.J. Todaro. 1981. Properties of a sarcoma-growth-factor-like peptide from cells transformed by a temperature-sensitive sarcoma virus. *J. Cell. Physiol.* **109:** 143.

Deuel, T.F., J.S. Huang, R.T. Proffitt, J.U. Baenziger, D. Chang, and B.B. Kennedy. 1981. Human platelet-derived growth factor. *J. Biol. Chem.* **256:** 8896.

Doolittle, R.F., M.W. Hunkapillar, L.E. Hood, S.G. Devare, K.C. Robbins, S.A. Aaronson, and H.N. Antoniades. 1983. Simian sarcoma virus *onc* gene, v-*sis*, is derived from the gene (or genes) encoding a platelet-derived growth factor. *Science* **221:** 275.

Frolik, C.A., L.L. Dart, C.A. Meyers, D.M. Smith, and M.B. Sporn. 1983. Purification and initial characterization of a type β transforming growth factor from human placenta. *Proc. Natl. Acad. Sci.* **80:** 3676.

Heldin, C., B. Westermark, and Å. Wasteson. 1981. Platelet-derived growth factor. *Biochem. J.* **193:** 907.

Marquardt, H. and G.J. Todaro. 1982. Human transforming growth factor: Production by a melanoma cell line, purification, and initial characterization. *J. Biol. Chem.* **257:** 5220.

Marquardt, H., M.W. Hunkapiller, L.E. Hood, D.R. Twardzik, J.E. De Larco, J.R. Stephenson, and G.J. Todaro. 1983. Transforming growth factors produced by retrovirus-transformed rodent fibroblasts and human melanoma cells: Amino acid sequence homology with epidermal growth factor. *Proc. Natl. Acad. Sci.* **80:** 4684.

Roberts, A.B., C.A. Frolik, M.A. Anzano, and M.B. Sporn. 1983a. Transforming growth factors from neoplastic and non-neoplastic tissues. *Fed. Proc.* **42:** 2621.

Roberts, A.B., M.A. Anzano, C.A. Meyers, J. Wideman, R. Blacher, Y.E. Pan, S. Stein, S.R. Lehrman, J.M. Smith, L.C. Lamb, and M.B. Sporn. 1983b. Purification and properties of a type β transforming growth factor from bovine kidney. *Biochemistry* **22:** 5692.

Sporn, M.B. and G.J. Todaro. 1980. Autocrine secretion and malignant transformation of cells. *New Engl. J. Med.* **303:** 878.

Sporn, M.B., D.L. Newton, A.B. Roberts, J.E. De Larco, and G.J. Todaro. 1981. Retinoids and suppression of the effects of polypeptide transforming factors—A new molecular approach to chemoprevention of cancer. In *Molecular actions and targets for cancer chemotherapeutic agents* (ed. A.C. Sartorelli et al.), p. 541. Academic Press, New York.

Sporn, M.B., A.B. Roberts, J.H. Shull, J.M. Smith, J.M. Ward, and J. Sodek. 1983. Polypeptide transforming growth factors isolated from bovine sources and used for wound healing *in vivo*. *Science* **219:** 1329.

Todaro, G.J., C. Fryling, and J.E. De Larco. 1980. Transforming growth factors produced by certain human tumor cells: Polypeptides that interact with epidermal growth factor receptors. *Proc. Natl. Acad. Sci.* **77:** 5258.

Twardzik, D.R., G.J. Todaro, H. Marquardt, F.H. Reynolds, and J.R. Stephenson. 1982. Transformation induced by Abelson murine leukemia virus involves production of a polypeptide growth factor. *Science* **216:** 894.

Waterfield, M.D., G.T. Scrace, N. Whittle, P. Stroobant, A. Johnsson, Å. Wasteson, B. Westermark, C. Heldin, J.S. Huang, and T.F. Deuel. 1983. Platelet-derived growth factor is structurally related to the putative transforming protein p28[sis] of simian sarcoma virus. *Nature* **304:** 35.

The Amount of EGF Receptor Is Elevated on Squamous Cell Carcinomas

G. Cowley,* J. A. Smith,* B. Gusterson,* F. Hendler,† and B. Ozanne‡

*Ludwig Institute of Cancer Research, Sutton, England; Departments of Internal Medicine† and Microbiology,‡ Southwestern Medical School, University of Texas Health Science Center, Dallas, Texas 75235

Epidermal growth factor (EGF) is a polypeptide hormone originally isolated from the mouse submaxillary gland (Cohen 1962). Related molecules that are functionally indistinguishable from murine EGF have been isolated from a variety of species, including man (for review, see Carpenter and Cohen 1979). When injected into newborn mice, EGF elicits, among other things, the proliferation and differentiation of the epidermis. In tissue culture, EGF induces the proliferation of a wide variety of cell types, including keratinocytes (Rheinwald and Green 1975) and fibroblasts (Carpenter and Cohen 1979). The effect of EGF is mediated through a cell-surface receptor. The number of receptors expressed on the cell surface varies with the cell type. For instance, human fibroblasts express approximately 4×10^4 receptors per cell while keratinocytes have 2×10^5 receptors per cell. The receptor is a plasma-membrane phosphoglycoprotein of 170 kd with a tyrosine-specific protein kinase activity (Ushiro and Cohen 1980). The gene for the human receptor is located on chromosome 7 (Waterfield et al. 1983a).

After binding EGF, the receptor-ligand complex sets off a complex chain of events that in responsive cells results in DNA synthesis and, ultimately, cell division (Carpenter and Cohen 1979). Perhaps the first consequence of EGF binding is the stimulation of the receptor's enzymatic activity (Cohen et al. 1980). Within seconds of EGF addition there is an increase in the total amount of cellular phosphotyrosine (Cooper and Hunter 1981; Erikson et al. 1981; Hunter and Cooper 1981). The receptor itself appears to be autophosphorylated at tyrosine residues (Ushiro and Cohen 1980). Also, there is a rapid increase in proteins phosphorylated at serine and threonine residues, which suggests the EGF receptor-ligand complex activates other protein kinases (Hunter and Cooper 1981).

The tyrosine specificity of the EGF receptor kinase is shared by the receptors for other peptide growth factors, platelet-derived growth factor (PDGF) (Cooper and Hunter 1982; Ek et al. 1982) and insulin, and protein kinases encoded by a variety of tumor virus–derived oncogenes (Hunter and Cooper 1983). The growth factor–activated receptor kinases share some substrates with the virally encoded tyrosine-specific protein kinases (Cooper and Hunter 1981; Erikson et al. 1981). One known intracellular substrate for the EGF receptor is a 36-kd protein. This 36-kd protein is also phosphorylated by the PDGF and insulin receptors at tyrosine and in cells transformed by tumor viruses whose oncogenes encode tyrosine-specific protein kinases (Cooper and Hunter 1981; Erikson et al. 1981; Hunter and Cooper 1983). The common enzymatic activities and substrates of the growth-regulating proteins (growth-factor receptors and oncogene-encoded protein kinases) suggest they may function through similar or related pathways.

The involvement of growth factors and their receptors in transformation is also apparent by the finding that many transformed cells produce transforming growth factors (TGFs) that interact with the receptors for serum-derived growth factors (DeLarco and Todaro 1978; Heldin et al. 1980; Dicker et al. 1981). For instance, sarcoma growth factor (SGF) is a TGF-α, which functions via the EGF receptor, and is produced by a variety of virally transformed fibroblasts and a human melanoma-derived cell line (DeLarco and Todaro 1978; Ozanne et al. 1980; Twardzik et al. 1982; Marquardt et al. 1983). Cells that produce SGF also produce a TGF-α synergistic factor, TGF-β, which greatly enhances the transforming activity of SGF and EGF on some test cells (Roberts et al. 1982). TGF-βs do not function through the EGF receptor. When normal rat kidney (NRK) cells, clone 49F, are exposed to TGFs (α and β) or rat-1 cells are treated with TGF-α or EGF, they transiently display many traits common to their virally transformed counterparts that produce TGFs. Their actin cytoskeleton becomes disorganized, and they transform morphologically (DeLarco and Todaro 1978; Ozanne et al. 1980). Also, their glucose transport is enhanced. They proliferate in low serum concentrations and in serum-free media and efficiently form colonies when suspended in soft agar (DeLarco and Todaro 1978; Ozanne et al. 1980).

SGF and the EGF receptor appear to play a role in maintaining the transformed phenotype through autocrine stimulation (Sporn and Todaro 1980; Kaplan et al. 1982). First, transformed cells that produce SGF do not bind EGF, presumably because of the saturation of the receptors by the ectopically produced SGF (DeLarco and Todaro 1978; Ozanne et al. 1980). Second, cells transformed by mutants of either Moloney murine sarcoma virus (Mo-MSV) (DeLarco et al. 1983) or Kirsten MSV (Ki-MSV) (Ozanne et al. 1980; Kaplan et al. 1982; DeLarco 1983) that are temperature sensitive for transformation only produce TGFs at the permissive temperature, at which the cells also fail to bind EGF. At the nonpermissive temperature, the cells bind EGF and respond mitogenically to both EGF and their own TGFs.

Third, growth of Ki-MSV-transformed rat-1 or NRK cells in serum-free media is dependent upon their own TGFs (Kaplan et al. 1982). Finally, variants of 3T3 cells, NR6, that lack the EGF receptor when transformed by Ki-MSV fail to express fully the transformed phenotype, whereas Ki-MSV transformed wild-type 3T3 cells do. Although both types of cells are morphologically transformed by the virus and produce TGFs, only the parental 3T3 cells are capable of growth in serum-free media and semisolid agar (P.L. Kaplan and B. Ozanne, unpubl.).

Although all this implicates the EGF receptor in maintaining transformation of cells that produce TGF-α, the final proof of autocrine stimulation of the transformed phenotype depends upon the demonstration that cells that produce TGF-α synthesize a functional EGF receptor. In this volume Kamata and Feramisco report experiments that conclusively demonstrate that Ki-MSV-transformed NRK cells synthesize functional EGF receptors. The recent reports that one viral oncogene, v-sis, encodes a protein closely related to PDGF (Doolittle et al. 1983; Waterfield et al. 1983b) further supports the suggestion that TGFs act in an autocrine fashion to maintain the transformed phenotype.

Another system in which the EGF receptor may be playing an important role in the expression of the malignant character of a cell is in squamous cell carcinomas (SCCs). It is well established that A431 cells derived from an epidermoid carcinoma of the cervix express high levels of EGF receptor (2×10^6 to 3×10^6/cell) compared with most other cell types (10^4–10^5/cell) (Todaro et al. 1976) and are tumorigenic in nude mice. The involvement of EGF with the growth of the epidermis and of keratinocytes in culture and the high level of expression of the EGF receptor in A431 cells suggest that the receptor might play a role in the malignant growth of SCCs. Below we present three lines of evidence that support the above supposition.

Methods

Squamous carcinoma cell lines
The cell lines HN-2, HN-5, HN-10, and HN-6Rr, derived from head and neck squamous carcinomas described by Easty et al. (1981), were maintained in Dulbecco's modified Eagle's medium (DMEM; Gibco) with 10% fetal calf serum (FCS; Sterile Systems) and supplemented with penicillin and streptomycin. A431 cells, EJ bladder carcinoma cells, and NRK cells were maintained in the same media.

Serum-free growth of A431 cells
Serum-free growth of A431 cells was performed as described for other cell types in Kaplan et al. (1982). Briefly, A431 cells were seeded at 2×10^4 cells/ml in a mixture of DMEM and Ham's F12 (Grand Island Biological Corporation), supplemented with 10 µg/ml transferrin and 1 µg/ml insulin (TI; Sigma) with varying concentrations of EGF (Collaborative Research). Cell counts were made every 3 days, when the media was changed.

EGF binding assays
EGF binding assays were as described by G. Cowley, B. Gusterson, and B. Ozanne (in prep.). EGF-R1, a murine monoclonal antibody to the human EGF receptor, was labeled with ^{125}I, using lactoperoxidase.

Metabolic labeling of the EGF receptor with [^{35}S]methionine and ^{32}P
The receptor was labeled with [^{35}S]methionine or ^{32}P according to Cooper and Hunter (1982). The labeled receptor was immunoprecipitated with EGF-R1 according to the method of Waterfield et al. (1983a).

Immunofluorescence with EGF-R1
Immunofluorescence assays were performed as described by Waterfield et al. (1983a), using the EGF-receptor-specific monoclonal antibody EGF-R1.

Assay for EGF receptor in human tissue
Tumor specimens were obtained from patients undergoing lung biopsies or other diagnostic procedures. Tissues were then frozen in a cryostat, and cryosections (6–8 µm) were prepared. The sections were fixed with acetone/phosphate-buffered saline (PBS) and reacted with either ^{125}I-labeled EGF-R1 or an irrelevant monoclonal antibody, ^{125}I-labeled 2B, of identical idiotype, according to the technique of Yuan et al. (1982). Sections were incubated with 500,000 cpm for 3 hours at 25°C and washed with PBS-FCS. The sections were exposed to X-ray film (Kodak X-Omatic) for 16 hours before development. Sections were either stained with hematoxylin and eosin or dipped in NTB-6 photographic emulsion. Slides dipped in emulsion were exposed for 96 hours, developed, and stained with hematoxylin and eosin. Pathology and binding was independently reviewed by W. Kingsley (binding of antibody was determined by at least two independent observers).

Results
EGF stimulates growth of A431 cells
Although it has been demonstrated that the growth of A431 cells is inhibited by concentrations of EGF that stimulate the growth of normal fibroblasts and keratinocytes (Gill and Lazar 1981; Buss et al. 1982), their growth is enhanced by concentrations of EGF that are submitogenic for normal cells. As seen in Table 1, the growth of A431 cells in serum-free media is dependent upon the presence of concentrations of EGF that are not mitogenic for fibroblasts and is inhibited by concentrations that are mitogenic for fibroblasts and keratinocytes (Rheinwald and Green 1975; Kaplan et al. 1982). The dependence of A431 cells upon low levels of EGF in serum-free media suggests that the high level of expression of EGF receptors by A431 cells might allow the cells to proliferate under conditions where normal keratinocytes could not and in that fashion enhance their growth.

Squamous cell carcinomas
To determine if the high level of receptors on A431 cells was unique to that cell line or common to SCCs, we

Table 1 Response of A431 Cells to EGF

Additions	EGF (ng/ml)	Cells/plate after 9 days
10% FBS	0	6.8×10^5
1% FBS	0	2.5×10^5
TI	0	4.5×10^4
	0.1	2.8×10^5
	0.3	1.1×10^5
	1.0	7.0×10^4
	3.0	9.7×10^4
	10.0	1.75×10^4

A431 cells were grown in varying concentrations of EGF (see Methods). (TI) Transferrin and insulin.

Table 2 Number of EGF Receptors on SCC Cell Lines

Cell line	Receptors/cell[a]	SCC/[b] keratinocytes
A431	3.0×10^6	11.1
LICR-HN-2	1.0×10^6	3.7
LICR-HN-3	2.0×10^6	7.4
LICR-HN-4	2.3×10^6	8.5
LICR-HN-5	14.8×10^6	54.8
LICR-HN-6	1.2×10^6	4.4
LICR-HN-10	2.5×10^6	9.2
LICR-HN-6Rr	2.1×10^6	7.7
LICR-HN-6nL	1.0×10^6	3.7
Keratinocytes	2.7×10^5	1.0

[a]Determined by Scatchard analysis.
[b]Ratio of EGF receptors/cell on SCC lines to normal human keratinocytes.

have determined the number of EGF receptors expressed on eight SCC-derived cell lines (Easty et al. 1981). Binding studies with ^{125}I-labeled EGF or ^{125}I-labeled EGF-R1 reveal that all SCC cell lines studied expressed levels of receptor similar to A431 cells (Table 2). Binding studies on normal human keratinocytes in culture reveal that they express 2.7×10^5 receptors per cell. Thus, the tumor-derived cell lines express at least 3.7 times more receptors than do normal keratinocytes in culture, and one cell line, HN-5, has 54 times the number of receptors. The growth of these cell lines, like A431 cells, is inhibited by 3–10 ng/ml of EGF (data not shown). The high level of expression of receptor can also be demonstrated by using immunofluorescence with the monoclonal antibody EGF-R1. HN-5 cells, which express 50 times more receptor than normal keratinocytes, are brightly stained in a pattern reminiscent of that seen with A431 cells (Waterfield et al. 1983a), whereas normal human keratinocytes are barely visible (Fig. 1).

Incubation with [^{35}S]methionine and subsequent immunoprecipitation of the receptors indicated that their synthesis is similar to that observed for A431 cells (Fig. 2). The receptors on these SCC cell lines appear to be functional. The receptors are phosphorylated after the addition of EGF to the media (Fig. 3) and are down regulated by EGF (Table 3). From the characterization of these cell lines, we conclude that it is a general property of SCC-derived cell lines to express elevated levels of functional EGF receptors and to have their growth inhibited by concentrations of EGF that are mitogenic for normal keratinocytes or fibroblasts. The mechanism that results in this increased expression of EGF receptors remains unclear. The recent involvement of karyotypic rearrangements and/or gene amplification in the activation of cellular homologs of viral oncogenes suggests that these might be possible mechanisms. Al-

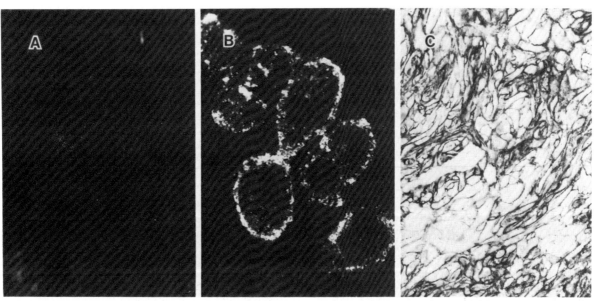

Figure 1 Indirect immunofluorescent staining of normal human keratinocytes (A) and HN-5 (B), using the EGF-receptor-specific monoclonal antibody EGF-R1. (C) Immunoperoxidase staining of the EGF receptor on a xenograft of HN-5 tumor grown in nude mice.

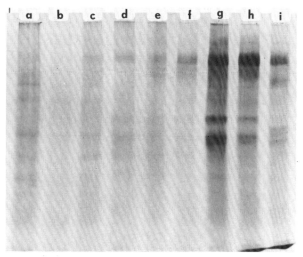

Figure 2 Immunoprecipitation of [³⁵S]methionine-labeled EGF receptors with EGF-R1. (Lane a) SVCBC, an SV40-transformed human fibroblast cell line; (lane b) NRK cells; (lane c) EJ bladder epithelial tumor cell line; (lane d) HN-2; (lane e) HN-10; (lane f) HN-6Rr; (lanes g,h) HN-5; (lane i) A431.

Table 3 Down Regulation of EGF Receptor

Cell line	% Down regulation	Receptors/cells
Keratinocyte	58.2	2.7×10^5
A431	62.3	3.0×10^6
HN-5	51.0	14.8×10^6

Receptors were down regulated by exposure to 50 ng/ml EGF for 12 hr. Cells were washed and EGF-binding assays performed.

ber of receptors might be the result of an adaptation to growth in tissue culture rather than a marker of malignant squamous cells. To determine if high levels of EGF receptors were present in SSCs before adaption to tissue-culture conditions, we have examined the ability of ¹²⁵I-labeled EGF-R1 to bind freshly frozen tissue obtained from squamous tumors and a variety of other tumors and normal tissues (Yuan et al. 1982). An autoradiograph of a representative assay is shown in Figure 4. The ¹²⁵I-labeled EGF-R1 binds very strongly to the two cryosections from patients with SCCs (Fig. 4b,d), weakly to the adenocarcinoma (Fig. 4a), and intermediately to the undifferentiated carcinoma (Fig. 4c). The specificity of the radioimmunobinding assay was demonstrated using emulsion autoradiography (Fig. 5). Silver grains are present in very high concentrations over the squamous tumor cells but not the stroma (Fig. 5a) and are absent from the section treated with similar concentrations of an irrelevant monoclonal antibody of the same idiotypic class (Fig. 5b). Specific binding of EGF-R1 is also demonstrated in a cryosection of normal skin (Fig. 5d) and a nonsquamous tumor (Fig. 5c) at one tenth to one hundredth of that observed in the squamous tumor. We have studied 10 squamous tumors of the lung and nasopharynx, and all have very high levels of EGF receptors. Fifty

though the 3.7-fold to 11-fold increase observed for most of the lines could have been achieved through increased transcription or decreased turnover of the receptors, metabolic labeling studies suggest the increased level is due to an increased rate of synthesis of the receptors, not a change in their half-life (data not shown).

EGF-receptor binding to human tumors

As increased levels of EGF receptors were observed solely in squamous tumor cell lines, the increased num-

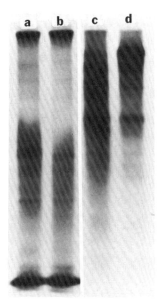

Figure 3 Immunoprecipitation of ³²P-labeled EGF receptors before and after stimulation with EGF. (Lane a) A431; (lane b) HN-5; (lane c) A431 + EGF; (lane d) HN-5 + EGF.

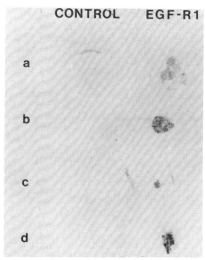

Figure 4 Binding of EGF-R1 to human lung cancer cells. Cryosections from biopsied specimens were prepared and reacted with either ¹²⁵I-labeled EGF-R1 or ¹²⁵I-labeled 2B (see Methods). (a) Tissues from a patient with adenocarcinoma; (b,c,d) specimens from a patient with SCC.

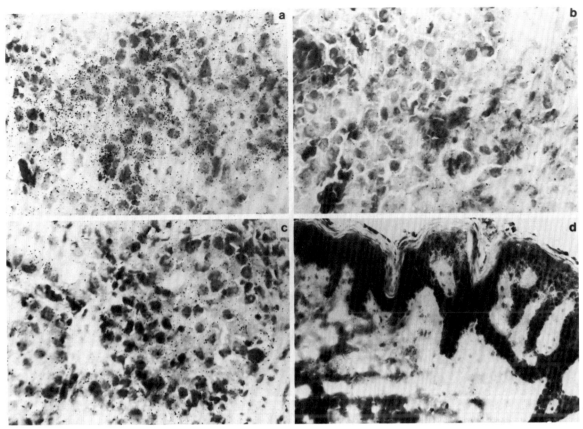

Figure 5 Emulsion autoradiography of EGF-R1 binding to human lung cancer. Cryosections from biopsied specimens were prepared and reacted with iodinated antibodies, and emulsion autoradiography was carried out (see Methods). Photomicrographs of the emulsions were made using Pan-X (Kodak) and a blue filter. (a) SCC of lung reacted with ^{125}I-labeled EGF-R1; (b) SCC of lung reacted with ^{125}I-labeled 2B; (c) undifferentiated carcinoma reacted with ^{125}I-labeled EGF-R1; (d) normal skin reacted with ^{125}I-labeled EGF-R1. Magnifications, 250×.

percent of undifferentiated lung tumors and two of eight adenocarcinomas of the lung have increased levels of EGF receptors. The level of EGF receptors in these tumors is rarely greater than that of normal skin. Thus, the presence of high levels of EGF receptors appears to be a consistent marker of squamous cell malignancies in vivo as well as in cell culture.

Discussion

The data presented above suggest that high-level expression of EGF receptors is a common property of SCCs and occurs sometime during the progression of normal epidermal cells to the malignant state. Further studies are required to determine at what stage this increased expression occurs and whether it is necessary for the development of the malignant phenotype. An alternative explanation for the high level of EGF-receptor expression on SCCs would be that these cells are a clone of a minor fraction of epidermal stem cells with high levels of receptors. These stem cells might naturally have high levels of receptors that decrease during differentiation. As a consequence of malignant transformation the differentiation could be blocked and the receptor levels remain elevated. If this were the mechanism for malignant squamous cells to have increased numbers of EGF receptors, the absence of cells with high numbers of EGF receptors in normal skin would suggest that these stem cells are very rare. Our goal is to ascertain whether elevated amounts of EGF receptors on SCCs contribute to tumorigenicity and tumor growth. Studies are underway to evaluate the EGF-receptor concentrations in premalignant states and to correlate the receptor concentrations in tumor specimens with the tumorigenicity in vivo.

Acknowledgments

We wish to thank D. Tucker for preparing the manuscript. Part of the work was supported by National Institutes of Health grants CA-32815 (to F.H.) and CA-23043 (to B.O.).

References

Buss, J.E., J.E. Kudlow, C.E. Lazar, and G.N. Gill. 1982. Altered EGF-stimulated protein kinase activity in variant A431 cells with altered growth responses to EGF. *Proc. Natl. Acad. Sci.* **79:** 2574.

Carpenter, G. and S. Cohen. 1979. Epidermal growth factor. *Annu. Rev. Biochem.* **48:** 193.

Cohen, S. 1962. Isolation of a mouse submaxillary gland protein accelerating incisor eruption and eyelid opening in the newborn animal. *J. Biol. Chem.* **237:** 1555.

Cohen, S., G. Carpenter, and L. King. 1980. Epidermal growth factor receptor-protein kinase interactions. *J. Biol. Chem.* **255:** 4834.

Cooper, J.A. and T. Hunter. 1981. Four different classes of retroviruses induce phosphorylation of tyrosine present in similar proteins. *Mol. Cell. Biol.* **1:** 394.

———. 1982. Similar effects of platelet-derived growth factor and epidermal growth factor on the phosphorylation tyrosine in cellular proteins. *Cell* **31:** 263.

DeLarco, J.E. 1983. Sarcoma growth factor and the transforming growth factors. In *Growth and maturation factors* (ed. G. Guroff), p. 193. Wiley, New York.

DeLarco, J.E. and G.J. Todaro. 1978. Growth factor from murine sarcoma virus-transformed cells. *Proc. Natl. Acad. Sci.* **75:** 4001.

Dicker, P., P. Pohjanpelto, P. Pettican, and E. Rozengurt. 1981. Similarities between fibroblast-derived growth factor and platelet-derived growth factor. *Exp. Cell Res.* **135:** 221.

Doolittle, R.F., M.W. Hunkapiller, L.E. Hood, S.G. Devane, K.C. Robbins, S.A. Aaronson, and H.N. Antoniades. 1983. Simian sarcoma virus onc gene, v-*sis*, is derived from the gene (or genes) encoding a platelet-derived growth factor. *Science* **221:** 275.

Easty, D.M., G.C. Easty, R.L. Carter, P. Monaghan, and L.J. Butler. 1981. Ten human carcinoma cell lines derived from squamous carcinomas of the head and neck. *Br. J. Cancer* **43:** 772.

Ek, B., B. Westermark, A. Wasteson, and C.-H. Heldin. 1982. Stimulation of tyrosine-specific phosphorylation by platelet-derived growth factor. *Nature* **295:** 419.

Erikson, E., D.J. Shealy, and R.L. Erikson. 1981. Evidence that viral transforming gene products and epidermal growth factor stimulate phosphorylation of the same cellular protein with similar specificity. *J. Biol. Chem.* **256:** 11381.

Gill, G.N. and C.S. Lazar. 1981. Increased phosphotyrosine content and inhibition of proliferation in EGF-treated A431 cells. *Nature* **293:** 305.

Heldin, C.-H., B. Westermark, and A. Wasteson. 1980. Chemical and biological properties of a growth factor from human cultured osteosarcoma cells: Resemblance with platelet-derived growth factor. *J. Cell Physiol.* **105:** 235.

Hunter, T. and J.A. Cooper. 1981. Epidermal growth factor induces rapid tyrosine phosphorylation of proteins in A431 human tumor cells. *Cell* **24:** 741.

———. 1983. A comparison of the tyrosine protein kinases encoded by retroviruses and activated by growth factors. *U.C.L.A. Symp. Mol. Cell Biol.* (in press).

Kaplan, P.L., M. Anderson, and B. Ozanne. 1982. Transforming growth factor(s) production enables cells to grow in the absence of serum: An autocrine system. *Proc. Natl. Acad. Sci.* **79:** 485.

Marquardt, H., M.W. Hunkapiller, L.E. Hood, P.R. Twardzik, J.E. DeLarco, J.R. Stephenson, and G.J. Todaro. 1983. Transforming growth factors produced by retrovirus-transformed rodent fibroblasts and human melanoma cells: Amino acid sequence homology with epidermal growth factors. *Proc. Natl. Acad. Sci.* **80:** 4684.

Monaghan, P., J. Knight, G. Cowley, and B. Gusterson. 1983. Differentiation of a squamous carcinoma cell line in culture and tumorigenicity in immunologically incompetent mice. *Virchows Arch. Cell Pathol.* **400:** 87.

Ozanne, B., R.J. Fulton, and P.L. Kaplan. 1980. Kirsten murine sarcoma virus-transformed cell lines and a spontaneously transformed rat cell line produce transforming factors. *J. Cell. Physiol.* **105:** 163.

Rheinwald, J.G. and H. Green. 1975. Serial cultivation of strains of human epidermal keratinocytes; the formation of keratinizing colonies from single cells. *Cell* **6:** 331.

Roberts, A.B., M.A. Anzano, L.C. Lamb, J.M. Smith, C.A. Frolik, H. Marquardt, G.J. Todaro, and M.B. Sporn. 1982. Isolation from murine sarcoma cells of novel transforming growth factors potentiated by EGF. *Nature* **295:** 417.

Sporn, M.B. and G.J. Todaro. 1980. Autocrine secretion and malignant transformation of cells. *New Engl. J. Med.* **303:** 878.

Todaro, G.J., J.E. DeLarco, and S. Cohen. 1976. Transformation by murine and feline sarcoma viruses specifically blocks bindings of epidermal growth factor to cells. *Nature* **264:** 26.

Twardzik, D.R., G.J. Todaro, H. Marquardt, F.H. Reynolds, and J.R. Stephenson. 1982. Transformation induced by Abelson murine leukemia virus involves production of a polypeptide growth factor. *Science* **216:** 894.

Ushiro, H. and S. Cohen. 1980. Identification of phosphotyrosine as a product of epidermal growth factor-activated protein kinase in A431 cell membranes. *J. Biol. Chem.* **255:** 8363.

Waterfield, M.D., E.L.V. Mayers, P. Stroobant, P.L.P. Bennett, S. Young, P.N. Goodfellow, G.S. Banting, and B. Ozanne. 1983a. A monoclonal antibody to the human epidermal growth factor receptor. *Cell. Biochem.* **20:** 149.

Waterfield, M.D., G.T. Scrace, N. Whittle, P. Stoobant, A. Johnsson, B. Westermark, G.H. Heldin, J.S. Huang, and T.F. Devel. 1983b. Platelet-derived growth factor is structurally related to the putative transforming protein P28^{sis} of simian sarcoma virus. *Nature* **304:** 35.

Yuan, D., F.J. Hendler, and E.S. Vitetta. 1982. Characterization of a monoclonal antibody reactive with a subset of human breast tumors. *J. Natl. Cancer Inst.* **68:** 719.

Is the *ras* Oncogene Protein a Component of the Epidermal Growth Factor Receptor System?

T. Kamata and J.R. Feramisco
Cold Spring Harbor Laboratory, Cold Spring Harbor, New York 11724

A number of human cellular oncogenes have been identified in part by transfection of DNA from tumor cell lines and tumor tissues into NIH-3T3 cells (Bishop 1983). Several of these oncogenes have been classified as members of the *ras* gene family because of their relatedness to the viral oncogenes of the Harvey (*v*-Ha-*ras*) or Kirsten (*v*-Ki-*ras*) murine sarcoma viruses (Ha-MSV or Ki-MSV). The human genes have been put into at least three classes, called *c*-Ha-*ras*, *c*-Ki-*ras*, or *c*-N-*ras*, depending upon the degree of homology to *v*-Ha-*ras* or *v*-Ki-*ras* oncogenes or the human cell line of initial description (i.e., neuroblastoma), respectively (Der et al. 1982; Parada et al. 1982; Shimizu et al. 1983b). All of the known *c*-*ras* oncogenes have normal or protooncogene forms that have only a single amino acid change from the corresponding activated oncogene forms (Reddy et al. 1982; Tabin et al. 1982; Capon et al. 1983; Shimizu et al. 1983a). In general, the proteins encoded by the *ras* family are about 21,000 daltons and are associated with the inner surface plasma membrane. The only known biochemical property common to all forms of the *ras* protein is the ability to bind guanine nucleotides (Papageorge et al. 1982). In the particular case of the *v*-Ha-*ras* protein, apparent autophosphorylation on threonine occurs (both in vivo and in vitro) (Shih et al. 1980). The guanine nucleotide–binding property is most likely an important property of the *ras* proteins since in a temperature-sensitive viral mutant of the *v*-Ki-*ras* gene, the virus is temperature sensitive for transformation and the *v*-Ki-*ras* protein is temperature sensitive for the guanine nucleotide–binding activity (Scolnick et al. 1979). Of the many alterations in the cellular phenotype associated with transformation by the *ras* oncogenes, the one relevant to the present work is the apparent loss of epidermal growth factor (EGF)-binding sites on the surfaces of the transformed cells (Todaro et al. 1976). This has been postulated to occur as a result of either down regulation of the receptors (Cherington and Pardee 1982) or occupancy of the receptors by the alpha-type transforming growth factor (TGF), which is produced by the transformed cells (Anzano et al. 1983). Because of the similar cellular locations of the *ras* oncogene proteins and the growth factor receptors, and the fact that several other hormone systems utilize guanine nucleotide–binding proteins as regulatory elements, we investigated the possible relationship of *ras* oncogene proteins to the EGF receptors. Here we describe findings that the normally low EGF-binding activity of *ras*-transformed cell membranes is stimulated by the addition of guanine nucleotides. By immunoadsorption of the *ras* proteins from the membranes, we found that this effect appears dependent upon the presence of the *ras* protein. In addition, we show that the phosphorylation of the *v*-Ha-*ras* protein and the guanine nucleotide–binding activity of the *v*-Ha-*ras* or *c*-Ha-*ras* proteins in membranes isolated from transformed cells are stimulated by the addition of EGF and MgGTP^{--}. Taken together, these results suggest a role of the *ras* oncogene protein in the EGF receptor system.

Results

Nucleotide effects on EGF binding to membranes from normal and transformed cells

A variety of nucleotides (Mg^{++} salts) were tested for their ability to affect the binding of ^{125}I-labeled EGF to membrane fractions isolated from normal and transformed cells. A viral nonproducer Ha-MSV transformed cell line (Ha-NRK) and a rat cell line (Q4) (Ruley 1983) cotransfected with the adenoviral E1A gene and the human *c*-Ha-*ras* (T24) oncogene were utilized in this study. (Although the biochemical basis for the need of the adenoviral E1A gene, or one of several other genes, as well as the *c*-Ha-*ras* oncogene to transform primary rat cells is not as yet known [Land et al. 1983; Ruley 1983], it would seem that expression of the E1A gene alone does not interfere with EGF binding or stimulate the production of transforming growth factors by, at least, KB cells [Fisher et al. 1983], which appear to be the only such cells tested in these properties to date.) As shown in Table 1, ^{125}I-labeled EGF binding to normal rat kidney cell (NRK) membranes is slightly stimulated by the addition of MgGTP^{--}, whereas the binding is increased twofold by the addition of MgATP^{--}. Other nucleotides had little effect on the ^{125}I-labeled EGF binding in NRK cells. In membranes isolated from rat cells transformed by either the *v*-Ha-*ras* (Ha-NRK cells) oncogene or the activated *c*-Ha-*ras* oncogene (Q4 cells), addition of MgATP^{--} caused an approximately fourfold stimulation in ^{125}I-labeled EGF binding. In these same membrane preparations from the *ras*-transformed cells, addition of MgGTP^{--} resulted in about a fivefold increase in the ^{125}I-labeled EGF binding. (As expected, however, the basal ^{125}I-labeled EGF-binding activity of the membranes from both of the transformed cell lines was very low compared with the basal binding activity of NRK cell membranes [see note to Table 1].) MgGDP$^-$ and MgGMP also had stimulatory effects on ^{125}I-labeled EGF binding to Ha-NRK cell membranes, although the

Table 1 Effect of Nucleotides on the ^{125}I-labeled EGF–binding Activity of Membranes from Normal and ras-transformed Cells

Additions	NRK	Ha-NRK	Q4
None	100	100	100
GTP	117	444	633
GDP	103	283	—
GMP	100	279	—
GppNHp	67	97	118
ATP	207	420	423
AppNHp	63	93	67
TTP	85	117	—
UTP	92	101	—
CTP	150	130	—

NRK cells, Ha-MSV-transformed cells, and Q4 cells (Ruley 1983) (primary rat cells cotransfected with T24 ras cDNA and adenoviral E1A DNA fragment provided by E. Ruley) were cultured in Dulbecco's modified Eagle's medium (DMEM) supplemented with 10% calf serum. The cells were harvested at a subconfluent stage of cell growth and washed twice with PBS. The cells were homogenized (dounce homogenizer) in 10 vol of buffer A (10 mM Tris · HCl, pH 7.4, 10 mM NaCl, 0.1 mM EDTA, 0.1 mM 2-mercaptoethanol, 1 mM phenylmethylsulfonylfluoride) and 10 vol of deionized H$_2$O. After pelleting the nuclei by low-speed centrifugation, the supernatant was centrifuged at 100,000g for 60 min at 4°C. The particulate fraction was solubilized in buffer A containing 1% Triton X-100. Aliquots of the solubilized membranes (2.5 × 10^6 cell equivalent) were adjusted to 0.1% BSA and 10 mM MgCl$_2$, and the indicated nucleotides (5 mM) were added. The mixtures were incubated with 10^5 cpm of ^{125}I-labeled EGF (150 μCi/μg; NEN) for 15 min at room temperature. ^{125}I-labeled EGF binding of membrane was carried out by filter-binding assay using 8.5% polyethylene glycol 6000 as described by Carpenter et al. (1979). Nonspecific binding was determined in the presence of unlabeled EGF (50-fold excess). Data represent the means of specific binding in two separate experiments. For NRK membranes, 100% binding was 3.2 fmoles/10^6 cells; for Ha-NRK cell membranes, 100% binding was 0.57 fmoles/10^6 cells; for Q4 cell membranes, 100% binding was 0.63 fmoles/10^6 cells.

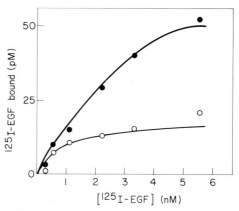

Figure 1 Effect of MgGTP^{--} on the saturation binding of EGF to the Ha-NRK cell membranes. The solubilized particulate fraction from subconfluent Ha-NRK cells (as described in Table 1) (2.5 × 10^6 cell equivalent/data point) was preincubated in the presence or absence of 5 mM GTP (with 10 mM MgCl$_2$) for 15 min at 4°C. Indicated amounts of ^{125}I-labeled EGF were then added for 15 min at room temperature and specific-binding activity was measured (as described in Table 1). Similar results were obtained in two separate experiments.

effects were much less than that caused by the addition of MgGTP^{--}. Nonhydrolyzable analogs of ATP and GTP were not stimulatory in the ^{125}I-labeled EGF–binding assay using membranes from normal or transformed cells. The optimal concentration for the stimulatory effects of ATP and GTP in transformed cell membranes was approximately 1–5 mM, and the effect required the presence of Mg^{++} (10 mM).

In membrane preparations from Ha-NRK cells, saturation-binding curves for ^{125}I-labeled EGF assayed in the presence or absence of MgGTP showed that MgGTP^{--} resulted in an increase in the maximum binding of ^{125}I-labeled EGF and a slight increase in the concentration of ^{125}I-labeled EGF required for half-maximal binding (Fig. 1).

Effect of anti-ras antibodies on the nucleotide stimulation of ^{125}I-labeled EGF–binding activity

To gain insight into the nature of the nucleotide effects on the ^{125}I-labeled EGF–binding activity of the ras-transformed cells, and in particular to determine if the ras protein is involved in the nucleotide effects, experiments in which the membranes were preincubated with antibodies directed against the ras protein and then assayed for ^{125}I-labeled EGF–binding activity in the presence and absence of nucleotides were performed.

First, varying amounts (0–5 μg of IgG) of monoclonal antibodies against the ras proteins were added to the solubilized membranes from Ha-NRK cells (10 min, 4°C) prior to the measurement of the ^{125}I-labeled EGF–binding activity. Compared with the addition of control antibodies (nonimmune rat or mouse IgG), anti-ras IgGs reduced the MgGTP^{--} stimulation of the ^{125}I-labeled EGF–binding to the membranes in a dose-dependent manner by about 70% (not shown). Second, if the immune complexes of the ras proteins were removed prior to the ^{125}I-labeled EGF–binding assay by the addition of rabbit anti-rat IgG and protein A–Sepharose, the MgGTP^{--} stimulation of the ^{125}I-labeled EGF–binding activity was completely abolished compared with control antibody treatment (Fig. 2). In contrast, the MgATP^{--} stimulation of the ^{125}I-labeled EGF–binding activity was reduced by about 50%. Interestingly, the antibodies to the ras proteins had no effect on the basal ^{125}I-labeled EGF–binding activity.

Stimulation of phosphorylation of the v-Ha-ras protein by EGF

One of the biochemical markers of the v-Ha-ras protein is its phosphorylation both in vivo and in vitro (Shih et al. 1980). Although it is not known what the biological consequences (if any) of this phosphorylation are, the

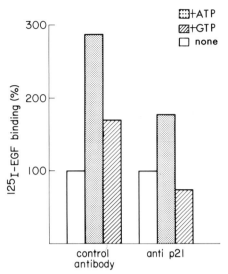

Figure 2 Effect of immunoadsorption of the ras proteins on the nucleotide stimulation of ^{125}I-labeled EGF binding to Ha-NRK cell membranes. Membranes were isolated from subconfluent Ha-NRK cells as described above. Membranes (10^6 cells) were preincubated with antibody no. 259 for 1 hr at 4°C and further incubated with anti-rat IgG and protein A–Sepharose for 1 hr. After the immune complexes were pelleted by centrifugation, the supernatant was mixed with 5 mM MgATP^{--} or MgGTP^{--} for 15 min at 4°C. Then 10^5 cpm of ^{125}I-labeled EGF was added for 15 min at room temperature, and the binding activity was measured as described above. Data show specific binding in duplicate experiments.

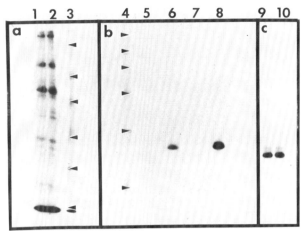

Figure 3 Stimulation of phosphorylation of v-Ha-ras p21 in the presence of EGF. Ha-NRK cells were cultured in Dulbecco's modified Eagle's medium (DMEM) supplemented with 10% calf serum. When the cells were confluent, they were rinsed once in DMEM and maintained in DMEM (serum-free) for 24–48 hr. The cells were harvested and washed with PBS, and membranes were prepared as described in Table 1. Aliquots of the solubilized membranes (2.5×10^6 cell equivalent) were incubated with EGF (500 ng/ml in 1 mg/ml BSA; BRL) or BSA alone for 10 min at 4°C. The phosphorylation reaction was initiated by the addition of 10 μCi of [γ-^{32}P]GTP (2500 Ci/mmole; NEN) and MgCl$_2$ to 5 mM. The reaction was carried out for 10 min at 30°C. The reaction mixtures were preadsorbed with rabbit anti-rat IgG (2 μl) and protein A–Sepharose (40 μl of 50% suspension). The v-Ha-ras protein was immunoprecipitated by the addition of a rat anti-p21 monoclonal antibody (no. 259) (4 μl). For control precipitations, a nonimmune antibody was used. After incubation for 1.5 hr at 4°C, 4 μl of rabbit anti-rat IgG and 40 μl of a 50% suspension of protein A–Sepharose were added (1 hr). The immunoprecipitates were analyzed by SDS-PAGE. Alternatively, Ha-NRK cells were labeled with [^{35}S]methionine (80 μCi/ml) for 18 hr, and the membranes were isolated as described above. After the membranes were treated with EGF or BSA as above in the presence of 5 mM MgCl$_2$ and GTP (100 μM) for 10 min at 4°C, v-Ha-ras p21 was immunoprecipitated by monoclonal antibody no. 259 as described above and analyzed by SDS-PAGE followed by fluorography (Bonner and Laskey 1974). (a) The total phosphoproteins detectable in the particulate fraction of Ha-NRK cells in the absence (lane 1) or presence (lane 2) of EGF analyzed by a 10% SDS-polyacrylamide gel; molecular-mass markers (lane 3) from top to bottom are 200,000, 94,000, 68,000, 40,000, 30,000, and 14,000 daltons. (b) The immunoprecipitated, phosphorylated v-Ha-ras proteins from membranes treated without (lane 6) or with (lane 8) EGF; (lanes 5,7) immunoprecipitations of the reaction mixtures without and with EGF, respectively, using a nonimmune antibody; molecular-mass markers on the 12.5% SDS-polyacrylamide gel are as indicated in a. (c) The [^{35}S]methionine-labeled v-Ha-ras proteins immunoprecipitated from reaction mixtures treated without (lane 9) or with (lane 10) EGF.

reaction appears to be one of autophosphorylation. It is possible, though, that this marker reaction reflects either the ras protein or the activity of an as yet unidentified protein. Because of the apparent effects of guanine nucleotides on the EGF-binding activity of cells transformed by the ras proteins, we examined the phosphorylation state of the v-Ha-ras protein in membranes as a function of added EGF.

Membranes isolated from Ha-NRK cells were preincubated with or without EGF, followed by the addition of MgGTP^{--} (γ-^{32}P labeled). Although no major changes in the total phosphoproteins in the membranes can be easily seen, immunoprecipitation of v-Ha-ras p21 showed that the phosphorylation of this protein was stimulated threefold to fivefold by the addition of EGF (Fig. 3). Platelet-derived growth factor (PDGF) had no effect on the phosphorylation of v-Ha-ras p21 under these conditions. Of importance in this reaction was the use of membranes isolated from serum-deprived Ha-NRK cells (24–48-hr deprivation), which seemed to lower the basal (EGF) phosphorylation activity of the v-Ha-ras protein. If membranes from cells growing in medium containing 10% calf serum were used, little or no stimulation of phosphorylation of v-Ha-ras was observed while the basal level of phosphorylation was high. This most likely reflects the presence of EGF (and perhaps other growth factors) in the serum.

The phosphorylation enhancement of v-Ha-ras p21 due to EGF was dependent upon the EGF concentration (Fig. 4) and was not due to an increase in the immunoprecipitable v-Ha-ras p21 over the course of the reaction period as seen by the levels of [^{35}S]methionine-labeled v-Ha-ras p21 recovered from the reaction in the presence or absence of added EGF (Fig. 3). Although the stoichiometry and the site(s) of the phosphorylation of the v-Ha-ras protein have as yet to be determined, the chemical nature of the reaction has been determined.

Figure 4 Effect of EGF concentration on the phosphorylation of v-Ha-ras p21. Ha-NRK cell membranes were isolated and solubilized as described in Fig. 3. After preincubation with varying concentrations of EGF (0–500 ng/ml), phosphorylation reactions were carried out in the presence of 10 μCi [γ-^{32}P]GTP and 5 mM MgCl$_2$. p21 was immunoprecipitated by anti-p21 (no. 259) as described in Fig. 3 and analyzed by SDS-PAGE.

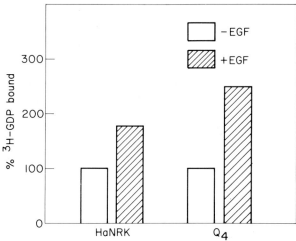

Figure 5 Stimulation of [^{3}H]GDP-binding activity of p21 by EGF. Membranes were isolated from serum-starved Ha-NRK or Q4 cells (~2.5 × 10^6 cell/data point) as described in Table 1 and preincubated with EGF (500 ng/ml) or BSA alone, then 3 μCi of [^{3}H]GDP (10 Ci/mmole; NEN) and 5 mM MgCl$_2$ were added (10 min at 4°C). [^{3}H]GDP bound to p21 was immunoprecipitated by anti-p21 antibody (no. 259) as described in Fig. 3 (Shih et al. 1980). The precipitates were washed four times with buffer A containing 5 mM MgCl$_2$ and analyzed by liquid scintillation counting. The 100% value was 0.15 pmole for Ha-NRK and 0.068 pmole for Q4 cells.

The immunoprecipitated phosphoproteins obtained from reactions done in the presence or absence of added EGF (as in Fig. 3) were treated with 1 M KOH for 2 hours at 55°C (Eckhart et al. 1979) and found to be completely labile, which indicates a lack of phosphotyrosine. In addition, partial acid hydrolysis of the proteins followed by two-dimensional phosphoamino acid analysis indicated the presence of phosphothreonine and phosphoserine but not of phosphotyrosine (data not shown).

Effect of EGF on the guanine nucleotide–binding activity of the ras proteins

One of the biochemical properties common to all of the ras proteins studied to date is the guanine nucleotide–binding activity (Papageorge et al. 1982). Either by immunoprecipitation assays or by direct measurements with the purified proteins (Shih et al. 1980), this property can be detected. The apparent K_D for GDP binding, for example, is 2×10^{-7} M, and the stoichiometry of binding is nearly 1 mole of bound nucleotide per mole of protein (Shih et al. 1980). We examined the GDP-binding activity of immunoprecipitated v-Ha-ras and c-Ha-ras proteins, using isolated membrane fractions from Ha-NRK cells and Q4 cells, respectively, and the effect of EGF on this activity. As shown in Figure 5, EGF causes about a twofold stimulation in [^{3}H]GDP-binding activity of the ras proteins present in both the Ha-NRK and Q4 membrane fractions. The stimulation depended upon the EGF concentration and was complete at about 200 ng/ml (not shown). We attempted to determine if the guanine nucleotide–binding activity of the ras proteins present in membranes isolated from normal human and rodent cell lines was also stimulated by EGF; however, with the extremely low binding activity present in normal cell membranes, it was unreliable to make such measurements.

Discussion

In attempting to determine the molecular function of the ras oncogenic proteins, several considerations must be taken into account. First, the protein resides in the plasma membrane and presumably exerts its effects at this location. Second, the ras family of proteins includes proto-oncogenic forms, "activated" oncogenic forms, and viral forms, all of which bind guanine nucleotides and are most likely involved in the same important cellular processes (Papageorge et al. 1982). And third, the ras activity(ies) is probably influenced by other cellular components (which follows from this observation that one of many single amino acid changes can convert the proto-oncogene forms into activated oncogene forms). Although we have little to say concerning the function of the ras proteins as yet, several interesting observations have been made and should be discussed.

As shown here, serum deprivation of v-Ha-ras-transformed cells (Ha-NRK) leads to a state of the ras protein in the membranes that is sensitive to EGF for the apparent ras autophosphorylation reaction. In addition, serum deprivation of cells transformed by either the c-Ha-ras or v-Ha-ras proteins (Q4 or Ha-NRK) leads to a state of the ras proteins that is sensitive to EGF in the guanine nucleotide–binding activity. These results seemingly indicate some type of biochemical interaction (either direct or indirect) between the EGF receptor system and the ras proteins.

In terms of the interaction of EGF with membranes from transformed cells, we have found that adenine and guanine nucleotides stimulate the measurable ^{125}I-labeled EGF binding to membranes from ras-transformed cells. Although MgATP^{--} stimulated the apparent EGF binding to membranes from both normal and ras-transformed cells, the MgGTP^{--} stimulated only the EGF binding to ras-transformed cell membranes. Removal of the ras proteins from the membranes prior to the measurement of the ^{125}I-labeled EGF–binding activity resulted in the abolishment of the MgGTP^{--} stimulation of the EGF-binding activity of the ras-transformed cell membranes but only resulted in a decrease by one half in the MgATP^{--}-stimulated EGF-binding activity of these same cells. Whether or not the nucleotide effects on EGF binding are mediated through phosphorylation reactions is not known; however, it would appear that the ATP and GTP effects are different, and the GTP effect is related to the presence of the ras proteins.

In this light, it is of interest that alterations in EGF binding have been found in a variety of transformed cells, including v-Ki-ras-, v-Ha-ras-, and c-Ha-ras-transformed cell lines (Todaro and DeLarco 1978; T. Kamata and J. R. Feramisco, unpubl.). The prevailing explanation for the reduced EGF binding to these cells is that the receptors are occupied by the alpha-type TGFs, which are produced by the transformed cells themselves and compete with EGF for binding to the receptors (Anzano et al. 1983). An alternative proposal is a disappearance of the receptors from the cell membrane through down regulation (Cherington and Pardee 1982). The observation presented here that EGF-binding sites in ras-transformed cell membranes can be uncovered by guanine and adenine triphosphates suggests that indeed EGF receptors are present in the transformed cells. The apparent number of EGF-binding sites can be increased in membranes from ras transformants by the addition of nucleotides to a number similar to that found in membranes from normal cells. Although the molecular mechanism is not as yet understood for the nucleotide effect of unmasking EGF-binding sites, it is possible that it involves either the dissociation of bound TGFs from the receptors or the appearance of new binding sites or other reasons.

It is of interest to discuss the known properties of several other hormone-receptor systems with regard to what has been presented here. In the well-known case of the alpha-, beta-adrenergic receptor system of adenylate cyclase, guanine nucleotides and guanine nucleotide–binding proteins regulate several features of the receptor system (for review, see Rodbell 1980). The guanine nucleotide–binding proteins, N_i and N_s, couple agonist receptor binding to second-messenger production (in this case, adenylate cyclase activity). In addition, guanine nucleotides diminish the binding of catecholamines, glucagon, and prostaglandin E to their receptors in membrane fractions. In yet three other hormone-receptor systems—the opiate-receptor system (Blume 1978), the insulin-receptor system (Rodbell 1980), and the IgE-receptor system (Comperts 1983)—guanine nucleotides again appear to play regulatory roles. In the first system, GTP in the presence of Mg^{++} or Mn^{++} increases the apparent number of high-affinity binding sites for enkephalins or clonidine. In the latter two systems, guanine nucleotide–binding proteins are thought to mediate or couple hormone-receptor occupancy to second-messenger production (i.e., Ca^{++}) in the case of the IgE receptor of mast cells or to a cyclic AMP–independent protein kinase activity (Walaas et al. 1979) and cyclic AMP–phosphodiesterase activity in the case of the insulin receptor (Heyworth et al. 1983).

In the case of the EGF-receptor system, the data presented here would seem to suggest a possible role of guanine nucleotides and, perhaps, the ras oncogene proteins in the activity of this receptor system. Considering that it appears as though growth factors alter the state of tyrosine phosphorylation of two other membrane-associated oncogene or proto-oncogene proteins (i.e., EGF stimulates polyoma middle-T-antigen phosphorylation [Segawa and Ito 1983] and PDGF stimulates pp60^{c-src} phosphorylation [R. Ralston, pers. comm.]), it may be a commonality to find interactions among various growth factors/receptor systems and the membrane-associated oncogene or proto-oncogene proteins. As for the present case of the ras proteins and the EGF-receptor system, the possible consequences of this interaction in terms of the relationship between EGF and TGF in transformed cells remains to be elucidated.

Acknowledgments

We thank J.D. Watson for his support of this work; Dr. Mike Wigler for his gift of cell lines and helpful discussions; Drs. Mark Furth and Ed Scolnick for their gift of monoclonal antibodies to p21; and Dr. Earl Ruley for his gift of transformed rat cell lines. This work was supported by National Institutes of Health grant GM-28277 and National Cancer Institute Cancer Center grant CA-13106.

References

Anzano, M.A., A.B. Roberts, J.M. Smith, M.B. Sporn, and J.E. De Larco. 1983. Sarcoma growth factor from conditioned medium of virally transformed cells is composed of both type α and type β transforming growth factors. *Proc. Natl. Acad. Sci.* **80:** 6264.

Bishop, J.M. 1983. Cellular oncogenes and retroviruses. *Annu. Rev. Biochem.* **52:** 301.

Blume, A.J. 1978. Interaction of ligands with the opiate receptors of brain membranes: Regulation by ions and nucleotides. *Proc. Natl. Acad. Sci.* **75:** 1713.

Bonner, W.M. and R.A. Laskey. 1974. A film detection method for tritium-labelled proteins and nucleic acids in polyacrylamide gels. *Eur. J. Biochem.* **46:** 83.

Capon, D.J., P.H. Seeburg, J.P. McGrath, J.S. Hayflick, U. Edman, A. Levinson, and D.V. Goeddel. 1983. Activation of Ki-ras 2 gene in human colon and lung carcinomas by two different point mutations. *Nature* **304:** 507.

Carpenter, G., L. King, and S. Cohen. 1979. Rapid enhancement of protein phosphorylation in A-431 cell membrane preparations by epidermal growth factor. *J. Biol. Chem.* **254:** 4884.

Cherington, P.V. and A.B. Pardee. 1982. On the basis for loss of the EGF growth requirement by transformed cells. *Cold Spring Harbor Conf. Cell Proliferation* **9**: 221.

Comperts, B.D. 1983. Involvement of guanine nucleotide-binding protein in the gating of Ca^{2+} by receptors. *Nature* **306**: 64.

Der, C.J., T.G. Krontiris, and G.M. Cooper. 1982. Transforming genes of human bladder and lung carcinoma cell lines are homologus to the *ras* genes of Harvey and Kirstein sarcoma viruses. *Proc. Natl. Acad. Sci.* **79**: 3637.

Eckhart, W., M.A. Hutchinson, and T. Hunter. 1979. An activity phosphorylating tyrosine in polyoma T antigen immunoprecipitates. *Cell* **18**: 925.

Fisher, P.B., M.R. Boersig, G.M. Graham, and I.B. Weinstein. 1983. Production of growth factors by type 5 adenovirus transformed rat embryo cells. *J. Cell. Physiol.* **114**: 365.

Heyworth, C.M., S. Rawal, and M.D. Houslay. 1983. Guanine nucleotides can activate the insulin-stimulated phosphodiesterase in liver plasma membranes. *FEBS Lett.* **154**: 87.

Land, H., L.F. Parada, and R.A. Weinberg. 1983. Tumorigenic conversion of primary embryo fibroblasts requires at least two cooperating oncogenes. *Nature* **304**: 596.

Papageorge, A., D. Lowry, and E. Scolnick. 1982. Comparative biochemical properties of p21 *ras* molecules coded for by viral and cellular *ras* genes. *J. Virol.* **44**: 509.

Parada, L.F., C.J. Tabin, C. Shih, and R.A. Weinberg. 1982. Human EJ bladder carcinoma oncogene is homologue of Harvey sarcoma virus *ras* gene. *Nature* **297**: 474.

Reddy, E.P., R.K. Reynolds, E. Santos, and M. Barbacid. 1982. A point mutation is responsible for the acquisition of transforming properties by the T24 human bladder carcinoma oncogene. *Nature* **300**: 149.

Rodbell, M. 1980. The role of hormone receptors and GTP-regulatory proteins in membrane transduction. *Nature* **284**: 17.

Ruley, H.E. 1983. Adenovirus early region 1A enables viral and cellular transforming genes to transform primary cells in culture. *Nature* **304**: 602.

Scolnick, E.M., A.G. Papageorge, and T.Y. Shih. 1979. Guanine nucleotide-binding activity as an assay for *src* protein of rat-derived murine sarcoma virus. *Proc. Natl. Acad. Sci.* **76**: 5355.

Segawa, K. and Y. Ito. 1983. Enhancement of polyoma virus middle T antigen tyrosine phosphorylation by epidermal growth factor. *Nature* **304**: 742.

Shih, T.Y., A.G. Papageorge, P.E. Stockes, M.O. Weeks, and E.M. Scolnick. 1980. Guanine nucleotide-binding and autophosphorylating activities associated with the $p21^{src}$ protein of Harvey murine sarcoma virus. *Nature* **287**: 686.

Shimizu, K., D. Birnbaum, M.A. Ruley, O. Fasano, Y. Suard, L. Edlund, E. Taparowsky, M. Goldfarb, and M. Wigler. 1983a. Structure of the Ki-*ras* gene of the human lung carcinoma cell line Calu-1. *Nature* **304**: 497.

Shimizu, K., M. Goldfarb, Y. Suard, M. Perucho, Y. Li, T. Kamata, J. Feramisco, E. Stavnezer, J. Fogh, and M.H. Wigler. 1983b. Three human transforming genes are related to the viral *ras* oncogenes. *Proc. Natl. Acad. Sci.* **80**: 2112.

Tabin, C.J., S.M. Bradley, C.I. Bargmann, R.A. Weinberg, A.G. Papageorge, E.M. Scolnick, R. Dhar, D.R. Lowry, and E.H. Chang. 1982. Mechanism of activation of a human oncogene. *Nature* **300**: 143.

Todaro, G.J. and J.E. DeLarco. 1978. Growth factors produced by sarcoma virus-transformed cells. *Cancer Res.* **38**: 4147.

Todaro, G.J., J.E. DeLarco, and S. Cohen. 1976. Transformation by murine and feline sarcoma viruses specifically blocks binding of epidermal growth factor to cells. *Nature* **264**: 26.

Walaas, O., E. Walaas, E. Lystad, A.R. Alertsen, and R.S. Horn. 1979. Effect of insulin and guanosine nucleotides on protein phosphorylations by sarcolemma membranes from skeletal muscle. *Mol. Cell. Endocr.* **16**: 45.

In Vitro Correlates of Tumorigenicity of REF52 Cells Transformed by Simian Virus 40

D.B. McClure, M. Dermody,* and W.C. Topp*†

The W. Alton Jones Cell Science Center, Old Barn Road, Lake Placid, New York 12946, and *Cold Spring Harbor Laboratory, Cold Spring Harbor, New York 11724

The properties of mammalian cells can become stably and inheritably changed following in vitro infection by a number of viruses, both DNA and RNA (Tooze 1981; Weiss et al. 1982). Many of these same viruses are oncogenic when injected into newborn hamsters (Diamandopoulis 1972, 1973, 1978), and a few are similarly, although more weakly, oncogenic in newborn mice and rats (Rabson et al. 1964; Yabe et al. 1964; Ibelgaufts et al. 1980). These virally induced tumors are almost always mesenchymal in origin and can be variously classified histologically as fibrosarcomas, reticulosarcomas, or angiosarcomas (Diamandopoulis 1973; Topp et al. 1981). The in vitro transformation of the properties of cultured mesenchymal cells is an accepted model for this viral tumor induction (Freedman and Shin 1974).

The virus/cell interaction, which is stabilized by the integration of the viral genome within the host chromosomes, resembles in many ways a perpetual abortive infection (Topp et al. 1982). A substantial amount of literature has accumulated over the years that addresses the question, which aspect or aspects of this virus/cell interaction are responsible for the oncogenic activity of the virus? A number of newly expressed cellular phenotypes have been found to correlate reasonably well with oncogenic growth potential as assayed by tumor formation in immune deficient nu/nu mice (Pollack et al. 1974, 1975a,b; Chen et al. 1976; Frisque et al. 1980). No one marker has proven to be sufficiently reliable that it is possible to conclude that the molecular mechanism by which that phenotype is acquired must be important in viral oncogenesis, although the ability of a cell to grow colonially in soft agar has repeatedly been found to correlate closely with oncogenic potential in the substantial majority of cases (Pollack et al. 1974, 1975b; Frisque et al. 1980).

A number of the properties of transformed cells, including growth in soft agar, growth to high population densities, loss of cytoplasmic actin microfilament bundles, and the production of a plasminogen activator, can all be induced in "normal" cells either by culture in a medium supplemented with higher than usual concentrations of serum or by the addition of physiological levels of factors that are similar to those present in serum (Holley and Kiernan 1968; Castor 1971; Dulbecco and Elkington 1973; Stoker 1973; Lee and Weinstein 1978;

O'Neill et al. 1979; Topp et al. 1979). Therefore it is possible that the use of culture media containing whole serum has variably modulated the behavior of the virally transformed cells and thus partially obscured our understanding of the virus/cell interaction.

We have chosen to reapproach the study of viral transformation by utilizing chemically defined cell-culture conditions. It is possible that in the absence of the uncontrolled and uncharacterizable influence of added serum, it will be possible to identify a reliable in vitro correlate of tumorigenicity. We have concentrated on three aspects of cellular behavior: the exact nature of the macromolecular requirements for unlimited monolayer proliferation, the ability of cells to grow in soft agar suspension, and the organization of the actin microfilaments. In a previous paper (McClure et al. 1982) we reported that a sub-tetraploid Fisher rat embryo–derived cell line, REF52, could be continuously cultured in a mixture of DMEM and F12 basal media supplemented with physiological levels of insulin, transferrin, hydrocortisone, epidermal growth factor (EGF), vasopressin, and the high-density lipoprotein complex (HDL). Simian virus 40 (SV40)–transformed variants of REF52 continued to depend on insulin and transferrin for exponential growth but showed a mildly toxic response to hydrocortisone. Unlike REF52, the transformants required either EGF and vasopressin (E/V) or HDL but not both. None of these lines were oncogenically transformed by the virus in the nude mouse assay.

Here we describe the isolation, by selection in soft agar followed by in vivo passage, of highly tumorigenic variants of these transformants. We report the additional alterations in growth factor requirements that accompany these selections, characterize the agar growth of the initial transformants and their variants both in serum-containing (SC) and serum-free (SF) medium, and compare these properties with the oncogenic growth potential of the cells. We conclude that agar growth in the *absence* of serum or added EGF correlates very well with growth potential in nu/nu mice, but that serum- or EGF-dependent agar growth does not.

Materials and Methods

REF52 is a sub-tetraploid, continuous cell line established essentially by a 3T3 protocol from a 14-day gestation Fisher rat embryo primary culture (Logan et al. 1981). The SV40-transformed variants have been described (McClure et al. 1982), and *dl*884 mutant SV40

†Present address: Otisville Biotech, Inc., Sanatorium Road, Otisville, New York 10963.

transformants were isolated in a similar manner. (d/884 is a deletion mutant of SV40 that has lost sequences encoding the viral small T protein [Shenk et al. 1976].) Soft agar sublines were established by removing single colonies of cells from soft agar plates with a pasteur pipette 3 weeks after plating. Tumor-derived lines were established by surgical excision of tumors from anesthetized nude mice subsequent to injection of 5×10^6 cells of the agar-derived sublines subcutaneously, followed by mincing and trypsin disaggregation (0.25% trypsin in phosphate-buffered saline [PBS] at 37°C with stirring for 15 to 30 min). All cells were cultured in a basal medium consisting of 25% Ham's F12 (Gibco 430-1700) and 75% DMEM (Gibco 430-2100) containing 5×10^{-9} M sodium selenite (Difco), 0.4 g/liter histidine, 50 mg/liter gentamicin sulfate, and 10^{-5} M 2-aminoethanol (all Sigma) supplemented with either 10% fetal calf serum (FCS; Reheis) or a combination of factors, including 100 ng/ml bovine insulin, 1 µg/ml hydrocortisone, 5 µg/ml transferrin, 100 ng/ml arginine vasopressin (all Sigma), 100 ng/ml epidermal growth factor (EGF; Collaborative Research), 1 µg/cm² fibronectin (Collaborative Research), and 25 µg/ml HDL (gift of D. Weinstein, U.C. San Diego). Where indicated, HDL was replaced by 50 mg/ml fatty acid–free bovine serum albumin (FAF-BSA; Sigma) conjugated by mixing with oleic acid (Sigma) freshly dissolved at 20 mg/ml in ethanol and used at a final concentration of 0.5 mg/ml FAF-BSA to 2.5 µg/ml oleic acid. The FAF-BSA was routinely treated with 1% w/w CAB-O-SIL (Packard) before use to remove any contaminating lipoprotein (Weinstein 1979). The basal medium was modified for use in 5% CO_2-gassed incubators by the inclusion of 1.2 g/liter sodium bicarbonate and 0.225 g/liter NaCl (Gibco Corp., pers. comm.) to maintain isotonic sodium concentrations. The basal medium was occasionally supplemented with 15 mM HEPES buffer, pH 7.4, for added pH stability. All solutions were prepared with reverse osmosis–purified water (at least 7 MΩ resistance; Millipore Corp.) and, where necessary, sterilized by filtration through 0.22-µm Nalgene or Millex filters. Cells were subcultured with 0.025% trypsin (Sigma) in PBS on ice or in a refrigerator; after use, trypsin was inhibited by the addition of soybean trypsin inhibitor (Sigma) to a final concentration of 1 mg/ml. Culture plates (Falcon) were routinely pretreated with 100 µg/ml poly-D-lysine (Sigma), followed by extensive rinsing with sterile water. Only disposable plasticware was used. These procedures have been more fully elaborated elsewhere (McClure et al. 1982).

Cloning assays were performed by plating 200 cells per 60-mm dish followed by formalin fixation and Giemsa staining 10 days later. Agar growth was assayed by plating 1×10^4 cells per 35-mm dish in 0.33% electrophoresis grade agarose (Sigma) over a layer of 0.9% agarose. Colonies (>0.1 mm) were fixed in formalin (10% v/v), stained with crystal violet (0.001% w/v), and counted after 3 weeks using a Wild dissecting microscope with a calibrated grid. Soft agar medium, factor addition, and colony staining were essentially as in McClure (1983). Tumorigenicity was assayed by subcutaneous injection of 5×10^6 cells into adult nu/nu mice. Actin stains were performed on cells grown directly on culture plates as described previously (McClure et al. 1982).

Results

Properties of transformed cells in serum-containing medium

As we reported previously, our REF52 cells, initially transformed by SV40 and selected for morphological variance, are only partially transformed by most standard criteria (McClure et al. 1982). Only one line, REF52 WT10, clones with an efficiency higher than 1% in SC soft agar and lacks appreciable cytoplasmic actin microfilament bundles, as visualized by immunofluorescence. None of the lines produce tumors when 5×10^6 cells are inoculated in nude mice, except WT10, which produced one tumor after a latency of 5 to 6 months. To obtain oncogenic lines, we established soft agar–derived sublines of each transformant by picking several of the few colonies that did appear in the initial cultures. Once expanded in monolayer culture, these sublines were tested for cloning efficiency in soft agar as well as actin cable patterns, and all were found to be essentially fully transformed by these two criteria (data not shown). Other cytoskeletal elements (vimentin- and tubulin-containing filaments) were unchanged. Either one or two soft agar sublines of WT1, WT2, WT4, and WT10 were assayed for tumorigenicity. These lines also failed to induce tumors within 2 months of injection of 5×10^6 cells. After 3 to 5 months, tumors did appear in roughly 50% of the injected animals surviving that length of time, and one or two of these tumors from each of two of the soft agar sublines (WT2 SA2.1 and WT10 SA10) were returned to in vitro culture. In the case of WT6, seven agar clones were isolated and injected into two nude mice each. In each case, after 2 months, one of two mice had nodules and/or tumors. After 3 months, all animals had either nodules or tumors, and two tumor-derived sublines were established, both from one agar subline. Once these tumor-derived sublines were expanded and reinjected at inocula of 5×10^6 cells, rapidly growing tumors were observed within 2 to 3 weeks. Again, all these lines were transformed according to the criteria of actin cable patterns and soft agar growth, and, in addition, the intermediate filament (vimentin) networks showed a partial disruption (data not shown). The properties of REF52 WT6 and the tumorigenic sublines derived from REF52 WT6 are listed in Table 1 and are completely representative of other lines similarly isolated.

Macromolecular growth factors required for SF colonial growth

The basal media DMEM and F12 were initially designed for the optimal growth of mesenchymal cultures (McClure 1983). The morphology as well as the HDL and vasopressin responsiveness of REF52 suggested that it might well be of vascular endothelial or smooth-muscle origin, so no further attempt was made to optimize the

Table 1. In Vitro Properties of REF52, REF52 WT6, and Soft Agar and Tumor-derived Variants in Serum Containing Medium

	REF52	+SV40 →	WT6	soft agar →	WT6 Ag6	nu/nu → mouse	WT6 Ag6 nu1
% CE in soft agar	0		0.5		16.6		35.0
Actin cables	++		+−		−		−
Tumorigenic in nu/nu mice	no		no		yes (2–3-mo latency)		yes (2–3-wk latency)

75:25 DMEM/F12 mix. Although we have reported that the several initial viral transformants differed one from another only in their requirement for oleic acid supplement, the degree of this effect varies from experiment to experiment, for reasons we have yet to determine. Except for this, the macromolecular growth factor requirements are constant within each type of transformant—initial, soft agar derived, and tumor derived—and are shown in Table 2 for the WT6 series. It is remarkable that the factor requirements correlate so well, both with selective pressure used to isolate the transformants and, by and large, with the phenotype of the transformants. Unlike the initial transformants, the soft agar–derived clones no longer require either E/V or HDL for unlimited colonial growth. Insulin, transferrin, and FAF-BSA/oleic acid are completely sufficient to maintain indefinite exponential growth on fibronectin/poly-D-lysine–coated plates. The initial morphological transformants retained a requirement for either E/V or HDL. We group EGF and vasopressin together as one requirement because REF52 will proliferate reasonably well in a medium supplemented with one or the other and there is only a marginal benefit to be derived by adding both at once, whereas the HDL requirement is absolute and unique (see McClure et al. 1982). No further change in requirements is seen when tumor-derived lines are established, except that hydrocortisone is no longer toxic, perhaps reflecting that the cells have been isolated for growth in the steroid-rich environment of the mouse.

There are specific cell-surface receptors for the transferrin molecule, and it is possible that this factor is stimulating growth by some other mechanism than as an iron-transport protein. We attempted to replace the transferrin with conalbumin (an iron-binding protein found in egg white) and were successful in doing so at 100 μg/ml iron-loaded conalbumin; this concentration partially replaced the transferrin requirement. We conclude that it is likely that the transferrin is serving simply to deliver iron to our cells and that the soft agar–and tumor–derived sublines are capable of unlimited monolayer culture in the absence of any mitogenic stimuli.

Because we observed a close relationship between the acquisition of soft agar growth potential and the complete obviation of a need for mitogenic factors, we thought it possible that the two effects might be related. Therefore we isolated six monolayer colonies that arose when REF52 WT6 was plated sparsely in SF medium containing only insulin, transferrin, and FAF-BSA/oleic acid. These WT6 variant colonies appeared at roughly the same frequency as did soft agar colonies of WT6 in SC medium. In four of the six sublines, the reduced factor requirement seemed to be stably acquired, but in only one of these was enhanced agar growth observed (percentage cloning efficiency [%CE] of 18.2 compared with %CE of 0.5 for WT6). Interestingly, only this line grew as a tumor in nu/nu mice, the tumors appearing after 2 or 3 months in only a few of the animals injected. This experiment will have to be expanded before firm conclusions can be drawn, but it is clear that the two properties, E/V plus HDL independence and soft agar growth, are not tightly linked.

Table 2 Growth Factor Requirements of REF52, WT6, WT6 Ag6, and WT6 Ag6 nu1 for Colony Formation in Serum-free Monolayer Culture

Additions to basal medium[a]	% Colony formation after 12 days			
	REF52[b]	WT6	WT6 Ag6	WT6 Ag6 nu1
F, I, T, O	0	2	20	20
F, I, T, O, E/V	0	25	25	22
F, I, T, O, HDL	1	32	35	20
F, I, T, O, E/V, HDL	30	25	30	20

[a] Abbreviations: (F) fibronectin; (I) insulin; (T) transferrin; (O) FAF-BSA/oleic acid; (E/V) EGF + vasopressin.
[b] Plus hydrocortisone.

Growth properties and factor requirements for REF52 transformed by the *dl*884 mutant of SV40

There are two proteins encoded within the transforming region of SV40, although only one, the large T protein, is required for lytic growth. Therefore it has been possible to isolate viable deletion mutants that are specifically mutated within the second gene, that encoding the small T protein (Shenk et al. 1976; Sleigh et al. 1978), and to use these mutants to evaluate the role of the small T protein in transformation. Past experience suggests that the small T protein is required for both growth in soft agar and for loss of actin microfilament bundles in established rat cell lines (Topp et al. 1979). We isolated five REF52 variants that were morphologically transformed by *dl*884 SV40, a small T mutant that not only has deleted substantial coding regions of the small T gene but also has lost sequences required to mature small T mRNA. This virus must lack essentially all small T protein activity.

As expected, none of the five transformants cloned at all in soft agar, and all retained well-articulated actin microfilament structures. Although the majority of the whole virus transformants also displayed an only partially transformed phenotype, the *dl*884 transformants were conspicuous by their essentially zero growth capacity in soft agar and the clear extent to which the actin structures were maintained. Clearly, the small T gene product is active in REF52 cells and participates in the development of the transformed phenotype.

However, in terms of alterations in growth factor requirements, the *dl*884 transformants were identical to the WT SV40 transformed lines (results not shown). It is likely that the initial reduction in growth factor requirement that we observe upon transformation of REF52 is the result of the activity of the large T gene product.

Phenotype of REF52, the transformed variants, and the sublines in SF culture

As elaborated above, there is no property of our REF52 transformants determined in SC culture that is unique to the highly oncogenic cells. The strongly tumorigenic tumor-derived lines possess a phenotype very similar to that of the weakly tumorigenic soft agar subclones. The only possible exception to this that we have found is the partial and not particularly striking diminution of the intermediate filament networks. Moreover, there is no further alteration in the macromolecular requirements for optimal, unlimited, SF monolayer proliferation that accompanies oncogenic transformation. We thus reexamined the phenotype of the various transformants in SF culture in the minimal medium determined for optimal monolayer proliferation (Table 3).

The microtubule and intermediate filament networks were essentially identical between SC and SF conditions for each of the four lines examined (WT2, WT4, WT6, and WT10), as were the growth rates, monolayer colony-forming efficiencies, and ultimate saturation densities (data not shown). A significant enlargement of cellular volume was observed in all SF cultures, but this might well represent simply an osmotic adjustment to the greatly reduced protein concentrations under SF conditions.

REF52 cells also behaved identically in SF and SC cultures with respect to soft agar growth (absolutely none under either condition) and the presence of pronounced actin-containing microfilament bundles. On the other hand, REF52 WT6, which gave a cloning efficiency of roughly 1% in SC soft agar cultures, reverted to behavior very similar to REF52 in the SF culture conditions that supported monolayer proliferation.

Importantly, the REF52 WT6 soft agar subline Ag6, which cloned with an efficiency of 10–20% in SC soft agar, also gave essentially no soft agar growth in the SF conditions that were optimal for monolayer proliferation. Only the nude mouse–derived subline REF52 WT6 Ag6 nu1 cloned in soft agar under minimally supplemented SF conditions (insulin, transferrin, hydrocortisone, and FAF-BSA/oleic acid). (Hydrocortisone, although slightly toxic to the in vitro transformants in monolayer culture, is essential for efficient soft agar growth and was included in all soft agar studies.) It would appear that the only correlate we can make between cellular properties in culture and tumorigenicity in nu/nu mice is the ability to clone in soft agar in the minimal medium required for monolayer proliferation.

Interestingly, the loss of cytoplasmic actin networks was also found to be highly dependent on culture conditions. Cell lines that had lost these structures in SC cultures tended to revert under SF conditions to the ordered cables characteristic of untransformed cells (Table 3). Only WT6 Ag6 nu1 is devoid of cables under all culture conditions. However, cable disruption was very dependent on cell density with all lines, tending toward dissociation as the cultures approached a monolayer. As a result, the interpretation of these results is

Table 3 Phenotype of REF52, WT6, WT6 Ag6, and WT6 Ag6 nu1 in the Minimal Serum-free Medium Necessary for Optimal Monolayer Colony Formation

Cell line (supplements)[a]	% CE soft agar	Actin cables
REF52 (F, I, T, H, O, E/V, HDL)	0	+ +
WT6 (F, I, T, O, E/V, or HDL)	0[b]	+ +
WT6 Ag6 (F, I, T, O)	0[b]	+ +
WT6 Ag6 nu1 (F, I, T, O)	18.0[b]	–

[a]Abbreviations: See legend to Table 2.
[b]Hydrocortisone (H) was included in soft agar assay (see text).

difficult, and, until clarified, these experiments are not being pursued.

Factors that modulate the transformed phenotype

REF52 WT6 and WT6 Ag6 are able to condition the medium over dense monolayer cultures with a factor or factors capable of enhancing the growth of WT6 in SC soft agar (Table 4). No detectable factor is produced by REF52, although previously a factor in REF52-conditioned medium has been found to alter somewhat the morphology of other established rat cell lines (M.A. Anderson and P. Kaplan, unpubl.). Medium (either SC or SF) conditioned by WT6 and WT6 Ag6 both seem to contain similar levels of WT6 soft agar growth–enhancing activity (data not shown). No significant enhancement in the SC soft agar growth of WT6 Ag6 or WT6 Ag6 nu1 is produced by any of the conditioned media. The addition of EGF, particularly in combination with hydrocortisone, to the SC cultures produces much the same effect as does conditioned medium on the agar growth of WT6. Platelet-derived growth factor (PDGF), endothelial cell growth supplement, fibroblast growth factor, and a variety of other factors had no effect (data not shown). This suggests that the SV40-transformed REF52 cultures might be producing a factor similar to EGF.

Both transformant-conditioned medium and EGF plus HC are capable of enhancing the soft agar growth of REF52 WT6 to the same level as is observed in the soft agar–derived subline REF52 WT6 Ag6 but are effective only in the presence of serum (Table 5). Serum factors were no longer necessary for the soft agar growth of the agar-derived sublines, in particular REF52 WT6 Ag6; however, EGF (not needed for monolayer proliferation) was definitely required, and even when present the colony size remained small compared with those seen in SC cultures. Apparently WT6 Ag6 continues to require a serum factor(s) for efficient soft agar growth that can be partially replaced by EGF. On the other hand, the tumor-derived subline WT6 Ag6 nu1 grew equally well in SC and in SF cultures and showed no apparent response to EGF or conditioned medium in either situation. Clearly, soft agar growth continues to require mitogenic stimulation by either EGF or conditioned medium in addition to a factor or factors present in serum under conditions where monolayer cultures show no response or

Table 4 Effect of Conditioned Medium on Soft Agar Growth of REF52 and Its SV40-transformed Derivatives

Cell line	Control	Conditioned medium[a] from		
		REF52	WT6	WT6 Ag6
REF52	0	0	0	0
WT6	0.8	1.4	14.0	14.5
WT6 Ag6	16.0	16.5	20.0	18.5
WT6 Ag6 nu1	35.0	32.0	38.0	40.0

Units represent percent colony formation in SC soft agar.
[a]SF medium was collected from high-density cultures after 24-hr incubation and assayed at a 1:3 dilution with fresh SC medium.

Table 5 Effect of Conditioned Medium or EGF on Soft Agar Growth under Serum-free Conditions

Cell line	% CE in soft agar		
	SF[a]	+CM[b]	+EGF
REF52	0	0	0
WT6	0	0	0.5
WT6 Ag6	0	6.0	7.4
WT6 Ag6 nu1	31.0	33.0	35.0

[a]SF: Serum-free medium containing I, T, H, O, HDL (see Table 2).
[b]CM: Serum-free conditioned medium from WT6.

dependence. This requirement is only obviated after a cell has undergone oncogenic transformation.

Both conditioned medium and EGF (plus HC) have a growth-promoting effect on soft agar cultures of WT6 and WT6 Ag6. As mentioned above, cell density and thus, potentially, medium conditioning have a profound effect on the actin structures in these two lines as well. However, no effect of EGF (plus or minus HC) has been found on SC or SF cultures of either WT6 or WT6 Ag6, which raises some questions as to the similarity of action of EGF and the conditioned medium.

Discussion

By controlling the use of whole serum in our cultures of REF52 cells and their transformed derivatives, we have been able to clarify the various roles played by the virus, serum factors, and cellular transforming factors in the development of the transformed phenotype.

The first conclusion that we can draw from these studies is that SV40-transformed variants of REF52 require fewer mitogenic stimuli for monolayer growth than do the untransformed parent cells and that selection for a maximally transformed phenotype, i.e., a high cloning efficiency in soft agar, produces a line that has completely obviated both the need for and the response to mitogenic stimuli for monolayer growth. Similar effects have been reported from a number of other laboratories (Stiles et al. 1980; Cherington and Pardee 1982; Sager et al. 1982). The loss of the mitogenic requirement for monolayer growth precedes, but may be a prerequisite for, the acquisition of the ability to grow as a tumor in nude mice.

On the other hand, when we characterized the agar growth and microfilament organization of the tumorigenic and nontumorigenic REF52 variants, a clear correlate did emerge. We can identify two distinct transformed phenotypes, one in which agar growth and the disruption of microfilament aggregates is dependent on serum or EGF in the culture medium and a second in which these are intrinsic properties of the transformants. The nontumorigenic and weakly tumorigenic variants of REF52 (e.g., REF52 WT6 and WT6 Ag6) are capable of substantial soft agar growth only in the presence of serum, and in the case of WT6, either EGF or medium conditioned with a potentially EGF-like factor is also required.

Our working hypothesis is that the role of the serum is to provide a "beta"-like factor reported by Sporn and colleagues (Roberts et al. 1981) to be necessary for EGF-induced soft agar growth, although the specific "beta" factor studied by that group has no effect on our cultures (D.B. McClure, unpubl.). Neither serum nor EGF has any apparent effect on the soft agar growth potential of the tumorigenic variant WT6 Ag6 nu1, which clones equally well in the mitogen-free SF medium and in 10% FCS.

Although both REF52 WT6 Ag6 and Ag6 nu1 would be characterized as maximally transformed by the standard criteria, behavior in 10% FCS–containing cultures, the correct correlate of cellular tumorigenicity in REF52 variants is the development of the transformed phenotype in SF and mitogen-free culture conditions. The great lengths to which we were forced before we ultimately obtained lines that both produced tumors in nude mice and possessed the factor-independent transformed phenotype are inconsistent with the generally good correlation that has been observed between soft agar growth and tumorigenicity. If our results can be generalized, then the strength of this correlation must rest upon the fact that the substantial majority of the cells that display a fully transformed phenotype in SC medium also would do so in SF conditions. It is possible that the difficulties we encountered may result from the fact that EGF is the principal mitogen for REF52, whereas almost all of the studies alluded to above employed fibroblastoid cultures for which the principal mitogen is PDGF. Nothing is known about the mechanism of action of either factor, and it is possible that the development of an EGF-independent phenotype is much more difficult. Were this so, a possible prediction would be that soft agar growth in SC cultures of epithelial cells might prove to be a substantially poorer correlate of tumorigenicity.

Our results regarding the potential role of cellular factors in the development of the transformed phenotype are also difficult to reconcile with the literature. The production of a factor by the REF52 variants that is capable of stimulating growth in soft agar does not correlate with oncogenic potential nor even with growth in SC soft agar. REF52 WT6 and WT6 Ag6 both produce roughly equivalent levels of an agar growth-promoting substance, yet WT6 grows only poorly in SC soft agar while Ag6 grows quite well. Two previous reports (Kaplan et al. 1981; McClure 1983) demonstrated a clear correlation between the level of factor production and the efficiency of soft agar growth. One possible explanation would be that the ability to grow in soft agar suspension is controlled either by the level of autostimulating factor or by the sensitivity of response to this factor. Perhaps WT6 Ag6 and Ag6 nu1 have been selected for progressively more efficient response to a set level of factor. Presumably, in this case the growth of WT6 Ag6 nu1 in SF and factor-free soft agar even at high cell dilutions must mean that within the microenvironment of a single cell, sufficient medium conditioning is possible for autostimulation. We find this unlikely, however; a definitive resolution of this situation must await some means of blocking either the factor or its receptor. On the other hand, it should again be noted that the previous reports have been concerned with the production of a PDGF-like factor by cells for which the primary mitogen is PDGF (Dicker et al. 1981; Kaplan et al. 1981), whereas we are dealing with an EGF-like factor and cells for which EGF is the mitogen. The importance of or even potential for autocrine stimulation could very well depend on the nature of the regulatory pathways involved. It is interesting to note that in this limited number of cases, the cellular factor produced is matched by the responsiveness of the cells transformed rather than determined by the nature of the transforming virus.

Acknowledgments

We would like to thank Gordon Sato, Margaret Hightower, Brad Ozanne, Paul Kaplan, Dan Rifkin, Marilyn Anderson, Fumio Matsumura, Jim Lin, Steve Blose, Robert Franza, and Jim Garrels for many helpful conversations during the course of these studies. We are also very grateful to Seung-il Shin for performing a number of the tumorigenicity assays, to Steve Blose and Jim Lin for the gift of the cytoskeletal antibodies, and to Anita Roberts for the gift of the purified beta factor.

References

Castor, L.N. 1971. Control of division by cell contact and serum concentration in cultures of 3T3 cells. *Exp. Cell Res.* **68:** 17.

Chen, L.B., P.H. Gallimore, and J.K. McDougall. 1976. Correlation between tumor induction and the large external transformation sensitive protein on the cell surface. *Proc. Natl. Acad. Sci.* **73:** 3570.

Cherington, P.V. and A.B. Pardee. 1982. On the basis for loss of the EGF growth requirement by transformed cells. *Cold Spring Harbor Conf. Cell Proliferation* **9:** 221.

Diamandopoulos, G.T. 1972. Leukemia, lymphoma, and osteosarcoma induced in the Syrian golden hamster by simian virus 40. *Science* **176:** 173.

———. 1973. Induction of lymphocytic leukemia, lymphosarcoma, reticulum cell sarcoma, and osteogenic sarcoma in the Syrian golden hamster by oncogenic DNA simian virus 40. *J. Natl. Cancer Inst.* **50:** 1347.

———. 1978. Incidence, latency, and morphological types of neoplasms induced by simian virus 40 inoculated intravenously into hamsters of three inbred strains and one outbred stock. *J. Natl. Cancer Inst.* **60:** 445.

Dicker, P., P. Pohjanpelto, P. Pettican, and E. Rozengurt. 1981. Similarities between fibroblast-derived growth factor and platelet-derived growth factor. *Exp. Cell Res.* **135:** 221.

Dulbecco, R. and J. Elkington. 1973. Conditions limiting multiplication of fibroblastic and epithelial cells in dense cultures. *Nature* **246:** 197.

Freedman, V.H. and S. Shin. 1974. Cellular tumorigenicity in nude mice: Correlation with cell growth in semi-solid medium. *Cell* **3:** 355.

Frisque, R.J., D.B. Rifkin, and W.C. Topp. 1980. The requirement for the large and small T proteins of SV40 in the maintenance of the transformed state. *Cold Spring Harbor Symp. Quant. Biol.* **44:** 325.

Holley, R.W. and J.A. Kiernan. 1968. "Contact inhibition" of cell division in 3T3 cells. *Proc. Natl. Acad. Sci.* **60:** 300.

Ibelgaufts, H., W. Doerfler, K.H. Scheidtmann, and W. Wechsler. 1980. Adenovirus type 12–induced rat tumor cells of neuroepithelial origin: Persistence and expression of the viral genome. *J. Virol.* **33:** 423.

Kaplan, P.L., W.C. Topp, and B. Ozanne. 1981. Simian virus

40 induces the production of transforming factor. *Virology* **108**: 484.

Lee, L.S. and I.B. Weinstein. 1978. Epidermal growth factor, like phorbol esters, induces plasminogen activation in HeLa cells. *Nature* **274**: 696.

Logan, J., J.C. Nicholas, W.C. Topp, M. Girard, T. Shenk, and A.J. Levine. 1981. Transformation by adenovirus early region 2A temperature sensitive mutants and their revertants. *Virology* **115**: 419.

McClure, D.B. 1983. Anchorage-independent colony formation of SV40 transformed Balb/c-3T3 cells in serum-free medium: Role of cell- and serum-derived growth factors. *Cell* **32**: 999.

McClure, D.B., M.J. Hightower, and W.C. Topp. 1982. Effect of SV40 transformation on the growth factor requirements of the rat embryo cell line REF52 in serum-free medium. *Cold Spring Harbor Conf. Cell Proliferation* **9**: 345.

Pollack, R.E., M. Osborn, and K. Weber. 1975a. Patterns of organization of actin and myosin in normal and transformed cultured cells. *Proc. Natl. Acad. Sci.* **72**: 994.

Pollack, R., R. Risser, S. Conlon, and D. Rifkin. 1974. Plasminogen activator production accompanies loss of anchorage regulation in transformation of primary rat embryo cells by simian virus 40. *Proc. Natl. Acad. Sci.* **71**: 4792.

Pollack, R., R. Risser, S. Conlon, V. Freedman, and S. Shin. 1975. Production of plasminogen activator and colonial growth in semisolid medium are in vitro correlates of tumorigenicity in the immune-deficient nude mouse. *Cold Spring Harbor Conf. Cell Proliferation* **2**: 885.

O'Neill, C.H., P.N. Riddle, and P.W. Jordan. 1979. The relation between surface area and anchorage dependence of growth in hamster and mouse fibroblasts. *Cell* **16**: 909.

Rabson, A.S., R.L. Kirschstein, and F.J. Paul. 1964. Tumors produced by adenovirus 12 in *Mastomys* and mice. *J. Natl. Cancer Inst.* **32**: 77.

Roberts, A., M. Anzano, L. Lamb, J. Smith, and M. Sporn. 1981. New class of transforming growth factors potentiated by epidermal growth factor: Isolation from non-neoplastic tissues. *Proc. Natl. Acad. Sci.* **78**: 5339.

Sager, R., F. Bennett, and B.L. Smith. 1982. Altered growth-factor requirements of transformed mutants and tumor-derived cell populations of CHEF cell origin. *Cold Spring Harbor Conf. Cell Proliferation* **9**: 231.

Shenk, T., J. Carbon, and P. Berg. 1976. Construction and analysis of viable deletion mutants of simian virus 40. *J. Virol.* **18**: 664.

Sleigh, M.J., W.C. Topp, R. Hanich, and J.F. Sambrook. 1978. Mutants of SV40 with an altered small-t protein are reduced in their ability to transform cells. *Cell* **14**: 179.

Stiles, C.D., W.J. Pledger, R.W. Tucker, R.G. Martin, and C.D. Scher. 1980. Regulation of the Balb/c-3T3 cell cycle: Effects of growth factors. *J. Supramol. Struct.* **13**: 489.

Stoker, M.G.P. 1973. Role of diffusion boundary layer in contact inhibition of growth. *Nature* **246**: 200.

Tooze, J., ed. 1981. *The molecular biology of tumor viruses*, 2nd edition, revised: *DNA tumor viruses*. Cold Spring Harbor Laboratory, Cold Spring Harbor, New York.

Topp, W.C., D.B. Rifkin, and M.J. Sleigh. 1981. SV40 mutants with an altered small-t protein are tumorigenic in newborn hamsters. *Virology* **111**: 341.

Topp, W.C., M.J. Hightower, M.B. Ramundo, D.M. Smith, and M.A. Anderson. 1982. Common features of transformation and tumor induction by DNA viruses. *Hepatology* **2**: 51S.

Topp, W.C., D. Rifkin, A. Graessmann, C. Chang, and M.J. Sleigh. 1979. The role of the early SV40 gene products in the maintenance of the transformed state. *Cold Spring Harbor Conf. Cell Proliferation* **6**: 361.

Weinstein, D.B. 1979. A single-step adsorption method for removal of lipoproteins and preparation of cholesterol-free serum. *Fed. Proc.* **59**: 204A.

Weiss, R., N. Teich, H. Varmus, and J. Coffin, eds. 1982. *The molecular biology of tumor viruses*, 2nd edition: *RNA tumor viruses*. Cold Spring Harbor Laboratory, Cold Spring Harbor, New York.

Yabe, Y., L. Samper, E. Bryan, G. Taylor, and J.J. Trentin. 1964. Oncogenic effect of human adenovirus type 12 in mice. *Science* **143**: 46.

Relationship between the Transforming Protein of Simian Sarcoma Virus and Platelet-derived Growth Factor

M.D. Waterfield,* G.T. Scrace,* N. Whittle,* P.A. Stockwell,*† P. Stroobant,*
A. Johnsson,‡ A. Wasteson,‡ B. Westermark,§ C.-H. Heldin,‡ J.S. Huang,**
and T.F. Deuel**

*Imperial Cancer Research Fund Laboratories, Lincoln's Inn Fields, London WC2A 3PX, England; ‡Institute of Medical and Physiological Chemistry and §Department of Pathology, University of Uppsala, Sweden; **Departments of Medicine and Biological Chemistry, Washington University School of Medicine, The Jewish Hospital of St. Louis, St. Louis, Missouri 63110

Studies of the growth properties of cells in culture have shown that for cells of mesenchymal origin the requirement for polypeptide mitogens in serum to optimize cell growth is mainly due to platelet-derived growth factor (PDGF; for recent reviews, see Deuel and Huang 1983; Westermark et al. 1983). The physiological role of PDGF remains rather unclear. However, since PDGF is released from platelet α-granules during the clotting process at the sites of damage to blood vessels, it is thought that it may be involved in the processes of tissue repair (Kohler and Lipton 1974; Ross et al. 1974; Westermark and Wasteson 1975; Ross and Glomset 1976; Witte et al. 1978; Kaplan et al. 1979; Scher et al. 1979). This concept is supported by the fact that PDGF is a potent chemoattractant for monocytes and neutrophils (Deuel et al. 1982) and for smooth muscle cells and fibroblasts (Grotendorst et al. 1981; Seppa et al. 1982; Senior et al. 1983) and that it is a mitogen for cells of the connective tissue (Kohler and Lipton 1974; Ross et al. 1974; Westermark and Wasteson 1975). PDGF has been shown to bind with a dissociation constant of 10^{-9}–10^{-11} M to specific saturable cell-surface receptors that vary in number from 40,000–400,000 on the surface of fibroblasts and glial cells (Heldin et al. 1981; Bowen-Pope and Ross 1982; Huang et al. 1982a). Following PDGF binding, protein kinase activity is stimulated, a protein thought to be the receptor becomes phosphorylated on tyrosine residues, and the radiolabeled protein can be detected as a polypeptide with an apparent molecular weight (app. m.w.) of about 185,000 on SDS-polyacrylamide gels (Ek and Heldin 1982; Ek et al. 1982; Nishimura et al. 1982; Heldin et al. 1983). The receptor for PDGF is clearly distinct from the receptor for epidermal growth factor (EGF; app. m.w. 175,000; Ek et al. 1982; Huang et al. 1982a), although interactions between these receptors occur since PDGF can reduce the affinity of the EGF receptor for EGF (Bowen-Pope et al. 1983; Collins et al. 1983). Rapid changes in ion fluxes follow PDGF binding to 3T3 cells (Mendoza et al. 1980) together with the accumulation of cAMP, which is mediated by the synthesis of a stable E-type prostaglandin (Coughlin et al. 1980; Shier 1980; Habenicht et al. 1981; Rozengurt et al. 1983). PDGF stimulates the phosphorylation of several intracellular proteins (Nishimura and Deuel 1981; Cooper et al. 1982) and alters the levels of specific mRNAs that may be correlated with changes in the amounts of certain proteins (Pledger et al. 1981; Smith and Stiles 1981; Cochran et al. 1983). PDGF also induces a series of morphological changes, including actin reorganization and membrane ruffling (Mellstrom et al. 1983). The relationship of all these changes to the mitogenic and chemotactic effects of PDGF remains to be elucidated.

Through comparative studies of the growth properties of normal and transformed cells, it has become clear that many transformed cells require lower levels of serum than normal cells for optimal growth (Temin et al. 1972; Vogel and Pollack 1974). It has been suggested that this reduced requirement may reflect the capacity of the transformed cell to secrete growth factors that replace the requirement for serum (De Larco and Todaro 1978a; Ozanne et al. 1980; Sporn and Todaro 1980). Progress in characterization of growth factors produced by transformed cells has been made for three groups of factors. The first group includes the multiplication-stimulating activities (MSAs or insulinlike growth factors [IGF-like]; Dulak and Temin 1973; De Larco and Todaro 1978b).

The second group is composed of the transforming growth factors (TGFs; De Larco and Todaro 1978a; Ozanne et al. 1980; Sporn et al. 1981), which have now been divided into two subgroups (Roberts et al. 1982). The α-TGFs compete with EGF for binding to the EGF receptor, and the β-TGFs do not bind to EGF receptors but may potentiate the growth-stimulating activity of α-TGFs in semisolid media (Anzano et al. 1982). Structural studies show that the MSA or IGF-like factors and the β-TGFs are distinct from PDGF (Rinderknecht and Humbel 1978; Antoniades and Hunkapiller 1983; Doolittle et al. 1983; Marquardt et al. 1983; Waterfield et al. 1983b). However, the relationship of the β-TGFs to PDGF is still unclear due to lack of structural data. Since PDGF may act like a β-TGF in that it will induce anchorage-independent growth in the presence of EGF (Dicker et

†Present address: Department of Biochemistry, University of Otago, Dunedin, New Zealand.

al. 1981), it is possible that a structural relationship may exist between these factors and PDGF.

The third group is made up of those factors thought to be related to PDGF. In the case of a growth factor secreted by osteosarcoma cells (ODGF; Westermark and Wasteson 1975; Heldin et al. 1980), a clear relationship between this factor and PDGF is established since antisera to PDGF recognize ODGF, which also shares various chemical and biological properties with PDGF. Another factor, fibroblast-derived growth factor (FDGF; Bourne and Rozengurt 1976; Burk 1976; Dicker et al. 1981; Rozengurt et al. 1982), secreted by SV40-transformed BHK cells, is indistinguishable from PDGF in many of its biological properties. For both ODGF and FDGF, no structural data is yet available.

To further analyze the structural and functional properties of PDGF and to clarify its relationships with other growth factors, we have determined a partial amino acid sequence of human PDGF. A comparison of this sequence with the sequences stored in the protein data bases of Doolittle (Newat; Doolittle 1981) and of the National Biomedical Research Foundation (the NBRF 1982 data base obtained from D.J. Lipman, pers. comm.) revealed a stretch of 104 amino acids that was virtually identical with that of the predicted putative transforming protein p28^{sis} of simian sarcoma virus (SSV; Devare et al. 1983). This observation was recently reported independently by two groups (Doolittle et al. 1983; Waterfield et al. 1983b). This homology suggests that SSV has acquired cellular sequences encoding a protein very similar to PDGF that could function as a growth factor, mediating the transforming function of SSV. Preliminary studies of a growth factor produced by SSV nonproducer cells would support this concept and are reported here (Deuel et al. 1983). These results suggest that autocrine growth regulation by growth factors may play an important role in the growth of transformed connective tissue cells and perhaps of other transformed cells as well.

Materials and Methods
Purification of PDGF
Two different purification procedures were employed.

1. PDGF from stocks of frozen outdated platelets was purified by the method of Deuel et al. (1981), using protease inhibitors in the early stages of the preparation. Material eluting from the final gel permeation column as PDGF II was used for structural study.
2. PDGF from fresh platelet lysates was purified by a modification of the procedure of Heldin et al. (1979), using the initial sulfadex step of Deuel et al. (1981) prior to the published protocol. The material eluting from the final gel permeation P-150 column was further fractionated on reverse-phase chromatography according to the procedure of Johnsson et al. (1982).

Fractionation of component polypeptide chains of PDGF
PDGF preparations were fully reduced and alkylated with [^{14}C]iodoacetamide (Amersham) by the procedure of Skehel and Waterfield (1975).

Gel permeation high-performance liquid chromatography (HPLC) in 6 M guanidine
Fully reduced and alkylated PDGF II prepared by the method of Deuel et al. (1981) was fractionated on a TSK 3000 column (LKB; 60 cm × 7 mm) at a flow rate of 0.5 ml/min and using solutions of 6 M guanidine·HCl (Schwarz-Mann) at pH 4.3 with 0.1 M potassium phosphate buffer. Effluent was monitored at 280 nm, and radioactivity on cysteine residues was measured for each fraction. Molecular weights were determined using fully reduced and alkylated bovine serum albumin, ovalbumin, chymotrysinogen, and lysozyme by plotting m.w.$^{0.555}$ against $K_d^{0.333}$ by the method of Ui (1979).

Reverse phase HPLC in guanidine
PDGF purified initially by the method of Heldin et al. (1979) and subsequently by the method of Johnsson et al. (1982), as described above, was fully reduced and alkylated, and the material was fractionated by reverse-phase HPLC in solutions of 1 M acetic acid containing 2 M guanidine·HCl, using a Lichrosorb RP8 column (0.4 × 2.5 cm, 5 µm particle size; Merck) that was developed using an n-propanol gradient. The column was monitored (following dialysis of samples against 0.1% trifluoracetic acid) by SDS-polyacrylamide gradient gel electrophoresis (as described below).

Reverse-phase HPLC in trifluoracetic acid
Peptides from gel-permeation separation in 6 M guanidine buffers or from reverse-phase separations in solutions of 2 M guanidine·HCl, 1 M acetic acid, and propanol were further fractionated by reverse-phase HPLC using 0.1% trifluoracetic acid solutions and linear acetonitrile gradients from 1% to 60% by a modification of the method of Bennett et al. (1980). HPLC was carried out on a Synchropac RPP column at a flow rate of 1 ml/min (0.45 × 7.5 cm; Synchrom, Linden, Ind.), and the column effluent was monitored at both 206 nm and 280 nm. Fractions were collected, and radioactivity on alkylated cysteine residues was measured by scintillation counting.

Amino acid sequence determination
Aminoterminal sequences were determined by using a gas-phase sequencer constructed and operated as described by Hewick et al. (1981). Acetonitrile was HPLC S grade, and ethyl acetate, butylchloride, heptane, phenylisothiocyanate, and trifluoracetic acid were sequencer grade, obtained from Rathburn Chemicals (Walkerburn, Scotland). Trifluoracetic acid was refluxed with chromic acid, distilled through a Widmer column, refluxed with dithiothreitol, and redistilled. Trimethylamine (British Drug Houses, Poole, England) was distilled through a column of phthalic anhydride (B.D.H.) before use. The glass fiber disks used in the sequencer were run for 10 precycles with polybrene and glycyl-glycine or cysteic acid before loading of samples.

The analysis of phenylthiohydantoin (PTH) amino acids was made using a Zorbax C8 column (4.6 mm × 25 cm; Dupont) at 43°C, developed at 2 ml/min using a 0.009 M sodium acetate buffer, pH 4.1, with a linear acetonitrile gradient over 8 minutes of 24–38% (Waterfield et al.

1983a). A Waters HPLC system with 2 M 6000A pumps, a Wisp autoinjector, and a system controller with a Beckman Model 160 detector were used. The amounts of PTH amino acids recovered were measured relative to standard PTH amino acids, using peak areas measured with a Waters Data Module. Yields of PTH serine and threonine were measured by peak height rather than by area since multiple peaks were obtained for these amino acids. Yield data for these PTH amino acids is thus semiquantitative.

SDS-polyacrylamide gel analysis
Samples were analyzed in the absence or presence of dithiothreitol on SDS-polyacrylamide gradient gels by the method of Laemmli (1970) or Blobel and Dobberstein (1975) and developed using Coomassie blue or the silver staining method of Morrissey (1981).

Results
Purification of PDGF
Two different PDGF preparations were used for structural characterization: PDGF II purified by the method of Deuel et al. (1981) or PDGF purified by the method of Heldin et al. (1979) with a final reverse-phase step as described by Johnsson et al. (1982) (see Materials and Methods). In both cases the purified PDGF migrated on nonreducing SDS-polyacrylamide gels with an apparent m.w. of 27,000–29,000 as a broad band of Coomassie blue. Following reduction of disulfide bonds, a series of five to seven polypeptides ranging in molecular weight from 17,500 to approximately 7,000 could be detected using gradient SDS-polyacrylamide gel electrophoresis. Aminoterminal sequence analysis of intact PDGF showed that both preparations contained at least five polypeptides with distinct sequences. To fractionate component polypeptide chains, the PDGF preparations were fully reduced and alkylated with [^{14}C]iodoacetamide and fractionated in guanidine solutions, using gel permeation HPLC or reverse-phase HPLC.

Sequence analysis of PDGF II
PDGF II purified by the method of Deuel et al. (1981) was fully reduced and alkylated and chromatographed on a TSK 3000 gel permeation column in 6 M guanidine solutions (SDS polyacrylamide gel analysis before and after reduction is shown in Fig. 1A). Four distinct fractions were obtained, as shown in Figure 1B. Using fully reduced and alkylated standard proteins for column calibration by the method of Ui (1979), the average molecular weights of the four fractions were found to be (1) 13,000, (2) 8,840, (3) 5,320, and (4) 3,454. Fraction 4 with average molecular weight of 3,454 was subfractionated by reverse-phase HPLC in 0.1% trifluoroacetic acid, using acetonitrile for an organic modifier (Fig. 1C), and six peptides were obtained.

Aminoterminal sequence analysis of material in fraction 2 showed that it contained at least two distinct polypeptides with the sequences shown in Figure 4 as peptides II and IV. Similar analysis of fraction 3 showed a single polypeptide with the sequence shown as peptide IV in Figure 5. Sequence analysis of the smaller peptides resolved by reverse-phase HPLC showed that the six peptides could be assigned to two groups. Peptides a and b differed by an aminoterminal lysine (the sequence data is shown in Fig. 3B), and their sequence is shown as peptide V in Figure 4. Peptides c, d, e, and f were found to be cleavage products derived from the amino terminus of peptide I.

Sequence analysis of PDGF
PDGF, prepared by the method of Heldin et al. (1979), was reduced, alkylated, and fratianated in 1 M acetic acid containing 2 M guanidine · HCl on a C8 reverse-phase HPLC column, using propanol as organic modifier. Fractions were dialyzed and analyzed by SDS-polyacrylamide gels, and results are shown in Figure 2. Certain fractions were further purified by gel-permeation HPLC in 6 M guanidine solutions or by reverse-phase HPLC using 0.1% trifluoroacetic acid and acetonitrile as organic modifier.

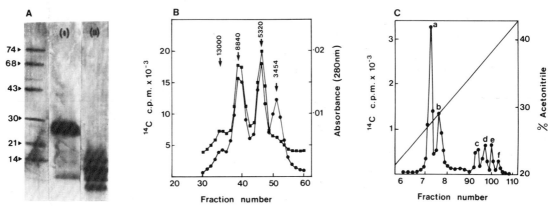

Figure 1 Fractionation of component polypeptides of PDGF II by HPLC. (*A*) SDS-polyacrylamide gel analysis of PDGF II in the absence (I) or presence (II) of dithiothreitol. Molecular-weight standards are shown on the left (m.w. × 10^{-3}). Gels were stained with Coomassie blue. (*B*) Gel-permeation HPLC in 6 M guanidine solutions of fully reduced and alkylated PDGF II chains. Average m.w.'s, determined by the method of Ui (1979), are shown for each peak. (●) Radioactivity; (■) absorbance. (*C*) Subfractionation of the material of average m.w. 3454 from *B* analyzed by reverse-phase HPLC in trifluoroacetic acid, acetonitrile. Details of analytical methods are given in the text. (Reprinted, with permission, from Waterfield et al. 1983b.)

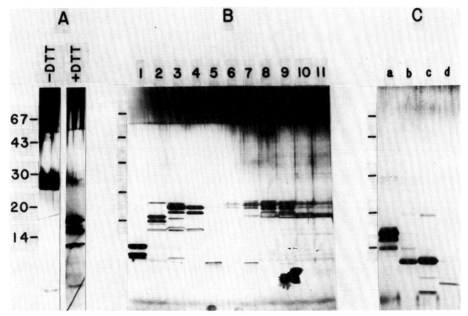

Figure 2 HPLC analysis of PDGF component polypeptides prepared by the methods of Heldin et al. (1979) and Johnsson et al. (1982). SDS-polyacrylamide gradient gel analysis of starting material in the absence and presence of dithiothreitol (*A*), of fractions from reverse-phase separation in 2 M guanidine and acetic acid, using propanol gradients (*B*), and subfractionation by gel permeation in 6 M guanidine of fractions 3 and 4 from *B* (*C*). Gels were developed with the silver stain of Morrissey (1981); full details are in the text. (Reprinted, with permission, from Waterfield et al. 1983b.)

Aminoterminal sequence analysis of the purified polypeptides was carried out. Material in fractions 10 and 11, for which sequence data is presented in Fig. 3C, was analyzed to give the sequence shown as peptide I in Figure 4. The sequence of peptide II (Fig. 4) was obtained from material in subfraction a (derived from gel-permeation HPLC of fractions 3 plus 4). Analysis of subfraction c revealed two sequences for which data are presented in Figure 3A: Sequences are shown as peptides II and III in Figure 4. Subfraction d contained material that gave a single sequence shown as peptide III in Figure 4.

The structure of PDGF

The partial amino acid sequences established for the component polypeptides of PDGF are shown in Figure 4 aligned with the sequence of the predicted translation product of the *v-sis* gene of SSV (Devare et al. 1983). The apparent molecular weight of the peptides sequenced has been used to calculate the approximate lengths of these peptides, which are indicated using dashed lines terminating with an asterisk for peptides I–V in Figure 4.

The aminoterminal sequences of peptides I and II are homologous but differ in length by 6 amino acids at the amino terminus. For the 25 amino acid stretch determined thus far, which is shared by these peptides, 12 residues are identical and 5 are conservative substitutions. The sequence of peptide II overlaps the sequences of peptides III and IV by 13 residues and gives a contiguous sequence of 69 residues. The difference between peptides III and IV at residue 2 may be due to polymorphism. Peptide V can be located carboxyterminal to peptides II, III, and IV by alignment with the predicted *v-sis* gene product; however, it is not known if peptide V is derived from cleavage of peptides I or II.

These results suggest that PDGF may contain two distinct but homologous polypeptides, together with various proteolytically derived cleavage products of these chains, which are held together by disulfide bonds.

Relationship between PDGF and p28^{sis}

The amino acid sequences of peptides I and II were used in a search of the 1983 protein data bases of Doolittle (Newat; Doolittle 1981) and a stripped down version of the 1982 NBRF data base, using the rapid search techniques of Wilbur and Lipman (1983). A stretch of 31 amino acid residues at the amino terminus of peptide I was found to be identical, except for 2 conservative substitutions, with the predicted amino acid sequence of the translation product p28^{sis} of the *v-sis* gene of SSV (Devare et al. 1983). Further searches with the sequences of peptides III, IV, and V showed that the region of virtual identity extends to the end of peptide V, providing a stretch of 104 amino acids that are shared by PDGF and by p28^{sis}. No clear relationship between PDGF and other protein sequences in the data bases was found. The partial amino acid sequence of peptide II was not encoded by *v-sis*.

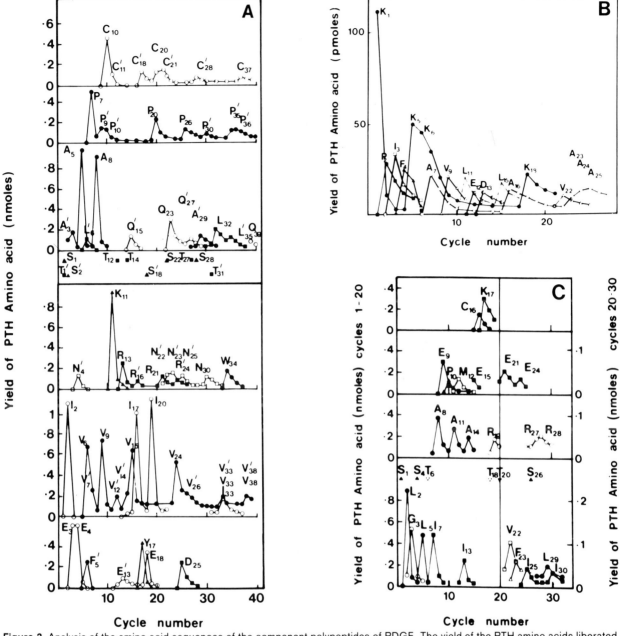

Figure 3 Analysis of the amino acid sequences of the component polypeptides of PDGF. The yield of the PTH amino acids liberated at each step of Edman degradation is plotted against the cycle number. (A) Analysis of peptides II and III sequenced as a mixture from subfraction c (see Fig. 2). The minor sequence is indicated by prime marks. (B) Analysis of peptide V. (C) Analysis of peptide I. The techniques used are described in Materials and Methods. (The results in A are reprinted, with permission, from Waterfield et al. 1983b.)

Growth factor production by SSV nonproducer cells

Investigations have now been completed in which a growth-promoting activity of sonicated lysates of SSV nonproducer 3T3 cells and uninfected 3T3 cells has been examined, using the assay of Huang et al. (1982b). A linear response occurs in incorporation of [^{3}H]thymidine with increasing concentrations of lysate similar to that observed for PDGF. Baseline activity only was observed with control 3T3 cell lysates. This growth factor activity can be partially blocked by addition of antisera specific for PDGF (Huang et al. 1983). Similar studies with SSV nonproducer rat cells, using a slightly different assay based on [^{3}H]thymidine incorporation into 3T3 cells, show that the medium of these cells grown in 1% PDGF-deficient fetal calf serum contains growth factor activity not found in that of noninfected cells (Deuel et al. 1983). Further purification of growth factors produced by these cells is in progress to investigate the possibility that the activity is due to p28^{sis} or a processed form of this protein.

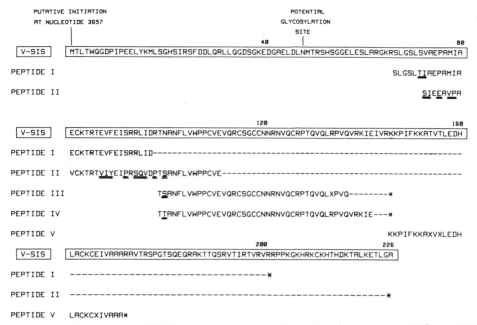

Figure 4 The partial amino acid sequences of PDGF peptides aligned with the predicted sequence of p28^{sis} from SSV. The sequence of the v-sis translation product p28^{sis} starting at the putative initiation codon 3657 and reading to the end of the open reading frame of v-sis is shown (Devare et al. 1983). The partial sequences of peptides I to V are aligned with the p28^{sis} sequence. Residues not determined are indicated by X, and the approximate size of each peptide is shown by dashed lines with an asterisk (*) at the possible carboxyl terminus.

Discussion

PDGF purified from either outdated platelets or from platelet lysates has an extremely basic pI (9.8–10.2) and is hydrophobic. Its mitogenic activity is stable to heat and resistant to SDS, and the unreduced active protein has an apparent molecular weight of 28,000–31,000 on SDS-polyacrylamide gels (for reviews, see Deuel and Huang 1983; Westermark et al. 1983). Surprisingly, PDGF appears to be extensively cleaved by proteases while retaining its mitogenic activity. The present study reveals the extent of such cleavages by analysis of the component polypeptide chains released following reduction of disulfide bonds (e.g., see Fig. 2a, b). Amino acid analysis of PDGF II suggests that PDGF could have nine disulfide bonds for a molecular weight of 28,500 (Deuel et al. 1981). The locations of some of the cysteine residues in PDGF have been established, but as yet we do not know the locations of the disulfide bonds. However, since all the component polypeptides purified contain cysteine residues, it is clear that they could all be linked by disulfides in the protease-cleaved PDGF molecule. The two largest component polypeptides found in PDGF are peptides I and II, which have apparent molecular weights of 14,500 and 17,500. It is possible that a PDGF species may exist that comprises two chains (A and B equivalent to peptides I and II), as suggested by Johnsson et al. (1982). The sequence data show that these chains would be homologous at least at their amino termini. Further cleavage of these two chains, for example, at basic residues homologous to those located in p28^{sis} at positions 99, 146, 172, and 201, could generate peptides III, IV, and V together with others not sequenced but observed in SDS-polyacrylamide gel analysis. Analysis of the carboxyterminal sequences of these peptides together with further sequencing studies should help to clarify the nature of the cleavages that take place. These studies are in progress, together with others to establish the extent of the homology between peptides I and II.

The relationship between PDGF and p28^{sis}, the putative transforming protein of SSV, was established during a computer search of the protein data bases of Doolittle and the NBRF for sequences homologous to the partial sequences determined for PDGF. Similar results were reported at the same time by Doolittle et al. (1983) using the Newat data base. The aminoterminal sequences of peptides I, III, IV, and V, which together cover a stretch of approximately 100 amino acids, are identical with the sequence of p28^{sis} between residues 67 and 171 with 3 differences (and 6 unidentified residues). The sequence of peptide I has as yet not been shown to be contiguous with that of peptides III or IV. Cleavage of partially pure peptide I with cyanogen bromide, followed by sequence analysis, showed that it contained a peptide having a sequence identical with peptides curred at the same tryptophane residue (presumably due to oxidation), and it was possible for them to determine a sequence that extends to the amino terminus of peptide V (see Fig. 4). In this case, as in our study, overlapping sequence data are needed to confirm the location of this sequence in peptides I or II.

The stretch of identical residues shared by PDGF and by p28^{sis} suggests that SSV has acquired cellular genomic sequences that encode part of PDGF or a closely related protein. Since v-sis cannot encode the amino terminus of peptide II, the simplest interpretation would be that SSV has acquired one of the polypeptide chains found in PDGF. However, since the precise details of the component polypeptide chain structure of PDGF remain unclear, several possibilities exist to explain our results. It is of course possible that SSV has rearranged the PDGF sequences, but further sequence data will be required to clarify this point.

Sequences related to v-sis have been located at a single locus (c-sis) in the human genome on chromosome 22 (Dalla Favera et al. 1982) and it seems probable that c-sis encodes PDGF. c-sis has five exons containing sequences related to v-sis (Dalla Favera et al. 1981; Wong-Staal et al. 1981; Josephs et al. 1983). The DNA sequence of the first exon of c-sis seems to encode amino acids 5–38 of the putative p28^{sis} protein (if initiation starts at nucleotide 3657 of v-sis in the SSV genome) (Josephs et al. 1983). The sequence of other exons related to v-sis will no doubt be determined soon and should help to clarify the structure of that part of the c-sis gene that is related to v-sis and to PDGF. Studies of human sarcoma cell lines (5/6), of glioblastoma cell lines (3/5), and of one T-cell leukemia cell line have shown that they contain a 4.2-kb transcript (together with a smaller transcript present in some cases at a lower level) related to v-sis that has not been detected in normal fibroblasts, in several carcinoma or melanoma cell lines, nor in SSV nonproducer cell RNA preparations (Eva et al. 1982). This 4.2-kb RNA could encode a much larger protein than PDGF, and we must await the cDNA sequence of this transcript before we can understand the coding potential of c-sis and the complete structure of PDGF. It has recently been reported (Besmer et al. 1983) that the Parodi-Irgens feline sarcoma virus has an oncogene related to that of SSV. However, sequence studies of the genome of this virus are not yet available.

Through the use of antisera raised against synthetic peptides of the putative p28^{sis}, it has been shown that a protein with an apparent molecular weight of 28,000 is produced in SSV-transformed cells (Robbins et al. 1982). Since p28^{sis} may be an env-sis fusion protein, it is possible that the sequences that allow membrane insertion of env could result in the secretion of p28^{sis}. Preliminary studies reported here suggest that SSV-nonproducer cells do make a growth factor that can stimulate DNA synthesis in 3T3 test cells. However, it remains to be seen if p28^{sis} or a processed product of this protein is responsible for this activity.

Abnormal expression of PDGF or a related protein that is a similar mitogen, through the process of transformation by SSV, would be expected to lead to uncontrolled growth of cells capable of responding to this mitogen. If the mitogen acts through the normal receptor pathway, then only cells having receptors would be expected to respond and become constitutively activated. Similar results might be expected from expression of the c-sis gene product if this product is a mitogen similar to PDGF. Sequence studies of the c-sis transcripts in normal and transformed cells will clarify these relationships and previous studies of growth factor production. The observation of transcripts related to v-sis and hence c-sis in cancer cell lines that would be expected to be capable of mounting a response to PDGF (Eva et al. 1982) together with their absence from normal fibroblasts and carcinoma and melanoma cell lines suggests that expression of the c-sis product could be important in the growth control of certain cancer cells.

Acknowledgments

We thank the Finnish Red Cross in Helsinki (Dr. G. Myllyla), the Department of Virology, University of Uppsala (Dr. G. Alm), and the American Red Cross Blood Banks in Chicago (R. Gilbert), Toledo (F. Courtwright and P. Lan), Tulsa (D. Kasprisin), Waterloo (J. Bender and T. Brown), and Fort Wayne (G. Drew and D. Dunfee) for generous supplies of platelets. We thank Dr. M.W. Hunkapiller for helpful discussions regarding preliminary sequence data established in his laboratory in the early stages of this work and Drs. T. Graf and S.A. Aaronson for SSV-nonproducer normal rat kidney and mouse 3T3 cells, respectively. We also thank N.F. Totty, R. Philp, S. Hermansson, B. Kennedy, and D. Chang for technical assistance. We are extremely grateful to Dr. R.F. Doolittle for providing the Newat data base and to Drs. W.J. Wilbur and D.J. Lipman for providing both their rapid search routines and for a stripped down version of the NBRF data base for which ICRF holds a license. We would also like to thank Dr. C. Rawlings of the ICRF Computer Laboratory for invaluable assistance with the computer studies and Mrs. A. Becket for preparation of this manuscript. The work was supported by Swedish MRC grant 4486, Swedish Cancer Society grants 689, 786, and 1794 (to A.W., B.W., and C.-H.H.) and National Institutes of Health grants CA-22409, HL-14147, and HL-22119 (to T.F.D.). P.A.S. is supported by a New Zealand Medical Research Council Fellowship.

References

Antoniades, H.N. and M.W. Hunkapiller. 1983. Human platelet-derived growth factor (PDGF): Amino-terminal amino acid sequence. *Science* **220**: 963.

Anzano, M.A. A.B. Roberts, C.A. Meyers, A. Komoriya, L.C. Lamb, J.M. Smith, and M.B. Sporn. 1982. Synergistic interaction of two classes of transforming growth factors from murine sarcoma cells. *Cancer Res.* **42**: 4776.

Bennett, H.P.J., C.A. Browne, and S. Solomon. 1980. The use of perfluorinated carboxylic acids in the reversed-phase HPLC of peptides. *J. Liq. Chromatogr.* **3**: 1353.

Besmer, P., H.W. Snyder, Jr., J.E. Murphy, W.D. Hardy, Jr., and A. Parodi. 1983. The Parodi-Irgens feline sarcoma virus and simian sarcoma virus have homologous oncogenes, but in different contexts of the viral genomes. *J. Virol.* **46**: 606.

Blobel, G. and B. Dobberstein. 1975. Transfer of proteins across membranes. *J. Cell Biol.* **67**: 835.

Bourne, H. and E. Rozengurt. 1976. An 18,000 molecular weight polypeptide induces early events and stimulates DNA synthesis in cultured cells. *Proc. Natl. Acad. Sci.* **73**: 4555.

Bowen-Pope, D.F. and R. Ross. 1982. Platelet-derived growth

factor. II. Specific binding to cultured cells. *J. Biol. Chem.* **257:** 5161.

Bowen-Pope, D.F., P.E. DiCorleto, and R. Ross. 1983. Interactions between the receptors for platelet-derived growth factor and epidermal growth factor. *J. Cell Biol.* **96:** 679.

Burk, R.R. 1976. Induction of cell proliferation by a migration factor released from a transformed cell line. *Exp. Cell Res.* **101:** 293.

Cochran, B.H., A.C. Reffel, and C.D. Stiles. 1983. Molecular cloning of gene sequences regulated by platelet-derived growth factor. *Cell* **33:** 939.

Collins, M.K.L., J.W. Sinnett-Smith, and E. Rozengurt. 1983. Platelet-derived growth factor treatment decreases the affinity of the epidermal growth factor receptors of Swiss 3T3 cells. *J. Biol. Chem.* **258:** 11689.

Cooper, J.A., D.F. Bowen-Pope, E. Raines, R. Ross, and T. Hunter. 1982. Similar effects of platelet-derived growth factor and epidermal growth factor on the phosphorylation of tyrosine in cellular proteins. *Cell* **31:** 263.

Coughlin, S.R., M.A. Moskowitz, B.R. Zetter, H.N. Antoniades, and L. Levine. 1980. Platelet-dependent stimulation of prostacyclin synthesis by platelet-derived growth factor. *Nature* **288:** 600.

Dalla Favera, R., E.P. Gelmann, R.C. Gallo, and F. Wong-Staal. 1981. A human onc gene homologous to the transforming gene (v-sis) of simian sarcoma virus. *Nature* **292:** 31.

Dalla Favera, R., R.C. Gallo, A. Giallongo, and C.M. Croce. 1982. Chromosomal localization of the human homolog (c-sis) of the simian sarcoma virus onc gene. *Science* **218:** 686.

De Larco, J.E. and G.J. Todaro. 1978a. Growth factors from murine sarcoma virus–transformed cells. *Proc. Natl. Acad. Sci.* **75:** 4001.

———. 1978b. A human fibrosarcoma cell line producing multiplication-stimulating activity (MSA)–related peptides. *Nature* **272:** 356.

Deuel, T.F. and J.S. Huang. 1983. Platelet-derived growth factor: Purification, properties, and biological activities. *Prog. Hematol.* **13** (in press).

Deuel, T.F., R.M. Senior, J.S. Huang, and G.L. Griffin. 1982. Chemotaxis of monocytes and neutrophils to platelet-derived growth factor. *J. Clin. Invest.* **69:** 1046.

Deuel, T.F., J.S. Huang, S.S. Huang, P. Stroobant, and M.D. Waterfield. 1983. Expression of a platelet-derived growth factor–like protein in simian sarcoma virus transformed cells. *Science* **221:** 1348.

Deuel, T.F., J.S. Huang, R.T. Proffitt, J.U. Baenziger, D. Chang, and B.B. Kennedy. 1981. Human platelet-derived growth factor. Purification and resolution into two active protein fractions. *J. Biol. Chem.* **256:** 8896.

Devare, S.G., E.P. Reddy, J.D. Law, K.C. Robbins, and S.A. Aaronson. 1983. Nucleotide sequence of the simian sarcoma virus genome: Demonstration that its acquired cellular sequences encode the transforming gene product p28^{sis}. *Proc. Natl. Acad. Sci.* **80:** 731.

Dicker, P., P. Pohjanpelto, P. Pettican, and E. Rozengurt. 1981. Similarities between fibroblast-derived growth factor and platelet-derived growth factor. *Exp. Cell Res.* **135:** 221.

Doolittle, R.F. 1981. Similar amino acid sequences: Chance or common ancestry? *Science* **214:** 149.

Doolittle, R.F., M.W. Hunkapiller, L.E. Hood, S.G. Devare, K.C. Robbins, S.A. Aaronson, and H.N. Antoniades. 1983. Simian sarcoma virus onc gene, v-sis, is derived from the gene (or genes) encoding a platelet-derived growth factor. *Science* **221:** 275.

Dulak, N.C. and H.M. Temin. 1973. Multiplication-stimulating activity for chicken embryo fibroblasts from rat liver cell conditioned medium: A family of small polypeptides. *J. Cell. Physiol.* **81:** 161.

Ek, B. and C.-H. Heldin. 1982. Characterization of a tyrosine-specific kinase activity in human fibroblast membranes stimulated by platelet-derived growth factor. *J. Biol. Chem.* **257:** 10486.

Ek, B., B. Westermark, A. Wasteson, and C.-H. Heldin. 1982. Stimulation of tyrosine-specific phosphorylation by platelet-derived growth factor. *Nature* **295:** 419.

Eva, A., K.C. Robbins, P.R. Anderson, A. Srinivasan, S.R. Tronick, E.P. Reddy, N.W. Ellmore, A.T. Galen, J.A. Lautenberger, T.S. Papas, E.H. Weston, F. Wong-Staal, R.C. Gallo, and S.A. Aaronson. 1982. Cellular genes analogous to retroviral onc genes are transcribed in human tumour cells. *Nature* **295:** 116.

Grotendorst, G.R., H.E.J. Seppa, H.K. Kleinman, and G.R. Martin. 1981. Attachment of smooth muscle cells to collagen and their migration toward platelet-derived growth factor. *Proc. Natl. Acad. Sci.* **78:** 3669.

Habenicht, A.J.R., J.A. Glomset, W.C. King, C. Nist, C.D. Mitchell, and R. Ross. 1981. Early changes in phosphatidylinositol and arachidonic acid metabolism in quiescent Swiss 3T3 cells stimulated to divide by platelet-derived growth factor. *J. Biol. Chem.* **256:** 12329.

Heldin, C.-H., B. Westermark, and A. Wasteson. 1979. Platelet-derived growth factor: Purification and partial characterization. *Proc. Natl. Acad. Sci.* **76:** 3722.

———. 1980. Chemical and biological properties of a growth factor from human cultured osteosarcoma cells: Resemblance with platelet-derived growth factor. *J. Cell. Physiol.* **105:** 235.

———. 1981. Specific receptors for platelet-derived growth factor on cells derived from connective tissue and glia. *Proc. Natl. Acad. Sci.* **78:** 3664.

Heldin, C.-H., B. Ek, and L. Ronnstrand. 1983. Characterization of the receptor for platelet-derived growth factor on human fibroblasts. Demonstration of an intimate relationship with a 185,000 dalton substrate for the PDGF dependent kinase. *J. Biol. Chem.* **258:** 10054.

Hewick, R.M. M.W. Hunkapiller, L.E. Hood, and W.J. Dreyer. 1981. A gas-liquid solid phase peptide and protein sequenator. *J. Biol. Chem.* **256:** 7990.

Huang, J.S., S.S. Huang, and T.F. Deuel. 1983. Human platelet-derived growth factor: Radioimmunoassay and discovery of a specific plasma-binding protein. *J. Cell Biol.* **97:** 383.

Huang, J.S., S.S. Huang, B. Kennedy, and T.F. Deuel. 1982a. Platelet-derived growth factor. Specific binding to target cells. *J. Biol. Chem.* **257:** 8130.

Huang, J.S., R.T. Proffitt, J.U. Baenziger, D. Chang, B.B. Kennedy, and T.F. Deuel. 1982b. Human platelet-derived growth factor: Purification and initial characterization. In *Differentiation and function of hematopoietic cell surfaces* (ed. V. Marchesi et al.), p. 225. A.R. Liss, New York.

Johnsson, A., C.-H. Heldin, B. Westermark, and A. Wasteson. 1982. Platelet-derived growth factor: Identification of constituent polypeptide chains. *Biochem. Biophys. Res. Commun.* **104:** 66.

Josephs, S.F., R. Dalla Favera, E.P. Gelmann, R.C. Gallo, and F. Wong-Staal. 1983. 5' Viral and human cellular sequences corresponding to the transforming gene of simian sarcoma virus. *Science* **219:** 503.

Kaplan, K.L., M.J. Broekman, A. Chernoff, G.R. Lesznik, and M. Drillings. 1979. Platelet α-granule proteins: Studies on release and subcellular localization. *Blood* **53:** 604.

Kohler, N. and N. Lipton. 1974. Platelets as a source of fibroblast growth-promoting activity. *Exp. Cell Res.* **87:** 297.

Laemmli, U.K. 1970. Cleavage of structural proteins during the assembly of the head of bacteriophage T4. *Nature* **227:** 680.

Marquardt, H., M.W. Hunkapiller, L.E. Hood, D.R. Twardzik, J.E. De Larco, J.R. Stephenson, and G.J. Todaro. 1983. Transforming growth factors produced by retrovirus-transformed rodent fibroblasts and human melanoma cells: Amino acid sequence homology with epidermal growth factor. *Proc. Natl. Acad. Sci.* **80:** 4684.

Mellstrom, K., A.S. Hoglund, M. Nister, C.-H. Heldin, B. Westermark, and U. Lindberg. 1983. The effect of platelet-derived growth factor on morphology and mobility of human glial cells. *J. Muscle Res. Cell Motil.* **4:** 589.

Mendoza, S.A., N.M. Wigglesworth, P. Pohjanpelto, and E. Rozengurt. 1980. Na entry and Na-K pump activity in murine, hamster, and human cells—Effect of monensin, serum, platelet extract, and viral transformation. *J. Cell. Physiol.* **103**: 17.

Morrissey, J.H. 1981. Silver stain for proteins in polyacrylamide gels: A modified procedure with enhanced uniform sensitivity. *Anal. Biochem.* **117**: 307.

Nishimura, J. and T.F. Deuel. 1981. Stimulation of protein phosphorylation in Swiss mouse 3T3 cells by human platelet-derived growth factor. *Biochem. Biophys. Res. Commun.* **103**: 355.

Nishimura, J., J.S. Huang, and T.F. Deuel. 1982. Platelet-derived growth factor stimulates tyrosine-specific protein kinase activity in Swiss mouse 3T3 cell membranes. *Proc. Natl. Acad. Sci.* **79**: 4303.

Ozanne, B., R.J. Fulton, and P.L. Kaplan. 1980. Kirsten murine sarcoma virus–transformed cell lines and a spontaneously transformed rat cell line produce transforming factors. *J. Cell. Physiol.* **105**: 163.

Pledger, W.J., C.A. Hart, K.L. Locatell, and C.D. Scher. 1981. Platelet-derived growth factor–modulated proteins: Constitutive synthesis by a transformed cell line. *Proc. Natl. Acad. Sci.* **78**: 4358.

Rinderknecht, E. and R.E. Humbel. 1978. The amino acid sequence of human insulin-like growth factor I and its structural homology with proinsulin. *J. Biol. Chem.* **253**: 2769.

Robbins, K.C., S.G. Devare, E.P. Reddy, and S.A. Aaronson. 1982. In vivo identification of the transforming gene product of simian sarcoma virus. *Science* **218**: 1131.

Roberts, A.B., M.A. Anzano, L.C. Lamb, J.M. Smith, C.A. Frolik, H. Marquardt, G.J. Todaro, and M.B. Sporn. 1982. Isolation from murine sarcoma cells of novel transforming growth factors potentiated by EGF. *Nature* **295**: 417.

Ross, R. and J.A. Glomset. 1976. The pathogenesis of atherosclerosis. *N. Engl. J. Med.* **295**: 369.

Ross, R., J. Glomset, B. Kariya, and L. Harker. 1974. A platelet-dependent serum factor that stimulates the proliferation of arterial smooth muscle cells in vitro. *Proc. Natl. Acad. Sci.* **71**: 1207.

Rozengurt, E., M. Collins, K.D. Brown, and P. Pettican. 1982. Inhibition of epidermal growth factor binding to mouse cultured cells by fibroblast-derived growth factor. *J. Biol. Chem.* **257**: 3680.

Rozengurt, E., P. Stroobant, M.D. Waterfield, T.F. Deuel, and M. Keehan. 1983. Platelet-derived growth factor elicits cyclic AMP accumulation in Swiss 3T3 cells: Role of prostaglandin production. *Cell* **34**: 265.

Scher, C.D., R.C. Shepard, H.N. Antoniades, and C.D. Stiles. 1979. Platelet-derived growth factor and the regulation of the mammalian fibroblast cell cycle. *Biochim. Biophys. Acta* **560**: 217.

Senior, R.M., G.L. Griffin, J.S. Huang, D.A. Walz, and T.F. Deuel. 1983. Chemotactic activity of platelet α-granule proteins for fibroblasts. *J. Cell Biol.* **96**: 382.

Seppa, H., G. Grotendorst, S. Seppa, E. Schiffmann, and G.R. Martin. 1982. Platelet-derived growth factor is chemotactic for fibroblasts. *J. Cell Biol.* **92**: 584.

Shier, W.T. 1980. Serum stimulation of phospholipase A2 and prostaglandin release in 3T3 cells is associated with platelet-derived growth-promoting activity. *Proc. Natl. Acad. Sci.* **77**: 137.

Skehel, J.J. and M.D. Waterfield. 1975. Studies on the primary structure of the influenza virus hemagglutinin. *Proc. Natl. Acad. Sci.* **72**: 93.

Sporn, M.B. and G.J. Todaro. 1980. Autocrine secretion and malignant transformation of cells. *N. Engl. J. Med.* **303**: 878.

Sporn, M.B., D.L. Newton, A.B. Roberts, J.E. DeLarco, and G.J. Todaro. 1981. Retinoids and suppression of the effects of polypeptide transforming factors—A new molecular approach to chemoprevention of cancer. In *Molecular actions and targets for cancer chemotherapeutic agents* (ed. A.C. Sartorelli et al.), p. 541. Academic Press, New York.

Smith, J.C. and C.D. Stiles. 1981. Cytoplasmic transfer of the mitogenic response to platelet-derived growth factor. *Proc. Natl. Acad. Sci.* **78**: 4363.

Temin, H.M., R.W. Pierson, Jr., and N.C. Dulak. 1972. The role of serum in the control of multiplication of avian and mammalian cells in culture. In *Growth nutrition and metabolism of cells in culture* (ed. G.H. Rothblatt and J. Cristofolo), vol. I, p. 50. Academic Press, New York.

Ui, N. 1979. Rapid estimation of the molecular weights of protein polypeptide chains using high-pressure liquid chromatography in 6 M guanidine hydrochloride. *Anal. Biochem.* **97**: 65.

Vogel, A. and R. Pollack. 1974. Methods for obtaining revertants of transformed cells. *Methods Cell Biol.* **8**: 75.

Waterfield, M.D., G. Scrace, and N. Totty. 1983a. Separation of PTH amino acids by HPLC. In *Practical protein biochemistry* (ed. A. Darbre and M.D. Waterfield). Wiley, New York. (In press).

Waterfield, M.D., G.T. Scrace, N. Whittle, P. Stroobant, A. Johnsson, A. Wasteson, B. Westermark, C.-H. Heldin, J.S. Huang, and T.F. Deuel. 1983b. Platelet-derived growth factor is structurally related to the putative transforming protein $p28^{sis}$ of simian sarcoma virus. *Nature* **304**: 35.

Westermark, B. and A. Wasteson. 1975. Response of cultured human normal glial cells to growth factors. *Adv. Metab. Disord.* **8**: 85.

Westermark, B., C.-H. Heldin, B. Ek, A. Johnsson, K. Mellstrom, M. Nister, and A. Wasteson. 1983. Biochemistry and biology of platelet-derived growth factor. In *Growth and maturation factors* (ed. G. Guroff), vol. 1, p. 73. Wiley, New York.

Wilbur, W.J. and D.J. Lipman. 1983. Rapid similarity searches of nucleic acid and protein data banks. *Proc. Natl. Acad. Sci.* **80**: 726.

Witte, L.D., K.L. Kaplan, H.L. Nossel, B.A. Lages, H.J. Weiss, and D.S. Goodman. 1978. Studies of the release from human platelets of the growth factor for cultured human arterial smooth muscle cells. *Circ. Res.* **42**: 402.

Wong-Staal, F., R. Dalla Favera, G. Franchini, E.P. Gelmann, and R.C. Gallo. 1981. Three distinct genes in human DNA related to the transforming genes of mammalian sarcoma retroviruses. *Science* **213**: 226.

Close Similarities between the Transforming Gene Product of Simian Sarcoma Virus and Human Platelet-derived Growth Factor

K.C. Robbins,* H.N. Antoniades,† S.G. Devare,* M.W. Hunkapiller,‡ and S.A. Aaronson*

*Laboratory of Cellular and Molecular Biology, National Cancer Institute, National Institutes of Health, Bethesda, Maryland 20205; †Center for Blood Research and Department of Nutrition, Harvard School of Public Health, Boston, Massachusetts 02115; ‡Division of Biology, California Institute of Technology, Pasadena, California 91125

Acute transforming retroviruses have arisen in nature by substitution of viral genes necessary for replication with discrete segments of host genetic information. When incorporated within the retroviral genome, such transduced cellular sequences, termed *onc* genes, acquire the ability to induce neoplastic transformation. Some of these same cellular genes or proto-oncogenes have also been implicated as important targets for genetic alterations that may lead normal cells to become malignant independent of virus involvement (Barth et al. 1982; Dalla Favera et al. 1982; Der et al. 1982; Harris et al. 1982; Parada et al. 1982; Rechavi et al. 1982; Santos et al. 1982; Shen-Ong et al. 1982; Taub et al. 1982).

Despite advances in identifying cellular genes with transforming potential, little is known about protooncogene function or how the altered counterparts of these genes disrupt normal growth regulation. Recently, however, studies of the *onc* gene of simian sarcoma virus (SSV), a primate transforming retrovirus, combined with investigations of platelet-derived growth factor (PDGF), a potent mitogen for connective tissue cells, have led to the discovery that the SSV transforming gene product and PDGF arise from the same or very closely related cellular genes. The implications of these findings with respect to normal growth regulation and neoplastic transformation are discussed.

Experimental Procedures

Nucleotide sequencing
Appropriate restriction fragments were labeled either at their 5' ends by using [γ-^{32}P]ATP (Amersham, 3000 Ci/mmole; 1 Ci = 3.7×10^{10} becquerels) and polynucleotide kinase (P-L Biochemicals) or at their 3' ends by using cordycepin 5'-[α-^{32}P]triphosphate (Amersham, 3000 Ci/mmole) and terminal deoxynucleotidyltransferase (P-L Biochemicals). End-labeled DNA fragments were digested with appropriate restriction endonucleases (New England BioLabs) and isolated by agarose or polyacrylamide gel electrophoresis. The nucleotide sequence was determined by the procedure of Maxam and Gilbert (1977).

Peptide antibodies
We chose for synthesis pentadecapeptides derived from the 5'- and 3'-terminal regions of *v-sis*. Rabbits were immunized with 100 μg of the peptide either alone or after conjugation with thyroglobulin. Thereafter, 100 μg of peptide was administered intraperitoneally at 14-day intervals. Animals were bled 1 week after each injection. The effectiveness of the immune response was monitored by the ability of serum from sequential bleedings to precipitate ^{125}I-labeled *sis* peptides.

Immunoprecipitation analysis
Subconfluent cultures (~10^7 cells per 10-cm petri dish) were labeled for 3 hours at 37°C with 4 ml of methionine-free Dulbecco's modified Eagle's minimal essential medium containing 100 μCi of [^{35}S]methionine (1200 Ci/nmole; Amersham) per milliliter. Labeled cells were lysed with 1 ml of a buffer containing 10 mM sodium phosphate, pH 7.5, 100 mM NaCl, 1% Triton X-100, 0.5% sodium deoxycholate, and 0.1 mM phenylmethyl-sulfonyl fluoride per petri dish, clarified at 100,000g for 30 minutes, and divided into four identical aliquots. Each aliquot was incubated with 4 μl of antiserum for 60 minutes at 4°C. Immunoprecipitates were recovered with the aid of *Staphylococcus aureus* protein A bound to Sepharose beads (Pharmacia) and analyzed by electrophoresis in sodium dodecyl sulfate (SDS)–14% polyacrylamide gels (Barbacid et al. 1980).

Amino acid sequencing
PDGF-A (18,000 daltons; 18 kD) was prepared by reduction, alkylation, and gel electrophoresis of active PDGF (Antoniades and Hunkapiller 1983). It consists of roughly equal amounts of PDGF-1 and PDGF-2. Treatment of the preparation with cyanogen bromide followed by gel electrophoresis yielded two protein bands (H.N. Antoniades and M.W. Hunkapiller, unpubl.). The larger of these (18 kD) contained only the PDGF-1 sequence, indicating that PDGF-1 was not cleaved by the cyanogen bromide treatment, but that PDGF-2 was. The smaller band (14 kD) contained a single sequence that must be derived from PDGF-2 by cleavage at two sites near the amino terminus of PDGF-2, one of which is presumably at Met$_{12}$ (methionine) of the PDGF-2 sequence. Amino acid sequence analysis was performed as described (Antoniades and Hunkapiller 1983) with the Caltech gas-phase protein sequenator.

Results

Biologic properties of SSV

SSV was isolated from a woolly monkey fibrosarcoma (Thielen et al. 1971). This virus induces fibrosarcomas and glioblastomas in susceptible hosts and transforms fibroblasts in tissue culture (Wolfe et al. 1971, 1972). Biological characterization of SSV initially presented difficulties because of its association with an excess of nontransforming helper virus, designated simian sarcoma–associated virus (SSAV) (Scolnick et al. 1972). Early investigations led to the isolation of cells nonproductively transformed by SSV (Aaronson 1973). The ability to rescue the transforming virus by superinfection with a type-C helper virus provided evidence of the replication-defective nature of the SSV genome (Aaronson 1973; Aaronson et al. 1975). Moreover, the development of clonal SSV transformants established the necessary biologic system for the subsequent molecular characterization of SSV structure and function.

An important feature of SSV initially noted in biologic investigations was the unusual morphology of cells transformed by the virus. Whereas transformed foci induced by Moloney or Kirsten murine sarcoma viruses were composed of rounded or spindle-shaped, highly refractile cells that often detached from the cell monolayer (Fig. 1A), SSV-transformed foci were much more mounded in appearance, and the transformed cells themselves were fibroblastic and not nearly as round or refractile (Fig. 1B). Subsequent studies have confirmed that SSV-induced transformants differ in these respects from those induced by other known transforming retroviruses of avian or mammalian origin.

Structural analysis of the SSV genome

To investigate the molecular organization and transforming functions of the SSV genome, we cloned the integrated provirus from DNA of an SSV-transformed nonproducer cell (Robbins et al. 1981). To determine whether our recombinant DNA clones possessed biological activity, they were analyzed by transfection of NIH-3T3 cells. Both of the clones that we obtained possessed transforming activity of approximately 10^4 foci/pmole viral DNA (Robbins et al. 1981). Focus formation was a linear function of the amount of DNA added, indicating that a single DNA molecule was able to induce a transformed focus. When individual transformed foci were selected by the cloning cylinder technique, grown to mass culture, and superinfected with amphotropic mouse type-C helper virus, focus-forming activity characteristic of SSV was rescued from each transformant tested.

Structural analysis of one clone, designated λ-SSV-11 Cl 1, revealed a 5.1-kb SSV genome containing 0.55-kb terminal repeats flanked by 0.45 kb and 0.25 kb of contiguous host-cell sequences. By R-loop analysis, the viral DNA molecule contained two regions of homology to SSAV, separated by a 1.0-kb nonhomologous region (Robbins et al. 1981). This SSV-specific sequence was designated *v-sis*.

v-sis is derived from a unique, highly conserved cellular gene

Like other *onc* genes, *v-sis* arose from a normal cellular gene. Using cloned probes, DNA fragments related to *v-sis* were detected within cellular DNAs of species as diverse as human and quail (Robbins et al. 1982a). Thus,

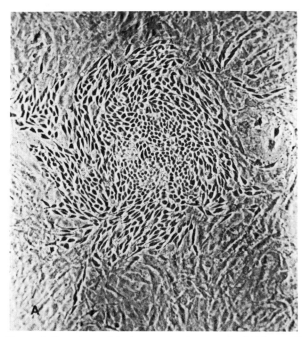
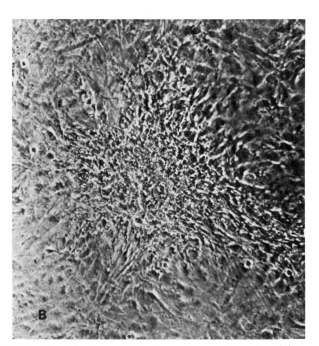

Figure 1 Transformed foci appearing in normal rat kidney cells at 7 days after infection with Kirsten murine sarcoma virus (*A*) or SSV (*B*). Magnification, 200×.

v-sis was not only derived from normal cells but was evolutionarily highly conserved. Such findings suggested that the normal cellular gene from which v-sis was derived must serve an important function.

It has been possible to demonstrate further that v-sis arose from the woolly monkey genome, by analysis of the extents of annealing in solution of different normal cellular DNAs with a single-stranded v-sis DNA probe prepared from cloned v-sis (Robbins et al. 1982a). The highest extents of annealing were exhibited by DNAs of New World primates with values ranging from 56% to 65%. Moreover, findings that the T_m of woolly monkey cellular DNA–sis DNA hybrids was identical to that of homologous v-sis DNA hybrids demonstrated that v-sis arose from within the woolly monkey genome.

v-sis is required for SSV transforming activity

To localize the region of SSV required for transformation, we constructed various deletion mutants from a molecular clone of SSV DNA and tested their ability to transform NIH-3T3 cells in a transfection assay (Graham and van der Eb 1973; Wigler et al. 1977). The intact viral genome (pSSV-11) exhibited a transforming activity of $10^{4.1}$ focus-forming units (ffu) per picomole of viral DNA (Fig. 2). A subgenomic clone, pSSV 3/1, from which the 3' long terminal repeat (LTR) was deleted, showed no reduction in biologic activity (Robbins et al. 1982b). In contrast, pSSV 3/2, a subclone that lacked the 3' LTR as well as all but 82 bp of v-sis, demonstrated no detectable transforming activity. These results indicated that v-sis was essential in SSV transformation. A mutant, pSSV I/1, which lacked an internal 1.8-kb BglII fragment, transformed NIH-3T3 cells with an efficiency of $10^{4.1}$ ffu/pmole viral DNA. However, a subclone, pSSV I/2, which lacked an additional stretch of 345 bp of SSAV sequences as well as the first 250 bp of v-sis, showed no transforming activity. These findings localized the SSV transforming gene to a region encompassing v-sis, along with 345 bp and 305 bp of flanking SSAV sequences, to the left and right of v-sis, respectively (Robbins et al. 1982b; Devare et al. 1983).

Nucleotide sequence of the SSV transforming gene

In an effort to better understand the structural organization of the SSV transforming gene as well as the molecular mechanisms involved in SSV transformation, we have determined the primary nucleotide sequence of the complete SSV genome and its rat cellular flanking sequences (Devare et al. 1983). Examination of the region of the SSV genome to which its transforming gene had been localized indicated the presence of a long open reading frame. This reading frame initiated at position 3657 within SSAV sequences 19 nucleotides to the left of the 5'-SSAV v-sis junction and terminated with an ochre codon within v-sis at position 4470 (Fig. 3). By sequence comparison with Moloney murine leukemia virus (Mo-MLV), the open reading frame to the left of v-sis was identified as initiating from the aminoterminal region of the SSAV env gene. Sequence comparison with Mo-MLV revealed 37% homology with SSV in this region. Moreover, sequence analysis of molecularly cloned SSAV confirmed that this region codes for the aminoterminal sequences of its env gene. The open reading frame of the SSAV env gene was identical from position 3657 to position 3810 at the 5' end of the sis open reading frame (Devare et al. 1982). This stretch of sequences contained two more ATG codons in the same reading frame. One of these corresponded to the ATG codon that has been proposed as the initiator codon of the Mo-MLV env gene product (Shinnick et al. 1981). This initiator codon could be used for synthesis of the v-sis gene product from a spliced mRNA analogous to that used for the env gene products (Mellon and Duesberg 1977; Rothenberg et al. 1978).

If the v-sis gene product were synthesized from the ATG at position 3657, a protein of approximately 33 kD containing 271 amino acids would result. A v-sis gene product synthesized from the second or third ATG would result in proteins of around 30 kD or 26 kD, respectively.

Detection of the SSV transforming gene product

To confirm our sequence analysis of v-sis as well as to develop antibodies capable of detecting the SSV transforming gene product, we synthesized peptides that represented amino- and carboxyterminal regions of the putative sis-coded protein (Devare et al. 1983). Such peptides can be used as haptens to elicit antibodies capable of recognizing translational products in vivo

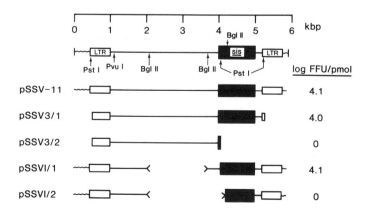

Figure 2 Construction and biologic analysis of SSV deletion mutants. Deletion mutants pSSV 3/1 and pSSV 3/2 were constructed by cloning products of a reaction in which purified SSV DNA was partially digested with PstI. The intact viral genome, pSSV-11, was obtained by cloning the λ-SSV-11 Cl 1 insert at the EcoRI site of pBR322. pSSV 1/1 and pSSV 1/2 were constructed by limited BglII digestion of pSSV-11 followed by religation. In each case, the structure of individual deletion mutants was determined by restriction enzyme and Southern blotting analysis. Transfection of NIH-3T3 cells with plasmids containing SSV wild-type or mutant DNAs was performed by the calcium phosphate precipitation technique (Graham and van der Eb 1973) as modified by Wigler et al. (1977). Transformed foci were scored at 14 to 21 days. (Reprinted, with permission, from Robbins et al. 1982b. Copyright 1982 by the AAAS.)

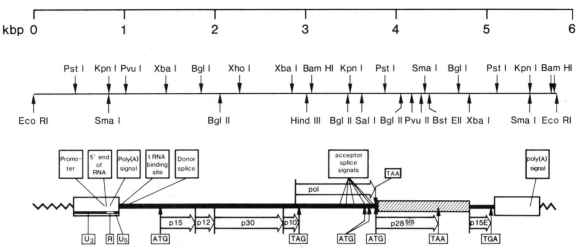

Figure 3 Summary of the major structural features of the SSV genome. The long open reading frame of v-sis, possible signals for promotion and polyadenylation, as well as donor and acceptor splice signals are illustrated.

(Sutcliffe et al. 1980; Walter et al. 1980). Serum against the aminoterminal sis peptide, designated anti-sis N, as well as the carboxyterminal peptide, designated anti-sis C, were capable of recognizing a protein of 28 kD specifically expressed in SSV-transformed cells (Robbins et al. 1982b; Devare et al. 1983). These findings provided conclusive evidence that the SSV transforming gene directed the synthesis of a protein of 28 kD (p28^{sis}) that corresponded in size to the sis open reading frame.

p28^{sis} is related to human platelet-derived growth factor

Independent investigations of the primary amino acid sequence of human platelet-derived growth factor (PDGF) in combination with knowledge of the predicted amino acid sequence of p28^{sis} have provided the first direct link between a transforming gene and a normal gene with known function (Antoniades and Hunkapiller 1983; Doolittle et al. 1983; Waterfield et al. 1983).

PDGF is a heat-stable (100°C), cationic (pI 9.8) protein (Antoniades et al. 1979). It circulates in blood stored in the α granules of platelets and is released into serum during blood clotting (Kaplan et al. 1979). It represents the major protein growth factor of human serum and is a potent mitogen for connective tissue and glial cells in culture (Ross et al. 1974; Scher et al. 1979). Unreduced, active PDGF exhibits multiple forms ranging in size from 28 kD to 35 kD (Antoniades et al. 1979; Antoniades 1981; Deuel et al. 1981; Heldin et al. 1981; Raines and Ross 1982). Reduction of PDGF produces inactive, smaller peptides ranging in size from 12 kD to 18 kD (Heldin et al. 1979; Antoniades 1981).

Amino acid sequence analysis of the aminoterminal portions of both active human PDGF and its inactive, reduced peptides revealed the presence of two homologous peptides (PDGF-1 and PDGF-2) in active PDGF preparations (Antoniades and Hunkapiller 1983). These peptides were identical at 8 of 19 positions near their amino termini, with no sequence gaps required for the homology alignment. Whether the active PDGF preparation is composed of a single protein formed by disulfide linkage of these two proteins, each of which consists of a disulfide-linked dimer of one of the peptides, is not yet known.

A computer comparison of the PDGF aminoterminal sequences with the sequence of p28^{sis} revealed a striking similarity between these proteins (Doolittle et al. 1983). This similarity was discovered by R.A. Doolittle during a search for sequence homology between the PDGF aminoterminal sequences and other protein sequences in the Newat sequence data base at the University of California, San Diego (Doolittle 1981). Sequence comparisons with additional PDGF primary sequences obtained from peptide fragments generated by cyanogen bromide cleavage of native PDGF have further strengthened this relationship (Doolittle et al. 1983; Waterfield et al. 1983).

The sequence alignments and matches derived from these searches are shown in Figure 4. The sequence comparisons are summarized as follows:

1. PDGF-1 matches p28^{sis} at 18 of 29 positions identified by protein sequencing.
2. PDGF-2 matches p28^{sis} at 26 of 31 positions at the PDGF-2 amino terminus and at 35 of 39 positions at the amino terminus of a 14-kD PDGF-2 cyanogen bromide fragment (total match at 61 of 70 identified positions).
3. No sequence gaps due to insertions or deletions were used in these alignments.

This extensive sequence homology left little doubt that the v-sis transforming gene was the result of a viral recombination with the host-cell gene or genes encoding PDGF or a very similar protein.

The match between the 70 identified PDGF-2 residues and the corresponding segment of p28^{sis} was 87.1%. Since the v-sis gene arose from within the genome of a woolly monkey (Wong-Staal et al. 1981; Robbins et al.

```
p28sis    1   M T L T W Q G D P I P E E L Y K M L S G H S I R S F D D L Q R L L Q G D S G K E D G A E L D L N M T   50

p28sis   51   A S H S G G E L E S L A R G K R S L G S L S V A E P A M I A E C K T R T E V F E I S R R L I D R T N   100
PDGF-2    1                                 S L G S L T I A E P A M I A E C K T R E E V F C I C R R L ? D R ? ?   34
PDGF-1    1                                             S T E E A V P A V C K T R T V T Y E I I S R R E L D I ? ? ?   28

p28sis  101   A N F L V W P P C V E V Q R C S G C C N R N V Q C R P T Q V Q L R P V Q V R K I E I V R K K P I F   100
PDGF-2   35   ? ? ? ? ? ? ? P P C V E V K A C T G C C N A N V K C A P S I Q V Q L A P ? Q V A K I E I V A K I [   80
PDGF-1   29   A N F L I [                                                                                          32

p28sis  151   K K A T V T L E D H L A C K C E I V A A A R A V T R S P G T S Q E Q R A K T T Q S R V T I R T V R V   200
PDGF-2                                                                                                         ]
PDGF-1                                                                                                         ]

p28sis  201   R R P P K G K H A K C K H T H D K T A L K E T L G A                                                  226
PDGF-2                                                                                                         ]
PDGF-1                                                                                                         ]
```

Figure 4 Sequence similarity between p28^{sis} and PDGF. Residue identity between the p28^{sis} and PDGF sequences is indicated by boxed sequences. A question mark indicates that no amino acid sequence assignment has yet been made for that position; the brackets indicate no sequence is yet available for the included segments. The box around p28^{sis} positions 65 and 66 indicates a possible proteolytic processing position for generation of a fragment of p28^{sis} corresponding to PDGF-2. Single letter abbreviations for the amino acid residues are: (A) alanine; (C) cysteine; (D) aspartic acid; (E) glutamic acid; (F) phenylalanine; (G) glycine; (H) histidine; (I) isoleucine; (K) lysine; (M) methionine; (N) asparagine; (P) proline; (Q) glutamine; (R) arginine; (S) serine; (T) threonine; (V) valine; (W) tryptophan; (Y) tyrosine. (Reprinted, with permission, from Doolittle et al. 1983. Copyright 1983 by the AAAS.)

1982a), a member of the family Cebidae (New World monkeys), and PDGF was isolated from human platelets, most, if not all, of the observed amino acid differences could represent species differences. This hypothesis is consistent with the known amino acid sequence similarity for myoglobin (90.8% identity) and fibrinopeptides A and B (70.0% identity) from humans and cebids (Doolittle 1981). Moreover, seven of the nine observed differences can be derived from single base changes.

Close structural and conformational similarity between the SSV transforming gene product and biologically active PDGF

Utilizing antibodies against the amino and carboxyl termini of the SSV transforming gene product, we were able to demonstrate that the primary translational product, p28^{sis}, of the SSV transforming gene undergoes cleavage to yield 11-kD and 20-kD polypeptides designated p11^{sis} and p20^{sis}, respectively (Fig. 5) (Robbins et al. 1983). The latter, which was shown to be derived from the carboxyl terminus of p28^{sis}, is very similar in size to the inactive, reduced 18-kD form of human PDGF-2. The sequence correspondence between PDGF-2 and the v-sis product begins at position 67. If cleavage of p28^{sis} were signaled by the two basic amino acids (lysine and arginine) at positions 65–66, a 160-residue protein would result, approximating p20^{sis} in its theoretical size. Thus, p20^{sis} closely corresponds in size and amino acid sequence to that observed for PDGF-2.

Unreduced, biologically active PDGF exhibits a dimeric structure and possesses multiple forms ranging in size from 28 kD to 35 kD. Under nonreducing gel conditions, p28^{sis} exhibited a molecular weight of 56,000. Our observations that this polypeptide was detectable with antibodies to both amino and carboxyl termini of p28^{sis} are most consistent with the possibility that p56^{sis} is a disulfide-linked dimer of p28^{sis}. Evidence that the carboxy- but not the aminoterminal sis antibodies also bound discrete sis-related polypeptides ranging in size from 28 kD to 46 kD implied further processing of p56^{sis} to smaller forms, some of which closely resembled the sizes of unreduced PDGF.

Anti-PDGF serum recognizes SSV transforming gene products

Striking evidence of the close conformational similarity between the two molecules derived from immunological studies. Anti-PDGF serum was shown to recognize the apparent p28^{sis} dimer as well as the other processed, unreduced forms of the protein recognized by the anti-

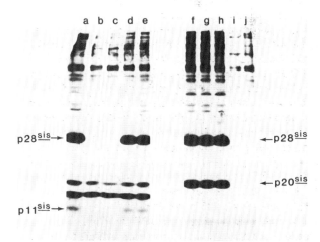

Figure 5 Identification of 20-kD and 11-kD SSV transforming gene products. Extracts of labeled SSV (SSAV)-transformed producer marmoset cells (HF/SSV) were immunoprecipitated with anti-sis-N (lanes a–e) or with anti-sis-C (lanes f–j) serum. In some cases antibodies were incubated prior to immunoprecipitation with 1 µg (lanes b,g) or 5 µg (lanes c,h) of sis-N or with 1 µg (lanes d,e) or 5 µg (lanes e,j) of sis-C peptide. Immunoprecipitates were analyzed by SDS-polyacrylamide gel electrophoresis. (Reprinted, with permission, from Robbins et al. 1983. Copyright 1983 by Macmillan Journals Limited.).

sis-C peptide serum (Robbins et al. 1983). In addition, anti-PDGF recognized a 24-kD protein not detected with either anti-sis-N or -C peptide sera. Pulse-chase analysis using anti-PDGF serum established p56^{sis} as a rapidly formed dimer of p28^{sis}. It was possible to demonstrate further processing in which p24^{sis} emerged as the most stable product of the SSV transforming gene detectable with the available antisera. Figure 6 represents a summary of these experiments. Our demonstration of cleavage at the amino terminus of the SSV primary translational product, p28^{sis}, implies that the failure of anti-sis-N peptide serum to detect processed, nonreduced forms smaller than p56^{sis} is due to cleavage of the domain from the molecule. Based upon its size and lack of detection with anti-sis-C peptide serum, p24^{sis} most likely reflects additional proteolytic cleavage at the carboxyl terminus. The dimeric nature of the SSV transforming gene product as well as its rapid proteolytic processing is consistent with the known dimeric structure and susceptibility to proteolysis of PDGF.

Discussion

The profound alterations in normal growth regulation induced by the cell-derived transforming genes of retroviruses have many similarities to the growth-promoting actions of certain hormones and growth factors. Each exerts pleiotropic effects on cellular metabolism and possesses the ability to induce sustained cell replication. Our studies have revealed a high degree of relatedness between the transforming gene product of a primate sarcoma virus and a potent growth factor for human fibroblasts, smooth muscle cells, and glial cells. Thus, the mechanism by which this onc gene transforms cells may involve the constitutive expression of functions similar to those of PDGF.

A major family of onc genes codes for protein kinases with specificity for phosphorylation of tyrosine residues (see Weiss et al. 1982); another has an associated guanosine triphosphate–binding activity (Shih et al. 1979). Moreover, previous computer searches have revealed distant homology between the src gene product and cyclic adenosine monophosphate kinase (Barker and Dayhoff 1982) and possibly between the B/ym gene product and transferrin (Goubin et al. 1983). Although these relationships are very distant, it is possible that other proteins with growth regulatory functions may be found to correspond to the products of known or as yet uncharacterized onc genes. The continued search for relationships of this kind will be of obvious interest and importance.

There have been reports of human osteosarcoma and glioblastoma cell lines that release PDGF-like polypeptides (Heldin et al. 1980; Nister et al. 1982; Graves et al. 1983). Moreover, sis-related transcripts have been detected frequently in human tumors of connective tissue origin but not in normal fibroblasts or in epithelial cell tumors (Eva et al. 1982). Thus, it is possible that activation of sis transcription alone or in combination with other genetic alterations affecting the sis structural locus may cause the sustained, abnormal proliferation of human cells responsive to the growth stimulatory effects of a PDGF-like molecule. If so, sis activation might be implicated as a step in the processes leading normal human cells of certain tissue types toward malignancy.

Acknowledgment

H.N.A. is supported by National Cancer Institute grant CA-30101.

References

Aaronson, S.A. 1973. Biologic characterization of mammalian cells transformed by a primate sarcoma virus. *Virology* **52**: 562.

Aaronson, S.A., J.R. Stephenson, S. Hino, and S.R. Tronick. 1975. Differential expression of helper viral structural polypeptides in cells transformed by clonal isolates of woolly monkey sarcoma virus. *J. Virol.* **16**: 1117.

Antoniades, H.N. 1981. Human platelet-derived growth factor: Purification of PDGF-I and PDGF-II and separation of their reduced subunits. *Proc. Natl. Acad. Sci.* **78**: 7314.

Antoniades H.N. and M.W. Hunkapiller. 1983. Human platelet-derived growth factor (PDGF): Amino terminal amino acid sequence. *Science* **220**: 963.

Antoniades, H.N., C.D. Scher, and C.D. Stiles. 1979. Purification of human platelet-derived growth factor. *Proc. Natl. Acad. Sci.* **76**: 1809.

Barbacid, M., A.L. Lauver, and S.G. Devare. 1980. Biochemical and immunological characterization of polyproteins coded for by the McDonough, Gardner-Arnstein, and Snyder-Theilen strains of feline sarcoma virus. *J. Virol.* **33**: 196.

Barker, W.C. and M.O. Dayhoff. 1982. Viral src gene products are related to the catalytic chain of mammalian cAMP-dependent protein kinase. *Proc. Natl. Acad. Sci.* **79**: 2836.

Barth, R., L. Hood, J. Prehn, and K. Calame. 1982. Mouse c-myc oncogene is located on chromosome 15 and translocated to chromosome 12 in plasmacytomas. *Science* **218**: 1319.

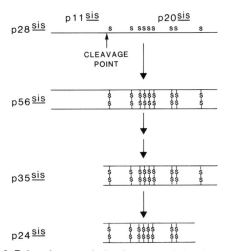

Figure 6 Pulse-chase analysis of posttranslational processing of p28^{sis}. HF/SSV cells were pulse labeled for 15 min with [^{35}S]-methionine and cysteine and chased as described (Robbins et al. 1983) for periods of 0 to 105 min. Immediately after the chase period, cells were washed on ice-cold phosphate-buffered saline and disrupted. Cell lysates were immunoprecipitated with anti-PDGF or normal rabbit serum under nonreducing conditions and immunoprecipitates were analyzed by SDS-polyacrylamide gel electrophoresis in the absence of reducing agent. A representation of the results from this experiment is shown.

Dalla Favera, R., M. Bregni, J. Erickson, D. Patterson, R.C. Gallo, and C.M. Croce. 1982. Human c-*myc onc* gene is located on the region of chromosome 8 that is translocated in Burkitt lymphoma cells. *Proc. Natl. Acad. Sci.* **79:** 7824.

Der, C.J., T.G. Krontiris, and G.M. Cooper. 1982. Transforming genes of human bladder and lung carcinoma cell lines are homologous to the *ras* genes of Harvey and Kirsten sarcoma viruses. *Proc. Natl. Acad. Sci.* **79:** 3637.

Deuel, T.F., J.S. Huang, R.T. Proffit, J.U. Baenziger, D. Chang, B.B. Kennedy. 1981. Human platelet-derived growth factor. Purification and resolution into two active protein fractions. *J. Biol. Chem.* **256:** 8896.

Devare, S.G., E.P. Reddy, J.D. Law, D.C. Robbins, and S.A. Aaronson. 1983. Nucleotide sequence of the siman sarcoma virus genome: Demonstration that its acquired cellular sequences encode the transforming gene product, $p28^{sis}$. *Proc. Natl. Acad. Sci.* **80:** 731.

Devare, S.G., E.P. Reddy, K.C. Robbins, P.R. Andersen, S.R. Tronick, and S.A. Aaronson. 1982. Nucleotide sequence of the transforming gene of simian sarcoma virus. *Proc. Natl. Acad. Sci.* **79:** 3179.

Doolittle R.F. 1981. Similar amino acid sequences: Chance or common ancestry. *Science* **214:** 149.

Doolittle, R.F., M.W. Hunkapiller, L.E. Hood, S.G. Devare, K.C. Robbins, S.A. Aaronson, and H.N. Antoniades. 1983. Simian sarcoma virus *onc* gene, v-*sis*, is derived from the gene (or genes) encoding a platelet-derived growth factor. *Science* **221:** 275.

Eva, A., K.C. Robbins, P.R. Andersen, A. Srinivasan, S.R. Tronick, E.P. Reddy, N.W. Ellmore, A.T. Galen, J.A. Lautenberger, T.S. Papas, E.H. Westin, F. Wong-Staal, R.C. Gallo, and S.A. Aaronson. 1982. Cellular genes analogous to retroviral *onc* genes are transcribed in human tumor cells. *Nature* **295:** 116.

Goubin, G., D.F. Goldman, J. Luce, P.E. Neiman, and G.M. Cooper. 1983. Molecular cloning and nucleotide sequence of a transforming gene detected by transfection of chicken B-cell lymphoma DNA. *Nature* **302:** 114.

Graham, F.L. and A.J. van der Eb. 1973. A new technique for the assay of infectivity of human adenovirus 5 DNA. *Virology* **52:** 456.

Graves, D.T., A.J. Owen, and H.N. Antoniades. 1983. Evidence that a human osteosarcoma cell line which secretes a mitogen similar to platelet-derived growth factor requires growth factors present in platelet-poor plasma. *Cancer Res.* **43:** 83.

Harris, L.J., P. D'Eustachio, F.H. Ruddle, and K.B. Marcu. 1982. DNA sequence associated with chromosome translocation in mouse plasmacytomas. *Proc. Natl. Acad. Sci.* **9:** 6622.

Heldin, C.H., B. Westermark, and A. Wasteson. 1979. Platelet-derived growth factor: Purification and partial characterization. *Proc. Natl. Acad. Sci.* **76:** 3722.

———. 1980. Chemical and biological properties of a growth factor from human-cultured osteosarcoma cells: Resemblance with platelet derived growth factor. *J. Cell. Physiol.* **105:** 235.

———. 1981. Demonstration of antibody against platelet-derived growth factor. *Exp. Cell Res.* **136:** 255.

Kaplan, D.R., F.C. Chao, C.D. Stiles, H.N. Antoniades, and C.D. Sher. 1979. Platelet alpha-granules contain a growth factor for fibroblasts. *Blood* **53:** 1043.

Maxam, A.M. and W. Gilbert. 1977. A new method for sequencing DNA. *Proc. Natl. Acad. Sci.* **74:** 560.

Mellon, P. and P.H. Duesberg. 1977. Subgenomic, cellular Rous sarcoma virus RNAs contain oligonucleotides from the 3′ half and the 5′ terminus of virion RNA. *Nature* **270:** 631.

Nister, M., C.H. Heldin, A. Wateson, and B. Westermark. 1982. A platelet-derived growth factor analog produced by a human clonal glioma cell line. *Ann. N. Y. Acad. Sci.* **397:** 25.

Parada, L.F., C.J. Tabin, C. Shih, and R.A. Weinberg. 1982. Human EJ bladder carcinoma oncogene is homologue of Harvey sarcoma virus *ras* gene. *Nature* **297:** 474.

Raines, E.W. and R. Ross. 1982. Platelet-derived growth factor I. High yield purification and evidence for multiple forms. *J. Biol. Chem.* **257:** 5154.

Rechavi, G., D. Givol, and E. Canaani. 1982. Activation of a cellular oncogene by DNA rearrangement: Possible involvement of an IS-like element. *Nature* **300:** 607.

Robbins, K.C., S.G. Devare, and S.A. Aaronson. 1981. Molecular cloning of integrated simian sarcoma virus: Genome organization of infectious DNA clones. *Proc. Natl. Acad. Sci.* **78:** 2918.

———. 1982a. Primate origin of the cell-derived sequences of simian sarcoma virus. *J. Virol.* **41:** 721.

Robbins, K.C., S.G. Devare, E.P. Reddy, and S.A. Aaronson. 1982b. In vivo identification of the transforming gene product of simian sarcoma virus. *Science* **218:** 1131.

Robbins, K.C., H.N. Antoniades, S.G. Devare, M.W. Hunkapiller, and S.A. Aaronson. 1983. Structural and immunological similarities between simian sarcoma virus gene product(s) and human platelet-derived growth factor. *Nature* **305:** 605.

Ross, R., J. Glomset, B. Kariya, and L. Harker. 1974. A platelet-dependent serum factor that stimulates the proliferation of arterial smooth muscle cells in vitro. *Proc. Natl. Acad. Sci.* **71:** 1207.

Rothenberg, E., D.J. Donoghue, and D. Baltimore. 1978. Analysis of a 5′ leader sequence on murine leukemia virus 21S RNA: Heteroduplex mapping with long reverse transcriptase products. *Cell* **13:** 435.

Santos, E., S. Tronick, S.A. Aaronson, S. Pulciani, and M. Barbacid. 1982. The T24 human bladder carcinoma oncogene is an activated form of the normal human homologue of BALB- and Harvey-MSV transforming genes. *Nature* **298:** 343.

Scher, C.D., R.C. Shepard, H.N. Antoniades, and C. D. Stiles. 1979. Platelet-derived growth factor and the regulation of the mammalian fibroblast cell cycle. *Biochim. Biophys. Acta* **560:** 217.

Scolnick, E.M., W.P. Parks, G.J. Todaro, and S.A. Aaronson. 1972. Immunological characterization of primate C-type virus reverse transcriptases. *Nat. New Biol.* **235:** 35.

Shen-Ong, G.L.C., E.J. Keath, S.P.E. Piccoli, and D.M. Cole. 1982. Novel *myc* oncogene RNA from abortive immunoglobulin-gene recombination in mouse plasmacytomas. *Cell* **31:** 443.

Shih, T.Y., M.O. Weeks, H.A. Young, and E.M. Scolnick. 1979. Identification of a sarcoma virus-coded phosphoprotein in nonproducer cells transformed by Kirsten or Harvey murine sarcoma virus. *Virology* **96:** 64.

Shinnick, T.M., R.A. Lerner, and J.G. Sutcliffe. 1981. Nucleotide sequence of Moloney murine leukemia virus. *Nature* **293:** 543.

Sutcliffe, J.G., T.M. Shinnick, N. Green, F.T. Liu, H.L. Niman, and R.A. Lerner. 1980. Chemical synthesis of a polypeptide predicted from nucleotide sequence allows detection of a new retroviral gene product. *Nature* **287:** 801.

Taub, R., I. Kirsch, C. Morton, G. Lenoir, D. Swan, S. Tronick, S. Aaronson, and P. Leder. 1982. Translocation of the c-*myc* gene into the immunoglobulin heavy chain locus in human Burkitt lymphoma and mouse plasmacytoma cells. *Proc. Natl. Acad. Sci.* **79:** 7837.

Theilen, G.H., D. Gould, M. Fowler, and D.L. Dungworth. 1971. C-type virus in tumor tissue of a woolly monkey (*Lagothrix* spp.) with fibrosarcoma. *J. Natl. Cancer Inst.* **47:** 881.

Walter, G., K.H. Scheidtmann, A. Carbone, A.P. Laudano, and R.F. Doolittle. 1980. Antibodies specific for the carboxy- and amino-terminal regions of simian virus 40 large tumor antigen. *Proc. Natl. Acad. Sci.* **77:** 5197.

Waterfield, M.D., G.T. Scrace, N. Whittle, P. Stroobant, A. Johnsson, A. Wasteson, B. Westermark, C.H. Heldin, J.S. Huang, and T.F. Deuel. 1983. Platelet-derived growth factor is structurally related to the putative transforming protein $p28^{sis}$ of simian sarcoma virus. *Nature* **304:** 35.

Weiss, R., N. Teich, H. Varmus, and J. Coffin, eds. 1982. *Molecular biology of tumor viruses*, 2nd ed.: *RNA tumor viruses*. Cold Spring Harbor Laboratory, Cold Spring Harbor, New York.

Wigler, M., S. Silverstein, L.S. Lee., A. Pellicer, Y. Cheng, and R. Axel. 1977. Transfer of purified herpes virus thymidine kinase gene to cultured mouse cells. *Cell* **11:** 223.

Wolfe, L.G., R.K. Smith, and F. Dienhardt. 1972. Simian sarcoma virus type 1 (*Lagothrix*): Focus assay and demonstration of nontransforming associated virus. *J. Natl. Cancer Inst.* **48:** 1905.

Wolfe, L.G., F. Deinhardt, G.J. Theilen, J. Rabin, T. Kawakami, and L.K. Bustad. 1971. Induction of tumors in marmoset monkeys by simian sarcoma virus, Type 1 (*Lagothrix*): A preliminary report. *J. Natl. Cancer Inst.* **47:** 1115.

Wong-Staal, F., R. Dalla Favera, E.P. Gelmann, V. Manzari, S. Szala, S.F. Josephs, and R.C. Gallo. 1981. The v-*sis* transforming gene of simian sarcoma virus is a new *onc* gene of primate origin. *Nature* **294:** 273.

The Platelet-derived Growth Factor Receptor Protein Is a Tyrosine-specific Protein Kinase

S.S. Huang, J.S. Huang, and T.F. Deuel
Departments of Medicine and Biological Chemistry, The Jewish Hospital of St. Louis at Washington University School of Medicine, St. Louis, Missouri 63110

The platelet-derived growth factor (PDGF) initiates DNA synthesis and cell division in Swiss mouse 3T3 cells and has been shown to be the major mitogen in human serum for smooth muscle cells, fibroblasts, and glial cells (Ross and Vogel 1978; Deuel and Huang 1983). PDGF is also a potent chemoattractant protein for human monocytes, neutrophils, fibroblasts, and smooth muscle cells (Grotendorst et al. 1981; Deuel et al. 1982b; Seppa et al. 1982; Senior et al. 1983). Recently, a partial amino acid sequence analysis of PDGF revealed near identity of the amino acid sequence of PDGF and that predicted for p28^{sis}, the product of the transforming gene of simian sarcoma virus (SSV) (Doolittle et al. 1983; Waterfield et al. 1983). This observation strongly suggested an important role for growth factor–like molecules in transformation by SSV and perhaps in cellular transformation by other agents. Therefore, PDGF as well as the related putative transforming protein p28^{sis} of SSV-infected cells (Robbins et al. 1982) not only may have a very important role in inflammation and repair, but also may play an important role in viral transformation or transformation by other agents. Precisely how PDGF or p28^{sis} interacts initially with cells to affect normal cellular growth, chemotaxis, or malignant transformation is not understood.

A PDGF-like growth factor activity has recently been found in lysates of SSV-transformed cells but not in lysates of noninfected cells. The growth factor activity is blocked by anti-PDGF antisera and has a mitogenic specific activity essentially identical with that of purified PDGF (Deuel et al. 1983). This growth factor activity is thus almost certainly the transforming protein of SSV, although this has not been conclusively established.

It is not yet established that the transforming protein of SSV is secreted to the external medium. We have therefore considered several models of how a transforming protein/growth factor activity potentially may mediate cellular transformation. First, the transforming protein may be expressed intracellularly and interact with intracellular receptors. Second, the transforming protein may be expressed and released to the extracellular environment, where it may interact with cell-surface receptors either on the secretor cells or on the nonsecretor neighboring cells, in each instance effecting the transformation event through a specific receptor. Third, the transforming protein may be interacting with both internal and external receptors at the same time. The striking near identity of the amino acid sequence predicted for p28^{sis} and that of PDGF suggests that each interacts with the same receptor. To better understand cellular events initiating DNA synthesis, the PDGF receptor protein from Swiss mouse 3T3 cells has now been purified.

We recently purified PDGF to apparent homogeneity and identified two equally active mitogenic proteins: PDGF I, with a molecular weight of 31,000 (31K), and PDGF II, with a molecular weight of 28,000 (28K) (Deuel et al. 1981a,b, 1982a; Huang et al. 1982b). Both PDGF I and II are glycoproteins and are highly basic (pI 10.2). Approximately 400,000 PDGF-specific, high-affinity ($K_d \simeq 0.7 \times 10^{-9}$ M) binding sites on the surface of Swiss mouse 3T3 cells were found (Huang et al. 1982a).

The PDGF receptor protein was initially identified in whole cells by crosslinking ^{125}I-labeled PDGF to the receptor. The receptor was then solubilized with retention of full binding activity, purified to near homogeneity by wheat germ lectin and PDGF affinity chromatographies, and shown to retain intrinsic and PDGF-stimulatable, tyrosine-specific protein kinase activity. The molecular weight of the purified receptor is approximately 180,000, which is in excellent agreement with the receptor molecular weight estimated by crosslinking ^{125}I-labeled PDGF with its receptor on intact 3T3 cells and with the molecular weight of the protein stimulated in tyrosine phosphorylation by PDGF. Copurification of intrinsic and PDGF-stimulated protein kinase activity was observed, further suggesting the PDGF receptor is a protein kinase that is activated by PDGF and capable of autophosphorylation and perhaps phosphorylation of other proteins as well.

Materials and Methods

Human PDGF II (28K) was isolated from outdated human platelet-rich plasma (Deuel et al. 1981a,b, 1982a; Huang et al. 1982b) and radiolabeled with ^{125}I, using the iodogen method as previously described (Huang et al. 1982a). Na^{125}I (17 Ci/mg) and [γ-^{32}P]ATP (2–10 Ci/mmole) were obtained from New England Nuclear. Swiss mouse 3T3 cells were obtained from the American Type Culture Collection and grown to confluence in Dulbecco's modified Eagle's medium (GIBCO), containing 10% fetal calf serum (Deuel et al. 1981a,b). For in vivo labeling of the receptor protein, cells were monolayer incubated in methionine-free media with [^{35}S]methionine (100 μCi/ml) added. Mouse epidermal growth factor (EGF) and bovine fibroblast growth factor (FGF) were obtained from Collaborative Research. Egg phosphotidylcholine, wheat germ lectin–Sepharose 6MB, cyanogen bromide–activated Sepharose 4B, human serum albumin, and salmon protamine sulfate were obtained from Sigma. Protein markers (myosin, β-galactosidase, phosphorylase B, and bovine serum albumin) were obtained from Bio-Rad. Ethylene glycol bis-succinimidyl succinate (EGS) and iodogen were obtained from Pierce. All other chemicals were reagent grade.

Membrane vesicles were prepared by a Parr nitrogen cavitation bomb. Purified membranes were obtained by centrifugal separation through 25% Ficoll (Whittenberger and Glaser 1977; Nishimura et al. 1982) and used for a protein kinase assay and for purification of the receptor. Protein was measured by a modified Lowry method (Lowry et al. 1951) or by the Bio-Rad method, including equal amounts of octylglucoside or Triton X-100 with the bovine serum albumin as a standard (Bradford 1976). To solubilize PDGF receptor activity, crude membrane pellets were suspended in 50 mM Tris, 200 mM NaCl, 2 mM $MgCl_2$, and 40 mM octylglucoside at a ratio of 6 mg of membrane protein per milliliter of detergent solution. The membrane pellet was dispersed with a syringe and stirred for 1 hour at 4°C. Nonsolubilized material was removed by centrifugation (35,000g, 30 min). The solubilized PDGF receptor activity was assayed using the methods of Schneider et al. (1979, 1980). Typically, 1 ml of membrane extract (0.2 mg protein, 40 mM octylglucoside) was diluted with 1.8 ml of the same buffer without octylglucoside. Six-tenths ml of phosphatidylcholine suspension (2.5 mg/ml) in Tris-$MgCl_2$-NaCl buffer (without detergent) was added to the solubilized membrane preparations. Two ml of acetone (−20°C) was added rapidly, the solution was centrifuged (20,000g, 20 min, 4°C), and the pellet was suspended in 5 mM HEPES, 0.15 M NaCl buffer (pH 7.4). PDGF receptor activity was assayed with ^{125}I-labeled PDGF (0–60 ng/ml) and 20 μg of solubilized membrane protein (acetone precipitate) in 5 mM HEPES (pH 7.4), 0.15 M NaCl, and human serum albumin (1 mg/ml) for 1 hour at room temperature. After centrifugation (10,000g, 5 min), the pellet was washed twice with 1 ml of phosphate-buffered saline (PBS) and radioactivity was determined. Nonspecific binding was measured with a 100-fold molar excess of unlabeled PDGF (Huang et al. 1982a).

Swiss mouse 3T3 cell membranes were incubated with 0.3 ml of 50 ng/ml ^{125}I-labeled PDGF in 5 mM HEPES, 0.15 M NaCl, and 0.1% human serum albumin (pH 7.4) for 1 hour at room temperature. EGS (20 mg/ml) in dimethylsulfoxide was diluted to 0.2 mg/ml in 50 mM dibasic potassium phosphate. Three hundred μl of this solution was immediately added to the binding buffer and the reaction continued for 15 minutes at 22°C. The membranes were pelleted and washed twice with 1 ml of PBS and dissolved in SDS buffer for electrophoresis. The gels were stained, destained, and dried. Autoradiography was developed using intensifying screens (DuPont).

Protein kinase activity in membranes was assayed as described previously (Nishimura et al. 1982). For solubilizing the protein kinase activity, the purified membranes (3 mg) were extracted with 1 ml of 1% Triton X-100 and 10% glycerol, 20 mM HEPES (pH 7.4), with occasional stirring at 4°C for 1 hour. The insoluble proteins were removed by centrifugation. The Triton X-100 extract was mixed with 1 ml of wheat germ lectin–Sepharose 6MB gels and rotated at 4°C overnight. The unretained fraction was removed by centrifugation. The gels were extensively washed with 20 mM HEPES, 0.15 M NaCl, and 0.1% Triton X-100 (pH 7.4). The gels were then eluted with 1 ml of 300 mM N-acetylglucosamine in the above buffer.

PDGF–Sepharose 4B gels were prepared as follows. One mg of purified PDGF II in 2 ml of 0.1 M $NaHCO_3$ (pH 8.2) was mixed with 1.5 ml of CNBr–Sepharose 4B (Sigma) at 4°C for 18 hours. The gels were washed with 0.1 M $NaHCO_3$ and then with 0.1 M acetate buffer (pH 4.0). The partially purified membrane receptors from wheat germ lectin affinity gels were mixed with 0.2 ml of PDGF affinity gels and rotated at 4°C overnight. The unabsorbed fraction was obtained by centrifugation. The gels were washed with the buffer, and an aliquot of PDGF affinity gels were directly assayed for protein kinase activity, and the gels subsequently were eluted with 0.5 M NaCl in 0.1% Triton X-100 and 20 mM HEPES (pH 7.4). Phosphotyrosine was measured after hydrolysis of protein extracted from gels by two-dimensional electrophoresis, using unlabeled phosphoamino acids as standards (Hunter and Cooper 1981; Nishimura et al. 1982).

Results

The molecular weight of the PDGF receptor on intact 3T3 cells was first estimated by crosslinking membrane-bound ^{125}I-labeled PDGF with 0.2 mM EGS. The ^{125}I-labeled PDGF–linked membranes were solubilized and analyzed by autoradiography of reduced SDS gels (Fig. 1). A band of radiolabeled protein is seen at ≃190K (Fig. 1, lane S). Specificity of ^{125}I-labeled PDGF binding was established by demonstrating that unlabeled PDGF (4 μg/ml) inhibited ^{125}I-labeled PDGF from binding to the tentatively identified membrane receptor (Fig. 1, lane C).

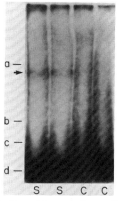

Figure 1 Affinity labeling of the membrane PDGF receptor from Swiss mouse 3T3 cells. ^{125}I-labeled PDGF (50 ng/ml) was crosslinked to the plasma membrane preparations (50 μg) of 3T3 cells with EGS in the presence (C) or absence (S) of a 100-molar excess of unlabeled PDGF. Protein markers: (a) myosin, 200K; (b) β-galactosidase, 116K; (c) phosphorylase B, 94K; (d) dye front. (I) The crosslinked complex of ^{125}I-labeled PDGF and the receptor.

Radiolabel is seen near the dye front; this activity is unreacted ^{125}I-labeled PDGF. Other than the single band at 190K, evidence of higher molecular weight multimers was not observed as it has been by others (Glenn et al. 1982; Heldin et al. 1983). If only one of the two polypeptide chains of ^{125}I-labeled PDGF is crosslinked to the receptor, the molecular weight of the receptor protein would be ~175K in reduced gels, in good agreement with the molecular weight of the 3T3 cell membrane protein phosphorylated in tyrosine residues by the PDGF-stimulatable protein kinase activity, as previously observed (Nishimura et al. 1982).

Membranes from 3T3 cells were then solubilized (see Materials and Methods), and the binding of ^{125}I-labeled PDGF was measured and compared with binding to intact 3T3 cells. ^{125}I-labeled PDGF bound to the solubilized PDGF receptor as a hyperbolic function of ^{125}I-labeled PDGF added (Fig. 2, left). Scatchard analysis (Fig. 2, right) revealed a single class of high-affinity binding sites with an apparent K_d of $\sim 1.2 \times 10^{-9}$ M, which is in good agreement with the apparent K_d of $\sim 0.7 \times 10^{-9}$ M measured for ^{125}I-labeled PDGF binding to intact 3T3 cells (Huang et al. 1982a). Nonspecific binding was ~40% of the total binding at 30 ng/ml ^{125}I-labeled PDGF (at ~K_d); approximately 2 pmoles of PDGF were bound per milligram of solubilized protein. The binding of ^{125}I-labeled PDGF to the solubilized receptor was ionic-strength dependent; ^{125}I-labeled PDGF binding was progressively reduced as the ionic strength was increased above physiological concentration. At 150 mM NaCl, ~70% binding was retained; at 1 M NaCl, less than 10% of ^{125}I-labeled PDGF binding was observed (data not shown). The specificity of binding was tested by seeking competition of ^{125}I-labeled PDGF with other proteins. The binding was specific; 100 ng/ml unlabeled PDGF reduced ^{125}I-labeled PDGF binding by ~50%; 600 ng/ml PDGF reduced ^{125}I-labeled PDGF binding to essentially nondetectable levels. EGF in concentrations to 3 µg/ml did not compete with ^{125}I-labeled PDGF for binding, whereas protamine sulfate, a competitive inhibitor of ^{125}I-labeled PDGF binding to intact 3T3 cells (Huang et al. 1982a), reduced binding to less than 50% at 800 ng/ml and to less than 5% at 3.3 µg/ml (data not shown). These results thus established that a single class of PDGF receptors had been solubilized in high yield; these receptors fully retained the reported characteristics and specificities of the PDGF receptor previously identified on intact 3T3 cells (Huang et al. 1982a); coupled with the 85% recovery of binding activity, these results establish that the receptor has been successfully solubilized.

The solubilized receptor and intact membranes were then tested for PDGF-stimulated protein kinase activity to see if the PDGF-stimulated kinase was solubilized with the PDGF receptor binding activity. When PDGF (0–1000 nM) and either membranes from 3T3 cells (Fig. 3A) or 1% Triton X-100 membrane extracts (Fig. 3B) were tested for PDGF-stimulated protein kinase activity, intrinsic and PDGF-stimulated phosphorylation of an ~180K protein were readily demonstrated in both preparations. The optimum PDGF concentrations stimulating protein kinase activity in intact membranes were found

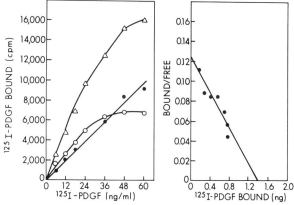

Figure 2 (*Left*) Concentration-dependent binding of ^{125}I-labeled PDGF to solublilized PDGF receptor from 3T3 cell membranes. The assay (0.3 ml) contained 23 µg of solubilized membrane protein and the indicated concentrations of ^{125}I-labeled PDGF (8600 cpm/ng). After incubation for 60 min at room temperature, bound ^{125}I-labeled PDGF was measured by the phosphatidylcholine/acetone precipitation procedure as described in the text. The nonspecific binding was determined in the presence of a 100-molar excess of unlabeled PDGF. (△) Total binding; (●) nonspecific binding; (○) specific binding. (*Right*) Scatchard plot of the specific binding data. The slope of the graph plotting the ratio of receptor-bound ^{125}I-labeled PDGF to free ^{125}I-labeled PDGF versus receptor-bound ^{125}I-labeled PDGF is equal to $1/K_d$. The molecular weight of 30K for PDGF was used to calculate K_d of PDGF at the binding sites—1.2×10^{-9} M.

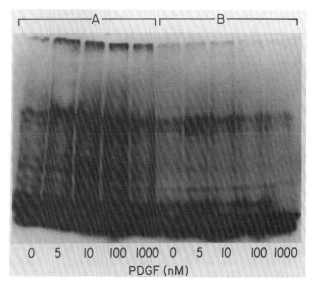

Figure 3 Effect of PDGF concentration on the phosphorylation of membrane (*A*) and Triton X-100–solubilized membrane (*B*) from 3T3 cells. Twenty µg of membrane protein or solubilized membrane protein were incubated with various concentrations of PDGF as indicated at 4°C for 30 min. Fifteen µM [γ-^{32}P]ATP was then added and incubation was continued for 10 min at 4°C. Five hundred µl of 10% TCA was added to stop the reaction. The protein was pelleted and washed twice with 10% TCA. The proteins were analyzed by 5% SDS-polyacrylamide gel electrophoresis and autoradiographed.

between 10 nM and 100 nM, whereas optimum phosphorylation in solubilized preparations occurred at ~5 nM PDGF. As PDGF concentrations were increased above optimum, progressive reductions in phosphorylation were observed. The solubilized receptor-containing fraction thus retained intrinsic and PDGF-activated protein kinase activity. Recovery of kinase activity exceeded 80% in 1% Triton X-100, in close agreement with the recovery of receptor binding activity.

The definitive identification of the PDGF receptor, the PDGF-stimulated protein kinase activity, and the ~180K phosphoprotein as arising from a single membrane protein or from very tightly associated activities depends upon the identification of each activity with a purified protein. Purification of the PDGF receptor protein was therefore attempted with wheat germ lectin chromatography and PDGF-Sepharose affinity chromatography (see Materials and Methods), using intrinsic protein kinase activity as assay. Recovery was ~50% of the activity eluted from wheat germ lectin chromatography and ~35% from PDGF-Sepharose affinity chromatography. PDGF stimulated phosphate incorporation into the ~180K protein 2.1-fold after elution from wheat germ lectin and 2.2-fold after elution from PDGF-Sepharose (see below).

The purity of the receptor preparation was estimated by SDS gel electrophoresis of receptor purified from cells grown with [^{35}S]methionine (Fig. 4). The principal protein band was found at ~180K. A minor protein band also was found at ~150K; this lower-molecular-weight, labeled protein is likely a proteolytic product of the PDGF receptor because it binds to PDGF-Sepharose, it is phosphorylated by intrinsic and PDGF-stimulated protein kinase activity, and because phosphopeptide map analysis of the PDGF-stimulated phosphorylated proteins revealed common proteolytic products of the two proteins (J. Nishimura et al., unpubl.). The SDS gels suggest the radiolabeled receptor protein is highly purified; no other [^{35}S]methionine-containing proteins were observed.

Purity was also tested by using intrinsic and PDGF-stimulated protein kinase activity (Fig. 5). The ~180K protein was phosphorylated in the absence of PDGF. PDGF stimulated incorporation of ^{32}P into this protein; 5 nM PDGF stimulated ^{32}P incorporation 2.1-fold above intrinsic (non-PDGF-stimulated) protein kinase activity, and 10 nM PDGF stimulated incorporation 2.2-fold above this mark. No evidence of other phosphoproteins was detected. Neither EGF, FGF, nor insulin stimulated protein kinase above the baseline, suggesting other receptor protein kinase activities were not purified with the PDGF receptor. Hydrolysis of the purified ~180K receptor phosphoprotein, followed by two-dimensional electrophoretic analysis of [^{32}P]phosphoamino acids, revealed only phosphotyrosine. No evidence for phosphoserine was obtained in this analysis (data not shown).

Discussion

PDGF is the major mitogen in serum for Swiss mouse 3T3 cells (Ross and Vogel 1978; Deuel and Huang 1983). A PDGF-like protein may also play an important role in cellular transformation by SSV, as suggested by the finding that a stretch of 104 amino acid residues of PDGF showed near complete identity with the sequence predicted for p28^{sis}, the putative transforming protein of SSV. The mechanisms whereby PDGF and the closely related p28^{sis} interact with cell-surface receptors and initiate DNA synthesis are thus of major importance to understand.

The binding of ^{125}I-labeled PDGF to the 3T3 cell surface has been previously investigated and found to be highly specific, of high affinity ($K_d \approx 0.7 \times 10^{-9}$ M), and

Figure 4 SDS gel electrophoresis pattern of [^{35}S]methionine-labeled receptor protein. Mouse 3T3 cells in petri dish (100 × 15 mm) were labeled with [^{35}S]methionine (100 μCi/ml) for 4 hr. The cells were solubilized with 20 mM HEPES containing 1% Triton and 10% glycerol, pH 7.4. The extracts were subjected to wheat germ lectin–Sepharose 6MB and PDGF–Sepharose 4B affinity chromatographies as described in Materials and Methods. The ^{35}S-labeled purified receptor was analyzed by SDS gel electrophoresis (7.5%) and fluorography. (→) The ^{35}S-labeled purified receptor with a molecular weight of ~180,000.

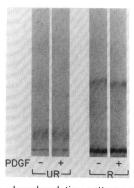

Figure 5 Protein phosphorylation patterns of unretained (UR) and retained (R) fractions from PDGF–Sepharose 4B affinity chromatography after wheat germ lectin–Sepharose 6MB affinity chromatography. One hundred μl (~1 μg) of a 0.5 M NaCl eluate of PDGF-affinity gel was assayed for protein kinase activity in the presence or absence of PDGF (10 nM) as described in the text. The autoradiograph was developed for 2 days at −70°C, using intensifying screens.

consistent with a single class of cell-surface receptors (Huang et al. 1982a). We have now identified the 3T3 cell PDGF receptor as an ~180K protein by crosslinking ^{125}I-labeled PDGF to the membrane receptor with EGS. We recently demonstrated than an ~180K protein is phosphorylated preferentially in tyrosine residues when PDGF, [γ-^{32}P]ATP, and membranes from 3T3 cells are incubated together (Nishimura et al. 1982). The molecular weight estimated by crosslinking ^{125}I-labeled PDGF to the receptor is consistent with the molecular weight of this phosphoprotein and the receptor solubilized from 3T3 cell membranes with ~80% recovery of receptor activity and with retention of the binding properties of the cell-surface receptor. The receptor protein purified to near homogeneity also has an apparent molecular weight of ~180,000 and is tentatively felt to be a single polypeptide chain on the basis of its migration in denaturant (SDS), reduced gels.

Glenn et al. (1982) estimated the molecular weight of the PDGF receptor on 3T3 cells to be ~165,000 on the basis of a molecular weight of ~193,000 observed for the ^{125}I-labeled PDGF receptor complex. Larger molecular weight bands were observed in these gels, which raised the possibility that ^{125}I-labeled PDGF may have crosslinked other proteins or formed multimers with additional receptor molecules. Heldin et al. (1983) similarly observed multiple labeled bands in crosslinking experiments with ^{125}I-labeled PDGF to intact human skin fibroblasts; the use of different methodologies in crosslinking may have accounted for the differences observed. Using similar assumptions to ours, a molecular weight of 180,000 was also estimated for the PDGF receptor.

Triton X-100 was used to solubilize the receptor. From the greater than 80% recovery of binding activity and from the demonstrated identical properties of the solubilized binding activity with those of the PDGF receptor on intact 3T3 cells, we conclude that the physiological receptor has been solubilized. The binding properties investigated include saturable binding, the K_d (1.2 × 10^{-9} M), evidence for only a single class of PDGF binding sites, inhibition by unlabeled PDGF and by protamine sulfate (a competitive inhibitor of PDGF) (Huang et al. 1982a), and the failure of EGF to compete for binding.

PDGF stimulates the incorporation of ^{32}P into tyrosine residues of an ~180K protein in 3T3 cell membranes. Cohen et al. (1980, 1982) showed copurification of EGF-binding activity and of the EGF-stimulated phosphorylation activity. The EGF receptor was thus identified as an intrinsic protein kinase that was activated by EGF binding, resulting in the autophosphorylation of the EGF receptor protein itself. We analyzed intrinsic and PDGF-stimulated protein kinase activity in 3T3 cell membranes, in detergent-solubilized receptor preparations, and after each subsequent purification step. Recovery was ~50% after wheat germ lectin chromatography and ~30% after PDGF-Sepharose chromatography. In each instance PDGF stimulated protein kinase activity, phosphorylating a protein essentially identical in molecular weight with that of the receptor protein identified by crosslinking ^{125}I-labeled PDGF. Wheat germ lectin chromatography was used to bind the PDGF receptor during purification, suggesting that the receptor is a glycoprotein. The EGF and insulin receptors also bind to wheat germ lectin–Sepharose (Cohen et al. 1982; Kasuga et al. 1982a,b). However, these receptors were not found in fractions retained by PDGF-Sepharose. The purified receptor protein retains PDGF-stimulated protein kinase activity (2.2-fold stimulation by PDGF) and is specifically phosphorylated in tyrosine residues by direct phosphoamino acid analysis. Because only a single class of binding sites was identified in the Triton X-100–solubilized fraction and because the binding activity shared identical properties with receptors on the surface of the 3T3 cell, it is clear that the binding activity being purified is the PDGF receptor. PDGF binding activity remained throughout the purification (i.e., PDGF-Sepharose binding, PDGF-stimulated protein kinase activity); the recovery of protein kinase activity was ~15–18% overall. These findings thus provide definitive evidence that the protein purified is the PDGF receptor and that the PDGF receptor protein is (or is very tightly associated with) a protein kinase that is activated by its specific ligand PDGF, activating autophosphorylation in tyrosine and presumably phosphorylation in tyrosine residues of other protein substrates. Evidence for extensive tyrosine phosphorylation of proteins in PDGF-stimulated intact 3T3 cells has been provided by Cooper et al. (1982) and may be of critical importance in mediating events leading to DNA synthesis and cell division.

EGF and insulin receptors also contain tyrosine-specific protein kinase activity (Cohen et al. 1980, 1982; Kasuga et al. 1982a,b). The establishment of similar protein kinase activity intrinsic to the PDGF receptor suggests the likelihood that many if not all polypeptide growth factors initiate signal transmission for DNA synthesis through stimulation of tyrosine-specific protein phosphorylation. Whereas the physiological protein substrates for additional phosphorylations are unknown, several proteins have been identified as substrates for tyrosine kinases in intact cells (Radke and Martin 1979; Hunter and Cooper 1981; Cooper et al. 1982). Although it is intriguing to believe that tyrosine-specific protein phosphorylation provides a common pathway for signal transmission in growth factor–stimulated cells, proof is lacking that protein phosphorylation is in fact the initial and necessary step for initiation of DNA synthesis.

Retrovirus-infected cells also have increased phosphotyrosine levels and tyrosine-specific protein kinase activity. This activity has been best defined in cells infected with Rous sarcoma virus (Erikson and Erikson 1980; Hunter and Sefton 1980; Levinson et al. 1981; Sefton et al. 1981). Phosphorylation of tyrosine residues is uncommon in normal cells (Hunter and Sefton 1980), suggesting that growth factor–stimulated cells and cells infected with retroviruses may share tyrosine-specific phosphorylations as immediate events requisite to the initiation of DNA synthesis and cell division. It is possible that the tyrosine-specific kinase activity may be expressed in some non-virus-infected human neoplasms, as well. Thus, as noted above (Doolittle et al. 1983; Waterfield et al. 1983), a striking homology has been observed between a partial amino acid sequence of PDGF and that predicted for p28^{sis}, the putative transforming protein of SSV (Devare et al. 1983). The cellular homolog of the viral transforming gene v-sis (c-sis) may thus be the structural gene for PDGF. Human c-sis is a gene present in single copy isolated by molecular cloning and localized to chromosome 22 (Dalla Favera et al. 1981, 1982). Eva et al. (1982) observed in five of six human sarcoma lines and in three of five human glioblastoma cell lines the expression of a transcript that hybridized to a v-sis probe; normal fibroblasts lacked expression of this gene. A human T-cell line (HUT 102) infected with the T-cell leukemia retrovirus HTLV also expresses c-sis (Westin et al. 1982). The product expressed by c-sis thus may contain the sequence required to interact with a PDGF receptor, activating tyrosine protein kinase activity. This activity has not yet been described, however, and the site of interaction of the v-sis transforming protein and the expressed product of the c-sis gene remained to be elucidated.

Heldin et al. (1983) recently presented a limited characterization of the PDGF receptor protein from human skin fibroblasts. Two different reagents were used to crosslink ^{125}I-labeled PDGF. Photoaffinity-labeled PDGF membrane receptor complex showed multiple crosslinked bands with molecular weights of 190,000 to 200,000; a low efficiency of labeling was encountered. Unlabeled PDGF competed for binding. When a bifunctional crosslinking reagent was used, bands of 200K–220K were observed; at higher concentrations of crosslinking reagent, additional complexes of ~400K were found. Heldin et al. (1983) also showed that ^{125}I-labeled PDGF bound to intact cells could be solubilized with Triton X-100; migration on Sepharose 6B was similar to a solubilized (PDGF-stimulated) phosphoprotein. This complex migrated at ~320K. An estimate of ~200K for the receptor protein was obtained after subtraction estimates of the contribution of Triton X-100 and PDGF. The 185K phosphoprotein had affinity for the same lectins as the putative PDGF receptor and bound to PDGF-Sepharose. A recovery of 10% of the material bound to PDGF-Sepharose was observed. These experiments thus suggested that the kinase activity and the PDGF receptor were related, but other interpretations were not excluded. Binding analyses of the solubilized receptor were not provided, having opened the possibility that the binding activity studied may be unrelated to the receptor protein. Pike et al. (1983) showed that PDGF stimulated the phosphorylation of a 170K protein in parental and variant 3T3 cell membranes and in membranes from human skin fibroblasts. Both phosphoserine and phosphotyrosine were found in this work. The intensity of protein phosphorylation paralleled the number of PDGF receptors identified by direct binding analysis of variant 3T3 cell lines and the number of receptors estimated in down-regulated 3T3 cells. These results thus provide additional support for the relatedness of the PDGF receptor and the protein kinase activity.

Conclusions

We conclude that the PDGF receptor protein is an ~180K protein with intrinsic and PDGF-stimulated (or, less likely, a very tightly associated) tyrosine-specific protein kinase activity that autophosphorylates the receptor protein and perhaps phosphorylates other intracellular proteins, as well. The PDGF receptor/protein kinase activity is similar to the receptors for EGF and insulin and is probably similar to receptors of other growth factors, as well; it is the site of initial cellular interaction of PDGF and perhaps of p28^{sis}.

Acknowledgments

We acknowledge with gratitude the cooperation of the American Red Cross Blood Banks in Chicago (R. Gilbert), Fort Wayne (G.D. Drew and D. Dunfee), Toledo (Dr. P. Lau and F. Courtwright), and Tulsa (Dr. D.O. Kasprisin) for generously supplying us with outdated human platelet packs. This work was supported by National Institutes of Health grants HL-14147 and HL-22119 and by the Monsanto Fund.

References

Bradford, M. 1976. Rapid and sensitive method for quantitation of microgram quantities of protein utilizing principle of protein-dye binding. *Anal. Biochem.* **72:** 248.

Cohen, S., G. Carpenter, and L. King, Jr. 1980. Epidermal growth factor-receptor-protein kinase interactions: Co-purification of receptor and epidermal growth factor–enhanced phosphorylation activity. *J. Biol. Chem.* **255:** 4834.

Cohen, S., H. Ushiro, C. Stoscheck, and M. Chinkers. 1982. A native 170,000 epidermal growth factor receptor–kinase complex from shed plasma membrane vesicles. *J. Biol. Chem.* **257:** 1523.

Cooper, J.A., D.F. Bowen-Pope, E. Raines, R. Ross, and T. Hunter. 1982. Similar effects of platelet-derived growth factor and epidermal growth factor on the phosphorylation of tyrosine in cellular proteins. *Cell* **31:** 263.

Dalla Favera, R., R.C. Gallo, A. Giallongo, and C.M. Croce. 1982. Chromosomal localization of the human homolog (c-sis) of the simian sarcoma virus onc gene. *Science* **218:** 686.

Dalla Favera, R., E.P. Gelmann, R.C. Gallo, and F. Wong-Staal. 1981. A human onc gene homologous to the transforming gene (v-sis) of simian sarcoma virus. *Nature* **292:** 31.

Deuel, T.F. and J.S. Huang. 1983. Platelet-derived growth factor: Purification, properties and biological activities. *Prog. Hematol.* **13:** 201.

Deuel, T.F., T. Maciag, and L.D. Witte. 1982a. Growth factors and the arterial cell wall. In *Differentiation and function of hematopoietic cell surfaces* (ed. V. Marchesi et al.), p. 217. A.R. Liss, New York.

Deuel, T.F., R.M. Senior, J.S. Huang, and G.L. Griffin. 1982b. Chemotaxis of monocytes and neutrophils to platelet-derived growth factor. *J. Clin. Invest.* **69:** 1046.

Deuel, T.F., J.S. Huang, R.T. Proffitt, D. Chang, and B. Kennedy. 1981a. Platelet-derived growth factor: Preliminary characterization. *J. Supramol. Struct. Cell. Biochem.* (suppl.) **5:** 128.

Deuel, T.F., J.S. Huang, S.S. Huang, P. Stroobant, and M.D. Waterfield. 1983. Expression of a platelet derived growth factor-like protein in simian sarcoma virus-transformed cells. *Science* **221:** 1348.

Deuel, T.F., J.S. Huang, R.T. Proffitt, J.U. Baenziger, D. Chang, and B.B. Kennedy. 1981b. Human platelet-derived growth factor: Purification and resolution into two active protein fractions. *J. Biol. Chem.* **256:** 8896.

Devare, S.G., E.P. Reddy, J.D. Law, K.C. Robbins, and S.A. Aaronson. 1983. Nucleotide sequence of the simian sarcoma virus genome: Demonstration that its acquired cellular sequences encode the transforming gene product p28^{sis}. *Proc. Natl. Acad. Sci.* **80:** 731.

Doolittle, R.F., M.W. Hunkapiller, L.E. Hood, S.G. Devare, K.C. Robbins, S.A. Aaronson, and H.N. Antoniades. 1983. Simian sarcoma virus *onc* gene, *v-sis*, is derived from the gene (or genes) encoding a platelet-derived growth factor. *Science* **221:** 275.

Erikson, E. and R.L. Erikson. 1980. Identification of a cellular protein substrate phosphorylated by the Avian sarcoma virus transforming gene product. *Cell* **21:** 829.

Eva, A., K.C. Robbins, P.R. Andersen, A. Srinivasan, S.R. Tronick, E.P. Reddy, N.W. Ellmore, A.T. Galen, J.A. Lautenberger, T.S. Papas, E.H. Westin, F. Wong-Staal, R.C. Gallo, and S.A. Aaronson. 1982. Cellular genes analogous to retroviral *onc* genes are transcribed in human tumour cells. *Nature* **295:** 116.

Glenn, K., D.F. Bowen-Pope, and R. Ross. 1982. Platelet-derived growth factor. III. Identification of a platelet-derived growth factor receptor by affinity labeling. *J. Biol. Chem.* **257:** 5172.

Grotendorst, G.R., H.E.J. Seppa, H.K. Kleinman, and G.R. Martin. 1981. Attachment of smooth muscle cells to collagen and their migration toward platelet-derived growth factor. *Proc. Natl. Acad. Sci.* **78:** 3669.

Heldin, C.-H., B. Ek, and L. Ronnstrand. 1983. Characterization of the receptor for platelet-derived growth on human fibroblasts. Demonstration of an intimate relationship with a 185,000-dalton substrate for the platelet-derived growth factor-stimulated kinase. *J. Biol. Chem.* **258:** 10054.

Huang, J.S., S.S. Huang, B.B. Kennedy, and T.F. Deuel. 1982a. Platelet-derived growth factor: Specific binding to target cells. *J. Biol. Chem.* **257:** 8130.

Huang, J.S., R.T. Proffitt, J.U. Baenziger, D. Chang, B.B. Kennedy, and T.F. Deuel. 1982b. Human platelet-derived growth factor: Purification and initial characterization. In *Differentiation and function of hematopoietic cell surfaces* (ed. V. Marchesi et al.), p. 225. A.R. Liss, New York.

Hunter, T. and J.A. Cooper. 1981. Epidermal growth factor induces rapid tyrosine phosphorylation of proteins in A431 human tumor cells. *Cell* **24:** 741.

Hunter, T. and B.M. Sefton. 1980. Transforming gene product of Rous sarcoma virus phosphorylates tyrosine. *Proc. Natl. Sci.* **77:** 1311.

Kasuga, M., Y. Zick, D.L. Blithe, M. Crettaz, and C.R. Kahn. 1982a. Insulin stimulates tyrosine phosphorylation of the insulin receptor in a cell-free system. *Nature* **298:** 667.

Kasuga, M., Y. Zick, D.L. Blithe, F.A. Karlsson, H.U. Haring, and C.R. Kahn. 1982b. Insulin stimulation of phosphorylation of the β subunit of the insulin receptor. Formation of both phosphoserine and phosphotyrosine. *J. Biol. Chem.* **257:** 9891.

Levinson, A.D., S.A. Courtneidge, and J.M. Bishop. 1981. Structural and functional domains of the Rous sarcoma virus transforming protein (pp60^{src}). *Proc. Natl. Acad. Sci.* **78:** 1624.

Lowry, O.H., N.J. Roseborough, A.L. Farr, and R.J. Randall. 1951. Protein measurement with the folin phenol reagent. *J. Biol. Chem.* **193:** 265.

Nishimura, J., J.S. Huang, and T.F. Deuel. 1982. Platelet-derived growth factor stimulates tyrosine-specific protein kinase activity in Swiss mouse 3T3 cell membranes. *Proc. Natl. Acad. Sci.* **79:** 4303.

Pike, L.J., D.F. Bowen-Pope, R. Ross, and E.F. Krebs. 1983. Characterization of platelet-derived growth factor-stimulated phosphorylation in cell membranes. *J. Biol. Chem.* **258:** 9383.

Radke, K. and G.S. Martin. 1979. Transformation by Rous sarcoma virus: Effects of *src* gene expression on the synthesis and phosphorylation of cellular polypeptides. *Proc. Natl. Acad. Sci.* **76:** 5212.

Robbins, K.C., S.G. Devare, E.P. Reddy, and S.A. Aaronson. 1982. In vivo identification of the transforming gene product of simian sarcoma virus. *Science* **218:** 1131.

Ross, R. and A. Vogel 1978. The platelet-derived growth factor. *Cell* **14:** 203.

Schneider, W.J., J.L. Goldstein, and M.S. Brown. 1980. Partial purification and characterization of the low-density lipoprotein receptor from bovine adrenal cortex. *J. Biol. Chem.* **255:** 11442.

Schneider, W.J., S.K. Basu, M.J. McPhaul, J.L. Goldstein, and M.S. Brown. 1979. Solubilization of low-density lipoprotein receptor. *Proc. Natl. Acad. Sci.* **76:** 5577.

Sefton, B.M., T. Hunter, E.B. Ball, and S.J. Singer. 1981. Vinculin: A cytoskeletal target of the transforming protein of Rous sarcoma virus. *Cell* **24:** 165.

Senior, R.M., G.L. Griffin, J.S. Huang, D.A. Walz, and T.F. Deuel. 1983. Chemotactic activity of platelet α-granule proteins for fibroblasts. *J. Cell Biol.* **96:** 382.

Seppa, H., G. Grotendorst, S. Seppa, E. Schiffman, and G.R. Martin. 1982. Platelet-derived growth factor is chemotactic for fibroblasts. *J. Cell Biol.* **92:** 584.

Waterfield, M.D., G.T. Scrace, P. Stroobant, A. Johnsson, A. Wasteson, B. Westermark, C.-H. Heldin, J.S. Huang, and T.F. Deuel. 1983. Platelet-derived growth factor is structurally related to the putative transforming protein p28^{sis} of simian sarcoma virus. *Nature* **304:** 35.

Westin, E.H., F. Wong-Staal, E.P. Gelmann, R. Dalla Favera, T.S. Papas, J.A. Lautenberger, A. Eva, E.P. Reddy, S.R. Tronick, S.A. Aaronson, and R.C. Gallo. 1982. Expression of cellular homologues of retroviral *onc* genes in human hematopoietic cells. *Proc. Natl. Acad. Sci.* **79:** 2490.

Whittenberger, B. and L. Glaser. 1977. Inhibition of DNA synthesis in cultures of 3T3 cells by isolated surface membranes. *Proc. Natl. Acad. Sci.* **74:** 2251.

Cell-cycle Genes Regulated by Platelet-derived Growth Factor

B.H. Cochran, A.C. Reffel, M.A. Callahan, J.N. Zullo, and C.D. Stiles
Department of Microbiology and Molecular Genetics, Harvard Medical School,
and Dana-Farber Cancer Institute, Boston, Massachusetts 02115

Circulating blood platelets contain a mitogen for connective tissue cells termed platelet-derived growth factor (PDGF). Human PDGF was purified to homogeneity several years ago (Antoniades et al. 1979; Heldin et al. 1979) and has since been characterized by several groups (Deuel et al. 1981; Raines and Ross 1982). Each circulating platelet contains about 1000 molecules of PDGF (Singh et al. 1982) sequestered within the alpha-granules (D.R. Kaplan et al. 1979; K.L. Kaplan et al. 1979; Gerrard et al. 1980). The contents of these alpha-granules are only released when blood clots; thus, clotted blood serum contains about 15–20 ng/ml PDGF whereas platelet-poor plasma contains less than 1 ng/ml (Singh et al. 1982; Westermark et al. 1983). Smooth muscle cells (Ross et al. 1974), fibroblasts (Scher et al. 1978), and glial cells (Heldin et al. 1979) require PDGF for optimum growth in vitro whereas epithelioid cells and hematopoietic cells do not (Stiles et al. 1979). This target specificity of PDGF is reflected in the presence or absence of PDGF-specific receptors in cultured cell lines (Heldin et al. 1981a; Bowen-Pope and Ross 1982; Huang et al. 1982; Singh et al. 1982).

Beyond its biologic activity as a connective tissue mitogen, PDGF makes a convenient teaching vehicle. There are growing indications these days that growth factors and oncogenes are relevant to each other. At least three candidate functions for cellular oncogenes may be discerned within the molecular biology of PDGF (Stiles 1983). The first candidate function lies at the onset of the mitogenic cascade wherein oncogenes may cause the production of a growth factor within cells that are responsive to that same growth factor. Simian sarcoma virus (SSV) appears to have acquired its transforming gene (v-sis) (Devare et al. 1983) from the structural gene for platelet-derived growth factor (PDGF) (Doolittle et al. 1983; Waterfield et al. 1983). The cellular homolog of v-sis (c-sis) is expressed in many connective tissue tumors (Eva et al. 1982) and these same tumors secrete a PDGF-like mitogen into cell-culture medium (Heldin et al. 1980, 1981b).

A second candidate function is at the receptor level. The receptor for PDGF is intimately associated with a tyrosine-specific protein kinase (Cooper et al. 1982; Ek and Heldin 1982; Nishimura et al. 1982; Heldin et al. 1983), an enzyme activity encoded by several viral oncogenes and their cellular homologs (Bishop 1983).

The third candidate function lies within the control of cell-cycle "early genes." Messenger RNA synthesis is required for traverse of G_1 phase of the cell cycle into S phase; however, the rate of RNA synthesis in exponentially growing cell cultures is not substantially greater than in quiescent cell cultures (see discussion in Rossini and Baserga 1978). This paradox could be reconciled by rare, unique-sequence genes whose transcription/translation products play a regulatory role in the mitogenic response. Indeed, analysis of mRNA:cDNA hybridization rates indicates that about 3% of the mRNA species found in exponentially growing fibroblast cultures are missing from quiescent cultures (Williams and Penman 1975). We have isolated gene sequences whose expression is modulated by PDGF (Cochran et al. 1983). The remainder of this article will focus on the structure, regulation, and possible function of these PDGF-regulated gene sequences.

Isolation of PDGF-regulated gene sequences

PDGF by itself is not an efficient mitogen for BALB/c-3T3 cells. A second set of growth factors contained in the platelet-poor-plasma fraction of serum functions synergistically with PDGF to promote the optimum mitogenic response (Pledger et al. 1977; Vogel et al. 1978). In the absence of these plasma growth factors, PDGF-treated cells remain growth arrested. Thus, biochemical changes observed following PDGF treatment do not occur merely as a consequence of growth and progression through the cell cycle. The ability to uncouple the immediate biochemical responses to PDGF from growth per se creates an opportunity for isolation of gene sequences whose expression is regulated directly by PDGF.

Figure 1 illustrates the strategy. A cDNA library was constructed from the mRNA of cells treated briefly with PDGF alone. Differential colony hybridization procedures were used to detect cloned gene sequences that reacted weakly with a cDNA probe prepared to mRNA from quiescent cells but that reacted strongly to a cDNA probe from PDGF-treated cells.

Figure 2 illustrates the actual appearance of one of our filter-screening assays. Sensitivity of the screening protocol can be calibrated by probing lysed bacterial colonies carrying a β-globin plasmid with a gradient dilution series of [^{32}P]cDNA directed against the globin insert (Fig. 2, top). The calibration data obtained with β-globin suggest that mRNAs represented at the 0.1% abundance level (~500 copies/cell) in the induced state would give a strong positive signal whereas message represented at the .01% abundance level or lower would give a weak or undetectable signal. Thus, by isolating colonies that gave a strong positive signal on PDGF-

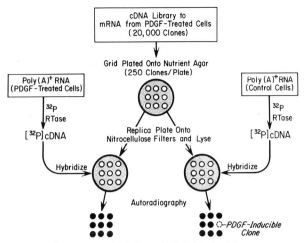

Figure 1 Strategy for isolation of PDGF-inducible genes.

treated filters and little if any signal at the corresponding position on filters probed with quiescent cDNA, we could expect to isolate gene sequences that were represented by low-abundance mRNAs in quiescent cells.

Frequency of PDGF-regulated gene sequences

More than 8000 cDNA clones were screened by the protocol illustrated in Figure 1. Because the mechanics of colony transfer to nitrocellulose filters generate some spurious signals, all colonies scored as PDGF-inducible in a first screening were twice retested. The results are summarized in Table 1. A total of 55 PDGF-inducible colonies were isolated from 8000 clones examined. Of these 55 clones, 46 of them represented only 5 independent gene sequences that had been isolated multiple times from the library. Nine additional clones resisted analysis for independence for technical reasons. Thus, at least 5 and no more than 14 independent PDGF-inducible genes were isolated from the library. Making the reasonable assumption that at least half of the 8000 clones screened contained unique gene sequences represented only once on an average in the library, we can estimate that PDGF regulates the expression of between 0.1% and 0.3% of the total genes transcribed in 3T3 cells. This estimate is in reasonably close agreement with that obtained by examining the frequency of PDGF-inducible proteins within 3T3 cells by two-dimensional gel electrophoresis (Pledger et al. 1981).

Abundance levels of PDGF-inducible mRNAs

Two of the PDGF-inducible gene sequences that we isolated were singled out for close inspection. The abundance level of the mRNAs corresponding to these gene sequences was quantitated by determining specific cDNA hybridization to cloned plasmids. The results show that within quiescent density-arrested monolayers, the KC and JE sequences correspond to low-abundance mRNAs (see Table 2). The levels found (70–100 copies/cell)

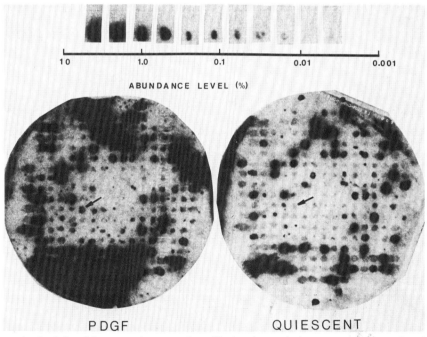

Figure 2 Sensitivity and selectivity of the screening procedure. The two large circles are autoradiographs of replica-plated colony grids that have been lysed and hybridized to 2.5×10^6 cpm of [^{32}P]cDNA from quiescent or PDGF-treated 3T3 cells as described in Cochran et al. (1983). (→) The location of a clone containing a PDGF-inducible gene sequence. (Top) Calibration strip showing autoradiographs of filters containing one colony each of pBR322 (upper) and pSV-β-globin (lower). Each filter was probed with twofold serial dilutions of β-globin [^{32}P]cDNA (100–200,000 cpm) under the same conditions as the large filters. The abundance level was calculated as (cpm ^{32}P–globin/2.5×10^6)%. The pBR322 colonies do not hybridize β-globin to any significant degree and therefore are not visible in the autoradiographs.

Table 1 Characterization of PDGF-inducible Clones

Plasmid	Insert size (bp)[a]	Times isolated from 8000-clone library[b]
pBC-JE	750	27
pBC-C1B	250	7
pBC-JB	1050	5
pBC-KC	820	4
pBC-CC	310	3
Others	—	9

[a]The size of the insert was determined by PstI digestion of the plasmid and electrophoresis on 5% polyacrylamide gels.

[b]Inserts were eluted from gels, nick-translated, and hybridized to all of the PDGF-inducible clones identified by differential colony hybridization.

correspond well to the steady-state mRNA levels for phenobarbital-regulated genes within rat liver that were measured by similar methods (Gonzalez and Kasper 1982). Following PDGF treatment, the mRNA content of JE and KC rises to the moderately abundant level. The higher level of JE within PDGF-treated cells relative to the level of KC (3000 copies/cell vs. 700 copies/cell) is consistent with their corresponding representation in the cDNA library (27 vs. 4 times) (see Table 1) and also with their relative fold induction as determined by Northern dot blots.

Specificity of the induction process

Quiescent BALB/c-3T3 cells were exposed to various growth factors for several hours. Total cytoplasmic RNA was extracted and the level of intracellular JE and KC mRNA was quantitated by dot-blot analysis. The data (Fig. 3) show that JE and KC respond preferentially if not specifically to PDGF. Up to 60-fold inductions of the JE sequence can be scored within 4 hours of PDGF treatment. Epidermal growth factor (EGF) sometimes gives a weak induction, whose significance is not understood at present. Insulin at pharmacologic concentrations has no detectable effect.

Superinduction of the JE and KC gene sequences in the presence of cycloheximide

Are the JE and KC gene sequences induced directly by PDGF or are they induced as a secondary event during the course of the mitogenic response? Clearly, the former type of control would be more interesting from the viewpoint of understanding the events that trigger the mitogenic response. The cDNA library from which JE and KC were isolated was directed to mRNA within cells that had been treated with PDGF alone. Since PDGF by itself does not efficiently promote progress through the cell cycle and replicative DNA synthesis, we should have discriminated against gene sequences induced as secondary events during the course of cell growth.

To confirm that JE and KC could be induced as a primary response to PDGF, we asked whether PDGF could induce KC and JE in the presence of cycloheximide. Inhibitors of protein synthesis block the progression of quiescent 3T3 cells through G_1 into S phase (Brooks 1977; Rossow et al. 1979); however, as shown in Figure 3, cycloheximide does not block the induction of these genes by PDGF. In fact, the KC sequence is superinduced 20-fold to 60-fold over the levels triggered by PDGF alone. Other inhibitors of protein synthesis, such as puromycin, give a similar superinduction response. The observations suggest that transcription of the KC sequence in particular may be feedback-inhibited by a subsequent translation product.

Time course of the induction process

The time course of the induction process is likewise consistent with the concept that these genes are regulated directly by PDGF. As shown in Figure 4, both the JE and KC sequences are induced nearly 20-fold within the first hour after exposure to PDGF, long before the onset of replicative DNA synthesis. In the case of KC, induction actually peaks at 1 hour and declines thereafter. The intracellular content of JE mRNA continues to increase within the first 2 to 4 hours of PDGF treatment so that final induction of greater than 60-fold can be measured. Figure 4 also illustrates that the induction of both JE and KC requires de novo RNA synthesis since

Table 2 Abundance of KC and JE Genes

Gene	Input (cpm)[a]	Total hybridization (cpm)[b]	pBR322 hybridization[b]	Percentage of total poly(A)+ RNA[c]	mRNA copies/ PDGF-treated cell[d]	mRNA copies/ quiescent cell[e]
JE	2.5×10^6	5225 ± 254	735 ± 101	0.9	3000	100
KC	2.5×10^6	1811 ± 107	735 ± 101	0.2	700	70

[a]Input is [^{32}P] cDNA made to poly(A)+ RNA from cells treated with PDGF for 4 hr.

[b]Hybridization was performed as described in Cochran et al. (1983). Each value is an average of three determinations.

[c]This value was calculated as described in Cochran et al. (1983). The hybridization efficiency was 20% as measured by determining the proportion of β-globin cDNA that hybridized to a lysed colony harboring a β-globin plasmid at several different levels of input cDNA. The efficiency was constant in the range of 1000–6000 cpm bound, indicating both that the hybridization had gone to completion and that the plasmid DNA was bound to the filter in excess of the input cDNA.

[d]This number was calculated by assuming approximately 350,000 mRNAs per cell (Lewin 1980).

[e]Copies of mRNA per cell in the induced state were determined by dividing the copy number per PDGF-treated cell by the fold induction as determined by dot blotting (see Fig. 3).

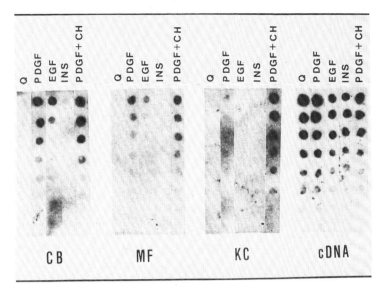

Figure 3 Quiescent (Q) BALB/c-3T3 cells were incubated at 37°C for 4 hr with PDGF (500 units/ml), EGF (50 ng/ml), insulin (INS) (5 μg/ml), or PDGF (500 units/ml) and cycloheximide (CH) (10 μg/ml). RNA was then extracted from the cells, taken through serial twofold dilution, and blotted onto nitrocellulose filter paper. The filters were probed with [^{32}P]cDNA corresponding to the plasmid inserts from three representative PDGF-inducible gene sequences, CB, MF, and KC. As a control, to demonstrate that roughly the same amount of RNA was present in each dilution series, one filter was probed with [^{32}P]-labeled total cellular cDNA. The filters were then processed by autoradiography. Details for these procedures are in Cochran et al. (1983).

the inhibitor, 5,6-dichloro-β-D-ribofuranosylbenzimidazole (DRB) prevents the response. The pronounced superinduction response of the KC gene sequence in the presence of cycloheximide is also illustrated in Figure 4.

Other mitogen-regulated gene sequences: A comparative analysis

Sequences responsive to SV40 T antigen (Schutzbank et al. 1982), EGF (Foster et al. 1982), and serum (Linzer and Nathans 1983) have been isolated by differential colony hybridization. Features of these other mitogen-inducible gene systems are summarized in Table 3 together with those of PDGF-inducible genes. Inspection of Table 3 reveals that each mitogen regulates only a small fraction of the total genes that are expressed in its target cell. The large T antigen of SV40 may exert the most sweeping effects on its target cells since 11 inducible clones were isolated from a cDNA library of only 430 colonies. The mRNA content corresponding to SV40, EGF, and serum gene sequences rises well before the onset of replicative DNA synthesis; however, there is yet no indication whether the induction is mediated directly by the mitogen or indirectly as a secondary event in the growth response. Induced levels in excess of 100-fold higher than basal state have been measured for the T-antigen-regulated sequences and for PDGF-regulated sequences when cycloheximide is used as a superinducer. One of the serum-responsive clones described by Linzer and Nathans (1983) is also responsive to PDGF and to SV40. Further relationships and interactions between these mitogen-regulated gene sequences may yet be awaiting explication.

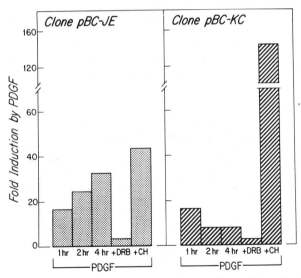

Figure 4 Time course of KC and JE induction and the effect of metabolic inhibitors. Quiescent 3T3 cells were treated with 500 units/ml of crude PDGF, and at the indicated time the cells were lysed and prepared for dot blots as described in Fig. 3. Each data point is an average of two experiments. DRB shows the effect of a 4-hr treatment of PDGF + 15 μg/ml 5,6-dichloro-β-D-ribofuranosylbenzimidazole. CH is a 4-hr treatment with PDGF + 10 μg/ml cycloheximide. DRB alone had no effect on gene induction. Cycloheximide, by itself, produced a onefold to fourfold increase in the level of gene transcripts.

Conclusions

Between 0.1% and 0.3% of the genes expressed in 3T3 cells are regulated by PDGF. This means that there are between 10 and 30 PDGF-inducible genes. Of these, 5 to 14 have been isolated by molecular cloning. The PDGF-inducible gene sequences correspond to low-abundance mRNAs in quiescent cells. Within 1 hour after PDGF addition, the abundance level of these mRNAs can be increased up to 20-fold.

Expression of PDGF-inducible genes may be feedback-inhibited by a labile repressor protein. By 12 hours after addition of PDGF to quiescent monolayer cultures (when the first cells have begun to enter S phase), the abun-

Table 3 Properties of Cloned Gene Sequences That Are Regulated by Mitogens

Stimulus	Percentage of mRNAs responding	Gene copy number	Abundance level of mRNAs (basal state)	Time of half-max induction	Maximum induction	References
PDGF	0.1–0.3%[a]	probably unique sequence[b]	low abundance (70–100 copies/cell)	<1 hr	60–120-fold	Cochran et al. 1983
T antigen	~5%[a]	N.S.	N.S.	6–8 hr[c]	100-fold	Schutzbank et al. 1982
EGF	N.S.	moderately repetitive (~100 copies/haploid genome)	N.S.	≥3 hr	10-fold	Foster et al. 1982
Serum	0.25%[a]	N.S.	N.S.	6–12 hr	5–20-fold	Linzer and Nathans 1983

[a]Calculated by dividing the number of responsive clones isolated from a cDNA library by 0.5 × the total number of clones screened. The correction of 0.5 assumes that about 50% of the colonies in a cDNA library correspond to high-abundance mRNAs, which are thus overrepresented.
[b]B.H. Cochran et al. (unpubl.)
[c]Measured following infection of normal fibroblasts with wild-type SV40. Individual clones varied in their induction kinetics.
(N.S.) Not specified.

dance level of PDGF-inducible messages has dropped considerably even with continual PDGF treatment. The abundance level of PDGF-inducible mRNAs in sparse, exponentially growing 3T3 cell cultures is moderately higher than that seen in quiescent monolayers but not dramatically elevated (B.H. Cochran et al., unpubl.); this minimum differential may reflect the fact that PDGF-inducible genes fire intermittently during the cell cycle. Thus, these PDGF-regulated gene sequences may be compared to the spark plug of an automobile engine, which fires intermittently during the power cycle.

Do PDGF-inducible mRNAs or their translation products play any role in the 10–12-hour chain of events that culminates in replicative DNA synthesis? There is now circumstantial evidence suggesting a functional role in mitosis but no direct evidence. Analysis of somatic cell fusion shows that (1) the mitogenic response to PDGF can be transferred from the cytoplasm of one cell to another and (2) de novo RNA synthesis is required before a PDGF-treated cell acquires the capacity to transfer the mitogenic response to another cell by fusion (Smith and Stiles 1981). Metabolic inhibitor data must be interpreted with caution; however, a simple view of the cell fusion data suggests that PDGF stimulates transcription of a new mRNA and that the translation product of this mRNA initiates the mitogenic response. That one or more of the PDGF-inducible genes described in this report mediates the growth response to PDGF is an exciting, but unproven, possibility.

It must be kept in mind that there are cellular responses to PDGF other than mitogenesis. PDGF regulates phospholipase activity (Shier 1980), lipid synthesis (Habenicht et al. 1981), and chemotaxis (Grotendorst et al. 1981). The PDGF-inducible genes described here could have a role in these nonmitogenic cellular functions.

A related question for future research is the relationship between the tyrosine-specific protein kinase activity of the PDGF receptor (Cooper et al. 1982; Ek and Heldin 1982; Nishimura et al. 1982; Heldin et al. 1983) and the coordinate induction of these gene sequences. It is possible that phosphoproteins serve as transducing elements in relaying the PDGF signal from the cytoplasmic membrane to chromatin.

Acknowledgments

Work from the authors' laboratory summarized here was supported by grants from the National Institutes of Health. B.H.C. is supported by a fellowship from the Noble Foundation. C.D.S. is supported by a Faculty Research Award from the American Cancer Society.

References

Antoniades, H.N., C.D. Scher, and C.D. Stiles. 1979. Purification of the human platelet-derived growth factor. *Proc. Natl. Acad. Sci.* **76:** 1809.

Bishop, J.M. 1983. Cancer genes come of age. *Cell* **32:** 1018.

Bowen-Pope, D.F. and R. Ross. 1982. Platelet-derived growth factor. II. Specific binding to culture cells. *J. Biol. Chem.* **257:** 5161.

Brooks, R.F. 1977. Continuous protein synthesis is required to maintain the probability of entry into S phase. *Cell* **12:** 311.

Cochran, B.H., A.C. Reffel, and C.D. Stiles. 1983. Molecular cloning of gene sequences regulated by platelet-derived growth factor. *Cell* **33:** 939.

Cooper, J.A., D.F. Bowen-Pope, E. Raines, R. Ross, and T. Hunter. 1982. Similar effects of platelet-derived growth factor and epidermal growth factor on the phosphorylation of tyrosine in cellular proteins. *Cell* **31:** 263.

Deuel, T., J.S. Huang, R.T. Proffitt, J.U. Baenziger, D. Chang, and B.B. Kennedy. 1981. Human platelet-derived growth factor. Purification and resolution into two active protein fractions. *J. Biol. Chem.* **256:** 8896.

Devare, S.G., E.P. Reddy, J.D. Law, K.C. Robbins, and S.A. Aaronson. 1983. Nucleotide sequence of the simian sarcoma virus genome: Demonstration that its acquired cellular sequences encode the transforming gene product p28^{sis}. *Proc. Natl. Acad. Sci.* **80**: 731.

Doolittle, R.F., M.W. Hunkapiller, L.E. Hood, S.G. Devare, K.C. Robbins, S.A. Aaronson, and H.N. Antoniades. 1983. Simian sarcoma virus *onc* gene, v-*sis*, is derived from the gene (or genes) encoding a platelet-derived growth factor. *Science* **221**: 275.

Ek, B. and C.H. Heldin. 1982. Characterization of a tyrosine-specific kinase activity in human fibroblast membranes stimulated by platelet-derived growth factor. *J. Biol. Chem.* **257**: 10486.

Eva, A., K.C. Robbins, P.R. Andersen, A. Srinivasan, S.R. Tronick, E.P. Reddu, N.W. Ellmore, A.T. Galen, J.A. Lautenberger, T.S. Papas, E.H. Westin, F. Wong-Staal, R.C. Gallo, and S.A. Aaronson. 1982. Cellular genes analogous to retroviral *onc* genes are transcribed in human tumour cells. *Nature* **295**: 116.

Foster, D.N., L.J. Schmidt, C.P. Hodgson, H.L. Moses, and M.J. Getz. 1982. Polyadenylated RNA complementary to a mouse retrovirus-like multigene family is rapidly and specifically induced by epidermal growth factor stimulation of quiescent cells. *Proc. Natl. Acad. Sci.* **79**: 7317.

Gerrard, J.M., D.R. Phillips, G.H.R. Rao, E.G. Plow, D.A. Waltz, R. Ross, L.A. Harker, and J.G. White. 1980. Biochemical studies of two patients with the Gray Platelet Syndrome. Selective deficiency of platelet alpha-granules. *J. Clin. Invest.* **66**: 102.

Gonzalez, F.J. and C.B. Kasper. 1982. Cloning of cDNA complementary to rat liver NADPH-cytochrome c (P-450) oxidoreductase and cytochrome P-4506 mRNAs. *J. Biol. Chem.* **257**: 5962.

Grotendorst, G.R., H.E.J. Seppa, H.K. Kleinman, and G.R. Martin. 1981. Attachment of smooth muscle cells to collagen and their migration toward platelet-derived growth factor. *Proc. Natl. Acad. Sci.* **78**: 3669.

Habenicht, A.J.R., J.A. Glomset, W.C. King, C. Nist, C.D. Mitchell, and R. Ross. 1981. Early changes in phosphatidylinositol and arachidonic acid metabolism in quiescent Swiss 3T3 cells stimulated to divide by platelet-derived growth factor. *J. Biol. Chem.* **256**: 12329.

Heldin, C.H., B. Ek, and L. Ronnstrand. 1983. Characterization of the receptor for platelet-derived growth factor on human fibroblasts. Demonstration of an intimate relationship with a 185,000 dalton substrate for the PDGF-stimulated kinase. *J. Biol. Chem.* **258**: 10054.

Heldin, C.H., B. Westermark, and A. Wasteson. 1979. Platelet-derived growth factor: Purification and partial characterization. *Proc. Natl. Acad. Sci.* **76**: 3722.

———. 1980. Chemical and biological properties of a growth factor from human cultured osteosarcoma cells: Resemblance with PDGF. *J. Cell. Physiol.* **105**: 235.

———. 1981a. Specific receptors for platelet-derived growth factor on cells derived from connective tissue and glia. *Proc. Natl. Acad. Sci.* **78**: 3664.

———. 1981b. Demonstration of an antibody against platelet-derived growth factor. *Exp. Cell Res.* **136**: 255.

Huang, J.S., S.S. Huang, B. Kennedy, and T.F. Deuel. 1982. Platelet-derived growth factor. Specific binding to target cells. *J. Biol. Chem.* **257**: 8130.

Kaplan, D.R., F.C. Chao, C.D. Stiles, H.N. Antoniades, and C.D. Scher. 1979. Platelet alpha-granules contain a growth factor for fibroblasts. *Blood* **53**: 1043.

Kaplan, K.L., J.M. Broekman, A. Chernoff, G.R. Lesznick, and M. Drillings. 1979. Platelet alpha-granule proteins: Studies on release and subcellular localization. *Blood* **53**: 604.

Lewin, B. 1980. *Gene expression 2*. Wiley, New York.

Linzer, D.I.H. and D. Nathans. 1983. Growth-related changes in specific mRNAs of cultured mouse cells. *Proc. Natl. Acad. Sci.* **80**: 4271.

Nishimura, J., J.S. Huang, and T.F. Deuel. 1982. Platelet-derived growth factor stimulates tyrosine-specific protein kinase activity in Swiss mouse 3T3 cell membranes. *Proc. Natl. Acad. Sci.* **79**: 4303.

Pledger, W.J., C.A. Hart, K.L. Locatell, and C.D. Scher. 1981. Platelet-derived growth factor-modulated proteins: Constitutive synthesis by a transformed cell line. *Proc. Natl. Acad. Sci.* **78**: 4358.

Pledger, W.J., C.D. Stiles, H.N. Antoniades, and C.D. Scher. 1977. Induction of DNA synthesis in BALB/c-3T3 cells by serum components: Reevaluation of the commitment process. *Proc. Natl. Acad. Sci.* **74**: 4481.

Raines, E.W. and R. Ross. 1982. Platelet-derived growth factor I high yield purification and evidence for multiple forms. *J. Biol. Chem.* **257**: 5154.

Ross, R., J. Glomset, B. Kariya, and L. Harker. 1974. A platelet-dependent serum factor that stimulates the proliferation of arterial smooth muscle cells in vitro. *Proc. Natl. Acad. Sci.* **71**: 1207.

Rossini, M. and R. Baserga. 1978. RNA synthesis in a cell cycle–specific temperature sensitive mutant from a hamster cell line. *Biochemistry* **17**: 858.

Rossow, P.W., V.G.H. Riddle, and A.B. Pardee. 1979. Synthesis of labile, serum-dependent protein in early G/1 controls animal cell growth. *Proc. Natl. Acad. Sci.* **76**: 4446.

Scher, C.D., W.J. Pledger, P. Martin, H.N. Antoniades, and C.D. Stiles. 1978. Transforming viruses directly reduce the cellular growth requirement for a platelet-derived growth factor. *J. Cell. Physiol.* **97**: 371.

Schutzbank, T., R. Robinson, M. Oren, and A.J. Levine. 1982. SV40 large tumor antigen can regulate some cellular transcripts in a positive fashion. *Cell* **30**: 481.

Shier, W.T. 1980. Serum stimulation of phospholipase A2 and prostaglandin release in 3T3 cells is associated with platelet-derived growth-promoting activity. *Proc. Natl. Acad. Sci.* **77**: 137.

Singh, J.P., M.A. Chaikin, and C.D. Stiles. 1982. Phylogenetic analysis of platelet-derived growth factor by radioreceptor assay. *J. Cell Biol.* **95**: 667.

Smith, J.C. and C.D. Stiles. 1981. Cytoplasmic transfer of the mitogenic response to platelet-derived growth factor. *Proc. Natl. Acad. Sci.* **78**: 4363.

Stiles, C.D. 1983. The molecular biology of platelet-derived growth factor. *Cell* **33**: 653.

Stiles, C.D., W.J. Pledger, J.J. Van Wyk, H.N. Antoniades, and C.D. Scher. 1979. Hormonal control of early events in the BALB/c-3T3 cell cycle: Commitment to DNA synthesis. In *Hormones and cell culture* (ed. G. Sato), p. 425. Cold Spring Harbor Laboratory, Cold Spring Harbor, New York.

Vogel, A., E. Raines, B. Kariya, M.-J. Rivest, and R. Ross. 1978. Coordinate control of 3T3 cell proliferation by platelet-derived growth factor and plasma components. *Proc. Natl. Acad. Sci.* **75**: 2810.

Waterfield, M.D., G.T. Scrace, N. Whittle, P. Stroobant, A. Johnsson, A. Wasteson, B. Westermark, C.H. Heldin, J.S. Huang, and T.F. Deuel. 1983. Platelet-derived growth factor is structurally related to the putative transforming protein p28^{sis} of simian sarcoma virus. *Nature* **304**: 35.

Westermark, B., C.H. Heldin, B. Ek, A. Johnsson, K. Mellstrom, M. Nister, and A. Wasteson. 1983. Biochemistry and biology of platelet-derived growth factor. In *Growth and maturation factors* (ed. G. Guroff), p. 75. Wiley, New York.

Williams, J.G. and S. Penman. 1975. The messenger RNA sequences in growing and resting fibroblasts. *Cell* **6**: 197.

Microfilaments in Normal and Transformed Cells: Changes in the Multiple Forms of Tropomyosin

J.J.-C. Lin, S. Yamashiro-Matsumura, and F. Matsumura
Cold Spring Harbor Laboratory, Cold Spring Harbor, New York 11724

Actin-containing microfilaments in cultured cells are found to show at least two distinctive supramolecular forms: microfilament meshworks (or microfilament gels) and microfilament bundles (or stress fibers or actin cables). Microfilament meshworks are located just under the plasma membrane in the cortical cytoplasm (Wolosewick and Porter 1979; Heuser and Kirschner 1980). By immunofluorescence microscopy, microfilament bundles have been shown to contain actin, myosin, filamin, α-actinin, tropomyosin, and vinculin. In normal cells, microfilament bundles are located primarily at the adherent side of the cells just under the surface membrane (Buckley and Porter 1967; Goldman 1975; Goldman et al. 1976). On the other hand, in various types of transformed cells, microfilament bundles become reduced in their size and number (McNutt et al. 1973; Pollack and Rifkin 1975; Goldman et al. 1976), and most microfilaments are found as a meshwork structure. Thus, after transformation, cells appear to rearrange microfilaments from a bundle state into a randomly interwoven meshwork. The mechanism of this rearrangement is not clearly understood; however, a direct comparison between protein compositions of microfilaments isolated from normal and transformed cells may generate some information about this mechanism.

To examine the protein composition of microfilaments, we have recently developed a rapid method to isolate tropomyosin-containing microfilaments from various types of cultured cells by using antitropomyosin monoclonal antibodies (Matsumura et al. 1983a). Using this method and two-dimensional gel electrophoresis, we have found changes in tropomyosin patterns of microfilaments between "normal" cells and cultured rat cells transformed by DNA or RNA viruses (Matsumura et al. 1983b). In this paper, we have extended this similar type of study to human and mouse cell lines with different types of transformation, such as transfection with tumor DNA, chemical carcinogen transformation, UV-irradiated transformation, and naturally occurring tumor cell lines. We have found that, in general, at least one of the major tropomyosin forms with a higher molecular weight appears to be decreased or missing, and the level of one of the tropomyosin forms with a lower molecular weight tends to be increased in all types of transformed cells. These changes in the pattern of tropomyosin expression may be responsible, in part, for the rearrangement of microfilamental bundles (or actin cables) into meshworks in the transformed cell.

Experimental Procedures

Cell culture and radioactive labeling

An established rat embryo cell line (REF52), a simian virus 40–transformed REF52 (REF4A), and a human type-5-adenovirus–transformed REF52 (Ad5D.1A) were kindly provided by W.C. Topp (Cold Spring Harbor Laboratory, New York). Normal rat kidney cells (NRK, CRL-1570) and Kirsten virus–transformed NRK cells (CRL-1569) were obtained from American Type Culture Collection. Mouse NIH/3T3 cells and NIH/3T3 secondary transformants derived from human bladder carcinoma (T24), neuroblastoma (SK-N-SH), and lung carcinoma (SK-LU-1) cells were kindly provided by M. Wigler (Cold Spring Harbor Laboratory, New York). A chemically transformed mouse epithelial cell line (MB49), a human bladder carcinoma cell line (EJ), a human breast adenocarcinoma cell line (MCF-7), and a human pancreatic carcinoma cell line (CRL-1420) were generous gifts from L.B. Chen (Sidney Farber Cancer Institute, Harvard Medical School). A mouse 3T3 cell (1-1), a chemically transformed 3T3 cell (BP-1), a UV-transformed 3T3 cell (UV-2), a human fibroblast (KD), and a chemically transformed KD cell (Hut-11) were generous gifts from T. Kakunaga (National Cancer Institute). All cell lines were maintained in Dulbecco's modified Eagle's medium (DMEM) containing 10% fetal calf serum (FCS) in an atmosphere of 5% CO_2 and 95% air at 37°C.

For [^{35}S] labeling of proteins, cells grown on 100-mm dishes at 90% confluence were incubated for 16 hours with 250 μCi of [^{35}S]methionine (1110 Ci/mmole) in 3 ml of methionine-free DMEM supplemented with 2.5% FCS. After labeling, cells were washed with phosphate-buffered saline (PBS: 137 mM NaCl, 2.7 mM KCl, 1.5 mM KH_2PO_4, 8 mM Na_2HPO_4, pH 7.3) and total cell lysates or microfilament isolates were prepared. Morphologies of cells were observed and photographed with a Leitz diavert microscope with phase-contrast illumination on Technical pan film.

Antibodies and immunoprecipitation

Monoclonal antibodies against tropomyosin (LCK16, JLH2, or JLF15) were prepared and characterized as described previously (Lin 1982). As reported previously (Matsumura and Lin 1982; Matsumura et al. 1983a,b; Lin et al. 1984), ascites fluids of these antibodies are equally effective in the isolation of microfilaments. Polyclonal rabbit antiserum against chicken gizzard tropomyosin was prepared and characterized as described previously (Matsumura et al. 1983a). Immunoprecipitation with rabbit antiserum against tropomyosin was carried out as described (Matsumura et al. 1983a) in the buffer containing 50 mM Tris, pH 8.0, 0.05% SDS, 0.5% Triton X-100, 1 mM EDTA, 1 mM phenylmethylsulfonylchloride (PMSF), 100 mM NaCl.

Isolation of microfilaments

Isolation of microfilaments from monolayers of cultured cells was accomplished using monoclonal antibodies against tropomyosin as described previously (Matsumura et al. 1983a). For the preparation of heat-stable proteins from microfilament fractions, the isolated microfilament was suspended in a small volume (30–50 μl) of PBS containing 5 mM $MgCl_2$ and 0.2 mM EGTA and placed in a boiling water bath for 10 minutes. After cooling on ice and centrifugation at 12,000g for 15 minutes, the supernatant containing heat-stable proteins was further analyzed on two-dimensional gels.

One- and two-dimensional gel electrophoresis

One-dimensional SDS-polyacrylamide gel electrophoresis was carried out according to Laemmli (1970) with a low bisacrylamide concentration (12.5% acrylamide and 0.104% bisacrylamide) (Blattler et al. 1972). Two-dimensional gel electrophoresis was performed by a modified procedure of O'Farrell (1975). The first-dimension tube gels contained 4% polyacrylamide and pH 5–7 ampholytes. The second-dimension gels were 12.5% SDS-polyacrylamide slab gels with a low bisacrylamide concentration. After electrophoresis, gels were stained with 0.15% Coomassie blue in 50% methanol/10% acetic acid and destained with 7.5% methanol/7.5% acetic acid. For fluorography, gels were processed according to the method of Bonner and Laskey (1974).

For measurement of ratios of tropomyosin/actin in isolated microfilaments, cells were first labeled with [^{35}S]methionine and microfilaments were isolated and analyzed on two-dimensional gels as described above. Gels were stained, destained, and dried on filter paper. The spots corresponding to actin and tropomyosin were cut out. Radioactive proteins were eluted by incubation of each gel slice for 24 hours in 0.2 ml of 2% SDS solution, and radioactivities were counted in 3 ml of Aquasol (New England Nuclear) with a liquid scintillation counter.

Results

Tropomyosin changes in "normal" and transformed rat cell lines

As described previously (Matsumura et al. 1983a,b), "normal" rat cell lines (REF52; fibroblastic NRK, CRL-1570; epithelial NRK, CRL-1571; and L6 myoblasts) contained five forms of tropomyosin: TM-1, 40K; TM-2, 36.5K; TM-3, 35K; TM-4, 32.4K; TM-5, 32K. Of these, three (TM-1, TM-2, and TM-4) were major tropomyosins and two (TM-3 and TM-5) were minor tropomyosins in "normal" rat cells. As can be seen in Figure 1, these proteins (indicated by numbers 1–5) were immunologi-

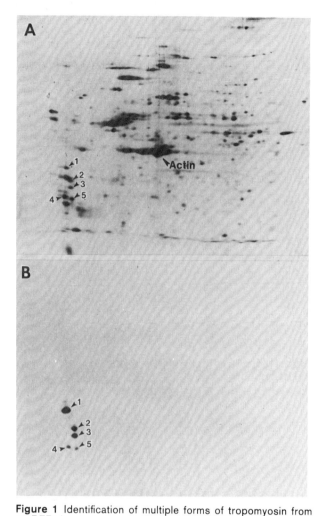

Figure 1 Identification of multiple forms of tropomyosin from rat REF4A cells by immunoprecipitation. Rat REF4A cells (one 100-mm dish at 90% confluence) were labeled in vivo with 250 μCi of [^{35}S]methionine (1110 Ci/mmole) for 16 hr in methionine-free DMEM containing 2.5% FCS. After washing three times with PBS, cells were scraped off and lysed in 0.1 ml of SDS gel sample buffer (2% SDS, 15% glycerol, 100 mM DTT, 80 mM Tris, 0.001% bromophenol blue, pH 6.8). The cell lysate was boiled at 100°C for 3 min and diluted (1:30) with buffer containing 50 mM Tris, pH 8.0, 0.5% Triton X-100, 1 mM EDTA, 1 mM phenylmethylsulfonyl fluoride, 100 mM NaCl. The lysate was used for immunoprecipitation by rabbit antiserum against gizzard tropomyosin as described in Experimental Procedures. The total cell lysate and the immunoprecipitates were analyzed by two-dimensional gel electrophoresis. (*A*) Total cell lysate of REF4A cells; (*B*) immunoprecipitate of cell lysate with rabbit antiserum against chicken gizzard tropomyosin. Multiple forms of rat tropomyosin (TM-1 to TM-5) are indicated by numbers 1–5. Note that TM-4 and TM-5 are not completely recovered in the immunoprecipitate with rabbit antiserum against chicken gizzard tropomyosin; this may be due to the lower cross-reactivity of the antisera to both proteins.

cally cross-reacted with rabbit antiserum directed against chicken smooth-muscle tropomyosin. In addition, they had the similar isoelectric point (pI) of 4.5 in two-dimensional gel analysis, which is the known pI value for muscle tropomyosin. These proteins were heat-resistant or heat-stable and have the ability to bind to F-actin filaments (Matsumura et al. 1983a). The identical five tropomyosins could be found in the microfilaments isolated from either cultured rat cell lines or primary cultures of rat embryo cells. The relative amounts of these five forms of tropomyosin found in microfilaments appeared to be unchanged throughout successive passages (at least 10) of rat embryo cells (data not shown). Therefore, it is likely that tropomyosin patterns in rats show no significant difference between primary and established cell lines.

Having the development of the microfilament isolation method and the identification of multiple forms of tropomyosin, we are able to examine the question as to whether tropomyosin patterns will change in accompaniment with cell transformation. Any change detected in the tropomyosin patterns may be, in part, responsible for the disruption of microfilament bundles (or stress fibers) found in transformed cells, because tropomyosin has been suggested to play a stabilizing role for microfilaments. As reported previously, we have found an apparent relation between changes in tropomyosin patterns and cell morphologies of "normal" and virally transformed rat cells (Matsumura et al. 1983b).

REF4A cells (SV40-transformed REF52 cells) showed a less spread morphology (Fig. 2D) when compared with a morphology of "normal" REF52 cells (Fig. 2B). In the microfilaments isolated from REF4A cells (Fig. 2C), the levels of both minor tropomyosins (TM-3 and TM-5) were found to be increased and the levels of two major tropomyosins (TM-1 and TM-2) were decreased. Ad5D.1A cells (human Ad5-transformed REF52 cells) showed a much more "rounded" morphology (Fig. 2F) than did REF4A cells (Fig. 2D). The changes in the tropomyosin patterns in Ad5D.1A cell microfilaments were much more drastic than those found in REF4A cells. The major tropomyosin TM-1 was entirely missing, and the minor tropomyosin TM-3 was greatly increased in amount.

In contrast with a well-spread morphology of "normal" rat kidney cells (CRL-1570, Fig. 2H), Kirsten RNA virus–transformed NRK cells (CRL-1569) showed a completely "rounded" morphology (Fig. 2J). In the microfilament isolated from CRL-1569, the level of major tropomyosin TM-1 was greatly decreased and another

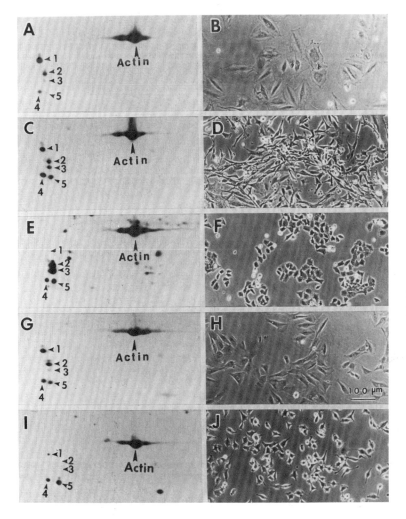

Figure 2 The relationship between tropomyosin patterns in microfilaments and cell morphologies. Microfilaments were isolated from [^{35}S]methionine-labeled cells by the method described previously (Matsumura et al. 1983a) and analyzed on two-dimensional gels. Fluorographs of a portion of the gels are shown on the left column with the acidic ends at left. The right column shows phase-contrast micrographs of cells. Multiple forms of rat tropomyosin (TM-1 to TM-5) are indicated by numbers 1–5. (A,B) "Normal" REF52 cells; (C,D) SV40-transformed REF52 cells (REF4A); (E,F) human Ad5-transformed REF52 cells (Ad5D.1A); (G,H) "normal" rat kidney cells (CRL-1570); (I,J) Kirsten virus–transformed rat kidney cells (CRL-1569). Note that the changes in tropomyosin patterns correlate well with the morphological changes.

major tropomyosin TM-2 was entirely missing. Although the level of minor tropomyosin TM-3 was not observed to be increased, unlike with DNA virus transformation, the level of the other minor tropomyosin TM-5 was greatly increased (Fig. 2I). Thus, the extent of changes in the tropomyosin patterns appears to be correlated well with the degree of morphological alteration upon transformation. This change in tropomyosin pattern is transformation-specific, because Rat-1 cells transformed with a temperature-sensitive mutant of Rous sarcoma virus show similar changes in tropomyosin patterns in conjunction with morphological alterations upon shifting down to the permissive temperature for transformation (Matsumura et al. 1983b).

Tropomyosin changes in "normal" and transformed mouse cell lines

"Normal" mouse cell lines also contained multiple forms of tropomyosin. Figure 3 shows the identification of mouse tropomyosins by two-dimensional gel analyses of the immunoprecipitate and the heat-stable proteins of the isolated microfilaments from NIH/3T3 cells. Rabbit antiserum against tropomyosin was able to immunoprecipitate at least five proteins: TM-1, TM-2, TM-3, TM-4, and TM-5 (Fig. 3B, indicated by numbers 1–5) from total cell lysates (Fig. 3A). The mouse TM-1 may be composed of two polypeptides. Mouse TM-1, TM-2, and TM-5 were considered as major tropomyosins, whereas TM-3 and TM-4 were minor forms. As can be seen in Figure 3C, these five forms of tropomyosin are identical with those in the heat-stable fraction of NIH/3T3 microfilament. It should be noted that although two protein spots, a and b in Figure 3C, possessed heat-resistant natures, they were not immunologically cross-reacted with rabbit antiserum against chicken gizzard tropomyosin.

We have further examined tropomyosin pattern changes in transformed mouse cells. One set of transformed cell lines was derived from transfection of NIH/3T3 cells with DNA prepared from human tumor cell lines (Perucho et al. 1981). Microfilaments were isolated from these cell lines and analyzed on two-dimensional gels (Fig. 4). When compared with the protein composition of NIH/3T3 microfilaments (Fig. 4A), tropomyosin patterns in the microfilaments from transformants derived from DNA transfection changed substantially (Fig. 4, B–D). One of the major tropomyosins, TM-2, with relatively higher molecular weight was missing, whereas the lower-molecular-weight TM-5 appeared to be slightly increased. In addition, TM-1 was found to be decreased, and TM-3 was also missing in the microfilaments from these transformed cells. It should be noted that the changes in tropomyosin patterns in these transformed cells were similar, regardless of the source (human bladder carcinoma, human neuroblastoma, and human lung carcinoma) of the DNA preparation used in DNA transfection.

Figure 5 shows the relationship between changes in tropomyosin patterns and morphologies of mouse cell lines. The microfilaments isolated from a well-spread

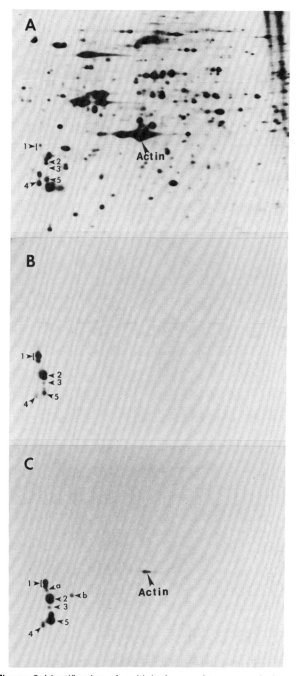

Figure 3 Identification of multiple forms of tropomyosin from mouse NIH/3T3 cells by immunoprecipitation and their heat-stable nature. Radioactive labeling of cells, preparation of total cell lysates, and immunoprecipitation were performed as described in the legend of Fig. 1. A heat-stable fraction of the isolated microfilaments from NIH/3T3 cells was prepared as described in Experimental Procedures. The total cell lysate (A), the immunoprecipitate (B), and the heat-stable fraction of the microfilaments (C) were analyzed by two-dimensional gel electrophoresis. Fluorographs of the gels are shown here with the acidic ends at left. Multiple forms of mouse tropomyosin (TM-1 to TM-5) are indicated by numbers 1–5. Heat-stable protein spots a and b did not cross-react with rabbit antiserum against chicken gizzard tropomyosin. TM-4 in total cell lysates was often contaminated by other nontropomyosin proteins in the two-dimensional gel.

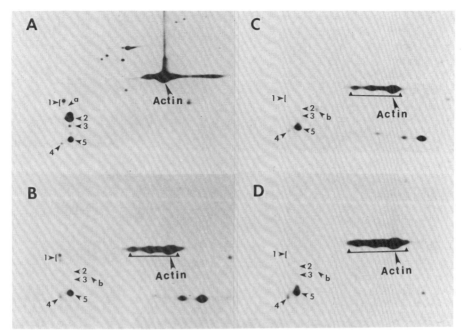

Figure 4 Two-dimensional gel analysis of the isolated microfilaments from "normal" mouse NIH/3T3 cells and the transformed cells derived from transfection of NIH/3T3 cells with DNA from various human tumors. Cells were labeled in vivo with [^{35}S]methionine and subjected to the microfilament isolation as described previously (Matsumura et al. 1983a). Isolated microfilaments were analyzed by two-dimensional gel electrophoresis. Fluorographs of a portion of gels are shown here with the acidic ends at left. Multiple forms of mouse tropomyosin (TM-1 to TM-5) are indicated by numbers 1–5. Protein spots a and b are heat-stable proteins but do not cross-react with rabbit antiserum against chicken gizzard tropomyosin. (A) Microfilaments isolated from NIH/3T3 cells; (B) microfilaments isolated from transformed cells derived from transfection of NIH/3T3 cells with human bladder carcinoma DNA; (C) microfilaments isolated from transformed cells derived from transfection of NIH/3T3 cells with human neuroblastoma DNA; (D) microfilaments isolated from transformed cells derived from transfection of NIH/3T3 cells with human lung carcinoma DNA. Note that the decreased amounts of TM-1, the entire absence of TM-2 and TM-3, and the increased amounts of TM-5 are found to be associated with all transformed cell microfilaments. The differences in actin patterns observed in this figure are not reproducible.

morphology of NIH/3T3 cells contained all five forms of tropomyosin (Fig. 5, A and B). In the microfilaments isolated from a transformed cell line derived from transfection of NIH/3T3 cells with human bladder cancer DNA, both TM-2 and TM-3 were missing and the level of TM-1 was reduced (Fig. 5C). The morphology of this transformed cell is spindle shaped and less spread than that of normal NIH/3T3 cells (Fig. 5D). In a chemically transformed mouse cell line (MB49), more-rounded morphology was observed (Fig. 5F). Again, TM-1 was entirely missing, and TM-5 appeared to be increased in the MB49 cell microfilament (Fig. 5E). Therefore, in transformed mouse cells, one of the higher-molecular-weight tropomyosins (either TM-1 or TM-2) was missing, and the tropomyosin with the lower molecular weight, TM-5, tended to be increased in amount.

We have also extended this type of study to another set of transformed cells. Normal mouse 3T3 cells (1-1) were transformed either by chemical carcinogen (derived BP-1 cells) or by UV irradiation (derived UV-2 cells). The morphologies of these transformed cells are less spread, but the degree of cell shape changes is less prominent than that observed in NIH/3T3 cells transformed by DNA transfection or MB49 cells. Microfilaments were isolated from these cell lines labeled with [^{35}S]methionine and analyzed on two-dimensional gels. Protein spots corresponding to actin and multiple forms of tropomyosin were cut and quantified as described in Experimental Procedures. Changes in tropomyosin patterns are also observed in these transformed cells, although the tropomyosin patterns change to a lesser extent. As shown in Table 1, the general tendency of tropomyosin changes is that a major tropomyosin, TM-2, with a higher molecular weight appeared to be decreased, and the other major tropomyosin, TM-5, with lower molecular weight tended to be increased in transformed cells.

Tropomyosin pattern changes in "normal" and transformed human cell lines

Like rat and mouse cell lines, human fibroblast (Hut-11) cells also contained multiple forms of tropomyosin. At least six polypeptides (TM-1 to TM-5, indicated by numbers 1–5 and spot a in Figs. 6 and 7) can be identified as tropomyosin by the following criteria: (1) They were immunologically cross-reacted with rabbit antiserum directed against chicken gizzard tropomyosin (Fig. 6B); (2) they possessed a heat-stable nature (Fig. 6C); (3) they had a pI value of 4.5; and (4) they were microfilament-associated proteins because they could be found in the isolated microfilament fraction from human KD fibroblasts (Fig. 7A). The human TM-1, like mouse TM-

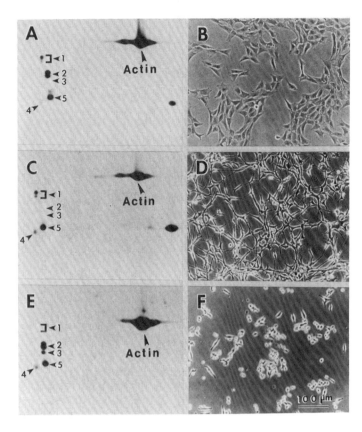

Figure 5 Changes in tropomyosin patterns of microfilaments and morphologies of "normal" NIH/3T3 and transformed mouse cells. Microfilaments were isolated from [^{35}S]methionine-labeled cells by the method described previously (Matsumura et al. 1983a). Isolated microfilaments were analyzed by two-dimensional gel electrophoresis. Fluorographs of a portion of the gels are shown in the left column with the acidic ends at left. The right column shows phase-contrast micrographs of cells. (A,B) "Normal" NIH/3T3 cells; (C,D) transformed cells derived from transfection of NIH/3T3 cells with DNA of human bladder carcinoma; (E,F) chemically transformed cells (MB49). Multiple forms of mouse tropomyosin (TM-1 to TM-5) are indicated by numbers 1–5. Tropomyosins TM-1 and TM-2, with higher molecular weights, were greatly decreased in both transformed cells, which showed altered morphologies (i.e., they were spindle shaped or rounded up).

1, consistently showed two spots on our two-dimensional gels that varied in molecular weight. Multiple forms of tropomyosins have also been reported in human fibroblasts by others (Giometti and Anderson 1981; Talbet and MacLeod 1983). At least four polypeptides having apparent molecular weights of 30K–36K were reported as tropomyosin in human cells. In our gel system, apparent molecular weights for TM-1 to TM-5 were 40K, 36.5K, 36K, 32.4K, and 32K, respectively. Of these, TM-1, TM-2, TM-3, and spot a were major tropomyosins, and TM-4 and TM-5 were minor tropomyosins in "normal" human KD cells.

Changes in tropomyosin patterns were examined in the isolated microfilaments of "normal" KD and chemically transformed Hut-11 cells and several cell lines derived from tumor tissues, such as human bladder carcinoma (EJ) cells, human pancreatic carcinoma (CRL-1420) cells, and human breast adenocarcinoma (MCF-7) cells. In Hut-11 and EJ cell microfilaments (Fig. 7, B and D, respectively), TM-5 appeared to be increased and TM-1 and spot a were decreased in amounts when compared with that in the KD cell microfilament (Fig. 7A). A greater change in tropomyosin pattern was found in the microfilaments isolated from CRL-1420 and MCF-7 cells (Fig. 7, C and E). In both transformed cell microfilaments, TM-1, TM-2, TM-3, and spot a diminished greatly and the level of TM-5 appeared to be increased. Although there were no normal counterparts for cell lines derived from tumor tissues to compare, it would be reasonable to use the KD cell tropomyosin pattern for the comparison because of the finding that normal rat cell lines, regardless of their source (myoblast, fibroblast, or epithelial), expressed a similar set of tropomyosins.

Discussion

In this report, we have shown that normal rat, mouse, and human cultured cells have multiple forms of tropomyosin and that multiple forms of tropomyosin are expressed differently in the microfilaments isolated from "normal" and transformed cell lines. A common phenomenon about the tropomyosin pattern change in transformed cells is that at least one major tropomyosin, which has a higher molecular weight, appears to be decreased or entirely missing and that the level of one tropomyosin with a lower molecular weight is increased. These changes in tropomyosin pattern are observed in all types of transformed cells tested so far, including DNA or RNA virus–transformed cells, chemical carcinogen–transformed cells, cells transformed by transfection with DNA from tumor cell lines, UV-irradiated transformed cells, and naturally occurring tumor cell lines. Moreover, the changes in tropomyosin patterns in transformed cells can be detected not only in the isolated microfilaments and total cell lysates, but also in the in vitro translation products of total mRNA (Matsumura et al. 1983b). Therefore, it is likely that the changes in tropomyosin patterns are regulated at the level of RNA rather than by

Table 1 Ratio of Radioactivities of [35S]Methionine Incorporated into Tropomyosins and Actin of Microfilaments Isolated from "Normal" Mouse 3T3 Cells (1-1), Chemically Transformed 3T3 Cells (BP-1), and UV-Transformed 3T3 Cells (UV-2)

	1-1	BP-1	UV-2
TM-1	2.5	2.1	1.6
TM-2	13.1	8.7	5.5
TM-3	1.0	1.0	0.7
TM-4	0.9	0.7	0.9
TM-5	4.9	6.0	8.4
Actin	100	100	100

Multiple forms of tropomyosin in the isolated [35S]methionine-labeled microfilaments were separated by two-dimensional gel electrophoresis. After electrophoresis, gels were stained, destained, and dried on filter papers. Protein spots corresponded to tropomyosin and actin (TM-1 to TM-5 were indicated by the numbers 1–5 in Figs. 3, 4, 5) were punched off and soaked in 0.2 ml of 2% SDS for 24 hr to elute the protein out of gels. The radioactivities were measured by a liquid scintillation counter.

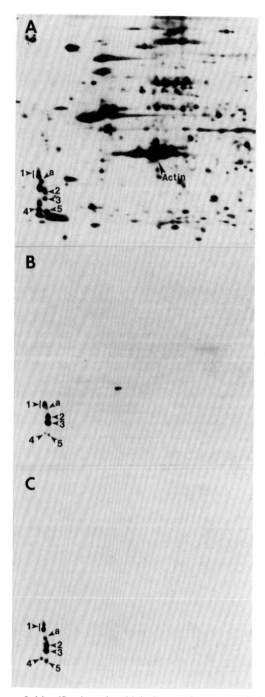

Figure 6 Identification of multiple forms of tropomyosin from human Hut-11 cells by immunoprecipitation and their heat-stable nature. Radioactive labeling of cells, preparation of total cell lysates, and immunoprecipitation were performed as described in the legend of Fig. 1. A heat-stable fraction of the isolated microfilaments from Hut-11 cells was prepared as described in Experimental Procedures. Total cell lysates (A), the immunoprecipitate (B), and the heat-stable fraction of the isolated microfilaments (C) were analyzed on two-dimensional gels. Fluorographs of the gels are shown here with the acidic ends at left. Multiple forms of human tropomyosin (TM-1 to TM-5) are indicated by numbers 1–5. In addition, protein spot a is also identified as a tropomyosin because of its immunological cross-reactivity and heat-stable nature.

posttranslational modification. The degree of tropomyosin changes, i.e., the increase of one lower-molecular-weight form of tropomyosin and the decrease of one major higher-molecular-weight form of tropomyosin, appears to correlate well with the extent of morphological alteration in transformed cells. This correlation is particularly obvious in rat and mouse cell lines (Figs. 2 and 5).

Changes in the synthesis and the total content of major tropomyosins in transformed cells have been reported by others (Paulin et al. 1979; Hendricks and Weintraub 1981; Leonardi et al. 1982). Mouse embryonal carcinoma (PCC3/A/1) cells induced to differentiate by hexamethylenebisacetamide lose their transformed morphology, develop actin microfilament bundles, and become well-spread, large cells. Protein analysis by two-dimensional gel electrophoresis showed the drastic increase in the synthesis of major tropomyosin during this induced differentiation (Paulin et al. 1979). In both chicken embryo fibroblasts transformed by Rous sarcoma virus and rat kidney cells transformed by Kirsten virus, the rate of synthesis and the total content of two major forms of tropomyosins have been shown to decrease or be entirely missing (Hendricks and Weintraub 1981; Leonardi et al. 1982). Therefore, it appears that a decrease in major tropomyosins accompanies cell transformation. However, in these reports only two protein spots on two-dimensional gels were identified as tropomyosins and examined with respect to changes in their amounts upon cell transformation. In this study, we have used our rapid microfilament isolation method together with two-dimensional gel electrophoresis and immunological methods to detect multiple forms (more than two) of tropomyosin in rat, mouse, and human cells. We have found that the levels of one tropomyosin with a lower molecular weight are consistently increased in transformed cells, in ad-

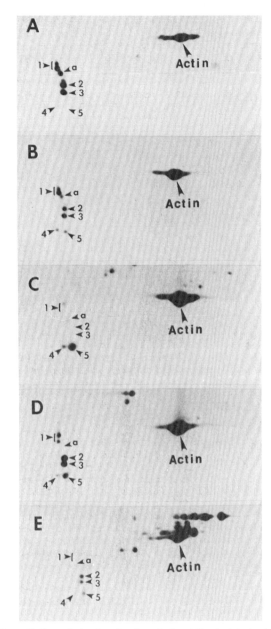

Figure 7 Two-dimensional gel analysis of the isolated microfilaments from "normal" human KD fibroblasts, their chemically transformed cells (Hut-11), and cell lines derived from various human tumors. Cells were labeled in vivo with [^{35}S]methionine and subjected to microfilament isolation as described previously (Matsumura et al. 1983a). The isolated microfilaments were analyzed by two-dimensional gel electrophoresis. Fluorographs of a portion of the gels, which contain tropomyosin and actin, are shown here with the acidic ends at left. (A) Microfilaments isolated from "normal" KD cells; (B) microfilaments isolated from chemically transformed Hut-11 cells; (C) microfilaments isolated from CRL-1420 human pancreatic carcinoma cells; (D) microfilaments isolated from EJ human bladder carcinoma cells; (E) microfilaments isolated from MCF-7 human breast adenocarcinoma cells. Multiple forms of human tropomyosin (TM-1 to TM-5) are indicated by numbers 1–5. Protein spot a is identified as a tropomyosinlike protein because of its immunological cross-reactivity and its heat-stable nature. Note that expression of major tropomyosins TM-1, TM-2, TM-3, and spot a of normal KD cells decreased and expression of minor tropomyosin TM-5 was increased in all transformed cells.

dition to the reduction of at least one major tropomyosin with a higher molecular weight.

Tropomyosin is a ubiquitous protein associated with the thin filaments of muscle cells and with the actin-containing microfilaments of nonmuscle cells. In the striated muscle, this protein together with the troponin complex confer the Ca^{++} regulation of actin-myosin interaction. However, as no troponinlike protein has been detected so far in smooth muscle or nonmuscle cells, tropomyosin may not function in this way.

One possible function of tropomyosin in nonmuscle cells is that tropomyosin may play a role in stabilizing the structure of F-actin filaments or microfilaments in the cell. This idea has been suggested from the observations that skeletal muscle tropomyosin apparently increases the rigidity of the F-actin filaments (Fujime and Ishiwata 1971), that by electron microscopy the tropomyosin-containing F-actin appears straighter (Kawamura and Maruyama 1970; Takebayashi et al. 1977), and that tropomyosin protects the F-actin from damage during fixation with OsO_4 or glutaraldehyde (Maupin-Szamier and Pollard 1978; Lehrer 1981). Furthermore, this idea has been strengthened by the two recent reports that tropomyosin inhibits the functions of gelsolin or actin-depolymerizing factor (ADF). Both muscle and nonmuscle tropomyosins appear to decrease the binding of gelsolin to actin filaments and to diminish the ability of gelsolin-Ca^{++} to fragment these filaments (Fattoum et al. 1983), and tropomyosin binding to F-actin protects the F-actin from disassembly by brain actin-depolymerizing factor (Bernstein and Bamburg 1982). It has been reported that muscle and nonmuscle tropomyosins are different in their properties. Nonmuscle tropomyosin purified from brain or platelet cells has a lower apparent molecular weight and shows the lower binding affinity to actin filaments relative to skeletal muscle tropomyosin (Fine et al. 1973; Cote and Smillie 1981). Tropomyosin purified from macrophages is less effective in protecting F-actin from the fragmentation by gelsolin-Ca^{++} than skeletal muscle tropomyosin (Fattoum et al. 1983). We have shown here that in transformed cells, one tropomyosin with a lower apparent molecular weight (called the transformed-cell tropomyosin) appears to be increased and the level of at least one of the major tropomyosins with a higher apparent molecular weight (called the normal-cell tropomyosin) is diminished. It is possible that the normal-cell tropomyosin and the transformed-cell tropomyosin may indeed be different with regard to stabilization of microfilaments in nonmuscle cells. The transformed-cell tropomyosin may show less affinity to actin or may be defective to regulate the functions of the actin-binding proteins such as gelsolin or ADF. These altered properties of tropomyosin may cause the transformed cell to rearrange the organization of microfilaments, such as reducing the number and size of microfilament bundles, and to show morphological alterations. However, the biochemical and functional comparison between the individual forms of tropomyosin is necessary to understand the meaning these changes in tropomyosin patterns have for cell transformation.

Acknowledgments

We would like to thank Dr. James D. Watson for his continuous and enthusiastic support of this work. We are grateful to Drs. S. Blose and J. Feramisco for their advice and criticism. We also thank Ms. Nancy Haffner for her competent technical assistance, Ms. Madeline Szadkowski for typing the manuscript, and Mr. Phil Renna for his photographic skills. This work was supported in part by a National Institutes of Health grant GM-31048, by National Cancer Institute grants CA-13106 and CA-35738, and by grants from the Muscular Dystrophy Association.

References

Bernstein, B.W. and J.R. Bamburg. 1982. Tropomyosin binding to F-actin protects the F-actin from disassembly by brain actin-depolymerizing factor (ADF). *Cell Motility* **2:** 1.

Blattler, D.P., F. Garner, K. Van Slyke, and A. Bradley. 1972. Quantitative electrophoresis in polyacrylamide gels of 2–40%. *J. Chromatogr.* **64:** 147.

Bonner, W.M. and R.A. Laskey. 1974. A film detection method for tritium-labelled proteins and nucleic acids in polyacrylamide gels. *Eur. J. Biochem.* **46:** 83.

Buckley, I.K. and K.R. Porter. 1967. Cytoplasmic fibrils in living cultured cells. A light and electron microscope study. *Protoplasma* **64:** 349.

Cote, G.P. and L.B. Smillie. 1981. The interaction of equine platelet tropomyosin with skeletal muscle actin. *J. Biol. Chem.* **256:** 7257.

Fattoum, A., J.H. Hartwig, and T.P. Stossel. 1983. Isolation and some structural and functional properties of macrophage tropomyosin. *Biochemistry* **22:** 1187.

Fine, R.E., A.L. Blitz, S.E. Hitchcock, and B. Kaminer. 1973. Tropomyosin in brain and growing neurones. *Nature* **245:** 182.

Fujime, S. and S. Ishiwata. 1971. Dynamic study of F-actin by quasielastic scattering of laser light. *J. Mol. Biol.* **62:** 251.

Garrels, J.I. 1979. Changes in protein synthesis during myogenesis in a clonal cell line. *Dev. Biol.* **73:** 134.

Giometti, C.S. and N.L. Anderson. 1981. A variant of human nonmuscle tropomyosin found in fibroblasts by using two-dimensional electrophoresis. *J. Biol. Chem.* **256:** 11840.

Goldman, R.D. 1975. The use of heavy meromyosin binding as an ultrastructural cytochemical method for localizing and determining the possible functions of actin-like mircofilaments in nonmuscle cells. *J. Histochem. Cytochem.* **23:** 529.

Goldman, R.D., M.-J. Yerna, and J.A. Schloss. 1976. Localization and organization of microfilaments and related proteins in normal and virus-transformed cells. *J. Supramol. Struct.* **5:** 155.

Hendricks, M. and H. Weintraub. 1981. Tropomyosin is decreased in transformed cells. *Proc. Natl. Acad. Sci.* **78:** 5633.

Heuser, J.E. and M.W. Kirschner. 1980. Filament organization revealed in platinum replicas of freeze-dried cytoskeletons. *J. Cell Biol.* **86:** 212.

Kawamura, M. and K. Maruyama. 1970. Electron microscopic particle lengths of F-actin polymerized in vitro. *J. Biochem.* **67:** 437.

Laemmli, U.K. 1970. Cleavage of structural proteins during the assembly of the head of bacteriophage T4. *Nature* **227:** 680.

Lehrer, S.S. 1981. Damage to actin filaments by glutaraldehyde: Protection by tropomyosin. *J. Cell Biol.* **90:** 459.

Leonardi, C.L., R.H. Warren, and R.W. Rubin. 1982. Lack of tropomyosin correlates with the absence of stress fibers in transformed rat kidney cells. *Biochim. Biophys. Acta* **720:** 154.

Lin, J.J.C. 1982. Mapping structural proteins of cultured cells by monoclonal antibodies. *Cold Spring Harbor Symp. Quant. Biol.* **46:** 769.

Lin, J.J.C., F. Matsumura, and S. Yamashiro-Matsumura. 1984. Tropomyosin-enriched and α-actinin-enriched microfilaments isolated from chicken embryo fibroblasts by monoclonal antibodies. *J. Cell Biol.* **98:** 116.

Matsumura, F. and J.J.C. Lin. 1982. Visualization of monoclonal antibody binding to tropomyosin on native smooth muscle thin filaments by electron microscopy. *J. Mol. Biol.* **157:** 163.

Matsumura, F., S. Yamashiro-Matsumura, and J.J.C. Lin. 1983a. Isolation and characterization of tropomyosin-containing microfilaments from cultured cells. *J. Biol. Chem.* **258:** 6636.

Matsumura, F., J.J.C. Lin, S. Yamashiro-Matsumura, G.P. Thomas, and W.C. Topp. 1983b. Differential expression of tropomyosin forms in the microfilaments isolated from normal and transformed rat cultured cells. *J. Biol. Chem.* **258:** 13954.

Maupin-Szamier, P. and T.D. Pollard. 1978. Actin filament destruction by osmium tetroxide. *J. Cell Biol.* **77:** 837.

McNutt, N.S., L.A. Culp, and P.H. Black. 1973. Contact-inhibited revertant cell lines isolated from SV40-transformed cells. *J. Cell Biol.* **56:** 412.

O'Farrell, P.H. 1975. High resolution two-dimensional electrophoresis of proteins. *J. Biol. Chem.* **250:** 4007.

Paulin, D., J. Perreau, H. Jakob, F. Jacob, and M. Yaniv. 1979. Tropomyosin synthesis accompanies formation of actin filaments in embryonal carcinoma cells induced to differentiate by hexamethylene bisacetamide. *Proc. Natl. Acad. Sci.* **76:** 1891.

Perucho, M., M. Goldfarb, K. Shimizu, C. Lama, J. Fogh, and M. Wigler. 1981. Human-tumor-derived cell lines contain common and different transforming genes. *Cell* **27:** 467.

Pollack, R. and D. Rifkin. 1975. Actin-containing cables within anchorage-dependent rat embryo cells are dissociated by plasmin and trypsin. *Cell* **6:** 495.

Takebayashi, T., Y. Morita, and F. Oosawa. 1977. Electron microscopic investigation of the flexibility of F-actin. *Biochim. Biophys. Acta* **492:** 357.

Talbet, K. and A.R. MacLeod. 1983. Novel form of non-muscle tropomyosin in human fibroblasts. *J. Mol. Biol.* **164:** 159.

Wolosewick, J.J. and K.R. Porter. 1979. Microtrabecular lattice of the cytoplasmic ground substance: Artifact or reality? *J. Cell Biol.* **82:** 114.

A Point Mutation and Other Changes in Cytoplasmic Actins Associated with the Expression of Transformed Phenotypes

T. Kakunaga,* J. Leavitt,†‡ T. Hirakawa,* and S. Taniguchi*
*Laboratory of Molecular Carcinogenesis, National Cancer Institute, National Institutes of Health, and †Food and Drug Administration, Bethesda, Maryland 20205

The most common biological characteristics of neoplastic cells are the alterations in cell morphology and motility. Since the cytoskeleton plays major roles in maintaining and controlling cell morphology, motility, and membrane fluidity, it is conceivable that the structure and function of the cytoskeleton are altered in transformed cells. In 1975, Pollack et al. reported that actin cable (or so-called stress fiber) was disorganized in transformed cells when the cells were stained with the fluorescently labeled antibody against actin. Since then, many observations have been reported that the diffusion of actin-cable structure generally correlates with the expression of transformed phenotypes such as the loss of contact inhibition of cell growth and motility, anchorage-independent cell growth, and tumorigenicity (Edelman and Yahara 1976; David-Pfuety and Singer 1980; Pollack 1980).

Although some cultured neoplastic cells were found to retain actin-cable network structure, they also showed normal morphology and cell-to-cell arrangement in culture, indicating again the close association of cytoskeletal structure and cell morphology and motility (Raz and Geiger 1982). On the other hand, transformed phenotypes consist of multiple and separable phenotypes. Alteration of the actin-cable structure may be related to only one or two phenotypic changes. The mechanisms by which the structure and function of the cytoskeleton are disrupted in transformed cells seem to vary depending on the developmental stages of the cells and the oncogenes expressed. Nonetheless they are hardly understood.

We have previously found a mutated β-actin in a chemically transformed human cell line (Leavitt and Kakunaga 1980). Although the mutated β-actin has been rarely found in other transformed cells, it has provided a clue in the investigation of the mechanism of disruption of actin cable and its relation to the expression of transformed phenotypes.

Methods and Materials

Cell cultures
HuT-11 and HuT-14 cells are in vitro–transformed human fibroblasts obtained by a single treatment of human diploid fibroblasts (KD) with 4-nitroquinoline-1-oxide (Kakunaga 1978). All the cells were cultured in Eagle's minimum essential medium (MEM) supplemented with 10% fetal calf serum (FCS).

Sublines defective in hypoxanthine-guanine phosphoribosyl transferase activity (HGPRT⁻) or resistant to ouabain (Ouaʳ) were obtained by isolating the cells from colonies grown in the presence of 6-thioguanine or ouabain, after exposing the cells to ultraviolet light (75 erg/mm^2). Both sublines were retested and confirmed for their resistance to 6-thioguanine or ouabain. Cell hybrids between HuT-14 and KD cells were isolated by fusing the HuT-14 cells (HGPRT⁻, Ouaʳ) with KD ((HGPRT⁺, Ouaˢ), using polyethyleneglycol, and by selecting the growing hybrid cells (HGPRT⁺, Ouaʳ) with HAT and ouabain in the culture medium.

Subconfluent and growing cultures were incubated in medium containing one tenth of the normal concentration of cold methionine and 0.15–0.5 mCi of [^{35}S]methionine for 4 hours, unless otherwise specified.

Preparation of Triton-soluble and- insoluble proteins and two-dimensional polyacrylamide gel electrophoresis
For the preparation of Triton-soluble proteins, the cell cultures were rinsed with cold phosphate-buffered saline (PBS) and then received 1 ml of cold Triton solution, consisting of 1.0% Triton X-100 (v/v), 10 mM Tris·HCl, pH 7.8, 10 mM MgCl$_2$, 30 mM KCl, 0.01 M 2-mercaptoethanol. The Triton solution was rinsed over the cell-layers (kept on ice) for 5 minutes, then thoroughly drained to the side of the dish, collected, and centrifuged. Triton-insoluble proteins were obtained from the residues attached to the dishes, after collection of Triton-soluble proteins, by scraping with a rubber policeman and suspending in Triton solution as described above, except that the concentration of Triton was reduced to 0.1%. The insoluble fraction was then pelleted by low-speed centrifugation and dissolved into lysis buffer A of O'Farrell (1975). The procedure for the two-dimensional gel electrophoresis was based on O'Farrell's methods (O'Farrell 1975) with a minor modification (Leavitt and Kakunaga 1980).

Isolation of mRNA and in vitro translation
The cells were lysed in hypotonic buffer containing 0.3% Nonidet-P40, and RNA was prepared from the cytoplasmic fraction by phenol/SDS extraction (Hamada et al. 1981).

‡Present address: Linus Pauling Institute, Palo Alto, California 94306.

RNA was translated into protein in the presence of [^{35}S]methionine by using a rabbit reticulocyte lysate translation kit purchased from New England Nuclear.

Isolation and purification of cytoplasmic actins
Actin was extracted and purified from 10^9–10^{10} HuT-14 or HuT-11 cells according to the method of Gordon et al. (1976) or Uyemura et al. (1978) with slight modifications. In some experiments, actin was extracted from 0.5–1.0 g of acetone powder with 50 times the volume of 2 mM imidazole, pH 8.0, 0.5 mM ATP, 0.2 mM DTT, 50 μM MgCl$_2$, 0.005% NaN$_3$, and 0.5 mM PMSF. The extract was concentrated to a few milliliters by ultrafiltration or salting out with ammonium sulfate and then loaded onto Sephacryl S-200 equilibrated with the extraction buffer. The fraction containing actin was further purified by DEAE-52 cellulose chromatography and ammonium sulfate fractionation in the same way as done by Uyemura et al. (1978).

In vitro polymerization of actins
Actins purified as above were dissolved in G buffer containing 5 mM Tris·HCl, pH 8.0, 0.1 mM CaCl$_2$, 0.2 mM ATP, 0.2 mM DTT, 0.01% NaN$_3$, added by appropriate concentration of KCl and MgCl$_2$, and then incubated for 3 hours at 30°C. The reaction mixture was then centrifuged at 27 psi in a Beckman Airfuge for 30 minutes. Polymerized actins were recovered as the pellet, suspended in lysis buffer, and subjected to isoelectric focusing. Gels were stained with Coomassie brilliant blue and scanned with a Gilford spectrophotometer. The relative ratio of the amounts of β-, γ-, and β^x-actins in the polymerized fraction was calculated from the microdensitometry of the corresponding stained bands by standardization to the amount of γ-actin present.

Results and Discussion

A mutated β-actin coexpressed with normal β- and γ-actins in a chemically transformed cell line

When the [^{35}S]methionine-labeled polypeptides were examined by the autoradiography of two-dimensional gels of normal human diploid fibroblasts and chemically transformed counterpart fibroblasts, the polypeptide species that disappeared or newly appeared in the transformed cells amounted in total to approximately 2% of approximately 1000 electrophoretically distinguishable polypeptide species on each autoradiograph (Fig. 1). Quantitative measurement of the polypeptide density on the autoradiograph, using a microdensitometer connected with computers, indicated that nearly 20% of the polypeptide species were modulated during the neoplastic transformation (Leavitt et al. 1982). Alterations of some polypeptides are common to many of the transformed and tumor cells tested. For example, as shown in Figure 1, a2 and a3 were detected in other chemically and virally transformed cells and the cells derived from human tumor tissues but not in normal cells. On the other hand, approximately half of the alteration of polypeptide species was unique to each transformed or tumor cell line. One of them, a large and dense spot, designated A^x in Figure 1B, was detected only in one of the cell lines (HuT-14) that were transformed by treatment with 4-nitroquinoline-1-oxide. This polypeptide (pI 5.2; m.w. 42,000) migrates very close to β- and γ-actin (pI 5.3; m.w. 43,000) on the two-dimensional gel and was found in both the Triton-soluble and -insoluble fractions of HuT-14 cells. The A^x polypeptide isolated from gels was immunoprecipitated with anti-actin antibody. The tryptic peptide pattern of the A^x polypeptide was very similar to those of β- and γ-actin polypeptides, indicating that the A^x polypeptide is a variant of cytoplasmic actin.

Two-dimensional gel electrophoresis of the translation products, obtained by in vitro translation in the presence of poly(A)$^+$ RNA extracted from HuT-14 cells, showed that A^x-actin (previously designated A' polypeptide) as well as β- and γ-actin were major translation products with either rabbit reticulocyte lysate or wheat germ extract (Hamada et al. 1981). On the other hand, poly(A)$^+$ RNA extracted from other transformed or normal human fibroblasts directed the synthesis of only β- and γ-actin, indicating the unique presence of mRNA encoding A^x-actin in the HuT-14 cells. The size of mRNA coding for A^x- and β-actin did not differ when it was determined either by fractionation of HuT-14 poly(A)$^+$ RNA on a sucrose gradient followed by a translation of each poly(A)$^+$ RNA fraction or by fractionation on an agarose gel followed by hybridization with nick-translated pcDd insert [^{32}P]DNA. pcDd insert DNA (i.e., actin cDNA complementary to actin mRNA of *Dictyostelium*) was found to hybridize specifically with A^x- and β-actin mRNA but not with γ-actin mRNA. These results suggested that A^x-actin is a variant of β-actin. The entire amino acid sequences of A^x- and β-actin isolated from HuT-14 cells, using DNase I affinity column and preparatory gel electrophoresis, were determined in collaboration with Vandekerckhove and Weber (Fig. 2). A^x-actin had an identical amino acid sequence to β-actin, except that at position 244, one amino acid, the glycine in normal β-actin, was substituted with aspartic acid in A^x-actin. This substitution corresponds to a GC→AT transition point mutation in the coding sequence. It is noteworthy that such a single amino acid substitution contributes to a significantly lower electrophoretic mobility in the SDS gel. The number of functional β-actin genes seems to be 1 copy per haploid human genome (T. Kakunaga et al., unpubl.). Since HuT-14 cells express both A^x and β-actin, there may be only one A^x-actin and one β-actin gene in the HuT-14 cells and a diploid β-actin gene set in other human cells. Thus, A^x-actin is considered to be a mutated β-actin and hereinafter designated β^x-actin.

Correlation between the expression of transformed phenotypes and actin mutation

A mutation of β-actin may be responsible for at least a part of the expression of transformed phenotypes of HuT-14 cells. Or, alternatively, the mutation may be a coincidence and have nothing to do with neoplastic transformation. To examine the role of a mutation in β-actin in the expression of the transformed phenotypes in HuT-14 cells, four different approaches have been taken. First, cell variants with the different transformed phenotypes were isolated from HuT-14 cells, and the alteration of

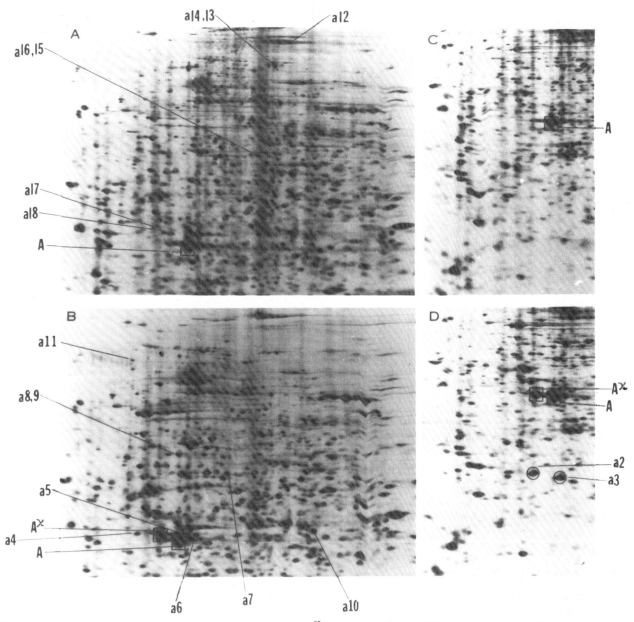

Figure 1 Autoradiographs of two-dimensional gels containing [^{35}S]methionine-labeled Triton-soluble polypeptides. Polypeptides were separated in the first dimension by isoelectric focusing between pH 4–7 range (right to left). (A,C) Polypeptides of KD cells; (B,D) polypeptides of HuT-14 cells. (A,B) 9% acrylamide gels; (C,D) 12.5% acrylamide gels. Polypeptide A is β- and γ-actin. Polypeptides A^x and a2 through a11 are polypeptide species present in only HuT-14 cells. Polypeptides a12 through a18 are those present in only KD cells. (Reprinted, with permission, from Leavitt and Kakunaga 1980.)

expression of β^x-actin was examined. Attempts to isolate the highly malignant variant cells by selecting them through serial passages of the cells in nude mice failed. The cells obtained from the tumors that were seriously transplanted several times into nude mice showed no increase in their abilities to grow in soft agar or to form tumors in nude mice, when compared with HuT-14 cells that were not passaged through animals. The cells also maintained similar amounts of expression of β^x-actin through many generations. These results indicate the considerable stability of the expression of both the transformed phenotypes and β^x-actin through generations of HuT-14 cells. In another attempt, HuT-14 cells were exposed to ultraviolet light and plated sparsely in culture. Subclones were isolated from the representative colonies showing a faster growth and a higher cell density. Although most of the subclones showed similar potentials to form colonies in soft agar and to produce tumors in nude mice, compared with HuT-14 cells, one subclone, HuT-14-3-1, showed significantly increased anchorage-independent cell growth and tumorigenicity in nude mice. Two-dimensional gel analysis of the polypeptides synthesized in the HuT-14-3-1 cells revealed a loss of β^x-actin and a gain of new polypeptide species. No other marked changes in the polypeptides synthesized, including normal β- and α-actin, were observed. One new polypeptide migrated the same distance as the β^x-actin in the SDS dimension but was more negatively

Figure 2 Part of amino acid sequence of human cytoplasmic β- and A^x-actin isolated from HuT-14 cells determined by Vandekerckhove et al. (1980). The only amino acid difference between β-actin and A^x-actin in position 244 is enclosed in the box. All other amino acid sequences (373 sequences) are identical between the actins.

charged than β^x-actin. The tryptic peptide pattern of this new polypeptide was very similar to β^x-actin and normal β-actin, indicating that the new polypeptide is a variant of actin and probably a variant of β^x-actin. Two-dimensional gel analysis of in vitro–translation products, synthesized in the presence of poly(A)$^+$ RNA extracted from HuT-14-3-1, suggests that the synthesis of the new actin variant is encoded within the messenger RNA. The new actin variant in HuT-14-3-1 was designated β^{xx}-actin.

Table 1 summarizes the relative ratios of normal β- and γ-actins and mutant β-actin in the soluble and insoluble cellular fractions of normal fibroblasts, HuT-14, HuT-14-3-1, and HuT-14-3-1T cells, and in the in vitro–translation products of poly(A)$^+$ RNA extracted from each cell strain or line. The insoluble fraction, under the conditions used, consists of cytoskeleton and nuclear matrix. The incremental increase in anchorage-independent cell growth and tumorigenicity of the HuT-14 sublines is associated with both an increase in the ratio of mutated β-actin (β^x- and β^{xx}-actin) to normal β-actin in the soluble fraction and a decrease in the ratio in the insoluble fraction. The decrease in the ratio of mutant β-actin in the insoluble fraction was much more marked in the sublines with higher malignancy. When we consider the high stability of the expression of actin in mammalian cells and of the expression of transformed phenotypes in HuT-14 progenies, it is likely that the alteration of β-actin is related to the expression of transformed phenotypes.

As the second approach to examining the role of the mutant β-actin in the expression of transformed phenotypes, we have isolated somatic cell hybrids between the transformed lines with mutant β-actin and the parental normal human fibroblasts and examined the expression of transformed phenotypes and actin in the hybrid cells. HuT-14 cells, in which the HGPRT$^-$ and Oua^r markers had been previously inserted, were fused with normal fibroblasts (HGPRT$^+$, Oua^s). The resulting cell hybrids (HGPRT$^+$, Oua^r) were selectively grown in the presence of hypoxanthine, aminoputerine, thymidine, and ouabain. Only 1:1 (transformed cell:normal cell) hybrids were selected, by examining their chromosome banding patterns, and examined for their morphology in culture, anchorage-independent growth, tumorigenicity in nude mice, and two-dimensional gel pattern of actin proteins. All of the hybrids showed transformed-like morphology, significant anchorage-independent growth, and the synthesis of mutant β-actin, except one hybrid that showed normallike morphology, low anchorage-independent growth, and no synthesis of mutant β-actin. All the hybrids, without exception, did not produce tumors in nude mice at a cell dose of 2×10^6 cells per animal, at which cell dose the parental transformed line produced tumors at 100% frequency. Suppression of tu-

Table 1 Stepwise Changes in the Cytoplasmic Actins and Transformed Phenotypes in the Variants of HuT-14 Cells

	K_D	HuT-14	HuT-14-3-1	HuT-14-3-1T
Plating efficiency in soft agar (%)	<0.0001	1.1	3.8	8.5
Tumorigenicity (TD_{50} in nude mice)	>10^7	3×10^5	3×10^4	<10^4
Relative ratio of actin[a]:				
in cellular soluble fraction				
β	1.0	1.0	NT^b	1.0
γ	1.2	1.2	NT	1.2
$β^x$ or $β^{xx}$	0	1.2	NT	1.7
in insoluble fraction (cytoskeleton)				
β	1.0	1.0	NT	1.0
γ	1.2	1.2	NT	1.2
$β^x$ and $β^{xx}$	0	0.6	NT	0.04
in in vitro translation				
β	1.0	1.0	1.0	1.0
γ	1.0	2.0	1.1	2.0
$β^x$ and $β^{xx}$	0	0.8	0.7	1.7

[a]Relative density ratio of the actin species to the β-actin (1.0) on the autoradiograph of two-dimensional gels containing [^{35}S]methionine-labeled proteins. ($β^x$) Mutant β-actin; ($β^{xx}$) new mutant β-actin.
[b]Not tested.

morigenicity in the hybrids is consistent with recent reports by others indicating an involvement of another unknown step in the acquisition of tumorigenicity in nude mice by human cells. The one hybrid line that showed normallike phenotype may be a segregant from a 1:1 cell hybrid, although no evidence to support this has been available. These results indicate that the synthesis of mutant β-actin and the expression of the in vitro–transformed phenotypes are dominantly expressed in the 1:1 cell hybrids and that the expression of mutant β-actin correlates with the expression of in vitro–transformed phenotypes.

Characterization of mutated β-actins

Abnormality observed in cells

As previously described, the ratio of newly synthesized mutant β-actins in the insoluble fraction to those in the soluble fraction was much less than that of newly synthesized, normal β-actins. This finding suggests that mutant β-actins are defective in incorporation into the cytoskeleton. On the other hand, the densitometry of Coomassie blue–stained two-dimensional gels indicated that the quantity of mutant β-actin relative to normal β- and γ-actins was reduced by approximately 95% in the highly malignant subline, HuT-14-3-1T cells, as compared with that in the mildly malignant HuT-14 cells. Since the [^{35}S]methionine incorporation into mutant $β^{xx}$-actin was markedly increased in HuT-14-3-1T cells (Leavitt et al. 1982), it is suggested that $β^{xx}$-actin is increased in both its synthesis and degradation. Pulse-labeling (8 or 15 min) with [^{35}S]methionine and chasing experiments showed that $β^{xx}$-actin in HuT-14-3-1T cells accumulated at a 50% faster rate than $β^x$-actin in HuT-14 cells and degraded more rapidly, with a half-life of approximately 1.5 hours, than $β^x$-actin or β-actin, whose half-lives are longer than 5 hours. Other polypeptides in HuT-14-3-1T cells showed half-lives similar to those in HuT-14 cells, indicating that the short half-life is not common to other polypeptides in HuT-14-3-1T cells.

In vitro polymerization

Cytoplasmic actins exist in nonmuscle cells in various forms: monomer, oligomer or polymer, associated or unassociated with other proteins, and bundled or unbundled. The dynamic shift between the various forms of actins is considered to play an important role in controlling cell shape, motility, and other cellular functions. One of the most crucial steps in the dynamic conversion of actin complexes is polymerization and depolymerization. The polymerization of actin monomers to form actin filaments is a highly cooperative condensation-polymerization reaction typical of the self-assembly processes of helical macromolecules (Korn 1982). To examine the polymerization kinetics of actins, it is essential to isolate and purify actins in a native form. In contrast to muscle actins, cytoplasmic actins are difficult to isolate and purify in a native form, especially from cultured mammalian cells, because of their instability, low yield, and limited cell sources. We have tried various methods to purify β- and γ-actins and found that none of the published procedures are satisfactory. We found, however, that regardless of the procedures used for isolation and purification, or the yields of actins, β^x-actin isolated from HuT-14 cells was always inefficient in polymerization (Table 2). A crude extract from HuT-14 cells contained β-, γ-, and β^x-actins at an approximately equal ratio. After purification through DEAE-cellulose chromatography, Sephadex G-150 chromatography, and/or ammonium sulfate precipitation, the relative ratios of β^x-actin and β-actin to γ-actin were reduced by 15–40% and 5–15%, respectively. Under the optimum conditions (i.e., in the presence of 100 mM KCl, 2 mM $MgCl_2$, 0.1 mM $CaCl_2$, and 0.2 mM ATP), β^x-actin showed a slightly reduced polymerization compared with β- and γ-actin. At lower KCl concentration, the decrease in polymerization of β^x-actin was marked. Similar results were obtained even by using an actin preparation that was previously purified by polymerization under optimum polymerization conditions (Exp. 2 in Table 2). When KCl and $MgCl_2$ were added to a supernatant that contained unpolymerized actins under low ionic concentration so that the ionic concentration of the supernatant was elevated to the optimum level, the β^x-actin polymerized as efficiently as the β^x-actin in the original actin preparation. Therefore the reduction of polymerization of β^x-actin by lowering ionic concentration is not ascribed to the denaturation of β^x-actin. These results indicate that the one amino acid substitution at position 245, glycine replaced by aspartic acid, results in the reduction of polymerization, particularly at a lower ionic concentration.

To examine in detail the ionic dependency of β-, γ-, and β^x-actins on polymerization, actins were isolated from HuT-14 cells, which synthesize β-, γ-, and β^x-actins, and from HuT-11 cells, which synthesize β- and γ-actin but not β^x-actin. Analysis of the actins polymerized at various KCl concentrations revealed that the polymerization of β-actin, as well as that of β^x-actin, was markedly reduced at low KCl concentration compared with that of γ-actin when β- and β^x-actins were present together in the reaction mixture (Fig. 3). On the other hand, β-actin polymerized as well as γ-actin in the absence of β^x-actin. These results indicate that mutant β^x-actin has a much greater dependency on KCl concentration for its polymerization than normal β- and γ-actins and that β^x-actin inhibits the polymerization of β-actin, particularly at low KCl concentration. The physiological and functional differences between β- and γ-actins are not known at all. The results described here indicate a very interesting possibility that β-actin and γ-actin polymerize separately and independently, and they provide the first evidence suggesting different functions of β- and γ-actins.

Role of mutation in actin in the neoplastic transformation

Our results described here suggest that a point mutation in β-actin is involved in the expression of transformed phenoyptes in one of the chemically transformed human fibroblast lines. The mutated β-actin is defective in polymerization from G-form to F-form and also partially inhibits polymerization of normal β-actin (Fig. 4). The defectiveness of mutant β-actin in polymerization observed in vitro may account for the low incorporation of mutant β-actin into cytoskeleton in vivo. Such a decrease in polymerization is likely to cause a disturbance in the dynamic shift of the actin state, i.e., conversion between monomer and polymer, which takes place in response to environmental and internal signals. Since the polymerization process of cytoplasmic actins seems to be involved in regulating cell shape, motility, membrane fluidity, and distribution and transportation of cellular components, alterations of which are an integral part of transformed phenotypes, the mutated β-actin may play a role in the expression of transformed phenotypes. It is also possible that the microfilaments, actin bundles, and cellular matrix that incorporate mutant β-actin are abnormal in their interaction with many proteins, including actin-associated proteins, in localization, in how they are modified, and in the mechanical functions that may result in the expression of some transformed phenotypes (Fig. 3). The further-altered mutant β-actin (β^{xx}-actin), which is present in the highly malignant subclone, is aggravated, at least in its defective incorporation into the cytoskeleton. The fact that β^{xx}-actin is extremely unstable and rapidly degraded in vitro makes it impossible to

Table 2 Relative Ratios of Actins Polymerized In Vitro

Polymerization	Actins polymerized				
	γ	:	β	:	β^x
Sample I	1.0	:	0.84	:	0.66
Exp. 1					
polymerized in 100 mM K^+ + 2 mM Mg^{++}	1.0	:	0.76	:	0.47
polymerized in 50 mM K^+	1.0	:	0.68	:	0.10
Exp. 2					
polymerized in 100 mM K^+ + 2 mM Mg^{++}	1.0	:	0.87	:	0.57
polymerized in 45 mM K^+	1.0	:	0.75	:	0.25
unpolymerized in 45 mM K^+	1.0	:	0.97	:	0.66
Sample II	1.0	:	0.79	:	0.45
polymerized in 100 mM K^+ + 2 mM Mg^{++}	1.0	:	0.70	:	0.32
polymerized in 50 mM K^+	1.0	:	0.77	:	0.18
(unpolymerized in 50 mM K^+) + (100 mM K^+ + 2 mM Mg^{++})	1.0	:	0.77	:	0.77

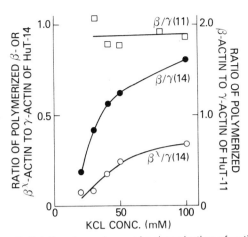

Figure 3 Relative dependency of polymerization of actins on KCl concentration. (□) Ratio of polymerized β-actin to γ-actin in the absence of $β^x$-actin; (●) ratio of polymerized β-actin to γ-actin in the presence of $β^x$-actin; (○) ratio of polymerized $β^x$-actin to γ-actin.

measure its biological function in vitro and even to determine its amino acid sequence. It is not difficult to imagine that $β^{xx}$-actin causes greater disturbance in the dynamic shift of the state of cytoplasmic actins and more profound abnormalities in the functions of actin complexes that integrate $β^{xx}$-actin than does $β^x$-actin. The aggravated malfunction of the cytoskeleton induced by $β^{xx}$-actin may be related to the increased malignancy of the subline that synthesizes $β^{xx}$-actin.

Although the above speculation is based on observations with only one transformed line and its subclones and hybrids, it may be generalized to many cases of transformation. Two-dimensional gel analysis will detect only some of the mutations occurring in actin molecules; therefore, many mutations will be left undetected. Furthermore, a cytoskeletal dysfunction similar to that caused by a mutation in actin molecules will also be induced by mutations in actin-associated proteins and other proteins regulating cytoskeletal functions. The probability of a mutation occurring in any one of these proteins is much higher than that in only two actins, β- and γ-actins. Further examination of these proteins may reveal many mutations.

On the other hand, the mechanisms of modification of the structural and/or regulatory proteins involved in the function of actin complexes may be altered in some cells, which would result in the expression of transformed phenoytpes. It has been reported that the tyrosine phosphorylation of one of the actin-associated proteins, vinculin, was increased in cells transformed by Rous sarcoma virus (Sefton et al. 1981). Although the role of actin mutation in the expression of the transformed state is still speculative, further somatic genetics studies, transfection studies of the relationship between the expression of mutant actin and transformed phenotypes, and studies on the basic function of cytoplasmic actins using mutant actins will shed more light on the mechanisms of expression of transformed phenotypes.

Acknowledgments

We thank Drs. S. Brenner and E.D. Korn for their valuable technical advice, Ms. J.D. Crow, C. Augl, M. Petrino, and G. Bushar for their expert technical assistance, and Ms. G.L. Wilson and L. Nischan for assistance in preparing the manuscript.

References

David-Pfuety, T. and S.J. Singer. 1980. Altered distributions of the cytoskeletal proteins vinculin and β-actin in cultured fibroblasts transformed by Rous sarcoma virus. *Proc. Natl. Acad. Sci.* **77**: 6687.

Edelman, G.M. and I. Yahara. 1976. Temperature-sensitive changes in surface modulating assemblies of fibroblasts transformed by mutants of Rous sarcoma virus. *Proc. Natl. Acad. Sci.* **73**: 2047.

Gordon, D.J., E. Eisenberg, and E.D. Korn. 1976. Characterization of cytoplasmic actin isolated from *Acanthamoeba* by a new method. *J. Biol. Chem.* **251**: 4778.

Hamada, H., J. Leavitt, and T. Kakunaga. 1981. Mutated β-

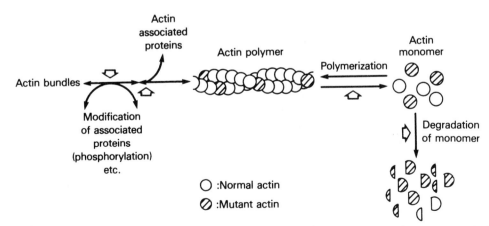

Figure 4 Scheme for the disruption of cytoskeletal functions induced by a point mutation in β-actin in a chemically transformed human fibroblast line, HuT-14. Mutant β-actins are defective in polymerization, stability, and interaction with actin-associated proteins.

actin gene: Coexpression with an unmutated allele in a chemically transformed human fibroblast cell line. *Proc. Natl. Acad. Sci.* **78:** 3634.

Kakunaga, T. 1978. Neoplastic transformation of human diploid fibroblast cells by chemical carcinogenesis. *Proc. Natl. Acad. Sci.* **75:** 1334.

Korn, E.D. 1982. Actin polymerization and its regulation by proteins from nonmuscle cells. *Physiol. Rev.* **62:** 672.

Leavitt, J. and T. Kakunaga. 1980. Expression of a variant form of actin and additional polypeptide changes following chemical-induced *in vitro* neoplastic transformation of human fibroblasts. *J. Biol. Chem.* **255:** 1650.

Leavitt, J., D. Goldman, C. Merril, and T. Kakunaga. 1982. Changes in gene expression accompanying chemically induced malignant transformation of human fibroblasts. *Carcinogenesis* **3:** 61.

O'Farrell, P.H. 1975. High resolution two-dimensional electrophoresis of proteins. *J. Biol. Chem.* **250:** 4007.

Pollack, R. 1980. Hormones, anchorage and oncogenic cell growth. In *Cancer achievements, challenges and prospects for the 1980's* (ed. I.J. Burchenal and H.F. Oettgen), p. 501. Grune and Stratton, New York.

Pollack, R., M. Osborn, and K. Weber. 1975. Patterns of organization of actin and myosin in normal and transformed nonmuscle cells. *Proc. Natl. Acad. Sci.* **72:** 994.

Raz, A. and B. Geiger. 1982. Altered organization of cell-substrate contacts and membrane-associated cytoskeleton in tumor cell variants exhibiting different metastatic capabilities. *Cancer Res.* **42:** 5183.

Sefton, B.M., T. Hunter, E.H. Ball, and S.J. Singer. 1981. Vinculin: A cytoskeletal target of the transforming proteins of Rous sarcoma virus. *Cell* **24:** 165.

Uyemura, D.G., S.S. Brown, and J.A. Spudich. 1978. Biochemical and structural characterization of actin from *Dictyostelium discoideum*. *J. Biol. Chem.* **253:** 9088.

Vandekerckhove, J., J. Leavitt, T. Kakunaga, and K. Weber. 1980. Coexpression of a mutant β-actin and the two normal β- and cytoplasmic actins in a stably transformed human cell line. *Cell* **22:** 893.

Mitochondria in Tumor Cells: Effects of Cytoskeleton on Distribution and as Targets for Selective Killing

L.B. Chen, I.C. Summerhayes, K.K. Nadakavukaren, T.J. Lampidis, S.D. Bernal, and E.L. Shepherd

Dana-Farber Cancer Institute and Department of Pathology, Harvard Medical School, Boston, Massachusetts 02115

It has been five years since we noted that mitochondria in living cells may be studied by cationic rhodamine dyes. Our work in this area up to 1981 is described in Chen et al. (1982). Here we summarize our findings on mitochondria in tumor cells.

Almost 50 years ago Otto Warburg proposed that defective mitochondria may be responsible for the high rate of aerobic glycolysis in cancer cells. Since then it has been a subject of controversy whether tumor mitochondria are indeed defective in respiration/oxidative phosphorylation. It is now widely "believed" that Warburg, who had revolutionized the field of biochemistry, was wrong in this particular issue. This "consensus" is most evident in the current lack of interest in tumor mitochondria; this is in contrast to the research of 15–20 years ago, when almost no cancer meeting went without a session on tumor mitochondria. The most persuasive evidence supporting the opponents of Warburg has been the in vitro comparison of mitochondria isolated from normal tissues and tumors, by which it was repeatedly demonstrated that tumor mitochondria are normal in respiration, ATP synthesis, P/O ratio, and so forth. However, it has been argued by proponents of the Warburg hypothesis that once mitochondria are removed from their native context, i.e., the living cells, comparison made in vitro should only be considered as the first step, not the last, toward the true. Unfortunately, the combination of the lack of proper tools to study mitochondria in the native context and the incredible popularity gained by the in vitro studies has made the field of tumor mitochondria dormant.

Rhodamine dyes seem to provide an opportunity to reconsider the issue of tumor mitochondria. Since it is a supravital dye, tumor mitochondria can be studied in their native context. Moreover, the amount of dye uptake reflects the electric potential across the mitochondrial membrane in living cells. Since this membrane potential is created essentially by proton pumps in the respiratory chain, it is possible to probe the most vital mitochondrial function by monitoring the rate and extent of dye uptake. Whether mitochondria in tumor cells are defective in respiration may thus be conveniently studied. In addition, other aspects of tumor mitochondria not directly related to respiration may also be studied since the detectability of these organelles in living cells is greatly improved over the conventional techniques, such as phase-contrast or Nomarski optics. These features include the number, length, size, volume, morphology, distribution, and motility of mitochondria. Although the role of any of these phenotypes in tumorigenesis is totally unknown, it is imperative that the information at least be collected.

It is possible that all these parameters may have biochemical meanings and that their alterations in tumor cells, if any, may contribute to the neoplastic state. For example, unusually low numbers of mitochondria may drive cells to utilize more aerobic glycolysis for the source of ATP. Abnormal mitochondrial distribution, as a manifestation of either defective mitochondrial motility or an aberrant cytoskeleton or both, may upset ATP distribution, which in turn could cause an imbalance in the reactions of ATP-utilizing enzymes, favoring those nearby the mitochondrion and slowing down those farther away.

For these reasons, using rhodamine dyes, we have systematically investigated mitochondria in more than 100 tumor-derived or tumorigenic cell lines. Whenever normal counterparts were available, an extensive comparison was made. Unexpectedly, we have encountered a long list of differences between tumor and normal mitochondria in one or a combination of the following: quantity (as reflected by number, length, size, and diameter), morphology, distribution, motility, membrane potential, pH gradient, electrochemical gradient, and retention time of rhodamine 123 (whose biochemical basis is still under investigation). Because of space limitations, we will mainly describe two aspects: abnormal distribution and unusually long retention of rhodamine 123 in some tumor cell lines. Most of our studies are still at the stage of phenomenology. Nonetheless, in one instance we have brought the discovery of the abnormality to the stage of practical application, i.e., we have exploited mitochondria that retain rhodamine 123 a long time as targets for chemotherapy in carcinoma cells. Mechanistic aspects of these abnormal phenotypes associated with mitochondria in tumor cells are now under active investigation.

Experimental Procedures

Cell culture

All fibroblasts, epithelial cells, and their transformants were grown in Dulbecco's modified Eagle's medium supplemented with 10% calf serum at 37°C, 5% CO_2, and 100% humidity. All cell types and lines used here were described in Summerhayes et al. (1982). Primary explant cultures of some human carcinomas were made according to Summerhayes et al. (1981).

Staining of mitochondria in living cells
Cells were grown on glass coverslips (12-mm d for round and 12 × 12 mm for square; Bellco) and placed on Parafilm, and a drop of rhodamine 123 (10 µg/ml in culture medium) was added. After 10 minutes at 37°C, cells were washed in dye-free medium for 2–5 minutes at room temperature, and the coverslip was mounted on a microscope slide with silicon rubber (North American Reiss Co.) as described in Johnson et al. (1980). For visualization of rhodamine 123, the filter system was that for fluorescein, not rhodamine, because of the unusual emission spectra of rhodamine 123.

The best lens for detecting mitochondria is Zeiss Planapo 40× or its equivalent, not the Zeiss Planapo 63× or its equivalent, which is commonly used for immunofluorescence. Because of the plasticity of living mitochondria, it is essential to use the shortest possible exposure time for photography. We used Kodak Tri-X at exposure index 1600 or 3200 and developed the film in Kodak HC-110, dilution B, for 12 or 15 minutes at 25°C. The optimal and inexpensive printing was made on Kodak Polycontrast resin-coated (RC) paper with a number 3 filter and processed by the Kodak Ektamatic Processor. The prints were fixed in Kodak Rapid Fixer for 2 minutes, rinsed in running water, and air-dried in a warm room.

Quantitation of rhodamine 123 uptake and retention
2-Butanol extraction of rhodamine 123 and its quantitation were described in Johnson et al. (1982). Fluorescence measurement of individual cells was according to Summerhayes et al. (1982).

Immunofluorescence staining of cytoskeletal components
Procedures and antibodies used are described in detail in Summerhayes et al. (1983). For localization of microtubules and intermediate filaments in tumor cells, the fixation procedures were developed by one of us (K.K.N.). Cells on coverslips were fixed in 0.025% glutaraldehyde and 0.2 mg/ml dithiobis [succinimidyl propionate] (DTSP; Pierce) in PBS at 37°C for 5 minutes, washed quickly in PBS, and incubated with 2% bovine serum albumin (BSA) in PBS for 3 minutes at 25°C. Cells were made permeable in 0.5% Triton X-100, 100 mM PIPES, pH 6.9, 1 mM EGTA, and 4% polyethylene glycol 8000 for 5 minutes at 37°C, rinsed in the same buffer without Triton X-100, and fixed in methanol at −20°C before application of antibodies.

Cell killing, clonogenic assay, and chemotherapy in mice
All methods and experimental protocols were previously described (Lampidis et al. 1983a; Bernal et al. 1982, 1983).

Results
Mitochondrial distribution in tumor cells
Despite the lack of quantitative means for "measuring" distribution, it is of interest to document the distribution of mitochondria in tumor cells, for two reasons. One is that the altered distribution may result from a more profound alteration hitherto unrecognized, such as those involved in cytoarchitecture. The other reason is that the altered distribution may have significant consequences with regard to the neoplastic state. Each time we stained a newly acquired cell line with rhodamine 123, the distribution of mitochondria was recorded. Over the years we have accumulated this information for more than 100 cell lines.

Figure 1 shows examples of mitochondrial distribution in some tumor cell lines. It is possible to categorize them into the following classes. Type R (denoted from random) has mitochondria so randomly distributed that some of them are located in a region normally devoid of this organelle (Fig. 1A). Type C (denoted from colchicine) has the majority of mitochondria distributed like those in colchicine-treated normal cells (Fig. 1B). The shape of these mitochondria also mimics those in cells with disrupted microtubule. Type N (denoted from nucleus) has most mitochondria clustered around the nucleus (Fig. 1C). Type U (denoted from unorthodox) has mitochondrial distribution unlike the above categories but clearly distinct from normal cells (Fig. 1D and Fig. 3); in the future we may be able to subclassify this type further. Some tumor cells express a combination of the above phenotypes. For instance, human adrenal cortex carcinoma has both type N and type C phenotypes. We denote this cell type NC. Likewise, type RC describes tumor cells, such as human Wilm's tumor, which have mitochondria distributed both randomly and in a pattern similar to colchicine-treated cells.

The mechanism responsible for the aberrant mitochondrial distribution is still unknown. Since it has repeatedly been suggested that the cytoskeleton is involved in the distribution of mitochondria, we have employed a variety of approaches to investigate this possible relationship. Normal cells were used in these studies because the best technique for studying such a relationship is epifluorescent microscopy, and cytoskeletal components of tumor cells are more difficult to visualize by this technique.

Mitochondrial distribution and cytoskeleton in normal cells
Photographic visualization of mitochondria with rhodamine 123 followed by processing of cells for immunofluorescent staining with actin, tubulin, or vimentin antibodies revealed a close correlation between the distribution of mitochondria and the distribution of microtubules or vimentin filaments, but not of microfilament bundles. The distribution and orientation of mitochondria within a single cell corresponded with the major networks of microtubules (Smith et al. 1975; Heggeness et

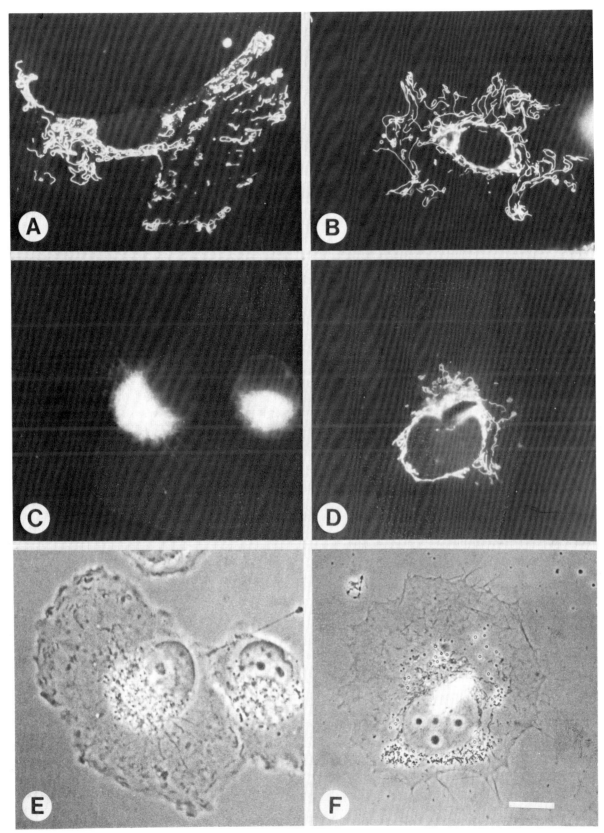

Figure 1 Mitochondria in living tumor cells as revealed by rhodamine 123. (*A*) Primary human bladder carcinoma; (*B*) primary human colon carcinoma; (*C*) primary human lung carcinoma; (*D*) human prostate carcinoma line PC-3; (*E* and *F*) phase-contrast micrographs of *C* and *D*, respectively. Bar, 25 μm.

al. 1978) as well as of intermediate filaments in normal fibroblasts (see Summerhayes et al. 1983). An example is shown in Figure 2.

In serially passaged, normal epidermal cells, where keratin is abundantly present, the correlation between mitochondrial distribution and keratin distribution is less clear cut since keratin is present virtually everywhere. In these cells, vimentin filaments are often not as prominent as in fibroblasts, and there is no certain correlation between the distribution of mitochondria and vimentin filaments (Summerhayes et al. 1983).

In primary bladder epithelial cells, for which we have previously reported the absence of vimentin (Summerhayes et al. 1981), mitochondria are distributed evenly throughout the cytoplasm. Whether they correlate with keratin is difficult to assess; mitochondria are present with an elaborate array of keratin filaments in all regions.

In some epithelial cell lines, such as canine kidney MDCK, where keratin is not as abundant as in epidermal cells, a good correlation between the distribution of keratin and that of mitochondria is often detected. However, if keratin is reduced to the extent that vimentin becomes the predominant component of the intermediate filament system, mitochondrial distribution correlates mainly with that of vimentin. The best example of this is with monkey kidney CV-1 cells.

It should be emphasized, however, that the correlation observed here is not of a single microtubule or inter-

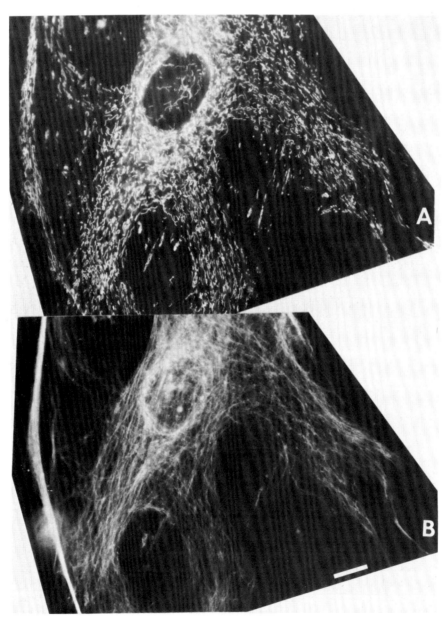

Figure 2 Mitochondria in a living gerbil fibroblast (A) and the corresponding vimentin filaments (B) in the same cell. Bar, 30 μm.

mediate filament to a single mitochondrion, although occasionally such a one-to-one correlation is detected. The correlation we have observed is with the overall distribution. Even with this qualification, however, the correlation is not absolute. Sometimes, although rarely, mitochondria are detected in a region devoid of major arrays of microtubules or intermediate filaments. Our impression is best summarized as follows: Mitochondria most often locate and migrate in the region of the cytoplasm abundant in microtubules and/or intermediate filaments, but these cytoskeletal components are not absolutely required for the presence of mitochondria. Mitochondria seem still to retain a certain degree of autonomy so that migration to unorthodox regions is not totally prohibited. This is most evident when a network of mitochondria migrate to the edge of a cell, as shown in Figure 11 in Chen et al. (1982).

Why do mitochondria favor the regions abundant in microtubules and intermediate filaments? Perhaps there is a nontrivial reason behind it, worthy of future exploration.

Mitochondrial distribution in normal cells with cytoskeletal alterations

Through prevention of repolymerization of microtubules, colchicine is able to make cells devoid of microtubules. Mitochondria in such cells become wavier in contour, and their active migration, as revealed by time-lapse video recording, is drastically curtailed. These results strongly suggest that microtubules are involved in the migration and therefore the distribution of mitochondria in untreated cells, either directly or indirectly. If microtubules promote mitochondrial migration, it is not surprising that these two systems are closely correlated in distribution. Moreover, if the correlation between vimentin filaments and mitochondria described above is true, then when vimentin is en route to the perinuclear region to form the characteristic nuclear whorl as a result of colchicine treatment, mitochondria should also follow these vimentin filaments. This is indeed the case (Summerhayes et al. 1983). Interestingly, in cell types in which vimentin filaments do not form the nuclear whorl after colchicine treatment, mitochondria seem to assume a more independent status, not following the vimentin filaments.

Why some cell types form the vimentin nuclear whorl with colchicine and others do not is still not understood, but it suggests that there may be a third cytoskeletal element involved in the complex interactions among microtubules, vimentin filaments, and mitochondria. Perhaps this third cytoskeletal element is responsible for both nuclear whorl formation and the entrapping of mitochondria in this massive structure (a good example of this entrapping may be found in Fig. 5 in Hynes and Destree [1978]). Human fibroblasts that do not form a nuclear whorl may be lacking this third element. Candidates for such a third system are the bridging structures described by Wang and Goldman (1978) and Smith et al. (1977).

It would be interesting to know how mitochondria would behave if microtubules were kept intact but vimentin distribution were altered. Until recently it has not been possible to alter vimentin organization without affecting microtubules. In collaboration with B.N. Fields and J.R. Murphy of Harvard Medical School, we have found two methods of achieving this in monkey kidney CV-1 cells. One is to inhibit protein synthesis by diphtheria toxin or cycloheximide, and the other is to infect the cells with reovirus. In both systems microtubule and microfilament bundles are intact, but vimentin filaments are drastically altered (Sharpe et al. 1980, 1982). Mitochondrial distribution in these cells is indeed dramatically altered, as shown in Figure 9 of Chen et al. (1982), Figure 7 of Summerhayes et al. (1983), and Figure 7 of Sharpe et al. (1982). A caveat for these experiments is that the altered mitochondrial distribution may result from the nonspecific inhibition in protein synthesis, rather than from the altered vimentin organization. At the moment the best, albeit imperfect, control for this possibility is that vimentin distribution in many other cell types is insensitive to the inhibition of protein synthesis. And in these cells mitochondrial distribution is unaltered. Thus, it is less likely that the effect seen in CV-1 cells is independent of the altered vimentin organization.

A third means of altering vimentin distribution without affecting microtubules has been developed by Lin and Feramisco (1981) at Cold Spring Harbor Laboratory. It involves the microinjection of monoclonal antibodies specific for a 95,000-dalton-protein associated with vimentin filaments (Lin 1981). Consistent with the initial observation made by Lin and Feramisco (1981), Figure 12 in Chen et al. (1982) and Figure 8 in Summerhayes et al. (1983) show that mitochondrial distribution is unaffected by the redistribution of vimentin filaments toward the nuclear region.

How can we reconcile all these seemingly contradictory results? Perhaps the simplest answer is, again, that there is a third element involved in effecting the migration of mitochondria, in addition to microtubules and vimentin filaments. It may be hypothesized that this third component is involved in a triangular interaction among microtubules, intermediate filaments, and mitochondria. When anti-95,000-dalton-protein antibodies are microinjected, because of the high specificity of these antibodies, it only leads to the retraction of vimentin filaments but does not affect the interactions of microtubules and the third element with mitochondria. Under such a circumstance, normal mitochondrial distribution can be maintained. When protein synthesis in CV-1 cells is inhibited, the third element may have a higher rate of turnover than microtubules/tubulin; thus, even though microtubules are intact, the third component may have been depleted. In the absence of the third component, it may then become difficult for microtubules alone to sustain the normal migration and distribution of mitochondria.

In epithelial cell lines such as PtK2 that express both keratin and a significant amount of vimentin, colchicine does not affect the keratin distribution but does redistribute vimentin filaments to form a nuclear whorl. In these cells mitochondrial distribution is unaffected (see Fig. 10 in Summerhayes et al. 1983). Evidently keratin exerts more influence on mitochondrial distribution than either microtubules or vimentin in these cell types. For vimentin-negative, keratin-abundant, primary mouse bladder epithelial cells, colchicine has absolutely no effect on mitochondrial distribution, as expected.

All the above experiments suggest that in normal cells mitochondrial distribution is under the strong influence of the two systems, microtubules and intermediate filaments. However, neither plays an absolute role. Mitochondria may retain some degree of autonomy. It is also possible there is a third cytoskeletal element, yet to be identified, that is involved in mitochondrial migration and hence distribution.

Is the cytoskeleton altered in tumor cells with aberrant mitochondrial distribution?

Immunofluorescent localization of cytoskeletal components in most tumor cells has been difficult. Using the available fixation procedures described in the literature for microtubules in normal cells, we were initially led to believe that tumor cells with type C mitochondrial distribution have altered microtubular distribution. However, we were skeptical and have continued to search for a better fixation procedure for microtubules in tumor cells. At the moment, the best fixation protocol appears to be 0.025% glutaraldehyde and 0.2 mg/ml DTSP, a cross-linking reagent, in PBS for 5 minutes at 37°C. This procedure has revealed that no tumor cells can be said to have altered microtubules. Tumor cells may be thicker or have unusual morphology, but microtubules are always abundant and intact and travel from the perinuclear region to the edge of the cells, as described by Osborn and Weber (1977). Examples are shown in Figure 3.

The above fixation protocol also turns out to be ideal for localizing intermediate filaments in tumor cells, even though it may not be always necessary. Methanol at

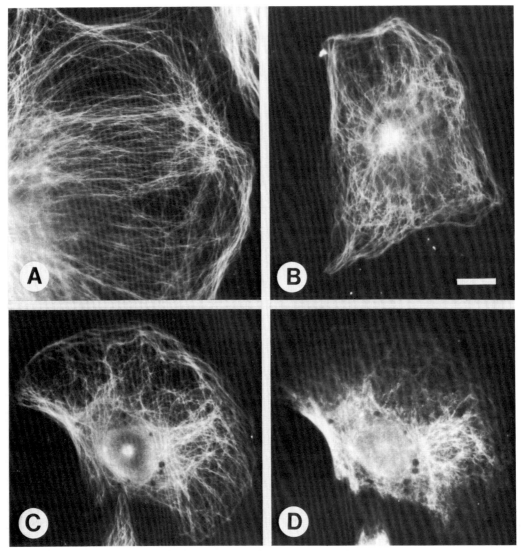

Figure 3 Microtubules in monkey kidney CV-1 cell (A), human pancreatic carcinoma PaCa-2 cell (B), and human Wilm's tumor line G-401 cell (C). (D) Vimentin filaments of the same cell in C. Bar, 25 μm.

−20°C seems to suffice for most cultured cells. For the results presented here, we have used both fixation procedures for vimentin and keratin. The majority of tumor cells do not have intermediate filament systems aberrant enough to be classified as grossly abnormal. However, among those tumor cells with types N, C, R, and U mitochondrial distribution, there are numerous examples of aberrant intermediate filament organization. Figures 3D and 4 illustrate examples of such findings. Most of these tumor cell lines also express reduced amounts of intermediate filaments, as initially reported in Summerhayes et al. (1981) and Summerhayes and Chen (1982). In the case of human adrenal cortex adenocarcinoma, which has type NC mitochondrial distribution, no vimen-

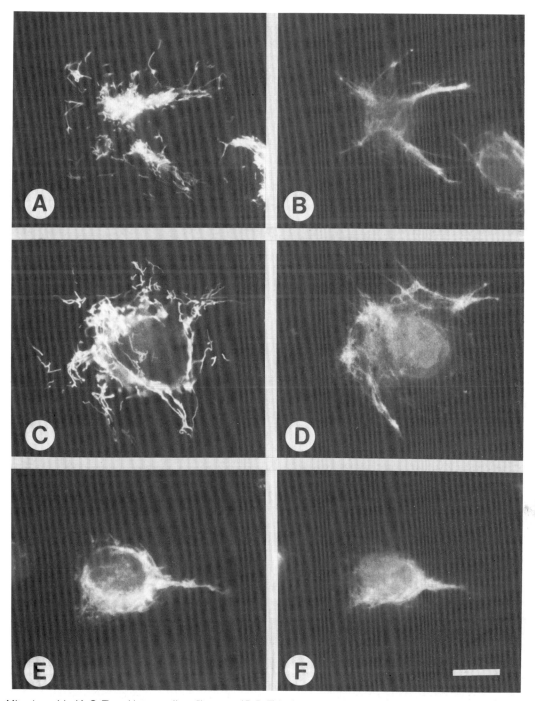

Figure 4 Mitochondria (A,C,E) and intermediate filaments (B,D,F) in the same respective tumor cells. (A,B) Human colon carcinoma line SW480 with vimentin staining; (C,D) human bladder carcinoma line EJ with keratin staining; (E,F) human lung carcinoma line LX-1 with keratin staining. Bar, 30 μm.

tin and keratin is detected. Since microtubules are still intact and abundant in this cell line, one is forced to think that the aberrant mitochondrial distribution may be related to the lack of intermediate filaments and possibly to the purported third element in this unusual tumor-derived cell line.

The triple stainings for mitochondria, tubulin, and vimentin in the same cell also support the notion that mitochondria tend to correlate better with the distribution of vimentin filaments than that of microtubules. An example of human Wilm's tumor line, G-401, with type NC mitochondrial distribution is demonstrated in Figure 5.

Obviously, we seem to have just scratched the surface of the problem. There is a long way to go before we understand the molecular mechanism governing mitochondria plasticity, migration, and distribution. We probably have only a first glimpse in observing that microtubules and intermediate filaments are involved, but these two systems do not explain all phenomena encountered. Hopefully, the variety of tumor cells with aberrant mitochondrial migration and distribution that we have identified can be used as "mutants" for unraveling the purported third element, and we can begin to answer the question, what is the molecular basis for the saltatory movement of mitochondria?

Unusual dye retention by mitochondria in carcinoma cells

It turns out that almost all these carcinoma cells with mitochondrial distribution of types N, C, U, R, and their combinations retain rhodamine 123 for much longer times than do normal epithelial cells (Tables 1 and 2) (see also Summerhayes et al. 1982). Several other carcinoma cell lines that do not express aberrant mitochondrial distribution also retain rhodamine 123. Thus, abnormal distribution is not a prerequisite for prolonged dye retention. The biochemical basis for dye retention is not fully understood. Mitochondrial membrane potential is necessary to keep dye inside mitochondria (Johnson et al. 1981), but other factors such as dye permeability, plasma membrane potential, and dye-binding sites in mitochondria may all be relevant (M.J. Weiss et al., unpubl.). However, if prolonged dye-retaining mitochondria in carcinoma cells also take up more dye than the normal cells, a higher electric potential across mitochondrial membrane may be one of the most important factors. It may be responsible for both increased uptake and prolonged retention, since we have previously shown that the dye uptake is mainly determined by the magnitude of the mitochondrial electric potential (Johnson et al. 1981). Whereas it is unlikely that all cases of prolonged retention result from this simplest of mechanisms, we have found thus far five out of five carcinoma cell lines that express long retention also take up more rhodamine 123 than do cells with short retention times (Fig. 6). These results do not preclude the possibility that plasma membrane potential and changes in lipid

Table 1 Rhodamine 123 Retention in Normal and Transformed Bladder Epithelial Cells After 24 Hours in Dye-free Medium

Cell type or lines	Fluorescence[a]
Primary bladder epithelial cells	
mouse	0
rabbit	0
Carcinogen-transformed	
MB48 (DMBA-mouse)	90
MB49 (DMBA-mouse)	90
BBN-6 (BBN-mouse)	70
SH264 (BBN-mouse)	62
SH257 (BBN-mouse)	0
RBC-1 (benzo[a]pyrene-rabbit)	84
Human bladder tumor lines	
EJ (transitional cell carcinoma)	90
J82 (transitional cell carcinoma)	15
RT4 (transitional cell carcinoma)	70
RT112 (transitional cell carcinoma)	75
HT-1376 (transitional cell carcinoma)	93

[a]The average fluorescence of rat cardiac muscle cells after 24 hr in dye-free medium was taken to be 100. Fluorescence of cells above and in Table 2 is expressed relative to this standard.

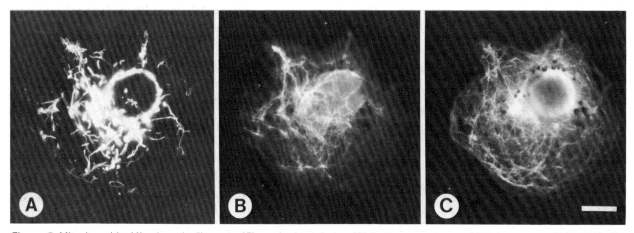

Figure 5 Mitochrondria (A), vimentin filaments (B), and microtubules (C) in the same human Wilm's tumor cell (G-401). Bar, 25 µm.

Table 2 Rhodamine 123 Retention of Human Tumor Lines After 24 Hours in Dye-free Medium

Carcinoma	Fluorescence[a]
MCF-7 (breast, adeno-)	95
T47D (breast, adeno-)	95
ZR-75-1 (breast, adeno-)	80
BT20 (breast, adeno-)	80
HBL-100 (breast, adeno-)	80
SW480 (colon, adeno-)	80
HeLa (cervix, adeno-)	70
CaSki (cervix, epidermoid)	90
A431 (vulva, epidermoid)	85
PaCa-2 (pancreas)	90
PC-3 (prostate)	90
OD562 (ovary)	80
OVCA433 (ovary)	70
SCC-4 (tongue, squamous)	60
SCC-9 (tongue, squamous)	65
SCC-12 (skin, squamous)	60
SCC-13 (skin, squamous)	50
SCC-15 (tongue, squamous)	40
SCC-25 (tongue, squamous)	20
SCC-27 (skin, squamous)	50
HUT23 (lung, adeno-)	75
HUT125 (lung, adeno-)	60
A549 (lung, adeno-)	60
U1752 (lung, squamous)	60
HUT157 (lung, large cell)	0
LX-1 (lung, poorly differentiated)	0
OH-1 (lung, oat cell)	0
HUT60 (lung, oat cell)	0
HUT128 (lung, oat cell)	0
HUT231 (lung, oat cell)	0
HUT64 (lung, oat cell)	0

[a]See Table 1 for definition of fluorescence units.

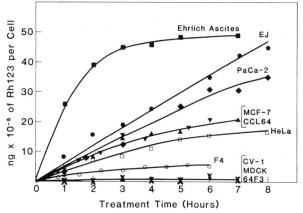

Figure 6 Rhodamine 123 uptake per cell by various cell lines. Dye was present continuously at 1 μg/ml. Cell lines represented are: Ehrlich ascites; human bladder carcinoma (EJ); human pancreatic carcinoma (PaCa-2); human breast carcinoma (MCF-7); human cervix carcinoma (HeLa); mink fibroblast (CCL64); FeSV- transformed CCL64 (64F3); monkey kidney epithelium (CV-1); dog kidney epithelium (MDCK).

composition may also contribute to the prolonged dye retention. We hope that in the near future the phenomenon can be explained fully.

Strong supporting evidence that dye-retaining carcinoma mitochondria also express a higher membrane potential came from studies of fusion between human breast carcinoma MCF-7 cells and nonretaining kangaroo rat epithelial PtK2 cells. In fused cells mitochondria from MCF-7 take up more rhodamine 123 than those of PtK2, despite the fact that they share the same cytoplasm and the plasma membrane has been fused (D. Chung et al., unpubl.). It is possible that these two types of mitochondria are intrinsically different. In vitro comparison of mitochondria may be warranted in this case.

Selective killing of dye-retaining carcinomas

Could the phenomenon described above be exploited for therapeutic purposes? The rationale is simple. If carcinoma mitochondria take up more rhodamine 123 and retain it longer, why cannot they also do the same with a poison that is also cationic and permeant, since it has been demonstrated previously that all fluorescent compounds with these two properties are taken up specifically by mitochondria (Johnson et al. 1981). If this is the case, a selective killing of dye-retaining carcinoma cells may be possible. We have tested a number of such poisons, such as rhodamine 6G and 3B, ethidium bromide, cyanines, and rhodanile blue. None of them exhibits selective toxicity between normal epithelial cells and carcinoma cells at the dose we have tested. Apparently these compounds are too toxic. A very small amount taken up by normal cells is sufficient to kill them.

Initially we had not considered the use of rhodamine 123 since in our earlier studies, out of many fluorescent dyes generally used for mitochondrial staining, this dye was selected because of its lack of toxicity to the fibroblasts and untransformed epithelial cells commonly used by cell biologists. Indeed, we have repeatedly suggested its use as a supravital dye (Johnson et al. 1980, 1981, 1982). Working with cardiac muscle cells, one of us (T.J.L.) noted serendipitously that prolonged exposure of these cells to a high dose of rhodamine 123 may be cytotoxic. Since cardiac muscle cells in culture also express prolonged retention of rhodamine 123, we immediately suspected that rhodamine 123 may be the drug we were looking for. All the subsequent experiments indeed showed this to be the case (Lampidis et al. 1982, 1983a). Figure 7 demonstrates the remarkable selective killing of human pancreatic and breast carcinoma cells with rhodamine 123. In the clonogenic assays, rhodamine 123 is also selectively toxic to the dye-retaining carcinoma cells (Bernal et al. 1982).

What is the basis for rhodamine 123 toxicity? Rhodamine 6G has previously been shown to inhibit oxidative phosphorylation in isolated mitochondria (Gear 1974). In collaboration with C. Salet and G. Moreno (Museum National d'Histoire Naturelle, Paris), we have found rhodamine 123 to inhibit the respiratory control ratio (RCR) of isolated rat liver mitochondria with a result similar to that of dinitrophenol (DNP) when glutamate or succinate was used as a substrate (Lampidis et al. 1983b). More recently we have collaborated with J. Aprille (Tufts University) to investigate further the mechanism leading to the inhibition of RCR. At this writing we do not have a

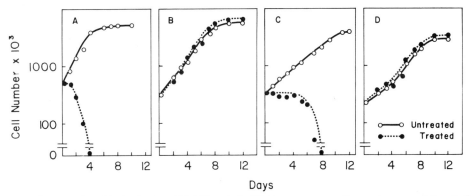

Figure 7 Killing of carcinoma cells with rhodamine 123 (10 μg/ml). (A) Human pancreatic carcinoma cell line PaCa-2; (B) monkey kidney epithelial cell line CV-1; (C) human breast carcinoma cell line MCF-7; (D) kangaroo rat epithelial cell line PtK-1. Each point represents the average cell count of triplicate cultures.

complete answer, but rhodamine 123 is definitely not an uncoupler like DNP; it does not inhibit nucleotide transport nor F1 ATPase (J. Aprille et al., unpubl.). Possibly it inhibits electron transport. Irrespective of the actual mechanism, rhodamine 123 would lead to the inhibition of mitochondrial ATP synthesis, as could be inferred from the inhibition of RCR. Thus, the basis for toxicity might be the inhibition of this vital mitochondrial function. In view of this possibility, it becomes evident that if one can deplete the other source of ATP production, cell killing by rhodamine 123 may be enhanced. Before testing anticarcinoma activity of rhodamine 123 in animals, it would also be advantageous if the carcinoma killing by rhodamine 123 could be accelerated by the inclusion of another nontoxic drug, since in culture it does take 4–7 days to kill most carcinoma cells (with the exception of human pancreatic adenocarcinoma, PaCa-2). The best candidate for such purpose is 2-deoxyglucose, which inhibits glycolysis, the nonmitochondrial source of ATP production, and has been used in human patients at high dosage without a significant side effect. Indeed, when rhodamine 123 and 2-deoxyglucose are combined, which should efficiently reduce the production of ATP, the carcinoma cell killing is enhanced. Moreover, the dose of rhodamine 123 needed to achieve sufficient killing can be substantially lower (Table 1 in Lampidis et al. 1983a). However, most remarkably, the doses of rhodamine 123 and 2-deoxyglucose used for killing carcinoma cells do not have any toxic effect on normal epithelial cells, untransformed epithelial cells, and a variety of normal cells, including fibroblasts, T and B lymphocytes, macrophages, chondrocytes, mast cells, endothelial cells, and smooth muscle cells.

Rational chemotherapy of some carcinomas in animals

Injection of a high dose of rhodamine 123 (100 mg/kg) intraperitoneally is indeed toxic to mice; most of the injected mice died the next day. Thus, there are normal cells that are sensitive in vivo to a high dose of rhodamine 123, despite the in vitro results. Surprisingly, the affected sites appear not to be the predicted areas, the heart and skeletal muscle. The fatal target responsible for death is still being investigated.

On days 1, 3, and 5, we chose to use doses of 15 mg/kg, which showed no acute toxicity nor latent side effects, such as tumor induction, after more than 18 months of follow-up observation. In combination with 2-deoxyglucose at 0.5 g/kg on days 1, 3, and 5, this dose of rhodamine 123 also did not produce acute toxicity or latent side effects.

Mice were injected intraperitoneally with 5×10^5 Ehrlich ascites tumor cells. These mice had a median survival time of 19 days (18–22 days). Treatment with rhodamine 123 at 15 mg/kg on days 1, 3, and 5 significantly prolonged survival to a median time of 50 days (T/C = 260%), with approximately 10% of the mice cured. However, treatment with the combination of 2-deoxyglucose and rhodamine 123 at the doses and regimen described above dramatically prolonged the survival time to a median of 80 days (T/C = 420%), with about 40% of the mice cured. 2-Deoxyglucose alone does not have any antitumor activity. These results are the average of five independent experiments. Results from one set of experiments are shown in Figure 8. A complete account of our first phase of animal experiments is documented in Bernal et al. (1983). These experiments demonstrate the feasibility of using rhodamine 123 and 2-deoxyglucose for the treatment of carcinoma cells expressing the characteristics of prolonged mitochondrial dye retention.

Discussion

Although we have not as yet resolved Otto Warburg's hypothesis in living cells with our new approaches, we have already uncovered that "different" mitochondria do exist in tumor cells. Earlier we had shown that FeSV-transformed mink fibroblasts have very low mitochondrial membrane potential compared with their normal counterparts (Johnson et al. 1982). Here we show that mitochondrial distribution can be very abnormal in tumor cells. In addition, we demonstrate that mitochondria with rhodamine 123 retention very different from normal ep-

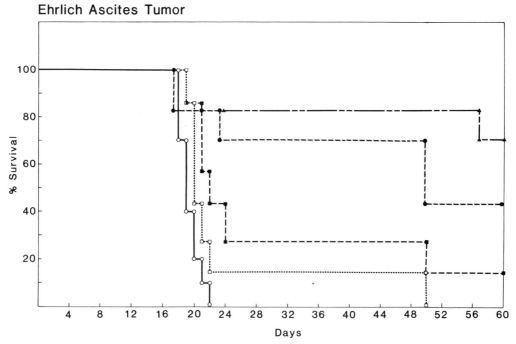

Figure 8 Antitumor activity of rhodamine 123 and 2-deoxyglucose in mice implanted with Ehrlich ascites tumor (carcinoma). (○) Untreated control, T/C = 100%; (□) rhodamine 123 (5 mg/kg, days 1,3,5,7,9), T/C = 105%; (■) rhodamine 123 (10 mg/kg, days 1,3,5,7,9), T/C = 115%; (●) rhodamine 123 (15 mg/kg, days 1,3,5), T/C = 263%; (▲) rhodamine 123 (15 mg/kg, days 1,3,5) and 2-deoxyglucose (0.5 g/kg, days 1,3,5), T/C = 400%. Each experiment included 10 BDF1 mice as untreated controls and 7 mice in each drug-treated group. The experiments illustrated above have been repeated five times, and the average is given in the text.

ithelial cells do exist in the majority of human carcinoma cell lines tested.

The biochemical consequence of aberrant mitochondrial distribution awaits further investigation. It is possible that the organization of cell structure and the distribution of organelles are involved in metabolism far more than we have expected. The ongoing marriage between cell biology and biochemistry may resolve this issue in near future.

The basis for the difference in mitochondrial rhodamine 123 retention between carcinoma cells and normal epithelial cells is still under investigation. However, it is likely one of the factors involved is the higher mitochondrial membrane potential in carcinoma cells. It is of interest to note that this is the reverse of what we have observed in normal and transformed mink fibroblasts. Despite the lack of understanding of the basic mechanism, this work confirms, in a partial and preliminary way, Warburg's intuition that paying attention to mitochondria in tumor cells may ultimately help the treatment of cancer.

Acknowledgments

This work has been supported by grants from the National Cancer Institute, the American Cancer Society, and the Council for Tobacco Research, U.S.A. L.B.C. is supported by an ACS Faculty Research Award. K.K.N. was supported by an NCI Training Grant to A.B. Pardee during 1981 and is currently supported by an NCI Postdoctoral Fellowship (CA-070407-01).

References

Bernal, S.D., T.J. Lampidis, R.M. McIsaac, and L.B. Chen. 1983. Anticarcinoma activity in vivo of a mitochondrial-specific dye, rhodamine 123. *Science* **222**: 169.

Bernal, S.D., T.J. Lampidis, I.C. Summerhayes, and L.B. Chen. 1982. Rhodamine 123 selectively reduces clonal growth of carcinoma cells in vitro. *Science* **218**: 1117.

Chen, L.B., I.C. Summerhayes, L.V. Johnson, M.L. Walsh, S.D. Bernal, and T.J. Lampidis. 1982. Probing mitochondria in living cells with rhodamine 123. *Cold Spring Harbor Symp. Quant. Biol.* **46**: 141.

Gear, A.R.L. 1974. Rhodamine 6G. A potent inhibitor of mitochondrial oxidative phosphorylation. *J. Biol. Chem.* **249**: 3628.

Heggeness, M.H., M. Simon, and S.J. Singer. 1978. Association of mitochondria with microtubules in cultured cells. *Proc. Natl. Acad. Sci.* **75**: 3863.

Hynes, R.O. and A.T. Destree. 1978. 10nm Filaments in normal and transformed cells. *Cell* **13**: 151.

Johnson, L.V., I.C. Summerhayes, and L.B. Chen. 1982. Decreased uptake and retention of rhodamine 123 by mitochondria in feline sarcoma virus-transformed mink cells. *Cell* **28**: 7.

Johnson, L.V., M.L. Walsh, and L.B. Chen. 1980. Localization of mitochondria in living cells with rhodamine 123. *Proc. Natl. Acad. Sci.* **77**: 990.

Johnson, L.V., M.L. Walsh, B.J. Bockus, and L.B. Chen. 1981. Monitoring of relative mitochondrial membrane potential in living cells by fluorescence microscopy. *J. Cell Biol.* **88**: 526.

Lampidis, T.J., S.D. Bernal, I.C. Summerhayes, and L.B. Chen. 1982. Rhodamine-123 is selectively toxic and preferentially retained in carcinoma cells in vitro. *Ann. N.Y. Acad. Sci.* **460**: 299.

———. 1983a. Selective toxicity of rhodamine 123 in carcinoma cells in vitro. *Cancer Res.* **43**: 716.

Lampidis, T.J., C. Salet, G. Moreno, and L.B. Chen. 1983b. Effects of the mitochondrial probe rhodamine 123 and related analogs on the function and viability of pulsating myocardial cells in culture. *Agents Actions* (in press).

Lin, J.J.-C. 1981. Monoclonal antibodies against myofibrillar components of rat skeletal muscle decorate the intermediate filaments of cultured cells. *Proc. Natl. Acad. Sci.* **78:** 2335.

Lin, J.J.-C. and J.R. Feramisco. 1981. Disruption of the *in vivo* distribution of the intermediate filaments in fibroblasts through the microinjection of a specific monoclonal antibody. *Cell* **24:** 185.

Osborn, M. and K. Weber. 1977. The display of microtubules in transformed cells. *Cell* **12:** 561.

Sharpe, A.H., L.B. Chen, and B.N. Fields. 1982. The interaction of mammalian reoviruses with the cytoskeleton of monkey kidney CV-1 cells. *Virology* **120:** 399.

Sharpe, A.H., L.B. Chen, J.R. Murphy, and B.N. Fields. 1980. Specific disruption of vimentin filament organization in monkey kidney CV-1 cells by diphtheria toxin, exotoxin A, and cycloheximide. *Proc. Natl. Acad. Sci.* **77:** 7267.

Smith, D.S., U. Jarlfors, and B.F. Cameron. 1975. Morphological evidence for the participation of microtubules in axonal transport. *Ann. N.Y. Acad. Sci.* **253:** 472.

Smith, D.S., U. Jarlfors, and M.L. Cayer, 1977. Structural cross-bridges between microtubules in central axons of an insect (*Periplaneta americana*). *J. Cell Sci.* **27:** 255.

Summerhayes, I.C. and L.B. Chen. 1982. Localization of a M_r 52,000 keratin in basal epithelial cells of the mouse bladder and expression throughout neoplastic progression. *Cancer Res.* **42:** 4098.

Summerhayes, I.C., D. Wong, and L.B. Chen. 1983. Effect of microtubules and intermediate filaments on mitochondrial distribution. *J. Cell Sci.* **61:** 87.

Summerhayes, I.C., Y.S.E. Cheng, T.T. Sun, and L.B. Chen. 1981. Expression of keratin and vimentin intermediate filaments in rabbit bladder epithelial cells at different stages of benzo[a]pyrene-induced neoplastic progression. *J. Cell Biol.* **90:** 63.

Summerhayes, I.C., T.J. Lampidis, S.D. Bernal, J.J. Nadakavukaren, K.K. Nadakavukaren, E.L. Shepherd, and L.B. Chen. 1982. Unusual retention of rhodamine 123 by mitochondria in muscle and carcinoma cells. *Proc. Natl. Acad. Sci.* **79:** 5292.

Wang, E. and R.D. Goldman. 1978. Functions of cytoplasmic fibers in intracellular movements in BHK-21 cells. *J. Cell Biol.* **79:** 708.

Movement of Nucleus and Centrosphere in 3T3 Cells

G. Albrecht-Buehler
Department of Cell Biology and Anatomy, Northwestern University Medical and Dental School, Chicago, Illinois 60611

If one compares a migrating animal cell to a moving vehicle, one would expect to find an engine (i.e., a transducer of chemical energy into mechanical energy), a frame (i.e., a propagator of mechanical force throughout the vehicle), a mechanism of traction (i.e., an interaction with the substrate), and a driver (i.e., a control mechanism for directionality and navigation). It appears that most scientists in the field of cell motility relate the engine to an actomyosin system, the frame to the cytoskeleton, and the mechanism of traction to the adhesion plaques and other junctional complexes. What mechanism corresponds to the driver?

From geometric patterns expressed in the tracks of migrating 3T3 cells, I have suggested that these cells are capable of storing programs of directional changes, reorienting their pathways in predictable ways after a collision, following and exploring alternatives of exogenous guidance, and coordinating each other's direction of migration (Albrecht-Buehler 1982a). These observations point to the existence of a directional control mechanism in cells that is able to assess the global properties of its environment.

The image of a tissue cell that emerged from this work is shown schematically in Figure 1. The cell is represented as an ordered mosaic of five major compartments, namely the nucleus, three cytoplasmic compartments, and the plasma membrane. The outermost compartment is the plasma membrane that mediates the interactions with the environment and provides junctions and adhesions. Immediately adjacent is the compartment that forms motile surface projections. It can be isolated in the form of autonomously moving microplasts (Albrecht-Buehler 1980). This compartment is actin-rich (Small 1981) and correspondingly sensitive to cytochalasin B. Figure 1 depicts the compartment divided into microplastlike domains, which may receive individual "instructions" for particular movements. The third compartment covers the space between the actin region and the nucleus. It contains the organelles, the inner membrane systems, and expresses liquid streams (see Rebhun 1972; Albrecht-Buehler 1982b). The dominant fibers of this compartment appear to be the intermediate filaments. Therefore, its movements and architecture are sensitive to colcemid and similar drugs. The fourth compartment is the centrosphere with the centrioles and a radial array of microtubules that traverse the organelle region and appear to be terminating at or looping through the actin region. The figure shows them terminating individually at a microplastlike domain. The microtubular centrosphere or the centrioles, are sensitive to colcemid and similar drugs. The fifth compartment is the nucleus.

This image of the cell is highly simplified, and the borderlines between some of the compartments are often ill-defined. However, the simplification may be helpful in discussing control mechanisms of cell movement.

This schematic suggests a spatial separation between the functional blocks of cell migration. In terms of this view, the membrane compartment provides traction, the actin compartment provides the engine, and the organelle compartment, with the intermediate filaments, represents the frame (Lazarides 1980). The nucleus has no direct involvement in cell migration, as demonstrated by the normal migration of cytoplasts (Goldman et al. 1973). Nevertheless, if one considers the morphogenetic movements of differentiating cells, a nuclear role in long-term programming of cell movement is almost certain. In addition, the shape of the nucleus is subjected to the body deformations of migrating cells, which may have effects on nuclear functions in general and feedback effects on migration (Folkman and Moscona 1978). Therefore, the compartment of the nucleus must be included in all considerations of cell migration.

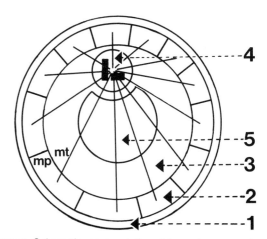

Figure 1 Schematic representation of five major compartments of a cell: (1) plasma membrane; (2) actin-rich compartment that produces motile surface projections; it is assumed to be divided into microplastlike domains (mp); (3) intermediate filament–rich compartment, containing inner membranes and organelles; (4) centrosphere with centrioles and radial microtubular array that traverses the cytoplasm; individual microtubules (mt) are assumed to terminate at or at least contact individual microplastlike domains; (5) nucleus.

Centrioles and microtubules appear to be directly involved in the control of cell polarity and are therefore instrumental in the control of cell migration. Consequently, the schematic view above considers the centrosphere, with its radial array of microtubules, as a candidate for the driver mechanism (Albrecht-Buehler 1981). The existence of such a function in cells would offer entirely new avenues of attack on cancer, focusing on the control mechanisms of migration and attempting to interfere with the locomotion of metastasizing cells.

Although we are far from understanding a control mechanism of cellular migration, initial steps toward its explanation are being taken that include observing the movements of the different compartments and attempting to relate them to each other. The movements of the membrane, of the actin-rich compartment, of organelles, and of liquid streams have been studied in considerable detail. Much less is known about nuclear movements, and so far the centrosphere was only visible in nonmoving cells. The following report suggests a method that allows for the first time the observation of the centrosphere in living, migrating cells. Therefore, one can now attempt to relate cell migration with movements of the centrosphere. In addition, the paper focuses on a particularly puzzling type of nuclear movement, namely, its occasional spinning within the cell. This strange nuclear behavior, which was first described 31 years ago (Pomerat 1953), has by its very rare occurrence escaped experimentation. However, by the same method that helped visualize the centrosphere, the frequency of nuclear rotation could be increased to make it more accessible for experimentation.

Methods

Near-infrared microscopy

The method that we developed to observe living cells in the near-infrared used a Zeiss standard RA microscope with normal phase-contrast lenses and condenser. The light source was a 20-W Osram halogen bulb that is standard equipment for the microscope. The color of the incident light was restricted to near-infrared by placing a 2-mm Corning 7-56 glass filter over the illumination aperture. The filter transmits light in the range between 800 nm and 4400 nm. The image, invisible to human eyes, was projected through a Zeiss photo-changer and a Zeiss straight tube for normal camera systems onto the vidicon tube of a Panasonic WV-1550 video camera, which is sensitive up to wavelengths of about 1100 nm. In this way, the video image was formed with wavelengths between 800 nm and 1100 nm. That close to the visible range (400–700 nm), the lenses produce still perfectly normal phase-contrast images. The image was displayed on a Hitachi VM-173U video monitor and recorded on a Panasonic VTR 8030 video time-lapse recorder. Still photographs were taken with an SLR camera from the video screen.

Cell culture and live-cell observations

3T3 cells were grown on Falcon plastic dishes in Dulbecco's modification of Eagle's medium (DMEM) supplemented with 10% calf serum in a 10% CO_2 atmosphere and saturated humidity. Cells were used up to 25 passages and then replaced with new cells at lower passage numbers. Experimental cells were grown on 25-mm circular glass coverslips (Bellco) and mounted in Dvorak-Stotler chambers. The temperature of 37°C was maintained on the microscope stage by a Sage 279 air-curtain incubator.

Indirect immunofluorescence

Cells grown on glass coverslips were fixed in 1:1 acetone:methanol at −20°C, then washed in phosphate-buffered saline (PBS) and incubated at 37°C for 15 minutes in monoclonal antibody against β-tubulin. The antiserum was a kind gift of S. Blose (Cold Spring Harbor Laboratory). Subsequently, the preparations were counterstained with FITC-labeled rabbit anti-mouse IgG at 37°C for 15 minutes and mounted in Gelvatol (Monsanto) and examined in a Zeiss Photomicroscope III with epifluorescence illumination.

Chemicals

Monensin (Calbiochem-Behring) was kept in 20 mM stock solutions in ethanol at 4°C. Incubation media were prepared by adding the appropriate amount of monensin to serum-free medium. Following sufficient mixing and dilution, the serum was added. Colcemid was purchased from Gibco.

Computer analysis

Nuclear rotation was measured by marking the position of nucleoli on the video screen and measuring their rotation throughout the recording. The data were entered into a Radio Shack TRS-80 Model III microcomputer with a program, written by the author, that calculated rotational speeds and statistical analysis and plotted the results on a flatbed plotter.

Results

Near-infrared microscopy of living, cultured tissue cells

In my experience and in the experience of many other students of the movements of cultured tissue cells, continuous microscopic observation in visible light at high magnification poses a problem for cell survival. To my knowledge, these problems have never been systematically examined and published. Nevertheless, the general experience seems to be that cultured cells under a 60× or a 100× objective lens respond to the necessarily high illumination density of the condenser with changes that suggest deterioration. For instance, wide lamellipodia with an equally wide front of ruffling tend to reduce the ruffles to smaller segments; mitoses rarely proceed past metaphase; and migrating cells often produce more spindly shapes and reduce the speed of locomotion. Frequently used countermeasures include the use of high-sensitivity video cameras or intermittent illumination. It is not yet clear which particular wavelengths are responsible for the changes in cell behavior and cell morphology upon continuous illumination. Obviously, the use of ultraviolet light instead of visible light would hardly improve the conditions of live-cell observation. The absorption of proteins and nucleic acids in the 200–300-Å range is likely to cause random chemical changes in the cells. The use of infrared would be unwise as well, because the heat radiation in the range of 5–20-μm wavelengths would not only cause heat effects in the cells, but the resolution of such a microscope would be reduced to objects larger than whole cell nuclei.

Consequently, I tested whether cells would tolerate illumination in the near-infrared. Using a normal phase-contrast microscope (which must not contain heat filters in the illumination pathway) with an appropriate filter, images with good contrast and resolution can be obtained using the invisible color range of 800–1100 nm (see Methods). I found that 3T3 cells exposed to continuous illumination under a 100× objective lens showed no visible changes of morphology and behavior over periods of 8 hours or longer. The line resolution of the microscope was better than 0.4 μm (Fig. 2). All subsequent observations were carried out with this microscope. As pointed out below, the observation of nuclear spinning was greatly facilitated by the use of the near-infrared microscope.

Effects of monensin on cell behavior

Monensin is a carboxylic ionophore with a 10-fold higher affinity for Na^+ than for K^+ ions (Pressman and Fahim 1982). For reasons not yet known, it seems to affect preferentially the membranes of the Golgi complex (Tartakoff and Vasalli 1978). Therefore, its major damaging effects on eukaryotic cells seem to be on secretory processes. For instance, the secretion of procollagen, fibronectin (Uchida et al. 1980), immunoglobulins (Tartakoff et al. 1977), and acetylcholinesterase (Similowitz 1979) are inhibited in the presence of monensin. With the exception of fibronectin (Yamada et al. 1976; Albrecht-Buehler and Chen 1977; Ali and Hynes 1978), none of these secreted proteins have been implicated in cell migration in vitro. Since the culture serum contains substantial amounts of fibronectin in the form of cold, insoluble globulin (CIG), one may not expect a major effect of monensin on cell locomotion in vitro. Indeed, at concentrations of 0.3 μM monensin and incubation times of 1 day, I found no inhibition of locomotion in 3T3 cells (Fig. 3) as judged by phagokinetic tracks (Albrecht-Buehler 1977) and observations of individual live cells. Within the first 24 hours of incubation, mitoses proceeded normally in 0.3 μM monensin. In cell growth experiments over 4 days, the proliferation of cells was reduced in a concentration-dependent manner. At

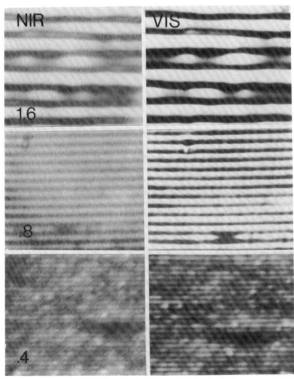

Figure 2 Line resolution in the near-infrared microscope (NIR) compared with visible light microscopy (VIS), using the same video display. The specimens are replicas of diffraction gratings. The line distance of 0.4 μm is still resolved in the NIR.

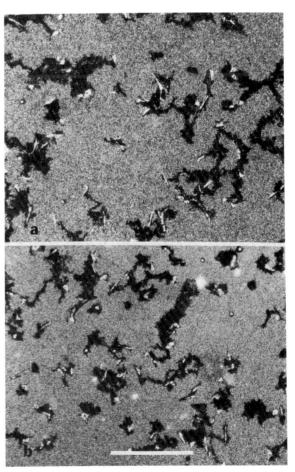

Figure 3 Phagokinetic tracks of 3T3 cells during 1 day of incubation in normal medium (a) and medium containing 0.3 μM monensin (b). Bar, 100 μm.

0.3 μM concentrations of monensin, growth was reduced by 60–80%.

Effects of monensin on cell morphology

In the presence of monensin, the cells underwent a number of morphological changes. After 1–2 days in 0.3 μM monensin, many cells appeared more rounded or spindle-shaped (Fig. 4d). The cell surfaces of mitotic cells seemed smoother than control cells, as judged by scanning electron microscopy (Fig. 4a,b), and the cytoplasm contained numerous vacuoles that were visible in phase-contrast microscopy and in favorable cases even in scanning electron microscopy (Fig. 4c). Although the cells showed a complete radial array of microtubules, the density of microtubules seemed to be reduced (cf. Figs. 5c, 6g,h). Transmission electron microscopy showed the expected vacuolarized morphology.

It should be emphasized, though, that these rather drastic changes appeared only after 1 day of incubation in 0.3 μM monensin. Most of the experiments about cell locomotion used incubation times of only a few hours, when most of these changes are not yet visible.

Live visualization of the centrosphere in migrating 3T3 cells

In nonmigrating cells (e.g., epithelial cells) in culture, the centrosphere is easily visible without any special treatment of the cells. It appears as a round area adjacent to the nucleus that is relatively free of granules and organelles. Mitochondria seem to be aligned along radii from this region. Most of the time the nuclei of such cells are bean-shaped and the centrosphere nests in the indentation of the nucleus. In fast-migrating fibroblasts, however, the nuclear shape changes often between a circular disk, a cigar shape, or a bean shape. In addition, organelles and vesicles appear to fill the entire perinuclear region. Therefore, the centrosphere is not visible in fast-migrating cells such as 3T3 cells.

Figure 5a shows live, confluent 3T3 cells that have been incubated in 1 μM monensin for 1 day. As a result the cells are rather vacuolarized. The vacuoles outline a round area adjacent to the nucleus that excludes the phase-dense granules and vacuoles (arrows in Fig. 5a,b). After acetone-methanol fixation (Fig. 5d) and staining against β-tubulin (Fig. 5c), the focus of the microtubules (i.e., the centrioles) was invariably found in the particular, specialized region (Fig. 6e,f,g,h). Since the specialized region of 3T3 cells after monensin treatment contains the centrioles and is relatively free of granules and organelles, I suggest to consider it as the centrosphere. The centrosphere can be observed continuously in living cells (Fig. 6a,b,c,d) until the morphological changes caused by monensin become unacceptable for the particular experimental question. Therefore, we are now for the first time in the position to monitor move-

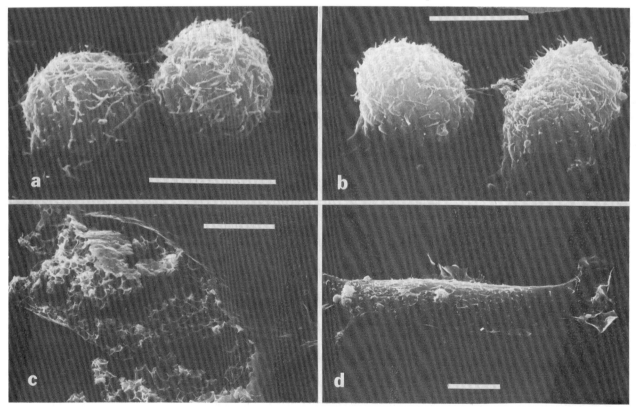

Figure 4 Scanning electron micrographs of 3T3 cells after 2 days of incubation in 0.3 μM monensin. Mitotic pairs (b) show a smoother surface than normal controls (a). Cells appear generally more spindly (d). The cytoplasm is filled with vacuoles, as shown by a cell that accidentally broke open (c). Bars, 10 μm.

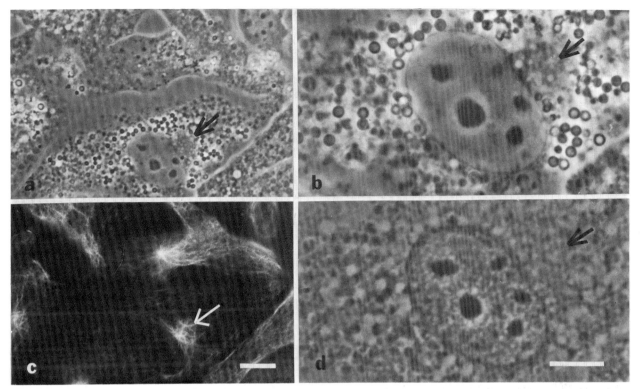

Figure 5 Visualization of the centrosphere in 3T3 cells incubated in 1 μM monensin for 1 day. (←) Centrosphere. (a) Living cells; (c) same cells fixed and stained with anti-β-tubulin, showing that the focal point of the microtubules coincides with the specialized area in a; (b) same cell as in a, live and at higher magnification; (d) same cell after methanol-acetone fixation. Bars, 10 μm.

ments of the centrosphere in migrating cells and to try to relate them with the locomotion of the cell. In particular, it will be important to determine whether dislocations of the centrosphere precede or follow directional changes of the cell. If the centrosphere is involved in a control function of migration, one would expect the centrospheric movements to precede a change of direction in migrating cells.

Movements of the centrosphere

The test of the above question will take considerable numbers of experiments before the statistical sample is large enough for an answer. At least it can be demonstrated that the centrosphere is able to migrate within 3T3 cells. Figure 7 shows a 3T3 cell in contact with another 3T3 cell. In the course of the experiment, the centrosphere migrates clockwise around the nucleus to the opposite (free) side of the cell, while the cell remains in contact with the other cell (upper left corner of Fig. 7). In this case the movement of the centrosphere precedes the change of direction that the cell will make in order to move away from the contact area. The centrosphere in Figure 7 is less well defined than in Figures 5 and 6 because the cell was incubated for only 1 hour in 1 μM monensin at the start of the experiment. It should be noted that the nucleus does not rotate synchronously with the centrosphere's circular movement around the periphery. Judging by the position of the nucleoli, the nucleus rotates clockwise by about 90 deg, followed by a counterclockwise rotation of about 30 deg.

Spinning of the nucleus

In the course of these experiments, such rotations of the nucleus occurred numerous times. The most pronounced rotations occurred after 1-day incubations of 3T3 cells in 0.3–0.5 μM monensin. The analysis of 55 experiments including controls showed that visual light decreased the average speed of rotation by about 50%. Therefore, all studies of nuclear spinning were carried out in the near-infrared microscope (Fig. 8).

Under these conditions average speeds reached up to 7 deg/min, with total rotations of up to 2.5 full revolutions of the nucleus. With the exceptions of early prophase "rolling" of the nucleus, all other nuclear rotations revolved strictly around axes perpendicular to the substrate. The spin could be counterclockwise or clockwise. In the cases of large rotations, the cells maintained the same direction of rotation throughout the entire episode, without wobbling.

The cytoplasmic granules, even those as close as the limit of resolution of the microscope did not participate in the rotation. As mentioned above, the centrosphere did not participate in the rotation either. Therefore, the nucleus in these particular cells appeared as a free, isolated compartment. The mechanism of rotation appeared to be either restricted to a very thin layer around the nuclear envelope, or to a mechanism that had no effect on granular saltations.

The nuclear rotations appear all the more remarkable because migrating 3T3 cells usually maintain the nucleus

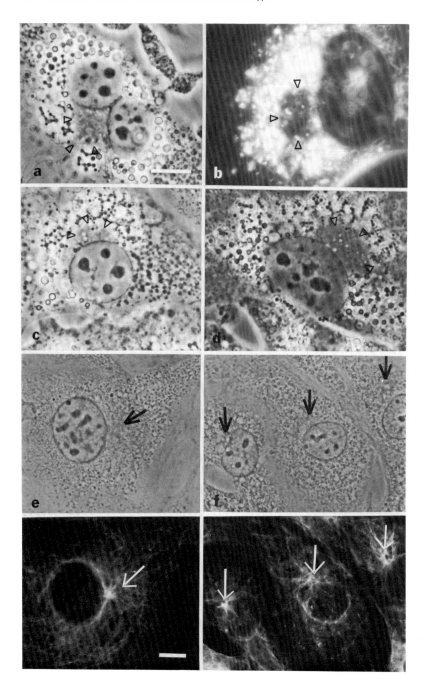

Figure 6 Examples of live-cell visualization of the centrosphere in mono- and binucleate 3T3 cells. Arrowheads outline the centrospheres; arrows point to the centrospheres in immunofluorescence micrographs. Incubation times in monensin: 0.3 μM for 5 hr (b); 1 μM for 3 days (a,c,d) and for 4 hr (e,f). (a,c,d) Live cells in phase-contrast microscopy; (b) live cell in darkfield microscopy, demonstrating the relative absence of light-scattering granules from the centrospheric region; (e,f) coincidence between specialized areas and focus of microtubules. Bars, 10 μm.

in a fixed orientation relative to the substrate despite drastic morphological changes, considerable deformations of their nuclei, and large directional changes.

Geometric factors

Rotating nuclei are extremely rare in normal 3T3 cultures. One reason may be found in the usually stretched shape of the nucleus during cell migration. Considering rotational speeds of less than 1 deg/min as background, the statistical analysis of 55 cases showed that rotation required almost circular outlines of the nucleus. If the ellipticity of the nucleus (largest dia/shortest dia) as measured in perpendicular projection of the microscope was larger than 1.3, only background rotation could be observed. Since the flattened cells squeeze their nuclei flat, the required shape for rotation of the nucleus is a circular disk. Consequently, the experimental protocol of studying nuclear rotation consists of three requirements:

1. 1-day incubation in 0.3 μM monensin;
2. use of near-infrared microscopy;
3. selection of cells with circular outlines of the nucleus.

Under these conditions we found a 1:3 chance of pre-

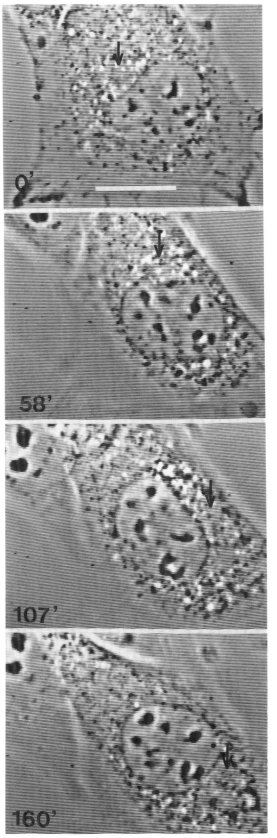

Figure 7 Movement of the centrosphere (↓) in a 3T3 cell, beginning after 2 hr in 1 μM monensin, as viewed with the near-infrared microscope. Bar, 10 μm.

dicting the rotation of the nucleus. This rate of prediction makes it feasible to experiment with at least single cells that express nuclear rotation.

Considering that bean-shaped nuclei do not rotate and that the bean shape may be caused by the centrosphere indenting the nuclear envelope, one may suspect that the position of the centrosphere is important for nuclear spinning. After 1 day of incubation in 0.3 μM monensin, 17 individual cells with rotating nuclei (rotation above background) were fixed and stained with anti-tubulin antibodies. All nuclei had circular outlines. In 60% of the cases, the centrioles appeared to the side of the nuclei without indenting them. In 40% of the cases, the centrioles were located inside the circular nuclear outlines, including 4 cases of central location above or below the nuclei. Random sampling of 154 other cells in these cultures yielded the same percentage of centriolar location, suggesting that the position of the centrioles is not related to the expression of nuclear spinning.

Inhibition of nuclear spinning

Within 30 minutes of incubation in 1 μg/ml colcemid, monensin-treated 3T3 cultures no longer contained cells with spinning nuclei. In addition to superficial screening of entire cultures, 9 individual cells with circular outlines of their nuclei, which had been incubated in 1 μg/ml colcemid for 1–4 hours, were observed for 1–2 hours in the near-infrared microscope. The rotational speeds were less than 0.2 deg/min. Other drugs or medium factors have not yet been tested.

Nuclear spinning under normal circumstances

Pomerat's original report of nuclear spinning did not expose the cells to drugs or special wavelengths of illumination (Pomerat 1953). Furthermore, as mentioned above, nuclei in cells may begin to roll and rotate during early prophase. In two cases I observed 3T3 cells with spinning nuclei in the absence of any drugs upon collision with another cell (Fig. 8c–d). Although the near-infrared microscope was used in these observations, it appears from the above results that nuclear spinning is not an artifact of monensin or near-infrared radiation. The extent of nuclear spinning seems to depend on the cell type. The primary epithelial cells studied by Pomerat expressed much higher rotational speeds of 288 deg/min with up to 16 full revolutions.

Discussion

Visualization and movement of the centrosphere

It is not yet clear why monensin mediates the visibility of the centrosphere in migrating cells. It is possible that the Na^+ influx into the Golgi apparatus caused by the ionophore inflates the Golgi cisternae and expels granules and organelles from the region. Since the Golgi apparatus is frequently found near the centrioles, the area above and below the centrosphere may become free of the phase-dense particles that hamper the vision.

The migration of the centrosphere within the cytoplasm poses again the unsolved question of the dynamics of microtubules. Is the microtubular network dragged

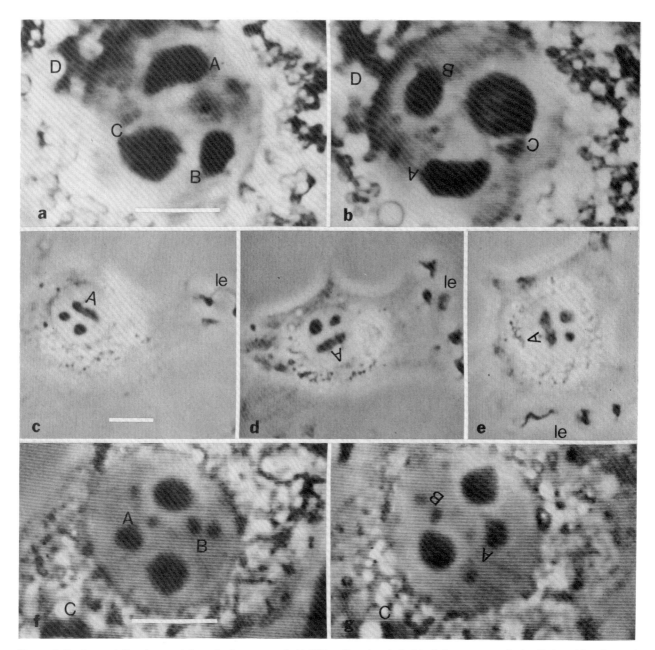

Figure 8 Nuclear rotation in near-infrared microscopy. (a,b) 3T3 cell preincubated in 0.3 μM monensin for 2 days (time interval between a and b is 48 min); (f,g) similar conditions as a and b, with a time interval of 62 min; (c,d,e) nuclear rotation in the absence of monensin in a 3T3 cell accompanying a turn of direction. (le) Leading lamellipodium. Times are 0 min (c), 42 min (d), and 134 min (e). Bars, 10 μm.

along with the moving centrioles or are they undergoing continuous assembly-disassembly processes? Interrupting centrospheric migration and staining the microtubules will be a possible way to answer the question.

Other new approaches to the problems of migrational control that are now possible include laser-microirradiation of centrospheres during locomotion of cells. If centrioles are involved in controlling the directionality of cells, laser ablation of centrioles should destroy directionality. Related experiments can be carried out using microneedles and microinjection of surplus centrioles.

Nuclear spinning: Possible interfaces of rotation

It is not yet clear between which interfaces the rotation occurs. Theoretically, it seems possible that the chromatin slides along the inner nuclear membrane. This possibility seems unlikely in view of the apparent intimate contact between heterochromatin and inner nuclear membrane. Alternatively, one may think of a sliding motion between the inner and outer nuclear membranes. This possibility appears very unlikely as well since the two membranes are bridged by numerous nuclear pores. Finally, it seems possible that the entire nucleus rotates relative to the surrounding cytoplasm. The objection to this possibility is the well-documented continuity between the nuclear envelope and the elaborate system of the endoplasmic reticulum.

The last objection seems to be the easiest to overcome because it is not certain that the endoplasmic reticulum is always connected to the nuclear envelope. The connection may be dynamic, i.e., it may reversibly open and break under certain conditions. Nuclear rotation occurs during early prophase when the nuclear envelope is about to disassemble. The increased frequency of occurrence of nuclear rotation in the presence of monensin may also point to a breakage of these membrane connections. The vacuolarization of the cytoplasm and the specific action of monensin on the Golgi apparatus seems consistent with a break between nuclear envelope and endoplasmic reticulum. Also, the occasional occurrence of nuclear rotation upon cell-cell collision might involve a break of the connection between nuclear envelope and endoplasmic reticulum, because the change of polarity of the cell may involve major cytoplasmic rearrangements.

Nuclear-cytoplasmic connections

The phenomenon of nuclear rotation poses the question of the mechanical ties between nucleus and cytoplasm. Not only is there the proven continuity between nuclear envelope and endoplasmic reticulum, but actin also seems involved somehow in positioning the nucleus within a cell, because cytochalasin B enucleates cells (Carter 1967), and actin is a major target for cytochalasin B (MacLean-Fletcher and Pollard 1980). There are reports of cages of intermediate filaments surrounding the nucleus, particularly during mitosis (Zieve et al. 1980). However, other cells showed only an equatorial torus of intermediate filaments in interphase and in mitosis (Blose and Chacko 1976) or no cages at all during mitosis (Aubin et al. 1980). Another report suggests that individual intermediate filaments attach to nuclear pores (Jones et al. 1982). There is also the claim of cytoplasmic fibers penetrating the nuclear envelope and forming a continuous nuclear-cytoplasmic matrix (Penman et al. 1982). From the report of an attachment between centrioles and nuclear envelope (Bornens 1977), one might think of a role of microtubules in the mechanical positioning of the nucleus.

The phenomenon of nuclear rotation does not contradict any one of these observations and claims, unless they interpret the mechanical ties between nucleus and cytoplasm as permanent. It seems difficult to reconcile the permanence of such connections with the observation of several full revolutions of the nucleus.

Possible significance of nuclear rotation

Assuming that there must be breaks of mechanical ties between nucleus and cytoplasm for nuclear rotation to occur, another necessary condition is the disk shape of the nucleus. The simplest explanation for this requirement may be that any other nuclear shape offers too much frictional resistance in the surrounding cytoplasm. Nuclei of disk shape are rare in cultured migrating cells, and therefore it is unlikely that nuclear rotation has a biological function of its own. Of course, the above in vitro observations cannot exclude that nuclear rotation occurs in vivo and fulfills an important biological function in the tissue. Nevertheless, the in vitro nuclear rotation points to the presence of an unknown interaction between nucleus and cytoplasm that may be present all the time but expresses itself as nuclear spinning only under the special conditions mentioned above.

Are saltatory movement and nuclear rotation related?

An exciting possibility of such interaction may be a mechanism that is related to the traffic of protein and RNA between nucleus and cytoplasm. In this respect, the inhibition of nuclear spinning by colcemid is interesting, because colcemid inhibits saltatory movements in the cytoplasm (Rebhun 1972). It is not yet clear whether saltatory movements and cytoplasmic traffic of macromolecules are related. Nevertheless, convulsive liquid streams exist in the cytoplasm of 3T3 cells and other cultured fibroblasts that may lead to blebbing or saltations depending on the flow resistance (Albrecht-Buehler 1982b). These streams cannot help but carry macromolecules between points of the cytoplasm, including areas bordering on the nuclear envelope.

How can the liquid streams lead to nuclear rotation? Assuming that they are involved in the macromolecular traffic between nucleus and cytoplasm, these streams must pass through the nuclear envelope (via pores?) in a presumably radial direction. Figure 9a depicts the situation schematically. If by accident one or more of the streams are deflected (Fig. 9b), a tangential frictional or electrical force component arises that may rotate the

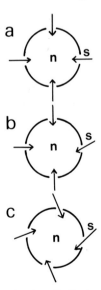

Figure 9 Schematic of a hypothetical mechanism to explain nuclear rotation by streams between nucleus and cytoplasm. (a) Normal radial direction of influxes; (b) accidental disturbance of one stream, giving rise to a tangential force; (c) slight turn of the nucleus in response to the tangential force, thus amplifying the tangential force component by deflecting all other streams.

nucleus by a small angle. Consequently, all other streams become deflected in the same way, thus amplifying the tangential force component (Fig. 9c). As a result the nucleus may now begin to be driven around in the accidental direction of the initial deflection, until other cytoplasmic events interrupt this self-reinforcing streaming pattern. This model considers only the streams flowing from the cytoplasm into the nucleus. The streams that flow out of the nucleus into the cytoplasm can propel the nucleus only through recoil like a jet. Otherwise, basic principles of mechanics require that their collective forces on the nuclear envelope cancel each other out. It is unlikely that there is a substantial recoil of the efferent streams but if there were, it would act parallel to the putative frictional or electrical force of the afferent streams, thus reinforcing their effect.

If this model can be experimentally supported, nuclear rotation may provide the first experimental handle on the spatial pathways of the macromolecular traffic between nucleus and cytoplasm. It is this hope that provided the reason to include these observations in a book concerned with basic aspects of cancer.

Acknowledgments

I wish to thank Mrs. Christiane Andrews for her valuable assistance. The work was supported by grant CA-30085 from the National Cancer Institute and a grant from the Searle Leadership Fund for the Life Sciences.

References

Albrecht-Buehler, G. 1977. The phagokinetic tracks of 3T3 cells. *Cell* **11**: 395.
———. 1980. The autonomous movements of cytoplasmic fragments. *Proc. Natl. Acad. Sci.* **77**: 6639.
———. 1981. Does the geometric design of centrioles imply their function? *Cell Motil.* **1**: 237.
———. 1982a. Control of tissue cell movement. *Natl. Cancer Inst. Monogr.* **60**: 117.
———. 1982b. Does blebbing reveal the convulsive flow of liquid and solutes through the cytoplasmic meshwork? *Cold Spring Harbor Symp. Quant. Biol.* **46**: 45.
Albrecht-Buehler, G. and L.B. Chen. 1977. Local inhibition of centripetal particle transport where LETS protein patterns appear on 3T3 cells. *Nature* **266**: 454.
Ali, I.U. and R.O. Hynes. 1978. Effects of LETS glycoprotein on cell motility. *Cell* **14**: 439.
Aubin, J.E., M. Osborn, W.W. Franke, and K. Weber. 1980. Intermediate filaments of the vimentin-type and the cytoskeleton-type are distributed differently during mitosis. *Exp. Cell Res.* **129**: 149.
Blose, S.M. and S. Chacko. 1976. Rings of intermediate (100 Å) filament bundles in the perinuclear region of vascular endothelial cells. *J. Cell Biol.* **70**: 459.
Bornens, M. 1977. Is the centriole bound to the nuclear membrane? *Nature* **20**: 80.
Carter, S.B. 1967. Effects of cytochalasins on mammalian cells. *Nature* **213**: 261.
Folkman, J. and A. Moscona. 1978. Role of cell shape in growth control. *Nature* **273**: 345.
Goldman, R.D., R. Pollack, and N.H. Hopkins. 1973. Preservation of normal behavior by enucleated cells in culture. *Proc. Natl. Acad. Sci.* **70**: 750.
Jones, J.C.R., A.E. Goldman, P.M. Steinert, S. Yuspa, and R.D. Goldman. 1982. Dynamic aspects of the supramolecular organization of intermediate filament networks in cultured epidermal cells. *Cell Motil.* **2**: 197.
Lazarides, E. 1980. Intermediate filaments: Integrators of intracellular space. *Nature* **283**: 249.
MacLean-Fletcher, S. and T.D. Pollard. 1980. Mechanism of action of cytochalasin B on actin. *Cell* **20**: 329.
Penman, S., A. Fulton, D. Capco, A. Ben Ze'ev, S. Wittelsberger, and C.F. Tse. 1982. Cytoplasmic and nuclear architecture in cells and tissue: Form, functions, and mode of assembly. *Cold Spring Harbor Symp. Quant. Biol.* **46**: 1013.
Pomerat, C.M. 1953. Rotating nuclei in tissue cultures of adult human nasal mucosa. *Exp. Cell Res.* **5**: 191.
Pressman, B.C. and M. Fahim. 1982. Pharmacology and toxicology of the monovalent carboxylic ionophores. *Annu. Rev. Pharmacol. Toxicol.* **22**: 465.
Rebhun, L.I. 1972. Polarized intracellular particle transport: Saltatory movements and cytoplasmic streaming. *Int. Rev. Cytol.* **32**: 93.
Similowitz, H. 1979. Monovalent ionophores inhibit acetylcholinesterase release from cultured chick embryo skeletal muscle cells. *Mol. Pharmacol.* **16**: 202.
Small, J.V. 1981. Organization of actin in the leading edge of cultured cells: Influence of osmium tetroxide and dehydration on the ultrastructure of actin meshworks. *J. Cell Biol.* **91**: 695.
Tartakoff, A. and P. Vassali. 1978. Comparative studies of intracellular transport of secretory proteins. *J. Cell Biol.* **79**: 694.
Tartakoff, A., P. Vassali, and M. Detraz. 1977. Plasma cell immunoglobulin secretion. Arrest is accompanied by alterations of the Golgi complex. *J. Exp. Med.* **146**: 1332.
Uchida, N., H. Similowitz, P.W. Ledger, and M.L. Tanzer. 1980. Kinetic studies of the intracellular transport of procollagen and fibronectin in human fibroblasts. *J. Biol. Chem.* **225**: 8638.
Yamada, K.M., S.S. Yamada, and I. Pastan. 1976. Cell surface protein partially restores morphology, adhesiveness, and contact inhibition of movement to transformed fibroblasts. *Proc. Natl. Acad. Sci.* **73**: 1217.
Zieve, G.W., S.R. Heidemann, and J.R. McIntosh. 1980. Isolation and partial characterization of a cage of filaments that surrounds the mammalian mitotic spindle. *J. Cell Biol.* **87**: 160.

Onco-fetal Genes and the Major Histocompatibility Complex

P.W.J. Rigby, P.M. Brickell, D.S. Latchman, D. Murphy, M.R.D. Scott*, K.-H. Westphal, V. Simanis, D.P. Lane, and K. Willison†

Cancer Research Campaign, Eukaryotic Molecular Genetics Research Group, Department of Biochemistry, Imperial College of Science and Technology, London SW7 2AZ, England; †Chester Beatty Laboratories, Institute of Cancer Research, London SW3 6JB, England

One viral or cellular gene product is generally sufficient to transform cells of an established line to the malignant phenotype (Tooze 1981; Cooper 1982; Weiss et al. 1982), whereas transformation of primary cell cultures requires at least two viral or cellular genes acting in concert (Rassoulzadegan et al. 1982; Land et al. 1983; Ruley 1983). Even in the latter case it seems unlikely that all of the biological and biochemical changes that distinguish the transformed cell from its normal parent can result from direct actions of the transforming proteins, and there must therefore be changes in cellular gene expression that contribute to the transformed phenotype. The first evidence that tumor cells may express genes not active in the corresponding normal cell came from the discovery of onco-fetal antigens (Schone 1906). A number of tumor cells express antigenic determinants normally restricted to embryonic cells, but such antigens are thought to be found only in tumors derived from the embryonic cell type in which their expression is appropriate (Alexander 1972). More recent work using nucleic acid hybridization techniques has shown that there are detectable differences between the cytoplasmic mRNA populations of an SV40-transformed human cell line and its normal parent (Williams et al. 1977) and that the transformation of primary chicken embryo fibroblasts by Rous sarcoma virus is accompanied by the activation of a large number of transcription units (Groudine and Weintraub 1980).

We have recently developed techniques that allow the cloning of cellular genes activated in transformed cells even when the corresponding mRNAs are of relatively low abundance in the transformed cell (Scott et al. 1983). We constructed cDNA libraries from the SV40-transformed line SV3T3 Cl38 and from the normal parental line BALB/c 3T3 A31 and then used preparative solid-phase and solution hybridization procedures to obtain a probe significantly enriched in transformed cell–specific sequences. This probe was used to screen the SV3T3 Cl38 cDNA library, and we isolated 42 clones that were divided into five sets on the basis of restriction enzyme mapping and cross-hybridization experiments (Scott et al. 1983; Rigby et al. 1984). Clones of a given set are thus derived from the same or closely related mRNAs.

The prototype cDNA clone of set 1, pAG64, contains a dispersed repetitive element present thousands of times in the mouse genome and hybridizes to three SV3T3 Cl38 mRNAs of 1.6 kb, 0.7 kb, and 0.6 kb (Scott et al. 1983). We have shown that pAG64 is a cDNA copy of the 1.6-kb mRNA and that it hybridizes to the two small RNAs, which are the products of a separate transcription unit(s), only because they also contain the repetitive element (Brickell et al. 1983). Elevated levels of the three mRNAs defined by set 1 are a general feature of mouse fibroblasts transformed not only by SV40, but also by other DNA tumor viruses, by retroviruses, and by chemical carcinogens (Scott et al. 1983). Moreover, elevated levels of these three mRNAs are found in a wide variety of transformed mouse cell lines and tumors of nonfibroblast origin (Brickell et al. 1983). These data are summarized in Table 1.

To understand the role that these genes may play in transformation, it would be of great value to know when they are expressed during the life cycle of the mouse. A

Table 1 Expression of Set-1-related mRNAs in Mouse Transformed Cell and Tumor Cell Lines

Cell line	1.6 kb	0.7/0.6 kb
BALB/c 3T3	−	−
SV3T3 Cl20	+	+/−
SV3T3 Cl26	+	+
SV3T3 Cl38	+	+
SV3T3 Cl49	+	+/−
SV3T3 ClH	+	+
SV3T3 ClM	+	+/−
Py3T3	+	+
Py*tsa*3T3	+	−
RSV3T3 TK3 BXB4	+	+/−
RSV3T3 BC6	+	+/−
ANN-1	+	+
BALB/c AMuLV A1R1	+	−
BALB/c 10ME HD A5R1	+	+
BALB/c 10Cr MC A2R1	+	+
AKR thymoma, BW5147.A11	+	+
Mastocytoma, P815	+++	+
Friend erythroleukemia GM0086D	++	+
Myeloma, SP2/0 Ag14	+	+
AMuLV-induced B-cell tumor	+++	−
Neuroblastoma, PLATT	+	+
Rat hepatoma, FAZA	+	+

*Present address: Department of Microbiology and Immunology, University of California Medical School, San Francisco, California 94143

1.6-kb mRNA that hybridizes to pAG64 is found at low levels in thymus, but we have not observed significant hybridization to cytoplasmic RNA extracted from a wide variety of other adult tissues (Murphy et al. 1983). However, when we analyzed RNA isolated from mouse embryos at 8.5 days and 9.5 days postcoitum, we found that pAG64 hybridizes to a large number of polydisperse RNAs (Murphy et al. 1983). The expression of these set-1-related RNAs is quantitatively regulated during development, increasing rapidly from 6.5 days to a peak at 9.5 days and then by 16 days declining to the low level observed in the adult organism (Murphy et al. 1983). We have also shown that a similar pattern of polydisperse, polyadenylated, cytoplasmic RNAs is found in pluripotential embryonal carcinoma (EC) and EK (Evans and Kaufman 1981) cells. When EK cells are induced to differentiate into embryoid bodies, there is an overall repression of the majority of the set-1-related mRNAs, although a small transcript(s) is considerably elevated. The nullipotential EC line F9 can differentiate only into endoderm. F9 cells exhibit a much more restricted pattern of set-1-related mRNAs, suggesting that partial repression of genes containing the set-1 repeat accompanies commitment. When such cells are induced to differentiate into visceral or parietal endoderm, almost complete repression of these genes occurs (Murphy et al. 1983).

The set-1 repeat thus identifies a large number of embryonic transcription units that are quantitatively regulated both during embryonic development in vivo and during the in vitro differentiation of embryonic cell lines. A subset of these genes are reactivated in all mouse transformed and tumor cell lines tested and thus define onco-fetal markers of great generality.

We present here an identification of one of the set-1 genes, discuss data that suggest that activation of this gene may be directly mediated by the viral transforming protein, and consider in some detail the relationships between genes activated in tumor cells, embryogenesis, and the immunological mechanisms involved in tumor rejection.

Experimental Procedures

Eukaryotic cells
The origin and growth of the eukaryotic cells used have been described previously (Rigby et al. 1980; Murphy et al. 1983; Scott et al. 1983).

Nucleic acids
The methods by which the cDNA clones were constructed and the procedures used in their analysis are described by Scott et al. (1983) and Murphy et al. (1983).

DNA sequencing
Nucleotide sequences were determined by the procedure of Maxam and Gilbert (1980), using DNA fragments terminally labeled with reverse transcriptase or terminal transferase.

Immunoprecipitation assay for SV40 large-T-antigen binding to DNA
The procedure was essentially as described in McKay (1981) and Fanning et al. (1982). Three 90-mm plates of subconfluent SV3T3 Cl38 or BALB/c 3T3 cells were extracted in 200 µl of buffer (50 mM Tris·HCl, pH 8, 120 mM NaCl, 1 mM $MgCl_2$, 0.5 mM DTT, 1 mM PMSF, 0.5% NP-40) for 15 minutes on ice. The cleared lysates were diluted 10-fold in binding buffer (20 mM MOPS, pH 7.2, 40 mM NaCl, 1 mM $MgCl_2$, 2.5% glycerol, 1 mM PMSF). Eighty µl of the dilute extract were incubated with 20 µl of end-labeled HinFI-digested pAG64 DNA (in binding buffer) for 15 minutes on ice, then 100 µl of a monoclonal anti-T antibody was added (pAb419; Harlow et al. 1981). After 30 minutes the immune complexes were harvested onto 25 µl of 10% (w/v) fixed Staphylococcus aureus cells. The complexes were washed twice in binding buffer, then twice in binding buffer containing 100 mM NaCl. The final pellet was eluted in 20 µl of 1% (w/v) SDS and 10 mM EDTA at 65°C for 30 minutes, and the eluate was loaded onto a 1.5% agarose gel. After electrophoresis, the gel was fixed in 10% acetic acid, dried, and autoradiographed at −70°C.

Results

pAG64 derives from a major histocompatibility complex gene
We have determined the complete nucleotide sequence of pAG64, the prototype clone of set 1, which corresponds to the 1.6-kb mRNA present at elevated levels in all mouse transformed and tumor cell lines tested. The sequence is given in Figure 1. To our considerable surprise, comparison of this sequence with known nucleotide sequences indicates that pAG64 is a cDNA copy of an mRNA encoding an antigen of the major histocompatibility complex (MHC). Indeed, a sequence almost identical to that of pAG64 has previously been reported by Lalanne et al. (1982), who assigned it as an H-2-like gene of the d haplotype. For reasons given in detail below, we believe that the gene resides in the Qa/Tla region of the MHC.

The coding region of pAG64 is highly homologous to all published sequences for MHC antigens. However, the set-1 repetitive element is found in only some MHC RNAs. It is absent from another clone isolated by Lalanne et al. (1982) and assigned by them as H-2K^d and from the H-2K^b sequences reported by Reyes et al. (1982b) and by Weiss et al. (1983). At least a part of the repetitive element is present in a third clone described by Lalanne et al. (1982), in pH-2II of Steinmetz et al. (1981b), in an H-2D^b sequence reported by Reyes et al. (1982a), and in an H-2L^d sequence reported by Moore et al. (1982). In these cases the published sequences do not extend to the 3' end of the mRNA, but it seems likely that the remainder of the repetitive element and the additional 3' noncoding sequences are present in the corresponding mRNAs. Steinmetz et al. (1981a) have characterized a genomic clone, called 27.1, which appears to be a pseudogene and which they have assigned to the Qa region

```
  1          10          20          30          40          50          60          70          80          90
AGA TAT GAG CCG CGG GCG CGG TGG ATA GAG CAG GAG GGG CCG GAG TAT TGG GAG CGG GAG ACA CGG AGA GCC AAG GGC AAT GAG CAG AGT
ARG TYR GLU PRO ARG ALA ARG TRP ILE GLU GLN GLU GLY PRO GLU TYR TRP GLU ARG GLU THR ARG ARG ALA LYS GLY ASN GLU GLN SER
                        100         110         120         130         140         150         160         170         180
TTC CGA GTG GAC CTG AGG ACC GCG CTG CGC TAC TAC AAC CAG AGC GCG GGC TCT CAC ACA CTC CAG TGG ATG GCT GGC TGT GAC GTG
PHE ARG VAL ASP LEU ARG THR ALA LEU ARG TYR TYR ASN GLN SER ALA GLY GLY SER HIS THR LEU GLN TRP MET ALA GLY CYS ASP VAL
                        190         200         210         220         230         240         250         260         270
GAG TCG GAC GGG CGC CTC CTC CGC GGG TAC TGG CAG TTC GCC TAC GAC GGC TGC GAT TAC ATC GCC CTG AAC GAA GAC CTG AAA ACG TGG
GLU SER ASP GLY ARG LEU LEU ARG GLY TYR TRP GLN PHE ALA TYR ASP GLY CYS ASP TYR ILE ALA LEU ASN GLU ASP LEU LYS THR TRP
                        280         290         300         310         320         330         340         350         360
ACG GCG GCG GAC ATG GCG GCG CAG ATC ACC CGA CGC AAG TGG GAG CAG GCT GGT GCT GCA GAG AGA GAC CGG GCC TAC CTA GAG GGC GAG
THR ALA ALA ASP MET ALA ALA GLN ILE THR ARG ARG LYS TRP GLU GLN ALA GLY ALA ALA GLU ARG ASP ARG ALA TYR LEU GLU GLY GLU
                        370         380         390         400         410         420         430         440         450
TGC GTG GAG TGG CTC CGC AGA TAC CTG AAG AAC GGG AAT GCT ACG CTG CTG CGC ACA GAT CCC CCA AAG GCC CAT GTG ACC CAT CAC CGC
CYS VAL GLU TRP LEU ARG ARG TYR LEU LYS ASN GLY ASN ALA THR LEU LEU ARG THR ASP PRO PRO LYS ALA HIS VAL THR HIS HIS ARG
                        460         470         480         490         500         510         520         530         540
AGA CCT GAA GGT GAT GTC ACC CTG AGG TGC TGG GCC CTG GGC TTC TAC CCT GCT GAC ATC ACC CTG ACC TGG CAG TTG AAT GGG GAG GAG
ARG PRO GLU GLY ASP VAL THR LEU ARG CYS TRP ALA LEU GLY PHE TYR PRO ALA ASP ILE THR LEU THR TRP GLN LEU ASN GLY GLU GLU
                        550         560         570         580         590         600         610         620         630
CTG ACC CAG GAA ATG GAG CTT GTG GAG ACC AGG CCT GCA GGG GAT GGA ACC TTC CAG AAG TGG GCA TCT GTG GTG GTG CCT CTT GGG AAG
LEU THR GLN GLU MET GLU LEU VAL GLU THR ARG PRO ALA GLY ASP GLY THR PHE GLN LYS TRP ALA SER VAL VAL VAL PRO LEU GLY LYS
                        640         650         660         670         680         690         700         710         720
GAG CAG AAG TAC ACA TGC CAT GTG GAA CAT GAG GGG CTG CCT GAG CCC CTC ACC CTG AGA TGG GGC AAG GAG GAG CCT CCT TCA TCC ACC
GLU GLN LYS TYR THR CYS HIS VAL GLU HIS GLU GLY LEU PRO GLU PRO LEU THR LEU ARG TRP GLY LYS GLU GLU PRO PRO SER SER THR
                        730         740         750         760         770         780         790         800         810
AAG ACT AAC ACA GTA ATC ATT GCT GTT CCG GTT GTC CTT GGA GCT GTG GTC ATC CTT GGA GCT GTG ATG GCT TTT GTG ATG AAG AGG AGG
LYS THR ASN THR VAL ILE ILE ALA VAL PRO VAL VAL LEU GLY ALA VAL VAL ILE LEU GLY ALA VAL MET ALA PHE VAL MET LYS ARG ARG
                        820         830         840         850         860         870         880         890         900         910
AGA AAC ACA GGT GGA AAA GGA GGG GAC TAT GCT CTG GCT CCA GTG TGA AGACAGCTGCCTAGTGTGGACTTGGTGACAGACAATGTCTTCACACATCTCCTGTG
ARG ASN THR GLY GLY LYS GLY GLY ASP TYR ALA LEU ALA PRO VAL ***
  920         930         940         950         960         970         980         990        1000        1010        1020        1030
ACATCCAGAGACCTCAGTTCTCTTTAGTCAAGTGTCTGATGTTCCCTGTGAGTCTGCGGGCTCAAAGTGAAGAACTGTGGAGCCCAGTCCACCCCTGCACACCAGGACCCTATCCCTGCA
 1040        1050        1060        1070        1080        1090        1100        1110        1120        1130        1140        1150
CTGCCCTGTGTTCCCTTCCACAGCCAACCTTGCTGCTCCAGCCAAACATTGGTGGACATCTGCAGCCTGTCAGCTCCATGCTACCCTGACCTTCAACTCCTCACTTCCACACTGAGAATA
 1160        1170        1180        1190        1200        1210        1220        1230        1240        1250        1260        1270
ATAATTTGAATGTGGGTGGCTGGAGAGATGGCTCAGCGCTGACTGCTCTTCCAAAGGTCCTGAGTTCAAATCCCAGCAACCACATGGTGGCTCACAACCATCTGTAATGGGATCTAACAC
 1280        1290        1300        1310        1320        1330        1340        1350        1360        1370        1380        1390
CCTCTTCTGCAGTGTCTGAAGACAGCTACAGTGTACTTACATATAATAATAAATAAGTCTTTAAAAAATTAATTTGAAAGTGACCTTGATTGTTAACATCTTGAGCTAGGGCTGATTTCT
 1400        1410        1420        1430        1440        1450        1460        1470        1480
TGTTAATTCATGGATTGAGAATACTTAGAGGTTTGTTTGTTTGATTGATTGATTTTTTTGAAGAAATAAATGGCAGATGAAGGAACTTCA
```

Figure 1 Nucleotide sequence of pAG64. The sequence is translated in the only long, open reading frame. Underlining indicates the location of the set-1 repetitive element within the 3' untranslated region.

of the MHC. In their sequence the set-1 repetitive element is not present in what would be the 3' untranslated region of the mRNA, but rather there is a copy of the repeat in the opposite orientation within one of the introns. These relationships are summarized in Figure 2.

The 0.7-kb and 0.6-kb mRNAs that hybridize to the set-1 repetitive element are not dissimilar in size to the mRNAs that encode β_2-microglobulin, the protein found in the cell membrane in association with MHC antigens. However, the sequence of β_2-microglobulin (Parnes and Seidman 1982; Parnes et al. 1983) does not contain the set-1 repetitive element.

Does SV40 large T antigen bind to the set-1 repetitive element?

The set-1 repetitive element is found in the 3' untranslated region of at least two mRNAs present at elevated levels in SV40 transformed cells. Moreover, we have isolated genomic clones homologous to pAG64 and shown that two more copies of this repetitive element are located upstream of the coding region (D.S. Latchman et al., unpubl.). cDNA clones of set 2 define mRNAs quite distinct from those of set 1, they do not contain the repetitive element, and they hybridize to a limited number of discrete genomic DNA fragments. However, in the set-2 genomic clone that we have isolated, a copy of the set-1 repeat is located downstream from the transcription unit (D.S. Latchman et al., unpubl.). It seemed to us to be unlikely that this association of the repetitive element with at least three transcription units activated by SV40 transformation was fortuitous. Analysis of the set-1 mRNAs in cells transformed by tsA mutants of SV40 has shown that the activation of the set-1 genes is tightly coupled to the expression of the transformed phenotype and that it appears to require the presence of functional SV40 large T antigen (Scott et al. 1983).

We have therefore asked whether large T antigen can bind to the repetitive element. Figure 3 shows the results of a McKay (1981) immunoprecipitation assay using pAG64 DNA and T antigen from SV3T3 Cl38. A single DNA fragment is immunoprecipitated, and this is the fragment containing the repetitive element. Figure 4 shows the sequence of the set-1 repetitive element from pAG64 and it can be seen that this sequence contains three copies of the consensus T-antigen-binding pentanucleotide identified by De Lucia et al. (1983). The set-

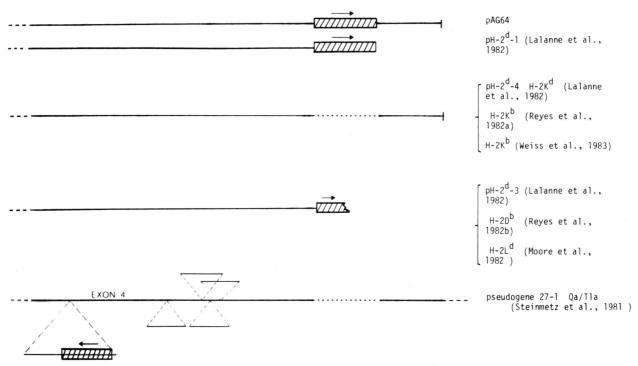

Figure 2 Relationships between pAG64 and DNA sequences from the MHC. (▨▨▨▨) The set-1 repetitive element; (———) deletions; (△) introns in the 27.1 pseudogene. Only the 3' portions of the various sequences are shown.

1 repetitive element also contains two copies of the enhancer core sequence (Hearing and Shenk 1983; Nordheim and Rich 1983), which are indicated in Figure 4. The juxtaposition of the T-antigen-binding sequences and of the enhancer core sequences within the set-1 repeat suggests that if this element is an enhancer, it may be regulated by large T antigen; experiments to test this possibility are in progress.

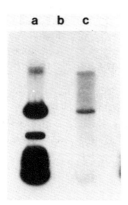

Figure 3 SV40 large-T-antigen binding to the set-1 repetitive element. 0.3 μg of HinfI-digested pAG64 end labeled with [α-^{32}P]TTP (3000 Ci/mmole; Amersham) and reverse transcriptase was incubated with either SV3T3 C138 or BALB/c 3T3 extract and immunoprecipitated with a monoclonal anti-T antibody as described in Experimental Procedures. (Lane a) Input DNA, 10 ng; (lane b) BALB/c 3T3 extract immunoprecipitate; (lane c) SV3T3 C138 extract immunoprecipitate.

Discussion

Identification of the 1.6-kb set-1 transcript as a class I MHC antigen mRNA

The nucleotide sequence of the cDNA clone pAG64, which was recovered from an SV40-transformed BALB/c 3T3 cell line, SV3T3 Cl38, (Scott et al. 1983), is almost completely homologous to clone pH-2^d-1 isolated from a lymphoid tumor, SL2, grown as ascites in DBA/2 mice (Lalanne et al. 1982). SL2 expresses H-$2D^d$, H-$2K^d$, H-$2L^d$, and TLA^d. The lack of complete amino acid sequence data for the H-$2K^b$ and H-$2K^d$ polypeptides and the fact that few of the molecularly cloned class I genes have been sequenced presently makes it impossible to identify the gene that encodes the transcript cloned in pAG64 and pH-2^d-1. Hybridization with the unique portion of pAG64 (i.e., only class I MHC antigen-coding sequences, not the repetitive element) to Northern blots of EC cell RNA (F9, PCC3, LT) shows a transcript of 1.6 kb (Murphy et al. 1983). This observation is inconsistent with the similar experiments of Morello et al. (1982) and Croce et al. (1981), particularly in the former case, since the pH-2^d-1 coding sequence was one of the probes used. It is possible that differences in hybridization stringency could account for this inconsistency.

It has been clearly demonstrated serologically that in vitro–cultured EC cells do not express classical H-2 antigens. EC cells of strain 129 origin (H-2^{bc}), F9, PCC3, PCC4, SIKR, were negative when tested with a variety of anti H-2^b reagents, some of broad and some of narrow specificity (Artzt and Jacob 1974; Stern et al. 1975; Ja-

```
5'
... GAATAATAATTTGAATGTGGGTGGCTGGAGAGATGGCTCAGCGCTGACTGC

    TCTTCCAAAGGTCCTGAGTTCAAATCCCAGCAACCACATGGTGGCTCACAA

    CCATCTGTAATGGGATCTAACACCCTCTTCTGCAGTGCTGAAGACAGCTAC

    AGTGTACTTACATATAATAATAAATAAGTCTTTAAA ...
```

Figure 4 Nucleotide sequence of the set-1 repetitive element from pAG64. The extent of the repetitive element was originally defined by comparing the sequences of pAG64 and pAG38. Four copies of the repeat have now been sequenced. Sequences related to the consensus large-T-antigen-binding pentanucleotide, G/ANGGC, are underlined. Sequences related to the consensus enhancer core sequence, TCAC^A/T^A/TCCA are indicated by dotted lines.

cob 1977). Conventionally, H-2 antigens are associated with β_2 microglobulin in the plasma membrane (reviewed by Ploegh et al. 1981) and, consistent with the absence of H-2 expression, EC cells appear to lack β_2 microglobulin in both serological and nucleic acid hybridization tests (Croce et al. 1981; Morello et al. 1982). Thus, it seems highly unlikely that we are detecting a conventional, left-end H-2 transcript when probing EC cell RNA with pAG64. Also, lowering the stringency of hybridization when probing SV3T3 Cl38 RNA allows the detection of another 1.6-kb transcript that is also present in the parental 3T3 line and presumably corresponds to the conventional H-2 transcript(s) found in most adult cells.

Recently, Winoto et al. (1983) showed that of the 36 class I genes (and pseudogenes) of the BALB/c mouse, only 5 map to the H-2 complex (H-2K, H-2D, H-2L). The remaining 31 map to the Tla complex, which is the distal part of the complex and includes the Qa-2,3, Tla, and Qa-1 genes (Fig. 5). This latter set of genes are differentiation antigens expressed on nucleated blood cells (Qa-2) and thymocytes (Tla) (for review, see Michaelson et al. 1984), unlike K, D, and L, which are expressed in most cell types, albeit at different levels, with the exception of EC cells, some cells of the early embryo, and sperm. The Tla complex shows both limited serological polymorphism and restriction enzyme site polymorphism. This is in stark contrast to the extensive polymorphism found for H-2K and D (see Klein et al. 1983). Transfection of cloned, class I antigen genes into mouse L cells can be used to identify genes in cases where antisera to the gene products are available. When all available antisera tested are unreactive, the ability of the cloned genes to raise the levels of surface β_2 microglobulin is taken as evidence for an active gene rather than a pseudogene. Of the five class I genes that map to H-2, three express the K, D, and L gene products and the other two are pseudogenes because they fail to increase β_2-microglobulin levels. Three of the Tla genes express serologically detectable gene products (Qa-2,3 and two Tl), and 10 more are active genes (Goodenow et al. 1982). Interestingly, and important to our argument, all 36 BALB/c class I genes and pseudogenes map to the MHC on chromosome 17 (Winoto et al. 1983).

We think it is likely that the 1.6-kb transcript(s) detected by the prototype sequence pAG64 in transformed cells (Scott et al. 1983) and EC cells (Murphy et al. 1983) are transcribed off H-2-like gene(s) mapping in Tla. An extensive amount of cDNA cloning and nucleotide sequencing of the 1.6-kb RNAs from different transformed cell lines will be required in order to assign the transcript(s) to specific class I genes in the Tla region.

Biological significance of the set-1-related transcripts

Transcripts containing the set-1 repetitive element are shared among cells of the early embryo, pluripotential EC cells, and transformed cells derived from cells of all three germ layers (Scott et al. 1983; Murphy et al. 1983). They therefore constitute a new class of general oncofetal marker that has not been described previously (see Alexander [1972] for discussion of onco-fetal antigens). It will clearly be of great interest to clone and sequence the numerous developmentally regulated, set-1-related transcripts present in early embryos.

The identification of pAG64 as a class I MHC antigen transcript raises important ideas about the possible functions of Tla genes in embryonic and tumor cells.

In the remaining discussion we shall make two major assumptions. First, the 1.6-kb transcript(s) detected by pAG64 is coded for by class I gene(s) mapping to Tla (i.e., the 31 distal MHC genes). Second, these transcripts are functional in the sense that polypeptides translated from them are expressed on the surface of cells.

The conventional H-2 complex, that region mapping between H-2K and H-2D and L, is involved in many immune phenomena. These range from allograft rejection to restriction of T-cell specificity. Zinkernagel and

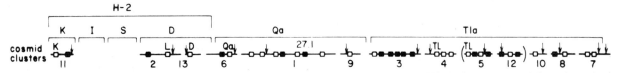

Figure 5 Map of the mouse MHC. The numbers below the map indicate cosmid clusters. Within the separate regions the order of the clusters with respect to one another is not known. Genes are indicated by boxes. Those shown as open boxes give elevated β_2-microglobulin expression in gene-transfer experiments; those indicated by closed boxes do not. (Reprinted, with permission, from Winoto et al. 1983.)

Doherty (1974a,b) showed that cytolytic T lymphocytes directed against cells infected with a virus in vivo could only recognize and kill other cells infected with the same virus in vitro if both infected cells were of the same H-2 genotype. This H-2 restriction of T-cell activity maps to the K and D regions. Since this discovery, a large body of work has confirmed and extended the finding, and a convincing case can be made for the physiological role of the H-2 complex: That is, the H-2 complex guides T lymphocytes in their function of distinguishing self from nonself (for review, see Klein et al. 1981). The Tla region was not included in Klein's thesis, probably because it was less well understood serologically and also because the gene-cloning studies showing Tla to be the larger genetic region (Winoto et al. 1983) had yet to be performed. We believe that our discovery of the expression of Tla genes in EC cells, embryos, and transformed cells can be accommodated within the Klein thesis and that it may explain some disparate observations in the field of tumor immunology.

Despite the fact that EC cells are serotypically negative for conventional H-2 antigens, they nevertheless express strong transplantation antigens, and several groups have investigated this allograft rejection by using various host/tumor genotypic combinations (Avner et al. 1978; Siegler et al. 1979; Ostrand-Rosenberg et al. 1980; Levine and Teresky 1981; Shedlovsky et al. 1981; Johnson et al. 1983). The genetic loci responsible for the rejection of EC tumors are named Gt (for graft of teratocarcinoma) and appear to map to both sides of H-2 on chromosome 17, Gt-1 mapping proximally (Levine and Teresky 1981; Shedlovsky et al. 1981), and Gt-2 mapping distally (Shedlovsky et al. 1981).

The Gt-1 loci behave like H-2 loci in the sense that F1 hybrid mice from tumor-accepting and tumor-rejecting parents accept the tumor in codominant expression. Similarly, like H-2 restriction, the rejection appears to be lymphocyte mediated. Recently it has appeared that previous data from one laboratory, based on apparent recombinants that were tumor sensitive but allogeneic for H-2 (i.e., differences in H-2 between EC cell line and recipient mouse), were the result of variants arising in the EC tumor lines (Johnson et al. 1983). Johnson et al. (1983) cannot separate Gt-1 and Gt-2 from H-2. We believe that the putative Tla RNA transcript(s) that we have detected in EC cells may in fact code for the target molecule(s) recognized by the cells that are cytotoxic for EC cells in vivo. What is the nature of the cytotoxic cells the activity of which is being measured in these allograft experiments? Stern et al. (1980) have shown that natural killer (NK) cells will lyse EC cells that lack H-2. NK cells differ from conventional H-2-restricted, cytotoxic T cells in that they show potent lytic activity against several tumor cell targets and also normal thymocytes and certain bone marrow stem cells (for review, see Kiessling and Wigzell 1979). Stern et al. (1980) suggested that the Gt loci may be regulatory genes for NK cell levels. However, Shedlovsky et al. (1981) argue against this idea on genetic grounds and suggest that Gt-1 and/or Gt-2 function in NK surveillance in the same way that H-2 does in T-cell surveillance. We further suggest that the 1.6-kb transcripts that we detect in EC cells code for Tla-like molecules that are Gt-1 and/or Gt-2 and that function as the targets for NK lysis of EC cells.

Furthermore, we note that numerous in vivo experiments show that prior challenge with 10-day embryo cells or EC cells confers weak protection against a subsequent challenge with virally or chemically transformed cells (Castro et al. 1974; Sikora et al. 1977; Medawar and Hunt 1978). It has been suggested by Medawar (1974) that cross-reactive antigens are responsible for these weak rejection reactions. The fact that EC cells, SV40-transformed cells, and methylcholanthrene-transformed cells all express the 1.6-kb Tla transcript provides an explanation for these phenomena.

The majority of the recently identified Tla genes are serologically undefined. The fact that sera reacting with these predicted tumor/embryonic antigens have not been described may not be surprising if they are involved in NK cell responses. We are presently using synthetic peptides predicted from the sequence of pAG64 to produce such sera.

Acknowledgments

D.M. and V.S. hold Research Studentships, and P.M.B. holds a Training Fellowship from the Medical Research Council. K.-H.W. is a Long-term Fellow of the European Molecular Biology Organisation. P.W.J.R. holds a Career Development Award from the Cancer Research Campaign. This research was paid for by grants from the Cancer Research Campaign and the Medical Research Council.

References

Alexander, P. 1972. Foetal antigens in cancer. *Nature* **235**: 137.

Artzt, K. and F. Jacob. 1974. Absence of serologically detectable H-2 on primitive teratocarcinoma cells in culture. *Transplantation* **17**: 632.

Avner, P.R., W.F. Dove, P. Dubois, J.A. Faillard, J.-L. Guenet, F. Jacob, H. Jakob, and A. Shedlovsky. 1978. The genetics of teratocarcinoma transplantation: Tumor formation in allogeneic hosts by the embryonal carcinoma cell lines F9 and PCC3. *Immunogenetics* **7**: 103.

Brickell, P.M., D.S. Latchman, D. Murphy, K. Willison, and P.W.J. Rigby. 1983. Activation of a Qa/Tla class 1 major histocompatibility antigen gene is a general feature of oncogenesis in the mouse. *Nature* **306**: 756.

Castro, J.E., R. Hunt, E.M. Lance, and P.B. Medawar. 1974. Implications of the fetal antigen theory for fetal transplantation. *Cancer Res.* **34**: 2055.

Cooper, G.M. 1982. Cellular transforming genes. *Science* **218**: 801.

Croce, C.M., A. Linnenbach, K. Huebner, J.R. Parnes, D.H. Margulies, E. Appella, and J.G. Seidman. 1981. Control of expression of histocompatibility antigen (H-2) and β_2-microglobulin in F9 teratocarcinoma stem cells. *Proc. Natl. Acad. Sci.* **78**: 5754.

De Lucia, A.L., B.A. Lewton, R. Tjian, and P. Tegtmeyer. 1983. Topography of simian virus 40 A protein DNA complexes: Arrangement of pentanucleotide interaction sites at the origin of replication. *J. Virol.* **46**: 143.

Evans, M.J. and M.H. Kaufman. 1981. Establishment in culture of pluripotential cells from mouse embryos. *Nature* **292**: 154.

Fanning, E., K.-H. Westphal, D. Brauer, and D. Corlin. 1982. Subclasses of simian virus 40 large T-antigen: Differential binding of two subclasses of T-antigen from productively infected cells to viral and cellular DNA. *EMBO J.* **1**: 1023.

Goodenow, R.S., M. McMillan, M. Nicolson, B.T. Sher, K. Eakle, N. Davidson, and L. Hood. 1982. Identification of the class I genes of the mouse histocompatibility complex by DNA-mediated gene transfer. *Nature* **300**: 231.

Gorman, C.M., L.F. Moffat, and B.H. Howard. 1982. Recombinant genomes which express chloramphenicol acetyltransferase in mammalian cells. *Mol. Cell. Biol.* **2**: 1044.

Groudine, M. and H. Weintraub. 1980. Activation of cellular genes by avian RNA tumor viruses. *Proc. Natl. Acad. Sci.* **77**: 5351.

Harlow, E., L.V. Crawford, D.C. Pim, and N.M. Williamson. 1981. Monoclonal antibodies specific for the SV40 tumor antigens. *J. Virol.* **39**: 861.

Hearing, P. and T. Shenk. 1983. The adenovirus type 5 *E1A* transcriptional control region contains a duplicated enhancer element. *Cell* **33**: 695.

Jacob, F. 1977. Mouse teratocarcinoma and embryonic antigens. *Immunol. Rev.* **33**: 3.

Johnson, L.L., L.J. Clipson, W.F. Dove, J. Feilbach, L.J. Maher, and A. Shedlovsky. 1983. Teratocarcinoma transplantation antigens are encoded in the *H-2* region. *Immunogenetics* **18**: 137.

Kiessling, R. and H. Wigzell. 1979. An analysis of the murine NK cell as to structure, function and biological relevance. *Immunol. Rev.* **44**: 165.

Klein, J., F. Figueroa, and C.S. David. 1983. *H-2* haplotypes, genes and antigens: Second listing. *Immunogenetics* **17**: 553.

Klein, J., A. Juretic, C.N. Baxeranis, and Z.A. Nagy. 1981. The traditional and a new version of the mouse *H-2* complex. *Nature* **291**: 455.

Lalanne, J.L., F. Bregegere. C. Delarbre, J.P. Abastado, G. Gachelin, and P. Kourilsky. 1982. Comparison of nucleotide sequences of mRNAs belonging to the mouse *H-2* multigene family. *Nucleic Acids Res.* **10**: 1039.

Land, H., L.F. Parada, and R.A. Weinberg. 1983. Tumorigenic conversion of primary embryo fibroblasts requires at least two cooperating oncogenes. *Nature* **304**: 596.

Levine, A.J. and A.K. Teresky. 1981. Teratocarcinoma transplantation rejection loci: Genetic localization of the *Gt-1* locus on chromosome 17 and the expression of alternate alleles. *Immunogenetics* **13**: 405.

McKay, R.D.G. 1981. Binding of a simian virus 40 T-antigen-related protein to DNA. *J. Mol. Biol.* **145**: 471.

Maxam, A.M. and W. Gilbert. 1980. Sequencing end-labeled DNA with base-specific chemical cleavages. *Methods Enzymol.* **65**: 499.

Medawar, P. 1974. Introduction to session on cross-reactive antigens. *Cancer Res.* **34**: 2053.

Medawar, P.B. and R. Hunt. 1978. Vulnerability of methylcholanthrene-induced tumors to immunity aroused by syngeneic foetal cells. *Nature* **271**: 164.

Michaelson, J., E.A. Boyse, M. Chorney, L. Flaherty, I. Fleisner, U. Hammerling, C. Reinisch, R. Rosenson, and F.W. Shen. 1984. The biochemical genetics of the Qa-Tla region. *Transplant. Proc.* (in press).

Moore, K.W., B.T. Sher, Y.H. Sun, K.A. Eakle, and L. Hood. 1982. DNA sequence of a gene encoding a Balb/c mouse L^d transplantation antigen. *Science* **215**: 679.

Morello, D., F. Daniel, P. Baldacci, Y. Cayre, G. Gachelin, and P. Kourilsky. 1982. Absence of significant *H-2* and β_2-microglobulin mRNA expression by mouse embryonal carcinoma cells. *Nature* **296**: 260.

Murphy, D., P.M. Brickell, D.S. Latchman, K. Willison, and P.W.J. Rigby. 1983. Transcripts regulated during normal embryonic development and oncogenic transformation share a repetitive element. *Cell* **35**: 865.

Nordheim, A. and A. Rich. 1983. Negatively supercoiled simian virus 40 DNA contains Z-DNA segments within transcriptional enhancer sequences. *Nature* **303**: 674.

Ostrand-Rosenberg, S., T.M. Rider, and A. Twarowski. 1980. Susceptibility of allogeneic mice to teratocarcinoma 402AX. *Immunogenetics* **10**: 607.

Parnes, J.R. and J.G. Seidman. 1982. Structure of wild-type and mutant mouse β_2-microglobulin genes. *Cell* **29**: 661.

Parnes, J.R., R.R. Robinson, and J.G Seidman. 1983. Multiple mRNA species with distinct 3' termini are transcribed from the β_2-microglobulin gene. *Nature* **302**: 449.

Ploegh, H.L., H.T. Orr, and J.L. Strominger. 1981. Major histocompatibility antigens: The human (*HLA-A,-B,-C*) and murine (*H-2K, H-2D*) class I molecules. *Cell* **24**: 287.

Rassoulzadegan, M., A. Cowie, A. Carr, N. Glaichenhaus, R. Kamen, and F. Cuzin. 1982. The roles of individual polyoma virus early proteins in oncogenic transformation. *Nature* **300**: 713.

Reyes, A.A., M. Schold, and R.B. Wallace. 1982a. The complete amino acid sequence of the murine transplantation antigen $H-2D^b$ as deduced by molecular cloning. *Immunogenetics* **16**: 1.

Reyes, A.A., M. Schold, K. Itakura, and R.B. Wallace. 1982b. Isolation of a cDNA clone for the murine transplantation antigen $H-2k^b$. *Proc. Natl. Acad. Sci.* **79**: 3270.

Rigby, P.W.J., W. Chia, C.E. Clayton, and M. Lovett. 1980. The structure and expression of the integrated viral DNA in mouse cells transformed by simian virus 40. *Proc. R. Soc. Lond. B.* **210**: 437.

Rigby, P.W.J., P.M. Brickell, D.S. Latchman, D. Murphy, K.-H. Westphal, and M.R.D. Scott. 1984. Oncogenic transformation activates cellular genes. In *Genetic manipulation. Impact on man and society* (ed. P. Starlinger and K. Illmensee). ICSU Press, Miami, Florida. (In press.)

Ruley, H.E. 1983. Adenovirus early region *1A* enables viral and cellular transforming genes to transform primary cells in culture. *Nature* **304**: 602.

Schone, G. 1906. Untersuchungen über carzinoninnunität bei Mäusen. *Muench. Med. Wochensch.* **51**: 2517.

Scott, M.R.D., K.-H. Westphal, and P.W.J. Rigby. 1983. Activation of mouse genes in transformed cells. *Cell* **34**: 557.

Shedlovsky, A., L.J. Clipson, J.C. Vandeberg, and W.F. Dove. 1981. Strong teratocarcinoma transplantation loci, *Gt-1* and *Gt-2*, flank *H-2*. *Immunogenetics* **13**: 413.

Siegler, E.L., N. Tick, A.K. Teresky, M. Rosenstraus, and A.J. Levine. 1979. Teratocarcinoma transplantation rejection loci: An *H-2*-linked tumor rejection locus. *Immunogenetics* **9**: 207.

Sikora, K., P. Stern, and E. Lennox. 1977. Immunoprotection by embryonal carcinoma cells for methylcholanthrene-induced murine sarcomas. *Nature* **269**: 813.

Steinmetz, M., K.W. Moore, J.G. Frelinger, B.T. Sher, F.W. Shen, E. Boyse, and L. Hood. 1981a. A pseudogene homologous to mouse transplantation antigens: Transplantation antigens are encoded by eight exons that correlate with protein domains. *Cell* **25**: 683.

Steinmetz, M., J.G. Frelinger, D. Fisher, T. Hunkapiller, D. Pereira, S.M. Weissman, H. Uehara, S. Nathenson, and L. Hood. 1981b. Three cDNA clones encoding mouse transplantation antigens: Homology to immunoglobulin genes. *Cell* **24**: 125.

Stern, P., G.R. Martin, and M.J. Evans. 1975. Cell surface antigens of clonal teratocarcinoma cells at various stages of differentiation. *Cell* **6**: 455.

Stern, P.L., M. Gidlund, A. Orn, and H. Wigzell. 1980. Natural killer cells mediate lysis of embryonal carcinoma cells lacking MHC. *Nature* **285**: 341.

Tooze, J., ed. 1981. *Molecular biology of tumor viruses*, 2nd edition, revised: *DNA tumor viruses*. Cold Spring Harbor Laboratory, Cold Spring Harbor, New York.

Weiss, E., L. Golden, R. Zakut, A. Mellor, K. Fahrner, S. Kvist, and R.A. Flavell. 1983. The DNA sequence of the $H-2K^b$ gene: Evidence for gene conversion as a mechanism for the generation of polymorphism in histocompatibility antigens. *EMBO J.* **2**: 453.

Weiss, R.A., N.M. Teich, H. Varmus, and J.M. Coffin, eds.

1982. *Molecular biology of tumor viruses,* 2nd edition: *RNA tumor viruses.* Cold Spring Harbor Laboratory, Cold Spring Harbor, New York.

Williams, J.G., R. Hoffman, and S. Penman. 1977. The extensive homology between mRNA sequences of normal and SV40-transformed human fibroblasts. *Cell* **11**: 901.

Winoto, A., M. Steinmetz, and L. Hood. 1983. Genetic mapping in the major histocompatibility complex by restriction enzyme site polymorphisms: Most mouse class I genes map to the *Tla* complex. *Proc. Natl. Acad. Sci.* **80**: 3425.

Zinkernagel, R.M. and P.C. Doherty. 1974a. Restriction of in vitro T-cell-mediated cytotoxicity in lymphocytic choriomeningitis within a syngeneic or semiallogeneic system. *Nature* **248**: 701.

———. 1974b. Immunological surveillance against altered self components by sensitised T lymphocytes in lymphocytic choriomeningitis. *Nature* **251**: 547.

Stimulation of Gene Expression by Viral Transforming Proteins

J. Brady, L.A. Laimins, and G. Khoury

Laboratory of Molecular Virology, National Cancer Institute, National Institutes of Health, Bethesda, Maryland 20205

Insight into the mechanisms controlling RNA transcription is essential to our understanding of gene regulation and transformation. In this regard, the small DNA tumor viruses have served as valuable model systems. An example of their usefulness is reflected in the recent discovery of viral enhancer elements. These sequences dramatically increase the transcriptional efficiency of an adjacent promoter in a fashion that is relatively independent of distance and orientation (Banerji et al. 1981; Benoist and Chambon 1981; Gruss et al. 1981; Moreau et al. 1981; see also Khoury and Gruss 1983). The cis-essential enhancer sequences were first discovered in the papovavirus simian virus 40 (SV40), where they control the level of early viral gene expression. The association of strong enhancer elements with a number of other early viral transcriptional units, including those of polyoma virus, BKV, bovine papillomavirus, adenovirus 2 (Ad2), and retroviruses, underscores their importance in initiating viral gene expression. Subsequently, additional groups of viral genes are transcriptionally activated, in trans, by some of these early viral proteins (Jones and Shenk 1979; Nevins 1981, 1982; Feldman et al. 1982). Since it seems likely that a similar stimulation of cellular gene expression contributes to the complex set of events that ultimately results in the transformed phenotype (Schutzbank et al. 1982; Scott et al. 1983), the mechanism by which these secondary genes are regulated is of considerable interest.

As an initial approach to this problem, we have examined the ability of SV40 T antigen and the Ad5 E1A protein to activate the expression of two unrelated genes, the SV40 late gene and the rat insulin I gene. We demonstrate that the early SV40 gene product, T antigen, stimulates late SV40 gene expression in the absence of viral DNA replication. The mechanism by which this induction may occur is discussed. Further studies demonstrate the ability of both T antigen and Ad5 E1A transforming proteins to induce the expression of transfected insulin genes but not their chromosomal counterparts.

Materials and Methods

DNA transfection and preparation of protein extracts
Transfection of either SV40 DNA or plasmid pSVs-L18 was carried out using the calcium phosphate precipitation method (Graham and van der Eb 1973). Whole-cell protein extracts were prepared following aspiration of growth media and washing of the cell monolayer once with phosphate-buffered saline (PBS). Buffer (0.5 ml) containing 50 mM Tris (pH 7.4), 50 mM NaCl, 5 mM $MgCl_2$, 0.5% NP40, 25 μg/ml Aprotinin, and 25 μg/ml N^α-tosyl-L-phenylalanyl-chloromethyl ketone (TPCK) was added to each 10-cm culture dish. DNase I was added to a final concentration of 50 μg/ml, and the dishes were incubated for 10 minutes at 25°C. One-tenth volume of 10% SDS–1.4 M β-mercaptoethanol was added, and incubation was carried out at 25°C for an additional 3 minutes. The soluble cell extract was transferred to a 1.5-ml Eppendorf tube and incubated at 95–100°C for 5 minutes. Protein extracts were stored at −20°C.

Insulin-containing plasmids were transfected into CV-1, COS-1, and HEK 293 cells, and the T24 human bladder cell line by the calcium phosphate method (Graham and van der Eb 1973). Cell lysates were prepared 48 hours posttransfection (p.t.) and analyzed for insulin expression by radioimmune assay (RIA kit from Amersham).

Western blot analysis of protein extracts
Synthesis of the SV40 late gene product, VP1, was determined by the Western blot method (Bittner et al. 1980). Protein extracts (40 μl) were denatured, electrophoresed on a 10% SDS-polyacrylamide gel, and transferred directly to nitrocellulose. The blot was then incubated in 1× TNE (50 mM NaCl, 2.5 mM EDTA in 10 mM Tris·HCl, pH 7.5) containing 2% bovine serum albumin (BSA) and 0.02% sodium azide (blocking buffer) for 5–6 hours at 25°C. A 1:100 dilution of antiserum in the same buffer was added and allowed to incubate overnight at 25°C. The antibody solution was removed and the blot was washed five times with 1× TNE. The blot was then incubated with ^{125}I-labeled protein A (5 μCi in 50 ml of 1× TNE, 2% BSA, 0.02% sodium azide) for 4–5 hours, washed 10–15 times with 1× TNE, dried, and exposed to X-ray film. Antiserum to either SDS-denatured SV40 virions or purified VP1, isolated from acrylamide gels, was produced in rabbits following the protocol of McMillan et al. (1977).

Cell cultures and plasmid constructions
BSC-1 cells are a continuous line of monkey kidney cells. COS-1 cells are SV40-transformed CV-1 monkey kidney cells obtained by the transformation and stable integration with an origin-defective, replication-negative SV40 DNA (Gluzman 1981). The integration pattern of the SV40 DNA allows the stable expression of both SV40

large T and small T antigens, but not VP1. C2, C6, and C11 are SV40-transformed monkey kidney cell lines obtained following infection of CV-1 cells with ultraviolet-irradiated SV40 virus (Gluzman et al. 1977). These cell lines constitutively synthesize altered SV40 T antigens, which do not complement SV40 tsA mutants. The integration pattern in each of the transformed cell lines interrupts the SV40 late genes, thus preventing expression. HEK-293 cells are a line of human embryonic kidney cells, transformed by the left end of Ad5, which express the E1A and E1B viral proteins (Graham et al. 1977).

Plasmid pSVs-L18 was constructed as follows. SV40 DNA was digested with restriction endonucleases BglI and BclI (Fig. 1). The larger restriction fragment (2770 bp), containing the SV40 late promoter and coding sequences, was purified and cloned into a pBR322 plasmid derivative, pML-2, at the BamHI restriction site. Following isolation of the plasmid, restriction enzyme digestion was used to verify the insertion and orientation of the SV40 late gene.

The 1.8-kb HincII-BamHI and 3.8-kb EcoRI-BamHI restriction fragments, containing the entire rat insulin I gene and varying lengths of upstream promoter sequences, were cloned into pBR322 at the BamHI and EcoRI endonuclease sites, respectively.

Results

SV40 late gene expression following transfection of BSC-1 and COS-1 cells with SV40 DNA and plasmid pSVs-L18

The efficiency of SV40 late gene expression was first determined following transfection of the monkey kidney cell lines BSC-1 and COS-1 with purified form-I (F_I) SV40 DNA. To exclude template amplification, cytosine arabinoside (25 μg/ml) was added to the culture media. Following transfection, whole-cell extracts were prepared and analyzed by Western blot analysis using anti-SV40 VP1 antiserum (Fig. 2). The level of SV40 late gene expression 30 hours p.t. with SV40 DNA is dramatically enhanced in COS-1 cells compared with control BSC-1 cells (Fig. 2, lanes 1 and 3). We find that VP1 synthesis is increased by 20-fold to 50-fold in COS-1 cells. No SV40 DNA replication could be detected in the presence of cytosine arabinoside, indicating that enhanced late gene expression is not due to template amplification.

The SV40 sequences cloned into plasmid pSVs-L18 consist of the larger (2770-bp) BglI/BclI restriction fragment (see Fig. 1). The 300 bp of upstream promoter sequences include the SV40 21-bp repeats, the 72-bp enhancer element, and the −25 late transcriptional control sequence, all of which have been shown to be important domains of the SV40 late promoter (Brady et al. 1982, 1984; Fromm and Berg 1982; Hansen and Sharp,

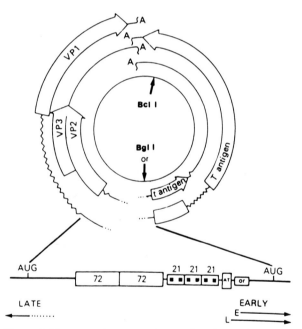

Figure 1 The genomic map and control region of SV40. The diagram presents the control region for expression of the SV40 early genes (large and small T antigens) and late genes (VP1, 2, and 3). The origin for viral DNA replication (or) is cleaved by the restriction enzyme BglI. To the late side of the origin are the early transcriptional control sequences, including the Goldberg-Hogness box (AT), the three 21-bp repeats (each containing two copies of a GC-rich hexanucleotide), and the tandem 72-bp enhancer element. The early transcripts are initiated predominately at position E early in infection and shift to position L after DNA replication. The late viral transcripts have heterogeneous 5′ ends, indicated by dots. The SV40 late promoter and coding sequences in mutant pSVs-L18 extend counterclockwise from the BglI site to the BclI site. (Reprinted, with permission, from Hamer and Khoury 1983.)

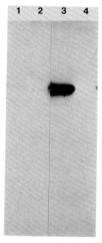

Figure 2 SV40 late gene expression in BSC-1 and COS-1 cells following transfection with SV40 DNA. Parallel cultures of BSC-1 and COS-1 cells (10-cm plate, ~60–70% confluent) were transfected with SV40 DNA (2 μg), using the calcium phosphate precipitation method (Graham and van der Eb 1973). Following transfection, cells were maintained in Dulbecco's modified Eagle's medium (DMEM) with 10% fetal calf serum (FCS) and 25 μg/ml cytosine arabinoside. At 30 hr p.t., whole-cell protein extracts were prepared and analyzed by Western blot analysis (see Materials and Methods). (Lane 1) jBSC-1 cells, SV40-transfected; (lane 2) BSC-1 cells, control; (lane 3) COS-1 cells, SV40-transfected; (lane 4) COS-1 cells, control.

1984; K. Subramanian, pers. comm.). This construction, which eliminates the SV40 DNA replication origin, allows us to evaluate the effect of T antigen on VP1 expression in the absence of inhibitors of DNA replication.

Western blot analysis of whole-cell protein extracts prepared from BSC-1 and COS-1 cells at 42-hours p.t. with plasmid pSVs-L18 are shown in Figure 3. Again, SV40 late gene expression is significantly enhanced in COS-1 cells, consistent with the results obtained after wild-type SV40 DNA transfection. These experiments exclude DNA replication or its initiation as a basis for the enhanced level of SV40 late gene expression in COS-1 cells. Although the SV40 sequences present in plasmid pSVs-L18 are sufficient to demonstrate a markedly different level of SV40 late gene expression in COS-1 cells compared with BSC-1 cells, the efficiency of late gene expression in this deletion mutant compared with intact SV40 DNA is reduced approximately 20-fold. These results may indicate an auxiliary requirement for sequences on the early side of the *Bgl*I site (Fig. 1) in the function of the late promoter.

SV40 late gene expression in SV40-transformed cells expressing altered T antigens

The above experiments demonstrate that a function associated with the early phase of a lytic infection, present in COS-1 cells but not in BSC-1 cells, is able to stimulate SV40 late gene expression. The obvious candidate for this function is T antigen. To confirm and extend this hypothesis, we transfected SV40 DNA in the presence of cytosine arabinoside into three additional monkey kidney cell lines that had been transformed with viruses, containing mutations in the T-antigen coding region (Gluzman et al. 1977). Production of VP1 after transfection of cell lines C2, C6, and C11 is compared with results in COS-1 cells (Fig. 4). None of these lines supports the same level of VP1 synthesis as COS-1 cells (Fig. 3, lane 2). The level of VP1 synthesis in C2 and C11 (Fig. 3, lanes 3 and 5), however, is at least 20 times as high as the level in C6 cells (Fig. 4, lane 4). These results correlate with the ability of the mutant T antigens in cell lines C2 and C11 to bind specifically to SV40 DNA, while the C6 T antigen is unable to bind (Scheller et al. 1982; C. Prives, pers. comm.).

The efficiency of SV40 late gene expression is directly comparable in COS-1, C2, C6, and C11 cells since we have determined by immunoprecipitation that each transformed cell line synthesizes approximately equal quantities of SV40 T antigen (data not shown). The level of T antigen is similar to that observed 14–20 hours after SV40 infection.

SV40 late gene expression in HEK-293 cells following transfection with SV40 DNA

Recent evidence from a number of laboratories suggests that certain DNA virus transforming proteins can induce the expression of cotransfected genes. In particular, the Ad5 E1A gene product, which is constitutively expressed in transformed human embryonic kidney cells (HEK-293), acts in *trans* to stimulate expression of several genes (Jones and Shenk 1979; Nevins 1981, 1982; Feldman et al. 1982). We wanted to know if this induction would extend to the late SV40 genes. Following transfection of control HEK, HEK 293, and COS-1 cells with SV40 F$_1$ DNA, whole-cell protein extracts were prepared (42 hr p.t.) and analyzed for VP1 synthesis by Western blot analysis. Cytosine arabinoside was added to the culture media to prevent DNA replication. The

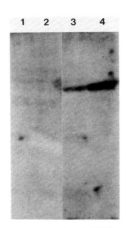

Figure 3 SV40 late gene expression in BSC-1 and COS-1 cells following transfection with plasmid pSVs-L18. Parallel cultures of BSC-1 and COS-1 cells (10-cm plate, ~60–70% confluent) were transfected with plasmid pSVs-L18 DNA (10 or 20 μg), using the calcium phosphate precipitation method (Graham and van der Eb 1973). Following transfection, cells were maintained in DMEM with 10% FCS. At 42 hr p.t., whole-cell protein extracts were prepared and analyzed by Western blot analysis (see Materials and Methods). (Lane *1*) BSC-1 cells, 10 μg pSVs-L18; (lane *2*) BSC-1 cells, 20 μg pSVs-L18; (lane *3*) COS-1 cells, 10 μg pSVs-L18; (lane *4*) COS-1 cells, 20 μg pSVs-L18.

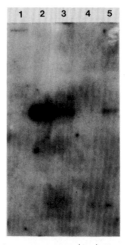

Figure 4 SV40 late gene expression in transformed cells containing mutant T antigens. Parallel cultures of BSC-1, COS-1, C2, C6, and C11 cells were transfected with SV40 DNA (2 μg) as described in Materials and Methods. Following transfection, cells were maintained in DMEM with 10% FCS and 25 μg/ml cytosine arabinoside. At 30 hr p.t., whole-cell protein extracts were prepared and analyzed by Western blot analysis. (Lane *1*) BSC-1 cells; (lane *2*) COS-1 cells; (lane *3*) C2 cells; (lane *4*) C6 cells; (lane *5*) C11 cells.

levels of VP1 synthesized in HEK, HEK 293, and COS-1 cells is shown in Figure 5, lanes 1, 2, and 3, respectively. Although we cannot exclude a minor stimulation, the level of VP1 synthesis in HEK-293 cells was at least 50-fold lower than that found in COS-1 cells under similar conditions. The low level of SV40 gene expression in HEK-293 cells is not due to poor transfection. VP1 is synthesized in HEK-293 cells, in the absence of cytosine arabinoside, at a level comparable to that found in BSC-1 cells (data not shown). In addition, we observed efficient expression from plasmids containing the Ad E2 gene (72K DNA binding protein) following transfection into HEK-293 cells (data not shown). We conclude that the efficient induction of SV40 late gene expression seen in the presence of T antigen in monkey kidney cells does not occur in the presence of E1A protein in HEK-293 cells.

Rat insulin I gene expression in COS-1, HEK-293, and T24 bladder carcinoma cells

We have also focused on the expression of one of two rat insulin genes, rI_1 (Lomedico et al. 1979). Two plasmids containing the entire rat insulin I gene with either 500 bp (1.8-kb insulin rI_1) or 2500 bp (3.8-kb insulin rI_1) of upstream sequence were used in the transfection studies. The expression level of the three plasmid constructions was first analyzed following transfection into CV-1 cells. As a control, the plasmid prI_2-SV40 was obtained from P. Lomedico (1982). This recombinant molecule contains the entire SV40 early promoter and enhancer, including the Goldberg-Hogness (TATA) box, the GC-rich 21-bp repeats, and the tandem 72-bp enhancer element ligated to the coding sequences of the rat insulin rI_2 gene. Its activity in CV-1 cells (100 μunits/10^6 cells) was scored as 100 (1 ng insulin/ml = 24 μunits/ml). At 48 hours p.t., cells were harvested and analyzed for insulin expression by radioimmune assay.

The average result of three experiments are shown in Table 1. In CV-1 cells, the upstream insulin promoter sequences present in either the 1.8-kb or the 3.8-kb plasmids, in the absence of the SV40 enhancer element, are insufficient for expression of detectable levels of insulin.

We next introduced the three plasmids into transformed COS-1 or HEK-293 cells. In contrast to the results obtained in CV-1 cells, we find efficient expression of the rat insulin I plasmids in COS-1 cells and HEK-293 cells in the presence or absence of the SV40 enhancer. The level of expression of either insulin plasmid, compared with the levels observed with plasmid prI_2-SV40 in CV-1 cells, was approximately 25% in COS-1 cells and 20% in HEK-293 cells. This represents a significant stimulation of the rat insulin gene expression (at least 20-fold) above levels observed in CV-1 cells. An enhanced expression of rI_1 was also seen when the 1.8-kb rI_1 plasmid was transfected into the human bladder tumor cell line T24 but not in normal human cells. This result suggests that a gene product, similar in function to either SV40 T antigen or Ad5 E1A, is present in the T24 tumor cell.

Discussion

Data are presented here that indicate that SV40 late gene expression is stimulated by T antigen. The use of DNA synthesis inhibitors indicates that the stimulation is not dependent upon DNA replication. Experiments with origin-deleted molecules, which were amplified in E. coli and introduced into monkey kidney cells by DNA transfection confirm this conclusion since they completely exclude low levels of DNA replication. Similar results have recently been obtained by J. Alwine and K. Subramanian and their colleagues (pers. comm.). Although these experiments clearly demonstrate that T antigen can stim-

Figure 5 SV40 late gene expression in HEK-293 cells following SV40 DNA transfection. Parallel cultures of HEK, HEK-293, and COS-1 cells were treated as described in the legend for Fig. 4. (Lane 1) HEK cells; (lane 2) HEK-293 cells; (lane 3) COS-1 cells.

Table 1 Insulin Expression in Transformed Cells

	CV-1	HEK-293	COS-1	T24
prI_2-SV40	100	54	66	40
1.8-kb Insulin rI_1	0	20	26	14
3.8-kb Insulin rI_1	0	18	24	n.d.*
Mock transfection	0	0	0	0

*Not determined.

Insulin expression is expressed as μunits/10^6 cells. The amount of insulin in cell lysates of transformed cells transfected by rat insulin–containing plasmids was determined by radioimmune assay at 48 hr p.t. COS-1 cells (SV40 T antigen–producing transformed cells), HEK-293 cells (adenovirally transformed human embryonic kidney cells synthesizing E1A), and T24 human bladder carcinoma cells were transfected with 25 μg of plasmids containing the rat insulin I gene rI_1 with 0.5 kb and 2.5 kb of upstream sequence. The plasmid prI_2-SV40, containing the rat insulin 2 gene (rI_2) coding sequences associated with SV40 transcriptional control sequences, was used as a control.

ulate late gene expression in the absence of viral DNA replication, they do not exclude the alteration and/or amplification of SV40 templates as contributory factors that further enhance the level of VP1 expression.

The mechanism by which T antigen induces late gene expression and whether it acts directly on the template or requires additional cellular protein(s) are not clear. However, some insight is gained from our examination of the ability of mutant T antigen in SV40-transformed cell lines to stimulate VP1 synthesis. Although the T antigens in transformed monkey cell lines C2, C6, and C11 are unable to initiate viral DNA synthesis (Gluzman et al. 1977), both C2 and C11 support rather efficient VP1 expression (~5-25% the level seen in COS-1 cells). On the other hand, T antigen from C6 cells does not stimulate VP1 expression. These results correlate well with the ability of T antigens from C2 and C11 cells, but not C6 cells, to bind specifically to the region around the SV40 origin of DNA replication (Scheller et al. 1982; C. Prives, pers. comm.). Thus, we suggest that the properties of T antigen required for DNA binding, but not those additionally required for initiation of DNA replication, are important for late gene expression.

Binding of SV40 T antigen to viral DNA occurs at three primary and adjacent sites (Tjian 1978; Tenen et al. 1982; DeLucia et al. 1983). All of binding site I and half of site II are missing in plasmid pSVs-L18. Since this plasmid actively expresses VP1 in COS-1 cells but not in BSC-1 cells, our results suggest that the remaining one and one-half T-antigen binding sites are sufficient for T-antigen-mediated stimulation of late gene expression. The overall decrease in late gene activity of pSVs-L18 compared with wild-type SV40 DNA, however, may indicate an auxiliary role for the deleted DNA sequences.

The SV40 72-bp tandem "enhancer" sequence is positioned between the early and late SV40 genes (Benoist and Chambon 1981; Gruss et al. 1981). Since this transcriptional unit functions equally well in both orientations (Banerji et al. 1981; Moreau et al. 1981), it is of interest that late SV40 gene expression is inefficient in the absence of T antigen. One possibility is that the late viral genes are under negative transcriptional control, which is mediated by a repressor. There are a number of experimental findings that could support such a model. Handa and Sharp reported that treatment of cells with protein synthesis inhibitors increased the expression of SV40 late RNA (Handa and Sharp 1980) early in the lytic cycle. Similarly, treatment of early transcriptional complexes with Sarkosyl, which releases most proteins other than preinitiated RNA polymerase, results in the synthesis of substantial amounts of late RNA (Birkenmeier et al. 1977; Ferdinand et al. 1977). Finally, Aloni and his colleagues have presented data suggesting that late SV40 transcription is attenuated (Hay et al. 1982). If such a repressor exists, it is likely to be cellular in origin and not associated with the infectious virions (Parker and Stark 1979). One might conceive of T antigen's role in late transcription to be the removal of this putative repressor.

Alternatively, T antigen may stimulate SV40 VP1 expression by positively activating the late transcriptional unit. This early viral protein may function by directing RNA polymerase molecules to a late promoter that, in its absence, is unresponsive. This is not unlike the mechanisms proposed for the adenoviral E1A protein in its activation of the other adenoviral early and late transcription units (Jones and Shenk 1979; Nevins 1981, 1982; Feldman et al. 1982).

Our results suggest that the Ad5 E1A protein, present in HEK-293 cells, is unable to stimulate SV40 late gene expression with an efficiency comparable to that of T antigen in COS-1 cells. This result is of interest, since the E1A-gene product induces expression of a number of transfected eukaryotic genes, including Ad E2, heat-shock (Jones and Shenk 1979; Nevins 1981, 1982; Feldman et al. 1982), and β-globin (M. Green et al., pers. comm.). We have shown in this manuscript that HEK-293 cells also stimulate rat insulin I gene expression. Although these results do not rule out a common mechanism for E1A and SV40 T-antigen-mediated gene activation, it is clear that T antigen, apparently through a specific interaction with SV40 DNA, is required for efficient SV40 late gene induction.

The SV40 T antigen, like the Ad5 E1A protein, can apparently activate other cellular genes which are transfected into cells. We have found that the rat insulin I gene is induced to express at high levels in transformed COS-1 cells but not in control CV-1 cells. The level of insulin expression is significant (25 μunits/10^6 cells), approximately 25% of that observed in CV-1 cells after transfection of a plasmid containing a rat insulin gene, under the control of the SV40 enhancer element. These experiments suggest that a function directly or indirectly associated with early SV40 gene expression, and present in COS-1 cells, is able to stimulate insulin production. In view of our results with SV40 late gene induction, we suggest that this function is SV40 T antigen. Experiments to confirm this suggestion and to define the sequences in rI, required for induction are in progress.

The ability to activate extrachromosomal insulin is not limited to transformed cells containing SV40 T antigen and adenoviral E1A. We find that human tumor cell line T24 also induces expression of the transfected rat insulin gene, suggesting that it contains an active function similar to either SV40 T antigen or adenoviral E1A. This conclusion is supported by recent data indicating that at least two genes may be required for cell transformation and that one of these is equivalent to adenoviral E1A (Land et al. 1983; Ruley 1983).

It should be noted that the endogenous insulin genes in both the control and transformed cell lines were not expressed. A similar observation was made by T. Maniatis and his co-workers for the endogenous β-globin genes (pers. comm.). The mechanism by which transforming proteins such as adenoviral E1A and SV40 T antigen stimulate the expression of transfected genes but not their endogenous cellular counterparts is obscure. The most likely explanation is that the cellular insulin genes are located in a chromosomal environment that is inaccessible to the viral proteins and/or their associated transcriptional machinery (Lacy et al. 1983). The heat-shock genes would then represent a set of cellular functions that reside in a susceptible chromosomal domain, readily inducible by either the appropriate functional stimuli or the alternative viral proteins. The observation that heat-shock genes are DNase hypersensitive, prior to activation by a thermal stimulus or E1A induction, suggests that they are preprogrammed for expression in most cells. One possibility is that a ubiquitous *trans*-acting factor interacts with heat-shock genes, preparing them for a second step that leads to expression. The chromosomal insulin gene may be provided with a similar set of factors only in the β-cell of the pancreas.

Acknowledgments

We thank Y. Gluzman for cell lines C2, C6, C11, and COS-1; P. Howley, G. Jay, R. Muschel, and N. Rosenthal for helpful advice; and M. Priest for preparation of this manuscript.

References

Banerji, J., S. Rusconi, and W. Schaffner. 1981. Expression of a β-globin gene is enhanced by remote SV40 DNA sequences. *Cell* **27**: 299.

Benoist, C. and P. Chambon. 1981. In vivo sequence requirements of the SV40 early promoter region. *Nature* **290**: 304.

Birkenmeier, E.H., E. May, and N.P. Salzman. 1977. Characterization of simian virus 40 *ts*A58 transcriptional intermediates at restrictive temperatures: Relationship between DNA replication and transcription. *J. Virol.* **22**: 702.

Bittner, M., P. Kupferer, and C.F. Morris. 1980. Electrophoretic transfer of proteins and nucleic acids from slab gels to diazobenzyloxymethyl cellulose or nitrocellulose sheets. *Anal. Biochem.* **102**: 459.

Brady, J., M. Radonovich, M. Thoren, G. Das, and N.P. Salzman. 1984. The SV40 major late promoter: An upstream DNA sequence required for efficient *in vitro* transcription. *Mol. Cell. Biol.* **4**: 133.

Brady, J., M. Radonovich, M. Vodkin, V. Natarajan, M. Thoren, G. Das, J. Janik, and N.P. Salzman. 1982. Site-specific base substitution and deletion mutants that enhance or suppress transcription of the SV40 major late RNA. *Cell* **31**: 625.

DeLucia, A., B. Lewton, R. Tjian, and P. Tegtmeyer. 1983. Topography of simian virus 40 A protein–DNA complexes: Arrangements of pentanucleotide interaction sites at the origin of replication. *J. Virol.* **46**: 143.

Feldman, C.T., M.J. Imperiale, and J.R. Nevins. 1982. Activation of early adenovirus transcription by the herpesvirus immediate early gene: Evidence for a common cellular control factor. *Proc. Natl. Acad. Sci.* **79**: 4952.

Ferdinand, F.-J., M. Brown, and G. Khoury. 1977. Characterization of early simian virus 40 transcriptional complexes: Late transcription in the absence of detectable DNA replication. *Proc. Natl. Acad. Sci.* **74**: 5443.

Fromm, M. and P. Berg. 1982. Deletion mapping of DNA regions required for SV40 early region promoter function *in vivo*. *J. Mol. Appl. Genet.* **1**: 457.

Gluzman, Y. 1981. SV40-transformed simian cells support the replication of early SV40 mutants. *Cell* **23**: 175.

Gluzman, Y., J. Davison, M. Oren, and E. Winocour. 1977. Properties of permissive monkey cells transformed by UV-irradiated simian virus 40. *J. Virol.* **22**: 256.

Graham, F.L. and A.J. van der Eb. 1973. A new technique for the assay of infectivity of human adenovirus 5 DNA. *Virology* **52**: 456.

Graham, F.L., J. Smiley, W.C. Russel, and R. Nairn. 1977. Characteristics of a human cell line transformed by DNA from human adenovirus type 5. *J. Gen. Virol.* **36**: 59.

Gruss, P., R. Dhar, and G. Khoury. 1981. Simian virus 40 tandem repeated sequences as an element of the early promoter. *Proc. Natl. Acad. Sci.* **78**: 943.

Hamer, D. and G. Khoury. 1983. Introduction. In *Enhancers and eukaryotic gene expression* (ed. Y. Gluzman and T. Shenk), p. 1. Cold Spring Harbor Laboratory, Cold Spring Harbor, New York.

Handa, H. and P.A. Sharp. 1980. Expression of early and late simian virus 40 transcripts in the absence of protein synthesis. *J. Virol.* **34**: 592.

Hansen, U. and P. Sharp. 1984. Sequences controlling transcription of SV40 promoters. *EMBO J.* (in press).

Hay, N., H. Skolnick-David, and J. Aloni. 1982. Attenuation in the control of SV40 gene expression. *Cell* **29**: 183.

Jones, N. and T. Shenk. 1979. An avenovirus 5 early gene product function regulates expression of other early viral genes. *Proc. Natl. Acad. Sci.* **76**: 3665.

Khoury, G. and P. Gruss. 1983. Enhancer elements. *Cell* **33**: 313.

Lacy, E., S. Roberts, E.P. Evans, M.D. Burtenshaw, and F.D. Costantini. 1983. A foreign β-globin gene in transgenic mice: Integration at abnormal chromosomal positions and expression in inappropriate tissues. *Cell* **34**: 343.

Land, H., L.F. Parada, and R.A. Weinberg. 1983. Tumorigenic conversion of primary embryo fibroblasts requires at least two cooperating oncogenes. *Nature* **304**: 596.

Lomedico, P.L. 1982. Use of recombinant DNA technology to program eukaryotic cells to synthesize rat proinsulin: A rapid expression assay for cloned genes. *Proc. Natl. Acad. Sci.* **79**: 5798.

Lomedico, P., N. Rosenthal, A. Efstratiadis, W. Gilbert, R. Kolodner, and R. Tizard. 1979. The structure and evolution of the two nonallelic rat prepro-insulin genes. *Cell* **18**: 545.

McMillan, J. and R.A. Consigli. 1977. Immunological reactivity of antisera to sodium dodecyl sulfate–derived polypeptides of polyoma virus. *J. Virol.* **21**: 1113.

Moreau, P., R. Hen, B. Wasylyk, R. Everett, M.P. Gaub, and P. Chambon. 1981. The SV40 72-base-pair repeat has a striking effect on gene expression both in SV40 and other chimeric recombinants. *Nucleic Acids Res.* **9**: 6047.

Nevins, J.R. 1981. Mechanism of activation of early viral transcription by the adenovirus E1A gene product. *Cell* **26**: 213.

———. 1982. Induction of the synthesis of a 70,000-dalton mammalian heat-shock protein by the adenovirus E1A gene product. *Cell* **29**: 913.

Parker, B.A. and G.R. Stark. 1979. Regulation of simian virus 40 transcription: Sensitive analysis of the RNA species present early in infections by virus or viral DNA. *J. Virol.* **31**: 360.

Ruley, H.E. 1983. Adenovirus early region 1A enables viral and cellular transforming genes to transform primary cells in culture. *Nature* **304**: 602.

Scheller, A., L. Covey, B. Barnet, and C. Prives. 1982. Small subclass of SV40 T antigen binds to the viral origin of replication. *Cell* **29**: 375.

Schutzbank, T., R. Robinson, M. Oren, and A.J. Levine. 1982. SV40 large tumor antigen can regulate some cellular transcripts in a positive fashion. *Cell* **30**: 481.

Scott, M.R.D., K.-H. Westphal, and P.W.J. Rigby. 1983. Activation of mouse genes in transformed cells. *Cell* **34**: 557.

Tenen, D.G., L.L. Haines, and D.M. Livingston. 1982. Binding of an analog of the simian virus 40 T antigen to wild-type and mutant viral replication origins. *J. Mol. Biol.* **157**: 473.

Tjian, R. 1978. The binding site on SV40 DNA for a T-antigen-related protein. *Cell* **13**: 165.

Changes in Specific mRNAs Following Serum Stimulation of Cultured Mouse Cells: Increase in a Prolactin-related mRNA

D.I.H. Linzer and D. Nathans

Howard Hughes Medical Institute, Department of Molecular Biology and Genetics, The Johns Hopkins University School of Medicine, Baltimore, Maryland 21205

The events that determine whether a cell will enter a new round of cell division occur between the periods of mitosis (M) and DNA synthesis (S), during the G_1 phase of the cell cycle (Pardee et al. 1978). A useful system for analyzing the steps involved in committing a cell to move through G_1 and into S phase is the stimulation of quiescent tissue-culture cells to reenter the cell cycle. This stimulation can be accomplished, for example, by administration of serum to resting cells (Holley and Kiernan 1974; Rudland et al. 1974). Among the many changes elicited by serum stimulation of growth-arrested tissue-culture cells is an increase in the transcription (Hershko et al. 1971; Rudland et al. 1974) and processing (Johnson et al. 1975) of mRNA precursors. Only a few percent of these mRNAs present in G_1 are absent in quiescent cells (Williams and Penman 1975). That these newly appearing mRNAs are translated is suggested by the detection of new proteins in cells that transit G_1 (Fox and Pardee 1971; Salas and Green 1971; Becker and Stanners 1972; Gates and Friedkin 1978; Riddle et al. 1979; Rossow et al. 1979; Pledger et al. 1981; Thomas et al. 1981). It is likely that at least some of these new proteins are essential for the commencement of DNA synthesis, since the inhibition of protein synthesis during G_1 can prevent entry into S phase (Harris 1959; Powell 1962; Mueller et al. 1962; Taylor 1965). To characterize these G_1-specific elements involved in the cell's commitment to replicate its DNA, we have begun a study of individual mRNAs that are absent or at a very low level in resting mouse BALB/c 3T3 tissue-culture cells and are increased in amount prior to the onset of cellular DNA synthesis stimulated by serum.

Methods

Cell culture
BALB/c 3T3 (Todaro and Green 1963) and C3H 10T1/2 (Reznikoff et al. 1973) tissue-culture cells were grown in Eagle's minimal essential medium with Earle's salts, supplemented to 10 units/ml penicillin, 10 units/ml streptomycin, 2 mM glutamine, and 10% fetal calf serum (MEM-10). Cultures were rendered quiescent by starving in MEM containing 0.5% (MEM-0.5) or 2% (MEM-2) fetal calf serum for at least 2 days after confluence. Serum stimulation of resting cultures was accomplished by feeding with MEM containing 20% fetal calf serum (MEM-20). SV40-transformed BALB/c 3T3 cell lines SVB 10-1 and SVB 10-2 were selected as foci growing in MEM-10 and MEM-2, respectively, by P. Mounts. Krebs ascites carcinoma cells were grown in the peritoneal cavity of BALB/c mice and harvested 10 days after inoculation.

Nucleic acid purification
Total cellular RNA was prepared by centrifugation of guanidinium thiocyanate cell lysates (Chirgwin et al. 1979) through CsCl cushions (Glišen et al. 1974). Pelleted RNA was solubilized in water, extracted with phenol-chloroform, and precipitated with ethanol. Recombinant plasmid DNAs were purified from bacterial lysates by banding twice in CsCl–ethidium bromide gradients (Peden et al. 1982). Ethidium bromide was extracted with isobutanol, and CsCl was removed by dialysis or ethanol precipitation.

RNA filter hybridization
Denatured total cellular RNA was electrophoresed on formaldehyde agarose gels (Lehrach et al. 1977; Goldberg 1980) and transferred to nitrocellulose (Thomas 1980). After baking under vacuum for 2 hours at 80°C, filters were prehybridized for 3 hours at 42°C in formamide buffer (Fellous et al. 1982) and then hybridized in fresh formamide buffer containing 10 μg/ml denatured salmon sperm DNA, 5 μg/ml tRNA, and 1×10^6 dpm/ml of recombinant plasmid DNA nick-translated (Rigby et al. 1977) to 1×10^8 to 2×10^8 dpm/μg. Hybridizations were terminated after 36–48 hours, and the filters were washed (Thomas 1980), dried, and autoradiographed.

Results

Changes in RNA levels following serum stimulation
Previously, we have reported the construction of a cDNA library containing the sequences present in the mRNA population of serum-stimulated BALB/c 3T3 cells during late G_1 phase (Linzer and Nathans 1983). Approximately 0.5% of these clones contained cDNA inserts representing mRNAs that were in greater abundance in actively growing than in quiescent cells. Employing these rare cDNA clones to probe for the corresponding levels of RNA in resting cell cultures and at various times after serum stimulation yielded the results seen in Figure 1. Clone 18A2 detected a 700-nucleotide (nt) RNA species that reached at least a fourfold higher level 18 hours after serum stimulation than in resting cells, where it was

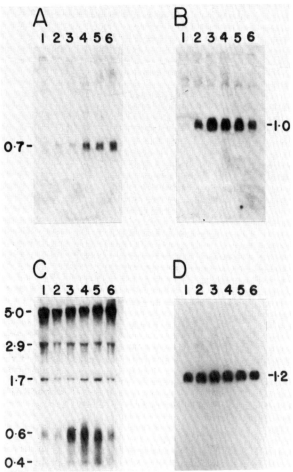

Figure 1 Time course of the levels of RNAs after serum stimulation. Total cellular RNA from cells that were resting (lane *1*) and from cells stimulated with serum for 6 hr (lane *2*), 12 hr (lane *3*), 18 hr (lane *4*), 24 hr (lane *5*), and 36 hr (lane *6*) was electrophoresed and transferred to nitrocellulose and then probed with cDNA clones 18A2 (*A*), 28H6 (*B*), 32A4 (*C*), or 31G8 (*D*). Approximate sizes in kilobases are given for each RNA species. (Reprinted from Linzer and Nathans 1983.)

concentration in cultures treated with 20% serum for 12 hours. The densitometric analyses of these data, along with that for a control clone that hybridized to an RNA whose abundance was not influenced by serum stimulation, are given in Figure 2.

RNA levels in other cell lines

The gene encoding the 28H6 RNA species appeared to be a good candidate for a gene involved in the process of cell growth, since this 1-kb RNA was absent in growth-arrested cell cultures and reached a peak level at the time of initiation of cellular DNA synthesis. In addition, we have demonstrated that this RNA is increased in amount after treatment of quiescent cells with purified simian virus 40 (SV40) or platelet-derived growth factor (Linzer and Nathans 1983). If this RNA has a crucial role in the transition from the resting to the growing state, then other fibroblastic mouse cell lines should also produce a change in the level of the 28H6 RNA in response to serum. To test this possibility, total cellular RNA was prepared from cultures of C3H 10T1/2 cells that were subconfluent and actively growing in MEM-10, from cultures that were fed with MEM-0.5 for 2 days after achieving confluence, and from these resting cultures that were stimulated with MEM-10 or MEM-20 for 12 hours. Analysis of BALB/c 3T3 cells grown under these conditions revealed a high level of 28H6 RNA in actively growing, subconfluent cultures, as well as an increase in the level of this RNA after serum stimulation of resting cultures

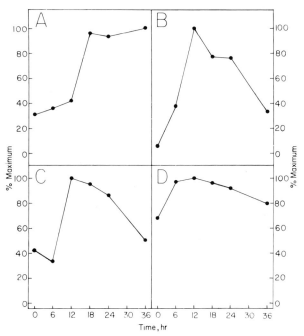

Figure 2 Quantitation of RNA levels after serum stimulation. The optical densities of the bands visualized in Fig. 1 were measured and normalized to the maximum hybridization for each RNA. These values were plotted versus duration of serum stimulation. (*A*) 0.7-kb 18A2 RNA; (*B*) 1.0-kb 28H6 RNA; (*C*) 0.6-kb 32A4 RNA; (*D*) 1.2-kb 31G8 RNA. (Reprinted from Linzer and Nathans 1983).

virtually absent. The peak in the amount of 18A2 RNA coincided with the peak of cellular DNA synthesis.

In contrast, clone 28H6 hybridized to a 1-kb RNA that attained a maximal level at the time of initiation of cellular DNA synthesis (12 hr after serum stimulation), and then decreased in quantity. This clone also hybridized to a less abundant, larger RNA species of approximately 4 kb that may be a precursor to the mature 1-kb mRNA. As with clone 18A2, no hybridization to the resting-cell RNA sample was detected with the cDNA clone 28H6 probe; the amount of the 1-kb RNA seen at 12 hours following stimulation represented at least a 15-fold to 20-fold increase above the resting-cell level.

A third clone, designated 32A4, detected several species of RNA, but only the two smallest RNAs changed in amount in response to serum stimulation. The 600-nt 32A4-specific RNA increased from a low, but detectable, level in resting cells to an approximately threefold greater that was dependent in magnitude on the concentration of serum (Linzer and Nathans 1983). The same results were obtained with the 10T1/2 cells, as shown in Figure 3. Actively growing and serum-stimulated cultures contained high levels of the 28H6 RNA compared with resting cell cultures, and the amount of this RNA was greater in cells treated with MEM-20 than with MEM-10. RNA was prepared from another mouse cell line, Krebs ascites carcinoma cells, that had been rapidly proliferating after intraperitoneal injection into BALB/c mice. Probing this RNA with cDNA clone 28H6 demonstrated a very high level of the 1-kb RNA and an easily detectable amount of the 4-kb RNA (Fig. 3).

The increase in the amount of 28H6 RNA after infection of quiescent cultures with SV40 suggested that SV40-transformed BALB/c 3T3 cells may maintain an elevated concentration of this RNA. Cultures of cell lines SVB 10-1 and 10-2 were grown from a low cell density (1×10^3 cells/cm^2) to a high density (2×10^5 cells/cm^2) in MEM-2 and MEM-10. RNA was prepared from cultures at these two densities as well as from two intermediate cell densities and examined with the 28H6 cDNA clone (Fig. 4). The SVB 10-1 cell line, which was originally selected as a focus growing in MEM-10, expressed a moderate amount of the 1-kb RNA when proliferating in MEM-10; this level of 28H6 RNA was independent of cell density but was considerably lower than that detected in actively growing or serum-stimulated BALB/c

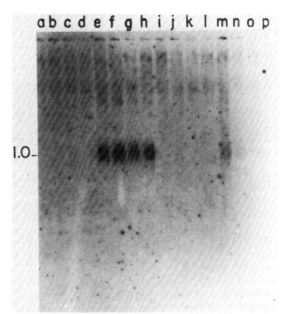

Figure 4 Analysis of 28H6 RNA levels of SV40-transformed cells. RNA samples prepared from SVB 10-1 cells grown in MEM-2 for 2 (*a*), 4 (*b*), 6 (*c*), and 8 (*d*) days after plating or in MEM-10 for 2 (*e*), 4 (*f*), 6 (*g*), and 8 (*h*) days after plating, along with RNA from SVB 10-2 cells grown in MEM-2 for 2 (*i*), 4 (*j*), 6 (*k*), and 8 (*l*) days after plating or in MEM-10 for 2 (*m*), 4 (*n*), 6 (*o*), and 8 (*p*) days after plating, were probed after electrophoresis and transfer to nitrocellulose with cDNA clone 28H6.

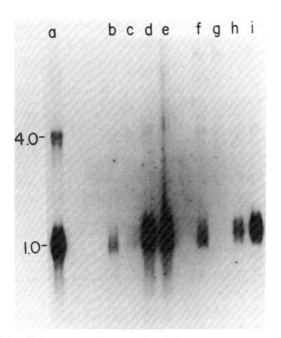

Figure 3 Analysis of 28H6 RNA levels in other mouse cell lines. cDNA clone 28H6 was used to probe total cellular RNA from Krebs ascites carcinoma cells (*a*), subconfluent C3H 10T1/2 cells growing in MEM-10 (*f*), resting 10T1/2 cultures in MEM-0.5 (*g*), and resting 10T1/2 cultures fed for 12 hr with MEM-10 (*h*) or MEM-20 (*i*). Included for comparison are BALB/c 3T3 RNA samples from cultures subconfluent and growing in MEM-10 (*b*), resting in MEM-0.5 (*c*), and fed with MEM-20 for 6 hr (*d*) or 18 hr (*e*).

3T3 cells. In contrast, SVB 10-1 dividing at approximately the same rate in MEM-2 failed to produce any observable 28H6 RNA. SVB 10-2, isolated as a focus in MEM-2, also gave no detectable 28H6 RNA in MEM-2 at any cell density tested; in MEM-10, these cultures contained a very low level of the 1-kb RNA at the lowest cell density, but this amount decreased rapidly below the level of detection as the cell density increased.

Identity of the 28H6 gene

Sequence analysis of the 28H6 cDNA insert, which has only the 3′ portion of the coding sequence, and systematic comparison with sequences in the Los Alamos Data Bank has revealed significant nucleotide and amino acid homology to mammalian prolactins, as noted earlier (Linzer and Nathans 1983). A summary of our current findings is given in Table 1; a complete presentation of the data will be reported elsewhere. From comparison of the degree of homology between 28H6 and prolactins to that between prolactins of different mammalian species, it seems very likely that the 28H6 RNA is not the mouse prolactin messenger. However, the sequence similarity between 28H6 and bovine prolactin is greater than the homology between bovine prolactin and bovine growth hormone, which is another member of the prolactin gene family. It is also significant that the three cysteine and the two tryptophan residues encoded by the portion of the 28H6 gene in the cDNA clone are found in the same locations in all of the mammalian prolactins so far sequenced (Miller and Eberhardt 1983).

Table 1 Relationship of 28H6, bPRL, and bGH

Proteins	Amino acid sequence comparisons		
	identical	related	total
28H6 × bPRL	35 (32%)	10 (9%)	45 (41%)
28H6 × bGH	28 (26%)	8 (7%)	36 (33%)
28H6 × bPRL or bGH	48 (44%)	16 (15%)	64 (59%)
bPRL × bGH	24 (22%)	8 (7%)	32 (30%)
bPRL × rPRL	67 (61%)	12 (11%)	79 (72%)

Abbreviations: (bPRL) bovine prolactin; (bGH) bovine growth hormone; (rPRL) rat prolactin.

Sequence data of bPRL, bGH, and rPRL are from Miller and Eberhardt (1983). The comparisons given in the table include only the region of each protein corresponding to the region of 28H6 that has been sequenced. Nucleotide sequence homology between 28H6 and rPRL in this region is 55%.

Discussion

By screening a cDNA library prepared from the RNA present in serum-stimulated BALB/c 3T3 cells during the late G_1 phase of the cell cycle, we have been able to isolate several clones that hybridize to RNA species that are at a very low concentration in quiescent cells and are elevated in amount after serum stimulation (Linzer and Nathans 1983). One clone in particular, designated 28H6, is a strong candidate for a gene involved in the process of cell growth, since (1) 28H6 RNA is absent or at extremely low levels in resting cells but is elevated in growing cells and in cells stimulated with serum, platelet-derived growth factor, or SV40; (2) the level of this RNA appears to reach a maximum at the onset of cellular DNA synthesis and then decreases in amount; and (3) several mouse cell lines express high levels of 28H6 RNA when actively growing.

The expression of this RNA in SV40-transformed BALB/c 3T3 cells was found to depend on both the conditions of the initial isolation of the specific cell line and on the growth conditions employed for the cell cultures during the preparation of RNA. The SV40-transformed cell line SVB 10-1 expressed 28H6 RNA independent of cell density when cultured in MEM-10 but did not contain detectable quantities of this RNA when proliferating in MEM-2; SVB 10-2 cells also failed to express measurable amounts of this RNA while growing in MEM-2 and revealed a cell-density-dependent level of 28H6 RNA in MEM-10-fed cultures. Thus, at least under certain circumstances, SV40 transformation enables a cell to grow in the absence of detectable 28H6 RNA. This indicates that the 28H6 gene probably does not encode a metabolic enzyme essential in cellular DNA replication nor a structural component of chromosomes or other cellular constituents. Rather, this conditional expression of the 28H6 gene suggests that its encoded protein may function as a regulatory factor in the process of normal cell growth that can be bypassed in SV40 transformants.

The tentative assignment of a regulatory role to the 28H6 gene is further suggested by its close relationship to the polypeptide hormone prolactin. Although the 28H6 mRNA sequence has not been completed, this relationship seems clear from (1) the nucleotide sequence homology between mammalian prolactin genes and the portion of the 28H6 gene included in the cDNA clone (approximately the 3' half of the mRNA sequence); (2) the amino acid homology between mammalian prolactins and the putative 28H6 protein; (3) the identical locations of the cysteine residues, which are known to form disulfide bridges in prolactin (Niall 1981); and (4) the size of the mRNAs for mammalian prolactin and 28H6 (the rat prolactin mRNA also being 1 kb; e.g., see Maurer 1981). The sequence homology strongly suggests that the 28H6 gene is a new member of the prolactin family, which is known to include growth hormone and placental lactogen. In fact, the homology between the mouse 28H6 gene and prolactin is greater than that between either growth hormone or placental lactogen and prolactin from the same species.

Among an astonishing array of functions attributed to it, prolactin has been implicated in the proliferation and differentiation of various glandular tissues (Nicoll 1974). The discovery of the activation of a prolactin-related gene during the transition of fibroblast cells from G_1 to S raises the possibility that a prolactinlike protein may have a more general function in cell proliferation. Further analysis of the 28H6 gene as well as of the structure, location, and action of the encoded protein will be essential to determine if the 28H6 gene product does indeed possess a hormonelike or autocrine activity involved in the regulation of mammalian cell growth.

Acknowledgments

We thank William Pearson for help with DNA sequence comparisons, Phoebe Mounts for supplying SV40-transformed lines, and Laura Sanders for expert technical assistance. This research was supported in part by grant 5-P01-CA-16519 from the National Cancer Institute to D.N. and by a postdoctoral fellowship (1-F32-CA-96699) from the National Cancer Institute to D.I.H.L.

References

Becker, H. and C.P. Stanners. 1972. Control of macromolecular synthesis in proliferating and resting Syrian hamster cells in monolayer culture. *J. Cell. Physiol.* **80:** 51.

Chirgwin, J.M., A.E. Przbyla, R.J. MacDonald, and W.J. Rutter. 1979. Isolation of biologically active ribonucleic acid from sources enriched in ribonuclease. *Biochemistry* **18:** 5294.

Fellous, M., U. Nir, D. Wallach, G. Merline, M. Rubinstein, and M. Revel. 1982. Interferon-dependent induction of mRNA for the major histocompatibility antigens in human fibroblasts and lymphoblastoid cells. *Proc. Natl. Acad. Sci.* **79:** 3082.

Fox, T.O. and A.B. Pardee. 1971. Proteins made in the mammalian cell cycle. *J. Biol. Chem.* **246:** 6159.

Gates, B.J. and M. Friedkin. 1978. Mid-G_1 marker protein(s) in 3T3 mouse fibroblast cells. *Proc. Natl. Acad. Sci.* **75:** 4959.

Glišin, V., R. Crkvenjakov, and C. Byus. 1974. Ribonucleic acid isolated by cesium chloride centrifugation. *Biochemistry* **13:** 2633.

Goldberg, D.A. 1980. Isolation and partial characterization of the *Drosophila* alcohol dehydrogenase gene. *Proc. Natl. Acad. Sci.* **77:** 5794.

Harris, H. 1959. The initiation of deoxyribonucleic acid synthesis in the connective-tissue cell, wtih some observations on the function of the nucleolus. *Biochem. J.* **72:** 54.

Hershko, A., P. Mamont, R. Shields, and G.M. Tomkins. 1971. The pleiotypic response. *Nat. New Biol.* **232:** 206.

Holley, R.W. and J.A. Kiernan. 1974. Control of the initiation of DNA synthesis in in 3T3 cells: Low-molecular-weight nutrients. *Proc. Natl. Acad. Sci.* **71:** 2942.

Johnson, L.F., J.G. Williams, H.T. Abelson, H. Green, and S. Penman. 1975. Changes in RNA in relation to growth of the fibroblast. III. Posttranscriptional regulation of mRNA formation in resting and growing cells. *Cell* **4:** 69.

Lehrach, H., D. Diamond, J.M. Wozney, and H. Boedtker. 1977. RNA molecular weight determination by gel electrophoresis under denaturing conditions, a critical reexamination. *Biochemistry* **16:** 4743.

Linzer, D.I.H. and D. Nathans. 1983. Growth-related changes in specific mRNAs of cultured mouse cells. *Proc. Natl. Acad. Sci.* **80:** 4271.

Maurer, R.A. 1981. Transcriptional regulation of the prolactin gene by ergocryptine and cyclic AMP. *Nature* **294:** 94.

Miller, W.L. and N.L. Eberhardt. 1983. Structure and evolution of the growth hormone gene family. *Endocrine Rev.* **4:** 97.

Mueller, G.C., K. Kajiwara, E. Stubblefield, and R.R. Rueckert. 1962. Molecular events in the reproduction of animal cells. *Cancer Res.* **22:** 1084.

Niall, H.D. 1981. The chemistry of prolactin. In *Prolactin* (ed. R.B. Jaffe), p. 1. Elsevier, New York.

Nicoll, C.S. 1974. Physiological actions of prolactin. In *Handbook of physiology*, section 7 (ed. E. Knobil and W.H. Sawyer), vol. 4, part 2, p. 253. American Physiological Society, Bethesda, Maryland.

Pardee, A.B., R. Dubrow, J.L. Hamlin, and R.F. Kletzien. 1978. Animal cell cycle. *Annu. Rev. Biochem.* **47:** 715.

Peden, K.W.C., P. Mounts, and G.S. Hayward. 1982. Homology between mammalian cell DNA sequences and human herpesvirus genomes detected by a hybridization procedure with high complexity probe. *Cell* **31:** 71.

Pledger, W.J., C.A. Hart, K.L. Locatell, and C.D. Scher. 1981. Platelet-derived growth factor–modulated proteins: Constitutive synthesis by a transformed cell line. *Proc. Natl. Acad. Sci.* **78:** 4358.

Powell, W.F. 1962. The effects of ultraviolet irradiation and inhibitors of protein synthesis on the initiation of deoxyribonucleic acid synthesis in mammalian cells in culture. *Biochim. Biophys. Acta* **55:** 969.

Reznikoff, C.A., D.W. Brankow, and C. Heidelberger. 1973. Establishment and characterization of a cloned line of C3H mouse embryo cells sensitive to postconfluence inhibition of division. *Cancer Res.* **33:** 3231.

Riddle, V.G.H., R. Dubrow, and A.B. Pardee. 1979. Changes in the synthesis of actin and other cell proteins after stimulation of serum-arrested cells. *Proc. Natl. Acad. Sci.* **76:** 1298.

Rigby, P.W.J., M. Dieckmann, C. Rhodes, and P. Berg. 1977. Labeling deoxyribonucleic acid to high specific activity in vitro by nick translation with DNA polymerase I. *J. Mol. Biol.* **113:** 237.

Rossow, P.W., V.G.H. Riddle, and A.B. Pardee. 1979. Synthesis of labile, serum-dependent protein in early G_1 controls animal cell growth. *Proc. Natl. Acad. Sci.* **76:** 4446.

Rudland, P.S., W. Seifert, and D. Gospodarowicz. 1974. Growth control in cultured mouse fibroblasts: Induction of the pleiotypic response by a purified growth factor. *Proc. Natl. Acad. Sci.* **71:** 2600.

Salas, J. and H. Green. 1971. Proteins binding to DNA and their relation to growth in cultured mammalian cells. *Nat. New Biol.* **229:** 165.

Taylor, E.W. 1965. Control of DNA synthesis in mammalian cells in culture. *Exp. Cell Res.* **40:** 316.

Thomas, G., G. Thomas, and H. Luther. 1981. Transcriptional and translational control of cytoplasmic proteins after serum stimulation of quiescent Swiss 3T3 cells. *Proc. Natl. Acad. Sci.* **78:** 5712.

Thomas, P.S. 1980. Hybridization of denatured RNA and small DNA fragments transferred to nitrocellulose. *Proc. Natl. Acad. Sci.* **77:** 3082.

Todaro, G.J. and H. Green. 1963. Quantitative studies of the growth of mouse embryo cells in culture and the development into established lines. *J. Cell Biol.* **17:** 299.

Williams, J.G. and S. Penman. 1975. The messenger RNA sequences in growing and resting mouse fibroblasts. *Cell* **6:** 197.

pp60$^{c\text{-}src}$ Expression in Embryonic Nervous Tissue: Immunocytochemical Localization in the Developing Chick Retina

L.K. Sorge,* B.T. Levy,† T.M. Gilmer,‡ and P.F. Maness*†

Departments of Biochemistry† and Microbiology‡ and Cancer Research Center,* University of North Carolina School of Medicine, Chapel Hill, North Carolina 27514

pp60$^{c\text{-}src}$ is a tyrosine-specific protein kinase that is highly homologous to the transforming protein of Rous sarcoma virus (RSV) but is encoded in the genome of normal cells. Species throughout the animal kingdom from human (Oppermann et al. 1979; Rohrschneider et al. 1979; Shealy and Erikson 1981), rat (Oppermann et al. 1979), chicken (Collett et al. 1978), Drosophila (Simon et al. 1983), to sponge (Schartl and Barnekow 1982) express low levels of pp60$^{c\text{-}src}$, yet the cellular role of this enzyme is unknown. The high degree of phylogenetic conservation of pp60$^{c\text{-}src}$ and the known ability of its homolog pp60$^{v\text{-}src}$ to induce cell proliferation in transformed cells (Hanafusa 1977) suggest that pp60$^{c\text{-}src}$ may be involved in growth or differentiation of multicellular organisms.

We have used the immune-complex protein kinase assay to examine the activity of pp60$^{c\text{-}src}$ in chick and human embryonic tissues (Levy et al. 1984). Both pp60$^{c\text{-}src}$ and pp60$^{v\text{-}src}$ phosphorylate immunoglobulin G (IgG) from rabbits bearing RSV-induced tumors at tyrosine residues on the heavy chains (Collett et al. 1978, 1979; Collett and Erikson 1978; Erikson et al. 1979; Karess et al. 1979; Maness et al. 1979; Oppermann et al. 1979; Rohrschneider et al. 1979; Hunter and Sefton 1980; Levinson et al. 1980; Maness and Levy 1983). Although IgG is an artificial substrate, this reaction has proved to be a useful tool for the study of the activity and localization of these two proteins because it is so highly specific. We found that pp60$^{c\text{-}src}$ kinase activity in extracts of chick embryo and human fetal brain increases dramatically during the mid to late organogenesis period, raising the possibility that pp60$^{c\text{-}src}$ is a product of a cell type that differentiates in the nervous system at this time. Similar results in the chick system have been reported by Cotton and Brugge (1983).

Identifying the cell type that expresses pp60$^{c\text{-}src}$ in the brain poses a formidable problem in view of the complexity of cell types present. In contrast, the neural retina, which develops as an outpocketing of the neural tube, offers the advantage of having fewer cell types. The identification of cell types in the neural retina is also facilitated by the organization of neurons into discrete layers according to a precise timetable during development of the eye. We have investigated the spatial and temporal expression of pp60$^{c\text{-}src}$ in the neural retina of the developing chick by immunocytochemistry. The results of these and other investigations (Sorge et al. 1984) show that pp60$^{c\text{-}src}$ is expressed in neurons at a time corresponding to the onset of differentiation.

Experimental Procedures

Preparation of extracts

Chick embryos negative for lymphoid leukosis virus (SPAFAS, Norwich, Conn.) were staged (Hamburger and Hamilton 1951) and frozen at −70°C until use. Embryos were dissected into two parts—trunk with appendages and heads. Trunks or heads from at least three embryos at the same developmental stage were homogenized in 4 volumes of lysis buffer (25 mM potassium phosphate, pH 7.1, 1.25 mM EDTA, 0.6% sodium deoxycholate, 1.25% Nonidet P-40, 12.5 mM NaF, 125 KIU/ml aprotinin, 2.5 mM dithiothreitol, 0.2 M KCl) by using 10 strokes of a Dounce tissue grinder fitted with a loose pestle and 10 strokes with a tight pestle, followed by sonication for 2 minutes using a Heat Systems–Ultrasonics Model 350 sonicator (Plainview, New York) at 65 W for 2 minutes at 4°C. Homogenates were centrifuged at 100,000g for 1 hour and the supernatants were retained. Pellets were reextracted in the same manner, and the resulting supernatants from both centrifugations were stored at −70°C. Of the total extracted protein, 95% was contributed by the first supernatant. In addition, the first supernatant contributed 97% of the total pp60$^{c\text{-}src}$ kinase activity. Extracts were prepared from organ tissues of the chick in the same manner. Samples from the frontal lobe of the cerebrum and the liver of a 32-week human fetus, which had been frozen at autopsy, were extracted similarly.

Immune-complex kinase assays

Immune-complex kinase assays for pp60$^{c\text{-}src}$ protein kinase were carried out as described previously (Collett and Erikson 1978). Briefly, pp60$^{c\text{-}src}$ was immunoprecipitated from embryo extracts using tumor-bearing rabbit (TBR) serum prepared by the method of Brugge and Erikson (1977). Antibodies in this serum were able to react with the cellular as well as the viral pp60^{src}. Three dilutions of extract were used in each assay to establish conditions of antibody excess. Duplicate measurements were made. Immune complexes were isolated from the extracts with protein A–Sepharose (Pharmacia), washed extensively, and incubated with [γ-^{32}P]ATP (2900 Ci/mmole; New England Nuclear) for 10 minutes at 37°C. IgG heavy chains (TBR IgG) were isolated on SDS-polyacrylamide gels, and their extent of phosphorylation was quantitated by counting dried gel slices for ^{32}P in Aquasol (New England Nuclear).

Protein was determined by the Lowry method (Lowry et al. 1951).

One unit of activity was defined as the amount of pp60^{c-src} kinase catalyzing the transfer of 1 femtomole of phosphate from ATP to the heavy chain of TBR IgG in 10 minutes at 37°C under the conditions of the assay.

Western blotting procedures
Neural retina extract containing 90–250 µg of protein were electrophoresed on an 8.5% SDS-polyacrylamide gel. Western blotting was carried out as described by Gilmer and Erikson (1983), using a 1:500 dilution of primary antiserum directed against pp60^{v-src} purified from bacteria expressing a molecularly cloned *src* gene (Gilmer and Erikson 1981).

Immunostaining procedures
Chick embryos (stages 21–32) or chick eyes (stages 36 to adult) were fixed for 24 hours in ice-cold 4% paraformaldehyde in 70 mM phosphate buffer at pH 6.8, dehydrated in an ascending series of alcohols and toluene, embedded in Paraplast-plus, and sectioned at 10 µm. Paraffin sections were mounted on glass slides and dried at 45°C for 1 week. After deparaffinizing, sections were treated with a 0.6 mg/ml phosphate-buffered saline (PBS) solution of bovine trypsin (40 units/mg protein; Boeringer Mannheim) for 10 minutes, then rinsed in two changes of PBS. Following a 10-minute preincubation with 1% normal sheep serum, the sections were incubated in primary antiserum directed against pp60^{v-src} purified from bacteria expressing a molecularly cloned *src* gene (Gilmer and Erikson 1981) at a dilution of 1:1000 to 1:4000 for 48 hours at 4°C. Two different preparations of antibody gave similar results. A modification (A.C. Towle et al., in prep.) of the avidin-biotin-complex technique (Hsu et al. 1981) was used to localize pp60^{c-src} immunoreactivity. To demonstrate competition, the primary antiserum was preincubated at a 1:2000 dilution with 30 µg/ml bacterially expressed pp60^{v-src} for 24 hours at 4°C prior to application to the section.

Results

Stage-specific expression of pp60^{c-src} kinase activity in chick embryos

The temporal expression of pp60^{c-src} in head and trunk regions of chick embryos from stages 21 to 37 (Hamburger and Hamilton 1951) was examined by the immune-complex kinase assay. These stages encompassed the major period of organogenesis. Head and trunk regions of embryos at all stages examined expressed pp60^{c-src} kinase activity with maximal activity in both regions present at stage 32 or approximately day 7.5 of incubation (Fig. 1). The head, which was predominantly neural tissue, showed a threefold increase in pp60^{c-src} kinase activity per milligram of protein from stages 21 to 32. The trunk region at stage 21 contained more pp60^{c-src} kinase activity per milligram than the head and showed only a slight increase as development approached stage 32. After stage 32, a rapid decline in activity occurred in

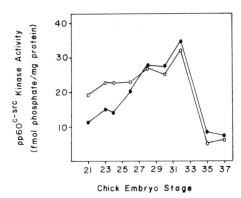

Figure 1 Stage-specific expression of pp60^{c-src} in head and trunk regions of the developing chick. Extracts were prepared from head and trunk regions of chick embryos, and pp60^{c-src} kinase activity was measured using the immune-complex kinase assay. Activity was expressed as fmoles of phosphate incorporated into IgG heavy chains per milligram of extract protein under the conditions of the assay as described in Experimental Procedures. Represented is pp60^{c-src} kinase activity in the head (●) and in the trunk with appendages (○).

both head and trunk extracts. During this developmental period, the embryo was engaged in rapid growth; the average protein content per embryo at stages 21, 32, and 37 was 1.1, 40, and 125 mg, respectively. Thus, the rapid drop in pp60^{c-src} kinase activity per milligram of protein after stage 32 may in part be due to increased synthesis of proteins during differentiation of the embryo rather than to decreased synthesis of pp60^{c-src}.

No evidence for an activator of pp60^{c-src} at stage 32 was found. Activity of pp60^{c-src} kinase in a mixture of stages 21 and 32 head extracts (3.65 units) was approximately the same as the sum of activity in individual extracts at stage 21 (0.60 units) and stage 32 (2.55 units). Selective degradation of ATP was not observed in immunoprecipitates of pp60^{c-src} from head extracts at stages 21, 26, 28, 32, and 37 when supernatants were analyzed by thin-layer chromatography (data not shown).

Tissue-specific expression of pp60^{c-src} kinase activity in chick and human embryos

The tissue distribution of pp60^{c-src} kinase was examined in the chick embryo at stage 32 (Table 1). Tissues were dissected from six stage-32 embryos, pooled for extract preparation, and assayed for pp60^{c-src} by the immune-complex kinase assay. The spinal region, which contained the notocord, spinal cord, and developing spinal ganglia, had the highest concentration of pp60^{c-src} per milligram of protein. The other neural ectoderm tissues, brain, and eye also had substantial concentrations. The expression of pp60^{c-src} kinase activity was not restricted to neural tissues, as significant levels were also found in the mesodermally derived chick embryo heart. In contrast, the liver, which is of endodermal origin, contained low levels of activity. Activity of pp60^{c-src} kinase per milligram of protein in stage-32 neural tissues was fourfold to eightfold greater than in secondary cultures of fibroblasts prepared from embryos at the same stage but

Table 1 Tissue Distribution of pp60$^{c\text{-}src}$ Kinase in Stage-32 Chick Embryos

	Activity/mg extract protein[a] (units/mg)	Total activity (units/embryo)
Spinal region	54	126
Brain	35	417
Heart	35	21
Carcass	27	157
Eye	25	157
Limb bud	23	70
Liver	5	6
Whole head	35	680
Whole trunk with appendages	38	633
Chick embryo fibroblasts	7	n.a.[b]
RSV-transformed chick embryo fibroblasts	200[c]	n.a.

[a] A unit of activity is defined as the amount of pp60$^{c\text{-}src}$ catalyzing the transfer of 1 fmole of phosphate from ATP to the heavy chain of TBR IgG in the immune complex in 10 min under the conditions of the assay.
[b] Not applicable.
[c] pp60$^{v\text{-}src}$ kinase activity.

lower than the level of pp60$^{v\text{-}src}$ in RSV-transformed fibroblasts.

To investigate the potential importance of pp60$^{c\text{-}src}$ in human development, we examined the levels of pp60$^{c\text{-}src}$ kinase activity in human fetal tissues. Extracts were prepared from frontal cerebral cortex and liver of a 32-week gestational-age human fetus and assayed for pp60$^{c\text{-}src}$ kinase activity in the immune-complex kinase assay. The kinase activity of pp60$^{c\text{-}src}$ per milligram of protein in the human fetal brain (57 units/mg) was eightfold greater than in the liver (7 units/mg). It is not known if the stage of human fetal tissue examined represented the maximal stage of expression of pp60$^{c\text{-}src}$ activity.

Biochemical and immunocytochemical analysis of pp60$^{c\text{-}src}$ in the developing chick retina

The chick neural retina was chosen as a model system for the immunocytochemical localization of pp60$^{c\text{-}src}$ in the developing nervous system because it is populated by a limited number of neuronal and nonneuronal cell types that organize to form functional layers during differentiation (Fig. 3D). Figure 2 demonstrates that the temporal expression of pp60$^{c\text{-}src}$ kinase activity in the developing neural retina is similar to that found in head extracts. Peak kinase activity occurred at stage 35. Neural retina extracts were subjected to Western blot analysis with primary antiserum directed against pp60$^{v\text{-}src}$ purified from bacteria expressing a cloned v-src gene (Gilmer and Erikson 1981). This antibody was shown to immunoprecipitate both native and denatured pp60$^{c\text{-}src}$ from chicken cells (Cotton and Brugge 1983). Extracts from neural retina at all stages exhibited a single protein species of 60,000 daltons on Western blots, indicating that immunocytochemical localization with this antibody was specific for pp60$^{c\text{-}src}$.

Immunocytochemical localization of pp60$^{c\text{-}src}$ was examined in neural retina tissues of the developing chick, using a modification (A.C. Towle et al., in prep.) of the avidin-biotin-complex technique (Hsu et al. 1981). Antibody dilutions equivalent to serum dilutions of 1:2000 or greater were used in all experiments. Early growth of the sensory layer is due to mitotic division of nuclei closest to the pigmented layer. Ganglion cell precursors cease DNA synthesis in the germinal zone and migrate through the central zone between stages 19 and 31 of development to reach their final destination at the vitreal surface of the retina (Romanoff 1960; Kahn 1973). Immunoreactive pp60$^{c\text{-}src}$ appears to be localized in these migrating postmitotic neuroblasts at stage 23 (day 4) (Fig. 3A). By stage 32 (day 7.5), the migrating cells reach the inner zone and organize to form a layer of neurons called the ganglion cell layer, which expresses pp60$^{c\text{-}src}$

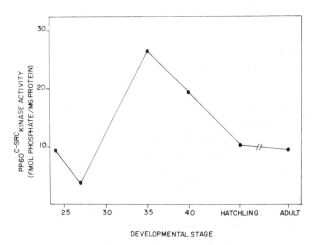

Figure 2 Stage-specific expression of pp60$^{c\text{-}src}$ in neural retina of developing chick. Extracts were prepared from neural retina of chick embryos, and pp60$^{c\text{-}src}$ kinase activity was measured using the immune-complex kinase assay as described in the text.

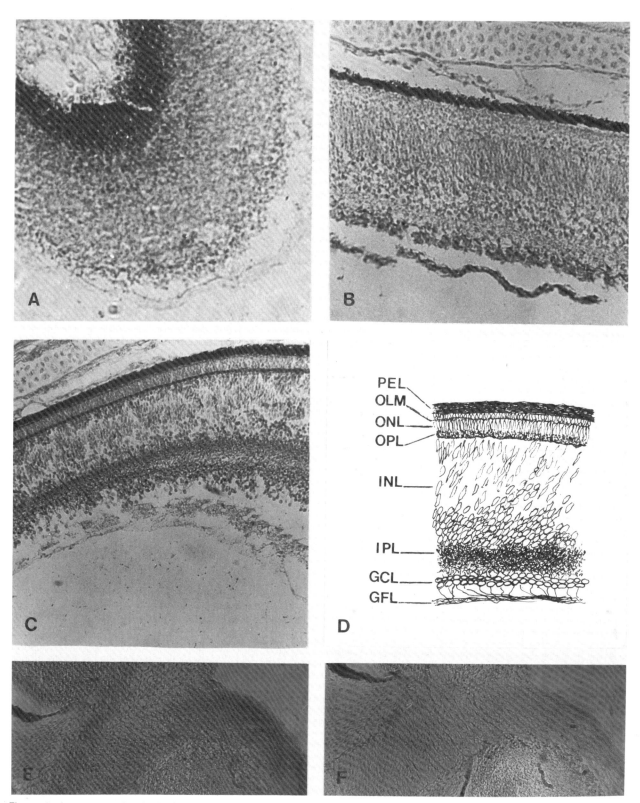

Figure 3 Immunocytochemical detection of pp60$^{c\text{-}src}$ in developing chick retina. (*A*) Immunocytochemical detection of pp60$^{c\text{-}src}$ showing staining in migrating neuroblasts at stage 23. (*B*) Staining of GFL, GCL, and OPL at stage 32. (*C*) staining of GFL, GCL, OPL, IPL, and OLM at stage 39. (*D*) Diagram of the fully differentiated chick retina. (PEL) Pigmented epithelium layer; (OLM) outer limiting membrane; (ONL) outer nuclear layer; (OPL) outer plexiform layer; (INL) inner nuclear layer; (IPL) inner plexiform layer; (GCL) ganglion cell layer; (GFL) ganglion fiber layer. (*E*) Staining of the optic nerve at stage 32. (*F*) Competition of optic nerve staining by preincubation of the antibody with pp60$^{v\text{-}src}$ as described in Experimental Procedures. All magnifications, 152×.

(Fig. 3B). The ganglion cells produce axons that form the prominent pp60^{c-src}-containing ganglion fiber layer. pp60^{c-src} was also localized within the developing inner plexiform layer consisting of ganglion cell dendrites and processes of the bipolar and amacrine cells of the inner nuclear layer. The optic nerve, which is an extension of the nerve fiber layer, shows pp60^{c-src} staining at stage 32 (Fig. 3E). pp60^{c-src} immunoreactivity at all stages examined was blocked by preincubation of the antibody with pp60^{v-src} purified from E. coli expressing the cloned v-src gene (Gilmer and Erikson 1981). Figure 3F demonstrates competition of stage-32 optic nerve staining; a 1:2000 dilution of antibody (15 μg/ml) was preincubated with 30 μg/ml of pp60^{v-src}.

The differentiation of rods and cones, starting at stages 36–38 (days 10–12), results in the formation of the outer nuclear layer, consisting of the cell bodies, and the outer plexiform layer, where the rods and cones form synapses with inner nuclear layer neurons. At stage 39 (day 13), in addition to the previously stained regions, immunoreactive pp60^{c-src} appeared in the outer plexiform layer (Fig. 3C). The outer limiting membrane containing junctional complexes between rods and cones and the apical ends of Müller glia also exhibited pp60^{c-src} staining. This is also the region where the inner segments of the rod and cone photoreceptors are located. Rod and cone cell bodies of the outer nuclear layer did not appear to stain as intensely for pp60^{c-src} as the outer plexiform layer. The pattern of pp60^{c-src} staining observed in stage-39 neural retinas persisted through stage-45 (days-19–20) embryos, hatchlings, and adults. A diagrammatic sketch of the fully differentiated chick retina is presented in Figure 3D.

Discussion

Elevated levels of the normal cellular protein pp60^{c-src} were found in neural tissues of the human and chick during embryogenesis. pp60^{c-src} activity was found to be expressed in a stage-specific manner in chick head regions and neural retina, implying that this protein is the product of a cell type that arises during embryogenesis and is important to the development of the nervous system. The amount of pp60^{c-src} kinase activity in neural tissues at the peak stage was significantly greater than from chick embryo fibroblasts prepared from chicks at the same developmental stage. This suggests that either the cell types expressing high levels of pp60^{c-src} in the embryo do not proliferate under standard tissue-culture conditions or the synthesis of pp60^{c-src} is turned off during differentiation of the cells in culture. pp60^{c-src} kinase activity in the neural retina, as well as in whole head and trunk regions of the embryo, showed a substantial decrease after stage 32 when expressed on a per-milligram-protein basis. This decrease was also observed on Western blots (data not presented). When analyzed in this way, the decrease in pp60^{c-src} activity may reflect proteins other than pp60^{c-src} being newly synthesized after stage 32 rather than a decrease in pp60^{c-src} synthesis. In support of the former interpretation, the immunostaining pattern of pp60^{c-src} seen in neural retinas of stage-36 embryos persisted in the hatchling and adult.

The time of appearance of pp60^{c-src} immunoreactivity in developing retina was correlated with the onset of differentiation of neuronal cells. The observed staining pattern of pp60^{c-src} in discrete neuronal cell layers is distinct from the distribution of Müller glia, which span the width of the sensory layer. Since Müller glia are the major nonneuronal cell type in the vertebrate retina (Bignami and Dahl 1979), it appears that pp60^{c-src} is a product of neuronal cell types. However, not all neuronal cells of the retina express observable amounts of pp60^{c-src}, as shown by the lack of staining in the inner nuclear layer of neurons. The ganglion fiber layer, inner plexiform layer, and outer plexiform layer appeared to express the highest concentrations of immunoreactive pp60^{c-src}. These layers are rich in neuronal cell processes. Nonetheless, pp60^{c-src} does not appear to be restricted to processes, because nerve cell bodies located in the ganglion cell layer clearly contained immunoreactive pp60^{c-src}. To localize pp60^{c-src} to discrete neuronal cell regions, a detailed immunocytochemical analysis of neural retina at the electron microscopic level is in progress.

Our results support a model in which pp60^{c-src} is the product of a developmentally regulated gene that is important in neuronal function. Other cellular oncogenes, including c-fos, c-abl, and c-ras^{Ha}, have recently been shown to be expressed in a stage- and tissue-specific manner during mouse development (Müller et al. 1982). Taken together, these results suggest that cellular oncogenes homologous to transforming genes of retroviruses may play a role in developmentally regulated processes of cell differentiation. The identification of a gene and its product, apparently important to neural development in the human and chick, provides a new opportunity for investigating the developing nervous system at the molecular level.

Acknowledgments

We gratefully acknowledge the generous assistance with immunostaining procedures provided by Drs. Andrew Towle and Jean Lauder (University of North Carolina).

This work was supported by grants from the National Science Foundation (PCM-8203857), National Cancer Institute (2-PO1-CA19014, 5T32-CA09156), and the American Cancer Society (BC-416).

References

Bignami, A. and D. Dahl. 1979. The radial glia of Müller in the rat retina and their response to injury. An immunofluorescence study with antibodies to the glial fibrillary acidic (GFA) protein. *Exp. Eye Res.* **28**: 63.

Brugge, J.S. and R.L. Erikson. 1977. Identification of a transformation-specific antigen induced by an avian sarcoma virus. *Nature* **269**: 346.

Collett, M.S. and R.L. Erikson. 1978. Protein kinase activity associated with the avian sarcoma virus src gene product. *Proc. Natl. Acad. Sci.* **75**: 2021.

Collett, M.S., J.S. Brugge, and R.L. Erikson. 1978. Characterization of a normal avian cell protein related to the avian sarcoma virus transforming gene product. *Cell* **15:** 1363.

Collett, M.S., E. Erikson, A.F. Purchio, J.S. Brugge, and R.L. Erikson. 1979. A normal cell protein similar in structure and function to the avian sarcoma virus transforming gene product. *Proc. Natl. Acad. Sci.* **76:** 3159.

Cotton, P.C. and J.S. Brugge. 1983. Neural tissues express high levels of the cellular *src* gene product pp60^{c-src}. *Mol. Cell. Biol.* **3:** 1157.

Erikson, R.L., M.S. Collett, E. Erikson, and A.F. Purchio. 1979. Evidence that the avian sarcoma virus transforming gene product is a cyclic AMP–independent protein kinase. *Proc. Natl. Acad. Sci.* **76:** 6260.

Gilmer, T.M. and R.L. Erikson. 1981. Rous sarcoma virus transforming protein, pp60^{src}, expressed in *E. coli*, functions as a protein kinase. *Nature* **294:** 771.

———. 1983. Development of anti-pp60^{src} serum antigen produced in *Escherichia coli*. *J. Virol.* **45:** 462.

Hamburger, V. and H.L. Hamilton. 1951. A series of normal stages in the development of the chick embryo. *J. Morphol.* **88:** 49.

Hanafusa, H. 1977. Cell transformation by RNA tumor viruses. In *Comprehensive virology* (ed. H. Fraenkel-Conrat and R.P. Wagner), vol. 10, p. 401. Plenum Press, New York.

Hsu, S.M., L. Raine, and H. Fanger. 1981. Use of avidin-biotin peroxidase complex (ABC) in immunoperoxidase techniques: A comparison between ABC and unlabeled antibody (PAP) procedures. *J. Histochem. Cytochem.* **29:** 577.

Hunter, T. and B. Sefton. 1980. Transforming gene product of Rous sarcoma virus phosphorylates tyrosine. *Proc. Natl. Acad. Sci.* **77:** 1311.

Kahn, A.J. 1973. Ganglion cell formation in the chick neural retina. *Brain Res.* **63:** 285.

Karess, R.E., W.S. Hayward, and H. Hanafusa. 1979. Cellular information in the genome of recovered avian sarcoma virus directs the synthesis of transforming protein. *Proc. Natl. Acad. Sci.* **76:** 3154.

Levinson, A.D., H. Oppermann, H.E. Varmus, and J.M. Bishop. 1980. The purified product of the transforming gene of avian sarcoma virus phosphorylates tyrosine. *J. Biol. Chem.* **255:** 11973.

Levy, B.T., L.K. Sorge, A. Meymandi, and P.F. Maness. 1984. pp60^{c-src} kinase is in chick and human embryonic tissues. *Dev. Biol.* (in press).

Lowry, O.H., N.J. Rosebrough, A.L. Farr, and R.J. Randall. 1951. Protein measurement with the Folin phenol reagent. *J. Biol. Chem.* **193:** 265.

Maness, P.F. and B.T. Levy. 1983. Highly purified pp60^{src} induces the actin transformation in microinjected cells and phosphorylates selected cytoskeletal proteins in vitro. *Mol. Cell. Biol.* **3:** 102.

Maness, P.F., H. Engeser, M.E. Greenberg, M. O'Farrell, W.E. Gall, and G.M. Edelman. 1979. Characterization of the protein kinase activity of avian sarcoma virus *src* gene product. *Proc. Natl. Acad. Sci.* **76:** 5028.

Müller, R., D.J. Slamon, J.M. Tremblay, M.J. Cline, and I.M. Verma. 1982. Differential expression of cellular oncogenes during pre- and postnatal development of the mouse. *Nature* **299:** 640.

Oppermann, H., A.D. Levinson, H.E. Varmus, L. Levintow, and J.M. Bishop. 1979. Uninfected vertebrate cells contain a protein that is closely related to the product of the avian sarcoma virus transforming gene (*src*). *Proc. Natl. Acad. Sci.* **76:** 1804.

Rohrschneider, L.R., R.N. Eisenman, and C.R. Leitch. 1979. Identification of a Rous sarcoma virus transformation-related protein in normal avian and mammalian cells. *Proc. Natl. Acad. Sci.* **76:** 4479.

Romanoff, A.L. 1960. *The avian embryo, structural and functional development*. Macmillan, New York.

Schartl, M. and A. Barnekow. 1982. The expression in eukaryotes of a tyrosine kinase which is reactive with pp60^{v-src} antibodies. *Differentiation* **23:** 109.

Shealy, D.J. and R.L. Erikson. 1981. Human cells contain two forms of pp60^{c-src}. *Nature* **293:** 666.

Simon, M.A., T.B. Kornberg, and J.M. Bishop. 1983. Three loci related to the *src* oncogene and tyrosine-specific protein kinase activity in *Drosophila*. *Nature* **302:** 837.

Sorge, L.K., B.T. Levy, and P.F. Maness. 1984. pp60^{c-src} is developmentally regulated in the neural retina. *cell* **36:** 249.

Gene Expression in Human Epidermal Basal Cells: Changes in Protein Synthesis Accompanying Differentiation and Transformation

J.E. Celis, S.J. Fey, P.M. Larsen, and A. Celis
Division of Biostructural Chemistry, Department of Chemistry, Aarhus University, DK-8000 Aarhus C, Denmark

One approach to searching for cellular proteins involved in the control of cell proliferation has been the analysis of the overall patterns of gene expression in pairs of normal and transformed cells by means of high-resolution, two-dimensional gel electrophoresis (O'Farrell 1975; O'Farrell et al. 1977; for references, see Celis et al. 1983). In this laboratory, we have hypothesized that neoplastic transformation is due to the abnormal expression of normal genes (Bravo and Celis 1980a, 1982b; Celis and Bravo 1981; Celis et al. 1983; see also other articles in this volume). Therefore, it was expected that a careful and detailed study of the proteins synthesized by normal and transformed cells under a variety of physiological conditions could lead to the identification of cellular proteins that may be involved in the control of cell proliferation. The ultimate goal of these studies is the elucidation of the pathway(s) that control(s) cell proliferation in normal cells and hence the determination of how alterations of this pathway lead to abnormal growth characteristics and/or to neoplastic transformation and cancer.

In general, the analysis of cultured somatic cells has shown that neoplastic transformation (virally, chemically, or spontaneously induced) results in changes in the relative proportions (rate of synthesis, modification, degradation) of proteins normally synthesized by somatic cells rather than in the appearance of new proteins (Leavitt and Moyzig 1978; Radke and Martin 1979; Forchhammer and Klarlund 1979; Bravo and Celis 1980a, 1982b; for further references, see Celis et al. 1983). Induction of a few protein(s) has been reported in virally transformed cells but with a few exceptions is not commonly observed in chemically or spontaneously transformed cells. Failure to detect new proteins, however, may be due to masking by other cellular proteins, low abundance, or both.

To date, a number of transformation-sensitive polypeptides common to various cell types have been identified in cultured human cells and tumors (Bravo and Celis 1982b; Forchhammer and Macdonald-Bravo 1983; Celis et al. 1983). The synthesis of many of these polypeptides is also sensitive to changes in growth rate (Bellatin et al. 1982), and a significant number have been identified in other species (Bravo and Celis 1980a; Celis et al. 1983). Of these, the nuclear polypeptide cyclin (IEF 49; M_r 36,000; HeLa protein catalog number, Bravo et al. 1981a,d; Bravo and Celis 1982a, 1983) is the most interesting as the rate of synthesis of this protein also increases in S phase of the cell cycle in HeLa cells (Bravo and Celis 1980b). The function of the transformation-sensitive proteins is as yet unknown, and work is now underway to prepare antibodies to some of these proteins. These will facilitate functional studies and should be of value in identifying these proteins in normal and tumor cells of various origins.

Presently, there is little information concerning the level of the transformation-sensitive polypeptides in human tumors and their corresponding normal counterparts (Forchhammer and Macdonald-Bravo 1983; Celis et al. 1983). The main obstacle in such studies is the heterogeneity of the tumor and normal tissue, which makes a direct comparison impossible. To approach this problem, we have chosen as a model system the human epidermis and its transformed counterparts (tumors and in vitro transformed cells). This system is of great interest because it allows an analysis of the interrelationship between cell proliferation and differentiation. As a first approximation, we have begun to prepare a battery of monoclonal and polyclonal antibodies to aid the identification and characterization of the epidermal basal cells. Here, we present a comparative study of the polypeptides synthesized by basal, differentiated, and SV40-transformed keratinocytes (Taylor-Papadimitriou et al. 1982). A preliminary analysis of basal cell carcinomas is also included. Besides revealing a few markers for differentiation, these studies have allowed us to identify some of the transformation-sensitive polypeptides in the SV40-transformed keratinocytes.

Experimental Procedures
Preparation of human epidermal cells
Strips of split skin from human leg (Fig. 1a) were washed three times in Hank's buffered saline and were placed in 10 ml of 0.25% trypsin in Hank's (Gibco 1:250) at 4°C for 15–17 hours. Following incubation, the strips were washed three times in Hank's, and the epidermis was detached from the dermis, using fine forceps. The epidermal samples were then washed twice in Hank's, minced finely with the aid of a scalpel, and incubated with stirring for 1 hour at 37°C in a 0.25% trypsin solution containing 0.02% EDTA. The cell suspension was filtered through gauze, and the cells were recovered by

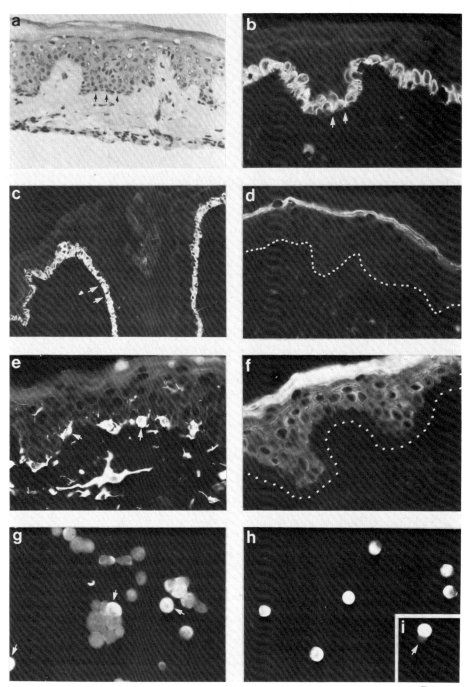

Figure 1 Immunofluorescent staining of human skin sections and epidermal cells with various antibodies. (a) Split skin section stained with hemotoxylin-eosin; (b,c) skin sections fixed with methanol:acetone and reacted with the basal cell monoclonal antibody; arrows in b and c (outer cell layer of the upper region of a hair follicle) indicate basal cells; (d) skin section fixed in methanol:acetone and reacted with a monoclonal antibody that stains stratum granulosum; (e) similar section as d but reacted with a monoclonal antibody that stains melanocytes and dermal fibroblasts; (f) similar section as d but reacted with a polyclonal antibody that stains suprabasal cells in the epidermis; (g) total epidermal cells fixed with methanol:acetone and reacted with the basal cell antibody; (→) basal cells; (h) purified preparation of basal cells reacted with the basal cell antibody; (i) same section as h but showing a suprabasal cell connected to a basal cell. The dotted lines in d and f denote the separation between dermis and epidermis. Magnifications: (a,c) 140×; (b,d-i) 360×.

centrifugation (5 min at 500g). They were then resuspended in Dulbecco's modified Eagle's medium (DMEM; Gibco, standard Ca^{++} concentration) containing 10% fetal calf serum and antibiotics (penicillin, 100 IU/ml; streptomycin, 50 µg/ml).

Preparation of a cell fraction highly enriched in basal cells

The suspension of epidermal cells was plated in microtiter plates (96 wells; Nunc) containing 9-mm² coverslips. After 3–4 hours at 37°C, the coverslips were removed from the wells and were washed eight times by dipping into wells containing DMEM. Under these conditions, only firmly attached round cells remained in the coverslips. Coverslips containing only a few cells (~50–100) were used for [^{35}S]methionine labeling.

The procedures for labeling cells with [^{35}S]methionine (Bravo et al. 1981c; Celis and Bravo 1981), two-dimensional gel electrophoresis (O'Farrell 1975; Bravo et al. 1982), and indirect immunofluorescence of cultured cells and of cryostat sections (Mose Larsen et al. 1983) have been described in detail elsewhere. The preparation of various monoclonal antibodies raised against formaldehyde-fixed total epidermal cells will be described elsewhere. Only the cells producing the basal cell antibody have been cloned so far (J.E. Celis et al., in prep.). The other two antibodies were used as supernatants from clones of cells obtained after fusion.

Results

Immunofluorescent characterization of cultured human epidermal basal and differentiated cells

The cells that attached to the 9-mm² coverslips 3–4 hours after plating were predominantly round, highly birefringent cells. Characterization of these cells was carried out by indirect immunofluorescence using monoclonal antibodies raised against formaldehyde-fixed epidermal cells. These include monoclonals that stain basal cells (Fig. 1b), including the outer cell layer of hair follicles (Fig. 1c) and eccrine sweat glands (not shown); stratum granulosum (Fig. 1d); and melanocytes and fibroblasts (Fig. 1e). Mouse polyclonal antibodies that stain suprabasal cells (Fig. 1f) and keratins in all layers of the epidermis (not shown) were also used.

Figure 1g shows immunofluorescence staining of a crude preparation of epidermal cells fixed with methanol:acetone and reacted with the basal cell antibody. A similar staining of a preparation of purified basal cells as used for [^{35}S]methionine labeling is shown in Figure 1h. Three hours after plating, about 89% of the cells stained positively with the basal cell antibody, whereas only 4% reacted with the suprabasal antibody (Fig. 2a). It should be stressed that the basal cell monoclonal antibody sometimes stains cells in suprabasal positions (not shown; J.E. Celis et al., in prep.), and these may be included in the analyzed population of cells (Fig. 1i). The preparation contained at the most 7% of melanocytes and/or fibroblasts (Fig. 2b; see also Fig. 1e). None of the cells reacted with the monoclonal antibody that stains the stratum granulosum (not shown; see Fig. 1d). Since we used epidermis from split skin to prepare the basal cells, it is likely that these are composed mainly of interfollicular cells, although basal cells from the upper part of the pilosebaceous tract and eccrine gland ducts should also be present (to be discussed later).

A similar preparation of basal cells was kept in culture for 12 days. After this time, the cells that reacted positively with the suprabasal cell antibody were usually stratified into a few cell layers. The proportion of cells staining with the suprabasal antibody could not be determined with accuracy since many cells detached during preparation of the coverslips for immunofluorescence. These cells, however, must comprise at least 30–35% of the total cell population as determined 30 hours after plating (Fig. 2c). Only a minor proportion of the flat cells that attached directly to the coverslips reacted with the suprabasal or the basal cell antibody (not shown). We rarely observed cells that stained with the basal cell antibody, and these were usually round (not shown).

Polypeptide patterns of basal and differentiated human epidermal cells

Figure 3a shows a two-dimensional gel pattern (isoelectric focusing; IEF) of [^{35}S]methionine-labeled proteins from a cell fraction (80–100 cells; labeled 4 hr after plating) containing 87% basal cells, 5% suprabasal keratinocytes, and 8% melanocytes and fibroblasts. Figure 3b shows a gel separation (IEF) of [^{35}S]methionine-labeled proteins from a similar preparation kept in culture for 12 days (DMEM, standard Ca^{++} concentration; 10% fetal calf serum).

The low amounts of vimentin observed both in the basal and differentiated cell populations (v in Fig. 3a,b) confirmed the immunofluorescence studies that showed that the basal cell preparation contained at the most an 8% contamination with melanocytes and fibroblasts. As far as we can judge from the two-dimensional gel patterns, the differentiated cell population was not dividing as the cells synthesized low amounts of cyclin (IEF 49) and high amounts of the tropomyosin-related polypeptide IEF 52 (Fig. 3b). Only very small amounts of cyclin could be detected in the basal cells (Fig. 3a), indicating that a significant proportion of these cells are also not cycling.

Proteins indicated with small asterisks in Figure 3, a and b, exhibited a higher relative proportion in the basal cells, whereas those indicated with large asterisks were higher in the cells after 12 days in culture. A few proteins were found predominantly in basal cells (termed b1 to b9), and these may serve as markers for these cells. The most prominent of these markers was a protein of $M_r \sim 43,000$ (b8) that could not be detected in the cell population containing differentiated cells (Fig. 4a,b; low exposure) and that was present in small amounts in gels of total epidermis labeled with [^{35}S]methionine (Fig. 4c; see also Fig. 5a). None of the basal cell markers could be detected in IEF gels of [^{35}S]methionine-labeled proteins from dermis pieces (Fig. 5b) obtained from split skin (Fig. 1a).

Of the polypeptides whose intensity increased in differentiated cells (large asterisks in Fig. 3a,b) we would like to mention first those indicated with a d (d1, M_r 124,000; d4, M_r 44,000; d9, M_r 29,500; d10, M_r 29,500; d11, M_r 28,500; d12 and d13, M_r not determined) because these may correspond to differentiation markers. The above polypeptides are important components of gels of [^{35}S]methionine-labeled epidermis (Fig. 5a), and except for d12 and 13 they are all abundant proteins, as judged by analysis of silver-stained gels (not shown).

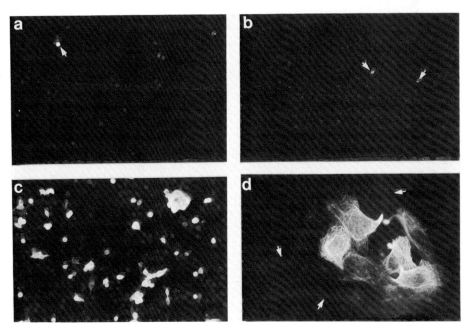

Figure 2 Immunofluorescent staining of cells stained with various antibodies. All cell preparations were fixed in methanol:acetone prior to immunofluorescence. (a) Basal cells fixed 3 hr after plating and reacted with the antibody that stains suprabasal cells; (b) similar preparation as a but reacted with the monoclonal antibody that stains melanocytes and fibroblasts; (c) epidermal cells fixed 30 hr after plating and reacted with the antibody that stains suprabasal cells; (d) SVK 14 cells reacted with the antibasal cell antibody. Arrows in a and b indicate positively stained cells; arrows in d indicate the positions of cells that do not react with the antibody. Magnifications: (a–c) 140×; (d) 400×.

With the possible exception of d11, none of these proteins could be observed in gels of dermis (Fig. 5b; not indicated), melanoma (not shown), transformed human amnion cells (AMA; not shown), and SV40-transformed human epidermal keratinocytes (SVK14; Fig. 7b) (Taylor-Papadimitriou et al. 1982).

Analysis of the keratins synthesized by the basal and differentiated cell populations revealed that the former exhibit mainly keratin K1 (Fig. 2a; M_r 44,000; IEF 44, HeLa protein catalog, Bravo and Celis 1982a; Fey et al. 1983; keratin 17 in the catalog of human cytokeratins, Moll et al. 1982b) and K2 (Fig. 2a; M_r 47,500; Fey et al. 1983; keratin 14 in the catalog of human cytokeratins, Moll et al. 1982b). Both keratins are present in gels of [^{35}S]methionine-labeled proteins of total epidermis (Fig. 5a; see also Fuchs and Green 1980; Fey et al. 1983), but only K2 is an abundant keratin, as determined by silver-staining (see also Fuchs and Green 1980; Moll et al. 1982a; Fey et al. 1983). Recently, Woodcock-Mitchell et al. (1982) and Skerrow and Skerrow (1983) have presented evidence indicating that basal cells contain both a 50-kilodalton (kD) and a 58-kD keratin. These results are in agreement with the observations of Fuchs and Green (1980) that also indicated the presence of a 58-kD keratin in the basal cell layer of the epidermis. Our IEF gel system, however, does not separate this keratin, and therefore we cannot comment on its occurrence. The relative proportion of both K1 and K2 increased considerably in the cell population containing differentiated cells, which also exhibited significant synthesis of a 46-kD keratin (K1a; Fig. 2a,b; keratin 16 in the catalog of human cytokeratins, Moll et al. 1982b).

The 46-kD keratin could only be weakly detected in heavily loaded, silver-stained gels of epidermis (Moll et al. 1982b; Fey et al. 1983), but it was readily detected in [^{35}S]methionine-labeled samples (Fig. 5a).

Interestingly, neither the basal nor the differentiated cells synthesized keratin K3 (d2, Fig. 5a; M_r 54,000; Fey et al. 1983), which sometimes is resolved in two components (Fig. 4c; termed K3 and K3a) of slightly different molecular weights. These proteins may correspond to keratins 10 and 11 in the catalog of human cytokeratins (Moll et al. 1982b). Since keratin K3 (d2) (Figs. 4c and 5a) may be expressed late in differentiation (Fuchs and Green 1980; Woodcock-Mitchell et al. 1982; Skerrow and Skerrow 1983), it would seem likely that the cultured, differentiated cells have not completed their differentiation program.

There are other putative differentiation markers that are not expressed in the differentiated cells discussed above. These correspond to proteins d3 (M_r 47,500), d5 (M_r 40,000), d6 (M_r 40,000), d7 (M_r 39,500), and d8 (M_r 30,000) (Fig. 5a). These proteins are not present in gels of dermis (cf. Fig. 5a,b) and could not be detected in a melanoma (not shown) or SVK14 cells (Fig. 7b, not indicated). Preliminary analysis of single colonies of epidermal cells obtained from independent experiments has revealed so far one case showing a more advanced pattern of differentiation (Fig. 6). This cell population (10 cells) expressed all the differentiation markers except d2 (keratin K3) and d8 (not shown in Fig. 6). The intensities of markers d9 through d11 was lower than that observed in Figure 3b, suggesting that these polypeptides are made transiently during differentiation. The cells synthe-

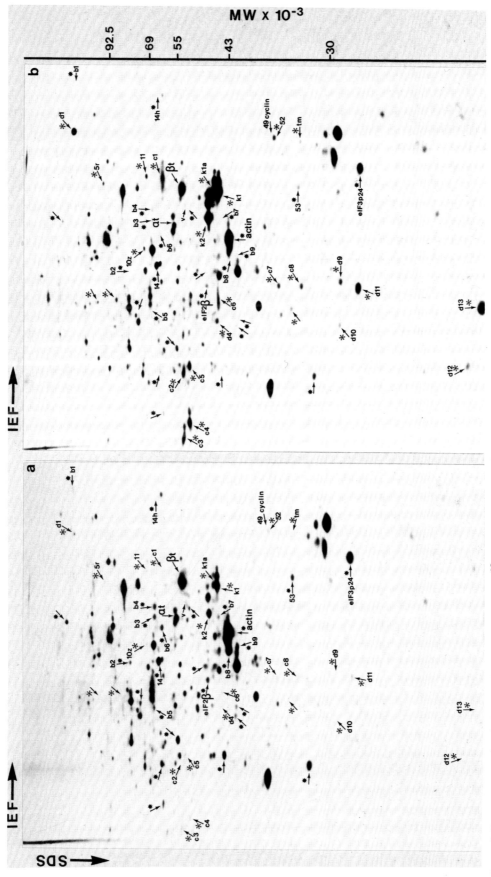

Figure 3 Two-dimensional gel electrophoresis (IEF) of total [^{35}S]methionine-labeled proteins from basal (a) and differentiated (b) cells. Cells were labeled with [^{35}S]methionine (1 mCi/ml) for 10 hr at 37°C. Large asterisks indicate polypeptides whose relative proportions increase in the differentiated cells. Polypeptides indicated with a d correspond to putative differentiation markers; those indicated with a c are most likely induced upon culturing. K1, K1a, K2, and K3 correspond to keratins. Polypeptides indicated with a b and a small asterisk correspond to basal cell markers. The positions of α- and β-tubulin (αt and βt) are indicated for reference.

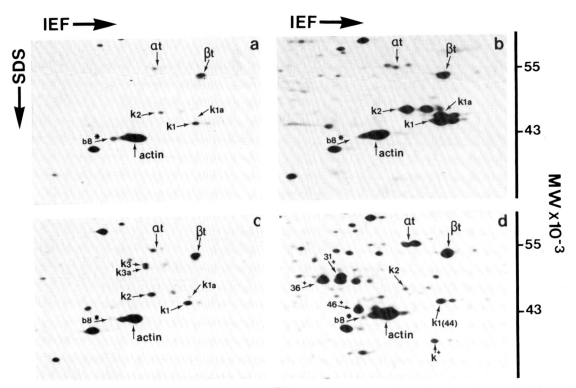

Figure 4 Two-dimensional gel electrophoresis (IEF) of total [^{35}S]methionine-labeled polypeptides from basal cells (a), basal cells after 12 days in culture (b), human epidermis (c), and SVK14 cells (d). Polypeptides are indicated as described in the legend of Fig. 3. The positions of α- and β-tubulin (αt and βt) and of actin are indicated for reference. In d the polypeptides indicated with plus signs over the arrows indicate keratins not present in basal cells.

sized little cyclin or IEF 52 and showed decreased synthesis of some of the identifiable subunits of protein synthesis initiation factors eIF2 and 3 (subunits, eIF2β; eIF3p24; see also Fig. 3a,b) (Duncan and Hershey 1983). Since these cells express keratin K2 but not keratin K3, it would seem likely that they are still in some stage of differentiation equivalent to the stratum spinosum in vivo (Fuchs and Green 1980).

Another set of interesting polypeptides whose relative proportion is higher in the cell population containing differentiated cells is that indicated with a c (c1–c8) in Figure 3, a and b (some of these are also indicated in Fig. 6). These polypeptides could not be detected in gels of total epidermis ([^{35}S]methionine, Fig. 5a; silver stain, not shown) or dermis (Fig. 5b; position not indicated), and thus their synthesis may be induced as a result of in vitro culturing. Whether these polypeptides arise from the differentiated cells or the cells that are negative for the basal and suprabasal antigen is still unknown. These polypeptides are not detected in preparations of total epidermal cells labeled during the first 20 hours but can be readily detected if the cells are labeled after that time (results not shown).

Comparison of the polypeptide synthesized by basal cells and SV40-transformed keratinocytes: Increased synthesis of cyclin

The SV40-transformed keratinocytes (SVK14; Taylor-Papadimitriou et al. 1982) used in these studies were kindly provided by E.B. Lane. The cells were originally isolated from SV40-infected human foreskin keratinocytes (Taylor-Papadimitriou et al. 1982) and were used in this study between passages 18 and 21. Methanol:acetone-fixed SVK14 cells reacted very weakly with the suprabasal antibody (not shown), but a significant number of cells (about 10%; Fig. 2d; arrows indicate the positions of unstained cells) stained brightly with the basal cell antibody to reveal a filamentous distribution of the antigen. The reason for the selective staining with the latter antibody is at present unknown. The basal cell antibody did not stain human cell lines AMA and A431 or monkey TC7 cells (not shown).

Figure 7, a and b, shows two-dimensional polypeptide patterns (IEF) of basal cells (Fig. 7a; same as shown in Fig. 3a) and sparsely growing SVK14 cells (Fig. 7b). Spots indicated with short arrows indicate polypeptides whose intensity increases considerably in the transformed cells. Polypeptides b1, b4, and b8 (Fig. 4d) correspond to basal cell markers expressed albeit at a reduced rate in the SVK14 cells.

Transformation-sensitive polypeptides whose relative intensities change in the SVK14/basal cell pair are indicated with their corresponding numbers in the HeLa protein catalog (Bravo and Celis 1982a,b). These include the following polypeptides that are present mainly in karyoplasts (Bravo et al. 1981b): IEF's 8z14 (M_r 84,000, increased in SVK14), 8m (phosphorylated, M_r 91,000, increased), 13e (M_r 67,000, increased), 13q (M_r

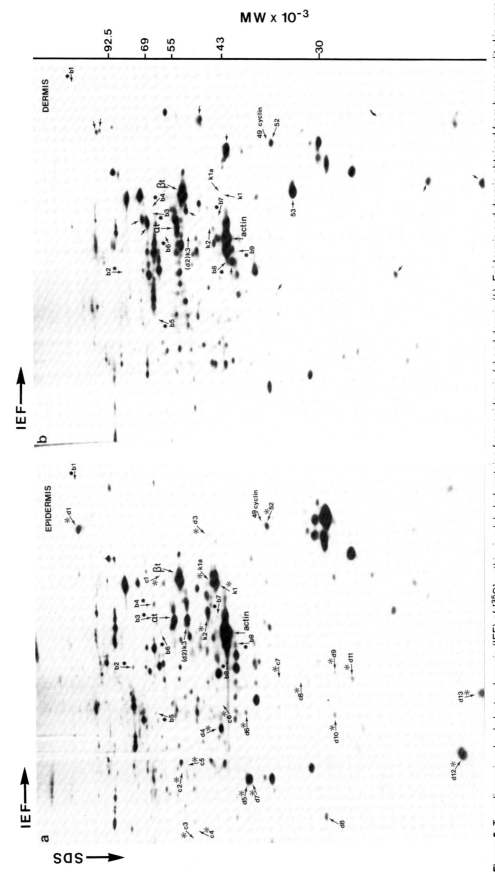

Figure 5 Two-dimensional gel electrophoresis (IEF) of [^{35}S]methionine-labeled proteins from epidermis (a) and dermis (b). Epidermis and dermis obtained from human split skin were labeled for 16 hr with [^{35}S]methionine. Short arrows in b indicate a few proteins present mainly in dermis. Other polypeptides are indicated as described in the legend of Fig. 3. Positions of α- and β-tubulin (αt and βt) as well as actin are indicated for reference. In b the positions of the epidermal keratins are indicated for comparison.

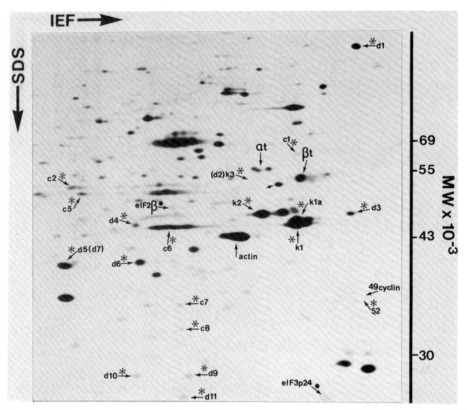

Figure 6 Polypeptides synthesized by a colony containing 8–10 differentiated epidermal cells. A single colony of differentiated cells was labeled with [^{35}S]methionine for 15 hr, 10 days after plating in microtiter wells containing 9-mm^2 coverslips. Polypeptides are indicated as described in the legend of Fig. 3. Positions of α- and β-tubulin (αt and βt) and actin are indicated for reference. The identity of the spot indicated with a short arrow is unknown.

66,500, increased), 14h (phosphoprotein, M_r 65,000, increased), 48z66 (M_r 38,500, increased), and 49 (cyclin, M_r 36,000, increased). Other transformation-sensitive polypeptides identified in SVK14 corresponded to IEF 12 (M_r 68,000, increased); 251 (M_r 55,000, increased), 33r (M_r 50,000, decreased), 39 (M_r 45,500, increased), and 52 (tropomyosin related, M_r 35,000, decreased) and 65 (M_r 16,000, increased).

The few new polypeptides observed in transformed cells that were not detected in basal cells corresponded to keratins (indicated with a plus sign over the arrows in Fig. 7). These proteins were identified by immunoprecipitation using a broad-specificity keratin antibody (not shown) and corresponded to HeLa keratins IEF 31 (M_r 50,000), 36 (M_r 48,500), and 46 (M_r 43,500) (Bravo and Celis 1982a; Bravo et al. 1982, 1983). These proteins correspond most likely to cytokeratins 7, 8, and 18 in the catalog of human cytokeratins (Moll et al. 1982b; see also Wu et al. 1982). SVK14 cells also synthesized a 40-kD keratin (K in Fig. 7a,b; keratin 19 in the catalog of human cytokeratins; Moll et al. 1982b; Lane 1982) that is not present in basal cells but is present in the epidermis only during early stages of embryonic development (Banks-Schlegel 1982). This keratin is also present in some squamous cell carcinoma lines derived from the epidermis and oral epithelium (Wu and Rheinwald 1981) and has been reported to be induced in cultured epidermal keratinocytes by high concentrations of vita-

min A (Fuchs and Green 1981). SVK14 cells synthesize, in addition to the keratins mentioned above, two other keratins (K1 and K2; Fig. 7b) that are present in basal cells. In general, these studies confirmed the results of Taylor-Papadimitriou et al. (1982) and Lane (1982) that first suggested the presence of a small keratin in SVK14 cells.

Up to now, it has not been possible to identify T antigen in the SVK14 cells. Crawford and O'Farrell (1979) have reported that T antigen focuses poorly, and E.B. Lane (pers. comm.) observed that T antigen is abnormal in SVK14 cells.

Careful analysis of the gel of SVK14 proteins further showed that these cells do not express (at least at a detectable level) any of the differentiation markers discussed earlier (Fig. 7b, markers are not indicated; cf. Figs. 3b, 5a, and 6), confirming earlier observations of Taylor-Papadimitriou et al. (1982) that showed that these cells are impaired in their ability to differentiate.

Preliminary analysis of basal cell carcinomas: Synthesis of cyclin

Our preliminary studies of basaliomas have been concentrated mainly on the two-dimensional gel electrophoretic analysis of unlabeled (silver staining) and [^{35}S]methionine-labeled proteins from tumor biopsies. Even though the interpretation of these data is compli-

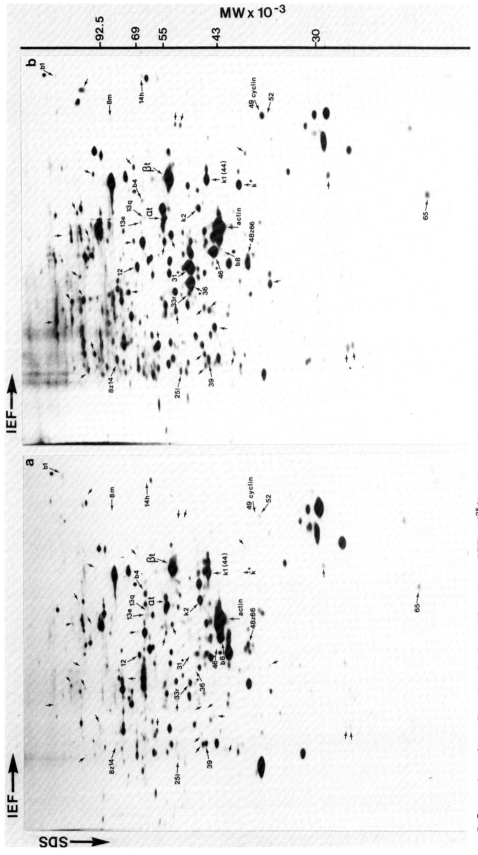

Figure 7 Comparison of two-dimensional polypeptide patterns (IEF) of [^{35}S]methionine-labeled proteins from basal (a) and sparsely growing SV40-transformed human keratinocytes (SVK14) (b). The pattern in a is the same as shown in Fig. 3a. Cells were labeled for 10 hr at 37°C with [^{35}S]methionine as described in the legend of Fig. 3. Polypeptides indicated with short arrows increased substantially in the transformed cells. Transformation-sensitive polypeptides are indicated with their numbers from the HeLa protein catalog. Polypeptides indicated with plus signs over the arrows indicate new keratins synthesized by the transformed cells. Other polypeptides are indicated as described in the legend of Fig. 3. Positions of α- and β-tubulin (αt and βt) and actin are indicated for reference.

cated by the heterogeneity of the tissue, we have obtained clear evidence for the increased synthesis of cyclin (IEF 49) in the tumor (Fig. 8a; [^{35}S]methionine labeling) compared with normal tissue (Fig. 8b). This increase is also observed in silver-stained gels (not shown), indicating that this transformation-sensitive polypeptide is synthesized in situ.

Experiments are now underway to analyze the polypeptide patterns of basal cell carcinoma cells by using microdissection of cryostat sections (10–20 μm in thick-

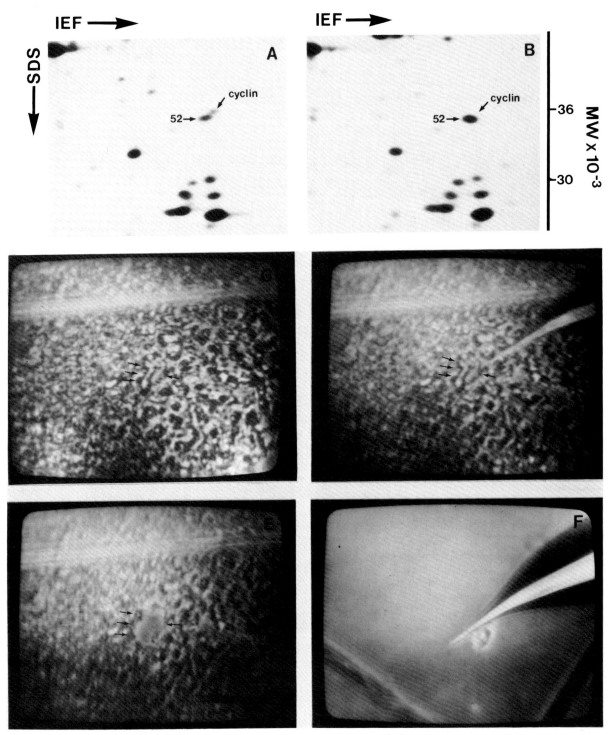

Figure 8 Synthesis of cyclin in a basal cell carcinoma: Microdissection of tumor cells. (a) IEF separation of [^{35}S]methionine-labeled proteins from a basal cell carcinoma labeled for 16 hr; (b) control tissue labeled under the same conditions; (c–f) microdissection of a small group of cells from a cryostat section (10 μm in thickness) of a human basalioma. (c and d) Phase-contrast photographs (as seen in the television monitor) showing the cells to be dissected (→). The arrows in d show the empty space left after dissection. d shows the piece of tissue after it has been placed in a 9-mm^2 coverslip. Magnifications: (c–f) 180×.

ness) obtained from tumor tissue prelabeled with [^{35}S]-methionine. Figure 8, c–f, shows the microdissection of a small island of basal cells from such a tumor (these cells also stain positively with the basal cell antibody; not shown). Both the apparatus and procedures for microsurgery are the same as used for microinjection and have been described in detail (Celis 1977; see also Moll et al. 1982a).

Discussion

Using a battery of antibodies, we have been able to identify and isolate a cell population that is highly enriched in basal cells. It is likely that this cell population is composed mainly of basal cells from the interfollicular epidermis, but it should also contain basal cells from the upper part of the pilosebaceous tract and eccrine gland ducts. Basal cells from the lower part of the hair follicles may not be present (at least in a significant proportion) since we have used epidermis from split skin as starting material. Also, as shown by Moll et al. (1982a), basal cells from the lower part of the hair follicles contain an abundant keratin (K1 in this study; keratin 17 in the catalog of human cytokeratins; Moll et al. 1982b) that we can barely detect in silver-stained gels of pure epidermis (Fey et al. 1983). We observed the synthesis of keratin K1 in [^{35}S]methionine-labeled samples (both in basal cells and epidermis), but this protein is only a minor component of the epidermis. We have previously suggested that this protein may exhibit a fast turnover (Fey et al. 1983).

The two-dimensional gel electrophoretic analysis (IEF) of the proteins synthesized by basal cells and by a cell population containing differentiated cells revealed various proteins that are putative markers for undifferentiated and differentiated cells. Among the first markers, polypeptide b8 (M_r 43,000) deserves further comment because it is the most highly labeled protein of this set. This polypeptide is not present in the differentiated cells or in the flat basal cells that fail to express both the basal and the suprabasal antigens. Since this polypeptide is a very minor component of the epidermis, it is possible that its synthesis is increased as a result of the in vitro culturing conditions. To answer this question, it will be necessary to compare the polypeptides synthesized by basal cells after different times in culture and in situ, using microdissection. These experiments will further allow us to compare the polypeptide patterns of basal cells obtained from epidermis and their appendages.

Thirteen putative differentiation-associated protein markers have been revealed so far, and some of these may be synthesized at different stages of differentiation. These markers add to those previously identified, which include involucrin (Rice and Green 1979; Banks-Schlegel and Green 1981; Green et al. 1982), keratins (Fuchs and Green 1980; for further references, see Green et al. 1982), and pemphigoid antigen (see Stanley and Yuspa 1983 and references therein). Preparation of antibodies to the differentiation markers may prove of value in dissecting the process of differentiation both in situ and in cultured cells.

As far as we can judge from the lack of expression of keratin K3, none of the differentiated cell populations that we have studied completed their differentiation program. These results are in agreement with various studies that have shown that large keratins are not synthesized in cultured epidermal keratinocytes (see Green et al. 1982 and references therein). Recently, however, Banks-Schlegel and Harris (1983) have shown the expression of a 67-kD keratin in cultured epidermal keratinocytes and suggested that variations in the serum composition could account for these observations. From this data it would be important to determine the components of the media that trigger and regulate differentiation (Hennings et al. 1980; for references, see Green et al. 1982). It should be stressed that only a fraction of the total cell population differentiated in DMEM containing standard Ca^{++} concentration. It is likely that the subpopulation of basal cells that differentiate may reflect the cell cycle stage of the original cells. Recently, Lavker and Sun (1982) have described two distinct populations of basal keratinocytes, one of which (nonserrated type) had the characteristics of stem cells. These cells, which are located at the tip of the deepest rete ridges, have primitive ultrastructural characteristics and exhibit low cycling kinetics. Whether these cells are related to the subpopulation that differentiates is at present unknown.

We observed a few polypeptides whose intensity increased in the cell population containing differentiated cells and that were not observed in whole epidermis or dermis. The origin of these polypeptides is unknown, although their synthesis may be stimulated as a result of the culturing conditions. Whether these polypeptides are derived from the differentiated cells, the flat cells that did not stain with either the basal or the suprabasal antibodies, or the few basal antigen-positive cells still present after 12 days in culture is still unknown. We are now determining the polypeptide pattern of the flat attached cells and of the loosely attached cells in an effort to answer this question.

Finally, the comparison between the two-dimensional protein patterns (IEF) of basal and SV40-transformed keratinocytes (SVK14) revealed a few transformation-sensitive polypeptides that have been previously shown to vary in various pairs of normal and transformed cells from various origins (Bravo and Celis 1982b; for references, see Celis et al. 1983). Of these, the nuclear polypeptide cyclin (IEF 49) has also been shown to increase in human basaliomas in contrast to normal tissue. We have also observed an increased synthesis of cyclin in melanomas (not shown). Experiments are now underway to prepare antibodies to some of the transformation-sensitive polypeptides. These should facilitate functional studies and will be of value in identifying these proteins in normal and tumor cells of various origins.

Note Added in Proof

Recently, M.B. Mathews, R.M. Bernstein, R. Franza, and J.I. Garrels (submitted) have shown that cyclin and the proliferating nuclear antigen (PCNA) are the same protein. Indirect immunofluorescent staining of basal and SV14 cells with PCNA antibodies kindly given by M.B. Mathews and R.M. Bernstein have confirmed the low and high levels of cyclin observed in these cells (Celis et al. 1984a,b).

Acknowledgments

We thank O. Jensen for photography and K. Dejgaard for assistance. We also thank the plastic surgery department of this University for providing the skin samples and E.B. Lane from the Imperial Cancer Research Fund for a gracious gift of the SV40-transformed keratinocytes.

S.F. is a recipient of a fellowship from the Danish Cancer Foundation. P.M.L. is a recipient of a fellowship from the Aarhus University. This work was supported by grants from Euratom, the Danish Cancer Foundation, the Danish Natural Science and Medical Research Council, the Carlsberg Foundation, NOVO, and Dir. Jakob Madsen and Wife Olga Madsen's Fund.

References

Banks-Schlegel, S. 1982. Keratin alterations during embryonic epidermal differentiation: A presage of adult epidermal maturation. *J. Cell Biol.* **93**: 551.

Banks-Schlegel, S. and H. Green. 1981. Involucrin synthesis and tissue assembly by keratinocytes in natural and cultured human epithelia. *J. Cell Biol.* **90**: 732.

Banks-Schlegel, S.P. and C.C. Harris. 1983. Tissue-specific expression of keratin proteins in human esophageal and epidermal epithelium and their cultured keratinocytes. *Exp. Cell Res.* **146**: 271.

Bellatin, J., R. Bravo, and J.E. Celis. 1982. Changes in the relative proportion of transformation sensitive polypeptides in giant HeLa cells produced by irradiation with lethal doses of X-rays. *Proc. Natl. Acad. Sci.* **79**: 4367.

Bravo, R. and J.E. Celis. 1980a. Gene expression in normal and virally transformed mouse 3T3B and hamster BHK21 cells. *Exp. Cell Res.* **127**: 249.

———. 1980b. A search for differential polypeptide synthesis throughout the cell cycle of HeLa cells. *J. Cell Biol.* **48**: 795.

———. 1982a. Updated catalogue of HeLa cell proteins: Percentages and characteristics of the major cell polypeptides labelled with a mixture of 16 [^{14}C]-amino acids. *Clin. Chem.* **28**: 766.

———. 1982b. Human proteins sensitive to neoplastic transformation in cultured epithelial and fibroblast cells. *Clin. Chem.* **28**: 949.

———. 1983. Catalogue of HeLa cell proteins. In *Two-dimensional gel electrophoresis of proteins: Methods and applications* (ed. J.E. Celis and R. Bravo), p. 445. Academic Press, New York.

Bravo, R., J. Bellatin, and J.E. Celis. 1981a. [^{35}S]-methionine labelled polypeptides from HeLa cells. Coordinates and percentage of some major polypeptides. *Cell Biol. Int. Rep.* **5**: 93.

Bravo, R., A. Celis, D. Mosses, and J.E. Celis. 1981b. Distribution of HeLa cells polypeptides in cytoplasts and karyoplasts. *Cell Biol. Int. Rep.* **5**: 479.

Bravo, R., S.J. Fey, P. Mose Larsen, N. Coppard, and J.E. Celis. 1983. Proteins IEF (Isoelectric focussing) 31 and IEF 46 are keratin-type components of the intermediate-sized filaments: Keratins of various human cultured epithelial cells. *J. Cell Biol.* **96**: 416.

Bravo, R., S.J. Fey, J.V. Small, P. Mose Larsen, and J.E. Celis. 1981c. Coexistence of three major isoactins in a single Sarcoma 180 cell. *Cell* **25**: 195.

Bravo, R., J.V. Small, S.J. Fey, P. Mose Larsen, and J.E. Celis. 1982. Architecture and polypeptide composition of HeLa cell cytoskeletons. Modification of cytoarchitectural proteins during mitosis. *J. Mol. Biol.* **159**: 121.

Bravo, R., S.J. Fey, J. Bellatin, P. Mose Larsen, J. Arevalo, and J.E. Celis. 1981d. Identification of a nuclear and of a cytoplasmic polypeptide whose relative proportions are sensitive to changes in the rate of cell proliferation. *Exp. Cell Res.* **136**: 311.

Celis, J.E. 1977. Injection of tRNAs into somatic cells. Search for in vivo sytems to assay potential nonsense mutations in somatic cells. *Brookhaven Symp. Biol.* **29**: 178.

Celis, J.E. and R. Bravo. 1981. Cataloguing human and mouse proteins. *Trends Biochem. Sci.* **6**: 197.

Celis, J.E., R. Bravo, P. Mose Larsen, S.J. Fey, J. Bellatin, and A. Celis. 1983. Expression of cellular proteins in normal and transformed human cultured cells and tumours. Two dimensional gel electrophoresis as a tool to study neoplastic transformation and cancer. In *Two dimensional gel electrophoresis of proteins: Methods and applications* (ed. J.E. Celis and R. Bravo), p. 307. Academic Press, New York.

Crawford, L.W. and P.Z. O'Farrell. 1979. Effect of alkylation on the physical properties of simian virus 40 T-antigen species. *J. Virol.* **29**: 587.

Duncan, R. and J.W.B. Hershey. 1983. Identification and quantitation of levels of protein synthesis initiation factors in crude HeLa cell lysates by two dimensional polyacrylamide gel electrophoresis. *J. Biol. Chem.* **258**: 7228.

Fey, S.J., P. Mose Larsen, R. Bravo, A. Celis, and J.E. Celis. 1983. Differential immunological crossreactivity of HeLa keratin antibodies with human epidermal keratins. Identification of a small epidermal keratin (k1) homologous to HeLa keratin IEF 44. *Proc. Natl. Acad. Sci.* **80**: 1905.

Forchhammer, J. and J. Klarlund. 1979. Changes in proteins from transformed cultures and tumours induced by sarcoma virus. In *Advances in medical oncology, research and education* (ed. P.G. Margison), vol. 1, p. 51. Pergamon Press, Oxford.

Forchhammer, J. and H. Macdonald-Bravo. 1983. Polypeptide synthesis in human sarcoma and normal tissue. In *Gene expression in normal and transformed cells* (ed. J.E. Celis and R. Bravo), p. 291. Plenum Press, New York.

Fuchs, E. and H. Green. 1980. Changes in keratin gene expression during terminal differentiation of the keratinocyte. *Cell* **19**: 1033.

———. 1981. Regulation of terminal differentiation of cultured human keratinocytes by vitamin A. *Cell* **25**: 617.

Green, H., E. Fuchs, and F. Watt. 1982. Differentiated structural components of the keratinocyte. *Cold Spring Harbor Symp. Quant. Biol.* **46**: 293.

Hennings, H., D. Michael, C. Cheng, P. Steinert, K. Holbrook, and S.H. Yuspa. 1980. Calcium regulation of growth and differentiation of mouse epidermal cells in culture. *Cell* **19**: 245.

Lane, E.B. 1982. Monoclonal antibodies provide specific intramolecular markers for the study of epithelial tonofilament organization. *J. Cell Biol.* **92**: 665.

Lavker, R.M. and T.T. Sun. 1982. Heterogeneity in epidermal basal keratinocytes: Morphological and functional correlations. *Science* **215**: 1239.

Leavitt, J. and R. Moyzig. 1978. Changes in gene expression accompanying neoplastic transformation of Syrian hamster cells. *J. Biol. Chem.* **253**: 2497.

Moll, R., W.W. Franke, B. Volc-Platzer, and R. Krepler. 1982a. Different keratin polypeptides in epidermis and other epithelia of human skin: A specific cytokeratin of molecular weight 46,000 in epithelia of the pilosebaceous tract and basal cell epitheliomas. *J. Cell Biol.* **95**: 285.

Moll, R., W.W. Franke, D.L. Schiller, B. Geiger, and R. Krepler. 1982b. The catalogue of human cytokeratins: Patterns of expression in normal epithelia, tumours and cultured cells, *Cell* **31**: 11.

Mose Larsen, P., S.J. Fey, R. Bravo, and J.E. Celis. 1983. Mouse mitochondrial protein IEF 24: Identification and immunohistochemical localization of mitochondria in various tissues. *Electrophoresis* **4**: 247.

O'Farrell, P.H. 1975. High-resolution two dimensional electrophoresis of proteins. *J. Biol. Chem.* **250**: 4007.

O'Farrell, P.Z., H.M. Goodman, and P.H. O'Farrell. 1977. High resolution two dimensional electrophoresis of basic as well as acidic proteins. *Cell* **12**: 1133.

Radke, K. and G.S. Martin. 1979. Transformation by Rous sarcoma virus: Effects of *src* gene expression on the synthesis and phosphorylation of cellular polypeptides. *Proc. Natl. Acad. Sci.* **76**: 5212.

Rice, R.H. and H. Green. 1979. Presence in human epidermal cells of a soluble protein precursor of the cross-linked envelope: Activation of the cross-linking by calcium ions. *Cell* **18**: 681.

Skerrow, D. and C.J. Skerrow. 1983. Tonofilament differentiation in human epidermis, isolation and polypeptide composition of keratinocyte subpopulations. *Exp. Cell Res.* **143**: 27.

Stanley, J.R. and S.H. Yuspa. 1983. Specific epidermal protein markers are modulated during calcium-induced terminal differentiation. *J. Cell Biol.* **96**: 1809.

Taylor-Papadimitriou, J., P. Purkis, E.B. Lane, I.A. Mckay, and S.E. Chang. 1982. Effects of SV40 transformation on the cytoskeleton and behavioural properties of human keratinocytes. *Cell Differ.* **11**: 169.

Woodcock-Mitchell, J., R. Eichner, W.G. Nelson, and T.T. Sun. 1982. Immunolocalization of keratin polypeptides in human epidermis using monoclonal antibodies. *J. Cell Biol.* **95**: 580.

Wu, Y.J. and R.G. Rheinwald. 1981. A new small (40 kd) keratin filament protein made by some cultured human squamous cell carcinomas. *Cell* **25**: 627.

Wu, Y.J., L.M. Parker, N.E. Binder, M.A. Beckett, J.H. Sinard, C.T. Griffith, and J.G. Rheinwald. 1982. The mesothelial keratins: A new family of cytoskeletal proteins identified in cultured mesothelial cells and non-keratinizing epithelia. *Cell* **31**: 693.

Transformation-sensitive Proteins of REF52 Cells Detected by Computer-analyzed Two-dimensional Gel Electrophoresis

B.R. Franza, Jr., and J.I. Garrels
Cold Spring Harbor Laboratory, Cold Spring Harbor, New York 11724

Although the genetic difference between a normal and a transformed cell can now in some cases be traced to the activity of one or a few viral genes or activated cellular oncogenes, the protein differences between normal and transformed cells remain a largely uncharted area. The products of the transforming genes are usually proteins of low abundance whose function is unknown. It is unlikely that any of these proteins can be directly responsible for the transformed phenotype. Changes in morphology, adhesion, and requirements for growth must be brought about by altered programs of protein synthesis and modification that result indirectly from the function of a transforming protein.

The proteins of normal and transformed cells have been analyzed by two-dimensional gel electrophoresis in a number of studies (Strand and August 1977; Leavitt and Moyzis 1978; Radke and Martin 1979; Bravo and Celis 1980; Brzeski and Ege 1980; Fransen et al. 1983). Typically between 5% and 10% of the 500 most abundant proteins are found to be quantitatively altered, and most studies have found that more proteins are repressed than are induced. Among the proteins reported to be reduced in rate of synthesis are the cytoskeletal proteins vimentin (Bravo and Celis 1980) and tropomyosin (Hendricks and Weintraub 1981; Leonardi et al. 1982; Matsumura et al. 1983b). Two transformation-induced proteins that have been discovered by these two-dimensional gel studies are the phosphoprotein of M_r 36,000 that is a substrate for tyrosine-specific protein kinases (Radke and Martin 1979) and the nuclear protein cyclin, which has been studied by Bravo and Celis (Bravo et al. 1981).

The viral transforming proteins, the products of cellular oncogenes, and the transformation-associated proteins such as the p53 T-antigen-associated protein are largely known through in vitro translation of selected mRNAs and by precipitation with specific antibodies. These important proteins have not been easily detected on two-dimensional gels of total cellular proteins, and most would not be scored in studies such as those mentioned above. Without the detection of low-abundance proteins, it is difficult to determine the significance of the changes found, and it is likely that protein changes critical to the transformation process will be missed.

To make possible much more detailed quantitative studies of proteins from unfractionated cells, we have developed improved methods of two-dimensional gel electrophoresis (Garrels 1979, 1983) and a system for computer analysis of two-dimensional protein patterns (Garrels et al. 1984). This system is being used to study more than 2000 proteins of the rat cell line REF52 and many of its transformed derivatives. The protein data base resulting from these studies can be used to compare the relative levels of any detected protein from normal and transformed cells grown and labeled under many different experimental conditions. The data base can also be used to search for proteins with specified patterns of induction or repression, and it can be used to judge the relatedness of cell lines based on the overall number and magnitude of protein changes. Finally, the data base can be used to record names and other properties for each of the detected proteins.

Here we will present identifications and preliminary quantitative data for a small number of proteins from the data base. These proteins include the actin and tropomyosin families, the transformation-induced protein cyclin, the transformation-associated protein p53, the adenovirally coded E1B-57K protein, and some of the products of the *ras* proto-oncogene family. These studies show that direct detection and quantitation of the minor transforming proteins will be feasible and suggest that altered synthesis of the more abundant proteins can reveal regulatory patterns characteristic of each transforming agent.

Experimental Procedures

Cell lines

REF52 cells, the SV40-transformed derivatives (REF52-WT2A, REF52-WT6A), and the type-5 adenovirus–transformed lines (Ad5D.4A and Ad5W.4A) were obtained from W.C. Topp (Cold Spring Harbor Laboratory). The origins of the normal and SV40-transformed lines and their serum requirements have been detailed previously (McClure et al. 1982). The adenovirus-transformed lines were isolated by J.S. Logan (State University of New York at Stony Brook) by DNA-transfection of REF52 cells with plasmids containing adenovirus E1 region DNA (see Matsumura et al. 1983b). Ad5W.4A is a member of the "W" series, whose cells do not express the E1B-57K tumor antigen. REF52 cells transformed by Kirsten murine sarcoma virus (REF52-KiMSV) were kindly provided by B. Ozanne (University of Texas).

REF52 cells and the transformed lines were grown in Dulbecco's modified Eagle's medium supplemented with

10% fetal calf serum. Lysates for two-dimensional gel analysis were prepared from cells labeled for 2 hours with 250–500 μCi/ml of [^{35}S]methionine (Amersham) in the logarithmic phase of growth, unless otherwise indicated. Lysates for immunoprecipitation were prepared from cells labeled for 24 hours with 500–1000 μCi/ml [^{35}S]methionine, using cells in late logarithmic growth, unless otherwise indicated.

Preparation of cell lysates for electrophoresis
Lysates of radiolabeled cells were prepared as described previously (Garrels 1979, 1983) except that the *Staphylococcal* nuclease step was omitted. Instead, the cells were scraped directly from the dish into a solution of 0.3% SDS, 1% β-mercaptoethanol, 20 mM Tris, pH 8.0, preheated to 100°C. The lysate was cooled on ice, digested with DNase I and RNase A, quick-frozen, and lyophilized as previously described.

Immunoprecipitation procedures
Monoclonal antibodies were generously provided by the following investigators: J. Lin (Cold Spring Harbor Laboratory), anti-actin and anti-tropomyosin; E. Harlow (Cold Spring Harbor Laboratory), anti-SV40 T antigen and anti-p53; M. Furth (Memorial Sloan-Kettering Cancer Center), anti-p21; and B. Stillman and R. McKay (Cold Spring Harbor Laboratory), anti-E1B-57K protein.

Lysates were prepared by addition of 1% Triton, 1% deoxycholate, in phosphate-buffered saline (PBS), pH 7.3, to labeled cells. After 15 minutes on ice, the cells were scraped from the dish and passed five times through a 28-gauge needle, and Protein A–Sepharose was added. After 30 minutes, the Protein A–Sepharose was removed by centrifugation, and the specific antibody was added to the supernatant. After 2 hours on ice, Protein A–Sepharose was added, and the supernatant was incubated for 30 minutes on ice and then pelleted by centrifugation. After washing three times with the above buffer and one time with the PBS, the pellet was resuspended in a small volume of 3% SDS, 10% β-mercaptoethanol and heated to 100°C for 5 minutes to dissociate the immune complexes. The supernatant was then removed, lyophilized, and resuspended in sample buffer for two-dimensional electrophoresis.

In the immunoprecipitations of the p21 proteins, a rabbit anti-rat IgG second antibody coupled to Protein A–Sepharose was used as described by Furth et al. (1982).

Two-dimensional gel electrophoresis and computer analysis
The electrophoretic procedures have been previously described (Garrels 1979, 1983). Most samples have been analyzed on gels containing LKB pH 3.5–10 ampholytes in the first dimension and 10% acrylamide in the second dimension. Narrow range (LKB pH 5–7 or LKB pH 6–8) ampholytes and other acrylamide concentrations have sometimes been used as indicated in the figure legends. All gels were processed for fluorography (Bonner and Laskey 1974) and exposed to Kodak XAR or XS film.

Films were scanned on an Optronics P-1000 scanner interfaced to a PDP-11/60 computer. The programs for spot detection, integration, pattern matching, and data management have been described elsewhere (Garrels et al. 1984). The quantitation of each protein involves conversion of film density to disintegrations per minute (dpm) per unit area, integration to obtain total dpm per protein spot, and finally normalization to express the quantitative value for each protein as a fraction of the total TCA-precipitable radioactivity. Data for each sample is stored in permanent data base records that can be addressed by sample number and standard spot number. From the protein data bases, one can determine the relative incorporation into any protein throughout the course of an experiment, one can compare relative intensities of proteins from various types of cells, and one can compare data from experiments done at different times.

Results

The properties of REF52 cells and three of its virally transformed derivatives are summarized in Table 1. The transformed lines have shorter cell cycle times and higher saturation densities than normal REF52 cells, and the transformed lines have a much more rounded morphology. The REF52 line has never been observed to form foci of transformed cells spontaneously, even after repeated passages, and it is not tumorigenic. The SV40-transformed line (REF52-WT2A) and the Ad5-transformed line (Ad5D.4A) used for this study have been selected for focus formation in monolayer culture. REF52-WT2A does not grow in soft agar and is not tumorigenic in nude mice. Ad5D.4A also does not grow in soft agar but is weakly tumorigenic. REF52 cells transformed by Kirsten murine sarcoma virus (REF52-KiMSV) grow well in soft agar and are tumorigenic in nude mice.

The proteins of normal and transformed REF52 cells have been radiolabeled and resolved on multiple two-dimensional gels. A typical pattern of REF52 proteins, separated using a broad-range pH gradient, is shown in Figure 1. Approximately 800 [^{35}S]methionine-labeled proteins can be detected on the film shown, and more than 2000 proteins are detectable on the film after a longer exposure. Many other proteins are not resolved on the gel shown but can be detected when the proteins are separated on narrower pH ranges or on slab gels with different second-dimension acrylamide concentrations. The locations of some of the proteins and protein families that have been partially characterized on this gel system are shown in Figure 1. Optimal resolution of some protein groups requires gels of a different pH range or acrylamide concentration and sometimes requires longer or shorter exposures to film. The members of the actin, tropomyosin, proliferating cell nuclear antigen (PCNA), p53, and p21 spot groups will be shown at optimal resolution in subsequent figures.

The group of actin proteins is shown on a narrow-range (pH 5–7) gel (Fig. 2). Normal REF52 cells, unlike most cultured cell lines, synthesize three forms of actin. In proliferating cells (Fig. 2A) allowed to incorporate

Table 1 Properties of REF52 and Transformed Derivatives

Cell line	Doubling time (hr)	Saturation density	Growth in soft agar	Tumorigenicity in nude mice
REF52	23	4×10^4	–	–
REF52-WT2	17	19×10^4	–	–
REF52-KiMSV	15	17×10^4	+	++
Ad5D.4A	17	19×10^4	–	+

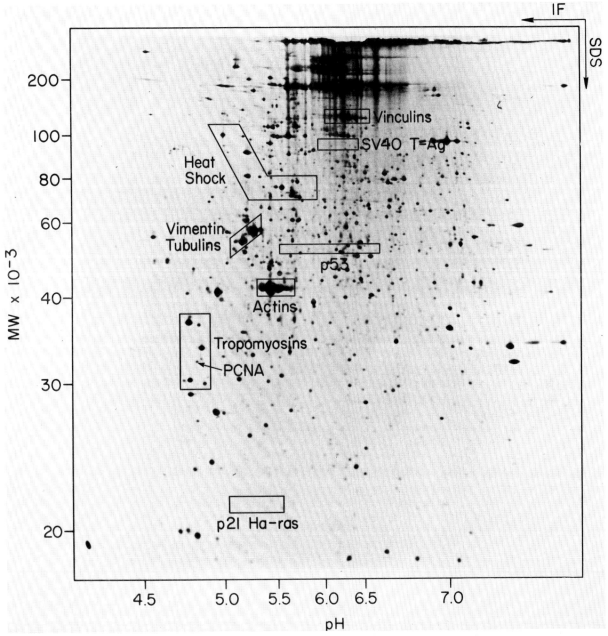

Figure 1 Identified protein groups of REF52 cells. The proteins of proliferating REF52 cells were resolved using pH 3.5–10 ampholytes in the first dimension and 10% acrylamide in the second dimension. This type of gel gives the best representation of the total protein pattern, although it does not provide optimal resolution in all areas. The regions of the identified spot groups are shown for reference. In subsequent figures, some of the individual proteins are identified using the pH range, acrylamide concentration, and exposure times optimal for each group. On gel shown, the p53 proteins, the SV40 T antigen, and the p21 proteins are not detected, but the regions of their migration are indicated. The locations of the individual heat-shock proteins in this gel system have been identified elsewhere (Welch et al. 1983). The pH and molecular-weight scales are based on a previous determination (Garrels 1979).

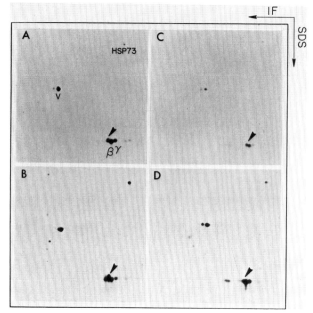

Figure 2 Actin isoforms in REF52 cells. Proteins of proliferating REF52 cells 4 days after plating (A) or nondividing confluent cells 12 days after plating (B) were resolved on gels containing pH 5–7 ampholytes in the first dimension and 10% acrylamide in the second dimension. Shown are the β- and γ-actins vimentin (V), the 73K heat-shock protein (HSP73), and the acidic form of smooth-muscle actin (▶). The tubulins are diffuse and not easily detected on this type of gel. Proteins immunoprecipitated from a lysate of REF52 cells (D) by the JLA20 anti-actin monoclonal antibody are shown in C.

[^{35}S]methionine for 2 hours, the most acidic form of actin is clearly detected, and in similarly labeled confluent, nondividing cultures (Fig. 2B), the most acidic form is the most predominant. To confirm that the growth-regulated protein is indeed a form of actin, we examined the proteins immunoprecipitated from lysates of REF52 cells by the JLA20 actin-specific monoclonal antibody (Lin 1981). Each of the actins is present in the immunoprecipitate (Fig. 2C), and the acidic actin isoform is enriched relative to the β- and γ-actins present in the lysate. The acidic actin isoform of REF52 cells is distinguishable on this gel system from the α-actin of skeletal muscle and comigrates with the most acidic of the smooth-muscle actins (Vandekerckhove and Weber 1978).

The tropomyosin proteins, shown on broad pH range gels, are identified in Figure 3. The tropomyosins, like the actins, have multiple isoforms (Matsumura et al. 1983a), some of which are specifically decreased or increased in transformed cells (Matsumura et al. 1983a, b; Lin et al., this volume). We wished to determine if REF52 cells contained any forms of tropomyosin that might be coregulated with the growth-regulated form of actin already identified. Each of the five known tropomyosins of REF52 cells was found in both dividing and confluent cultures, and only tropomyosin-1 (TM-1) was synthesized at a somewhat higher rate at confluence (Fig. 3A–C). However, one of the unknown proteins in the region of the gel was much more intensely labeled in confluent than in dividing cells, and we asked if this too could be a form of tropomyosin. We found that this protein, labeled TM-6 in Figure 3D, could be immunoprecipitated along with TM-1, TM-2, and TM-3 by a polyclonal anti-tropomyosin antibody. This antibody, raised against chicken smooth-muscle tropomyosin, is known to recognize the larger tropomyosin isoforms more strongly than the smaller tropomyosin isoforms (Matsumura et al. 1983a). A further indication that TM-6 is a form of tropomyosin is its apparent lack of proline. Parallel cultures of confluent REF52 cells were radiolabeled with either [^{35}S]methionine or [^{3}H]proline, and the proteins detected are shown in Figure 3, E and F, respectively. All forms of tropomyosin in REF52 cells were detected by methionine incorporation, but incorporation of proline could not be detected in any of the tropomyosin forms, including TM-6.

The relative rates of synthesis of the actins, tropomyosins, and several other major proteins have been determined for REF52 cells and for the three transformed lines described above. Most of the major cytoskeletal proteins are synthesized at lower rates, relative to total protein synthesis, in the transformed cells than in REF52 cells. As shown in Table 2, the β- and γ-actins and vimentin are synthesized at a lower rate relative to total protein synthesis in the transformed lines, whereas a major noncytoskeletal protein, the HSP73 heat-shock protein, is synthesized at a higher relative rate in the transformed cells. Tropomyosin synthesis in the transformed lines (Table 2 and Fig. 4) shows complex regulation. In agreement with studies of Matsumura et al. (1983b), TM-1 is reduced in REF52-WT2A and REF52-KMSV cells but is undetected in Ad5D.4A cells. TM-2 is reduced in REF52-WT2A and Ad5D.4A cells and is undetected in REF52-KiMSV cells. TM-3 is also undetected in REF52-KiMSV cells but is increased in amount in Ad5D.4A cells.

The growth-regulated forms of actin and tropomyosin are the cytoskeletal proteins most consistently and completely repressed by transformation of REF52 cells. Neither the smooth-muscle form of actin nor TM-6 could be detected in proliferating cultures of the three transformed lines (Table 2). In high-density cultures of transformed cells, a small amount of these proteins was detected in REF52-WT2A cells, but none was detected even at high levels of sensitivity in REF52-KiMSV or Ad5D.4A cells (data not shown).

Identification of a transformation-induced protein in REF52 cells

One protein in the tropomyosin region of the gels was found to be consistently elevated in transformed cells (Fig. 4 and Table 2). By an exchange of samples with the laboratory of J. Celis, we have identified this protein as IEF49, or cyclin (Bravo et al. 1981). These investigators have shown that cyclin is a nuclear protein, unrelated to tropomyosin, that is preferentially synthesized in proliferating and transformed cells.

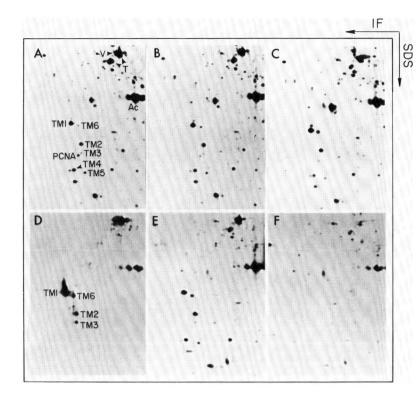

Figure 3 Tropomyosin isoforms in REF52 cells. Proteins of REF52 cells were labeled at 2 days (A), 4 days (B), or 12 days (C) after plating and were resolved on the same type of gel as shown in Fig. 1. Indicated in A are the tropomyosin isoforms (TM1–TM6), the proliferating cell nuclear antigen (PCNA), actin (Ac), vimentin (V), and the α- and β-tubulins (T). An immunoprecipitation of proteins from confluent REF52 cells by a polyclonal antitropomyosin antibody is shown in D. Proteins of confluent REF52 cells labeled for 24 hr with [^{35}S]methionine or [^{3}H]proline are shown in E and F, respectively.

In a different series of investigations, in collaboration with M. Mathews and co-workers, we have discovered that cyclin is identical to PCNA, an antigen recognized by autoimmune sera from a small fraction of patients with systemic lupus erythematosus (Miyachi et al. 1978). Human autoimmune sera with PCNA specificity can completely and specifically immunoprecipitate the cyclin spot from our gel patterns (M. Mathews et al., in prep.). PCNA has been shown by immunofluorescent staining of tissue sections to be present in the nuclei of dividing or transformed cells (Takasaki et al. 1981). We have retained the name PCNA because it was the name first applied to this protein.

Detection of minor transformation-associated proteins

It is known that both the SV40 large T antigen and the adenovirus E1B-57K protein bind to a cellular protein designated p53 (Lane and Crawford 1979; Linzer and Levine 1979; Sarnow et al. 1982). To determine the levels of the p53 protein in REF52 cells and in the transformed lines, we have used monoclonal antibodies (Harlow et al. 1981; Crawford and Harlow 1982) directed against SV40 T antigen and p53. The PAb-419 (L19) anti-T-antigen monoclonal antibody immunoprecipitates a faint streak from REF52-WT6A SV40-transformed cells, which is the T antigen itself, and it coprecipitates

Table 2 Quantitation of Proteins in REF52 and Transformed Derivatives

Protein	REF52	REF52-WT2A	REF52-KiMSV	Ad5D.4A
β-Actin	19,081[a]	14,575 (0.76)[b]	7,509 (0.39)	8,041 (0.42)
γ-Actin	9,947	4,819 (0.48)	4,101 (0.41)	3,729 (0.37)
sm-Actin	2,011	<92 (<0.046)	<68 (<0.034)	<92 (<0.046)
Vimentin	12,488	7,870 (0.63)	9,468 (0.76)	2,626 (0.21)
HSP73	2,289	3,272 (1.43)	2,981 (1.30)	3,031 (1.32)
TM-1	1,794	707 (0.39)	425 (0.24)	<6 (<0.0033)
TM-2	739	287 (0.39)	<5 (<0.007)	112 (0.15)
TM-3	121	140 (1.16)	<5 (<0.041)	258 (2.13)
TM-4	598	530 (0.89)	270 (0.45)	161 (0.27)
TM-5	240	311 (1.30)	234 (0.98)	181 (0.75)
TM-6	115	9 (0.078)	<5 (<0.04)	<6 (<0.05)
PCNA	242	555 (2.29)	359 (1.48)	1,111 (4.59)

All proteins were quantified from dividing cells 3 days after plating.
[a]Values for each protein indicate fraction of total incorporation of radiolabeled methionine into TCA-precipitable material ($\times 10^6$).
[b]Ratios of incorporation in transformed vs. normal REF52 cells are given in parentheses.

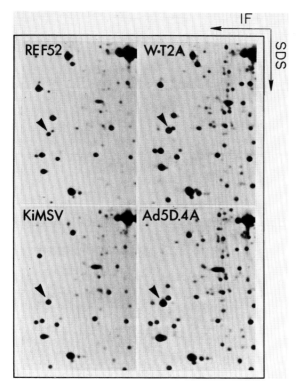

Figure 4 The tropomyosin/PCNA region from REF52 cells and transformed derivatives. An enlarged region of the day-2 REF52 pattern from Fig. 3A is compared with the same region of the three transformed cell patterns. All transformed cells were labeled at subconfluence 2 to 3 days after plating. The increased level of PCNA (▶) and the altered expression of the tropomyosins can be seen.

a heterogeneous series of spots at M_r 53,000 (Fig. 5A). The PAb-421 (L21) anti-p53 monoclonal antibody immunoprecipitates the same series of spots from REF52-WT6A cells (Fig. 5B) but immunoprecipitates only the most basic components of the series from Ad5W.4A adenovirally transformed cells (Fig. 5C). The latter components were also present among the proteins immunoprecipitated from adenovirally transformed cells by anti-E1B-57K monoclonal antibody (not shown).

Most of the members of the p53 complex have been difficult to detect on gels of total cell lysates because they are made in low amounts, are heterogeneous, and migrate in a crowded region of the gel. Two of the basic components, however, could be clearly identified among the proteins of unfractionated cell lysates. After labeling normal and transformed cells for 2 hours with [^{35}S]methionine, only the adenovirally transformed cells containing the E1B-57K protein showed elevated levels of these two spots (arrowheads in Fig. 6B). The levels found in REF52-WT6A cells (Fig. 6A) are typical of normal REF52 cells, of the other lines transformed by SV40 or KiMSV, and of REF52 cells transformed by adenovirus defective in the production of the E1B-57K protein (not shown). The E1B-57K protein itself is detectable on these gels as two spots (arrows on right half of Fig. 6B) that are adenovirus-specific and immunoprecipitated by the anti-E1B-57K monoclonal antibody. Of the proteins

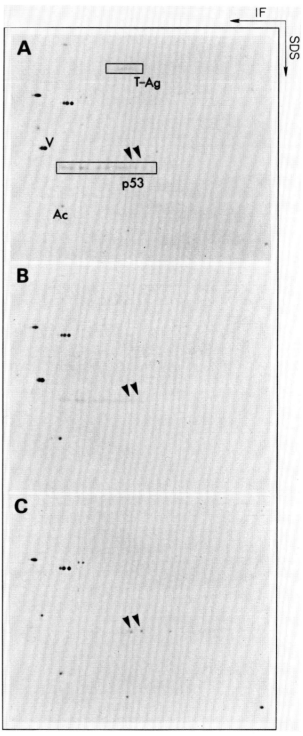

Figure 5 Immunoprecipitation of the p53 proteins from SV40- and adenovirally transformed cells. Proteins of REF52-WT6A cells were immunoprecipitated by the PAb-419 monoclonal antibody directed against SV40 T antigen (A) or by the PAb-421 monoclonal antibody directed against the p53 protein (B). The PAb-421 antibody was also used for immunoprecipitation from Ad5W.4A adenovirally transformed cells (C). The p53 protein components (enclosed by box) are identical in the anti-T and anti-p53 immunoprecipitations from REF52-WT6A cells. The arrows show the major p53 spots recognized in adenovirally transformed cells and indicate the alignment to minor components in A and B.

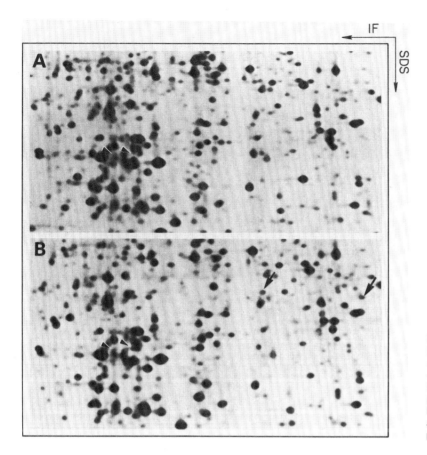

Figure 6 Identification of p53 forms in total cell lysates. REF52-WT6A cells (A) or Ad5D.4A cells (B) were labeled at near confluent density 4 days after plating. The p53 forms identified in these cells by alignment with the immunoprecipitated patterns are shown by arrowheads (in left half of each panel). The large arrows in B show two forms of the E1B-57K protein.

in the Ad5-transformed cells (Fig. 6B), the major p53 spot represents approximately 45 ppm and the E1B-57K spots each represent approximately 65 ppm of the TCA-precipitable radioactivity applied to the gel.

The ras gene products

Among the proteins of normal and transformed cells, we have also detected some of the p21 proteins, which are products of the ras proto-oncogenes. Immunoprecipitations using the Y13-259 anti-p21 monoclonal antibody (Furth et al. 1982) contain three protein spots at approximately M_r 21,000 that are detectable as minor spots in gels of total cell lysates (Fig. 7). Because these three spots are also recognized by the Y13-238 anti-p21 monoclonal antibody (not shown), which does not recognize products of the c-Ki-ras or c-N-ras genes (Furth et al. 1982), it is likely that all three spots represent products of the c-Ha-ras gene. The relationship between the p21 spots is not yet known. All three forms have been detected in REF52 cells and in each of the REF52 transformants examined. The major p21 spot detected in REF52 cells (Fig. 7) represents approximately 60 ppm of the TCA-precipitable radioactivity applied to the gel.

Discussion

We have begun the systematic, quantitative analysis of proteins in the rat cell line REF52 and its transformed derivatives, using methods of two-dimensional gel electrophoresis. The REF52 cell line has been chosen for these studies because (1) its growth in culture is tightly controlled, (2) it does not give rise to spontaneous transformants, (3) many of its growth factor requirements are known, and (4) a series of transformants with known history are already available. Although other lines, such as mouse 3T3 cells, have been used extensively in the past for studies of transformation, most of these lines do not meet all of the above criteria. A well-defined and stable cell system is especially important for studies such as ours in which protein data obtained by two-dimensional gel electrophoresis is being used to generate permanent protein data bases.

The cytoskeletal proteins are some of the most important to identify on two-dimensional gels because they are nearly ubiquitous markers on the gel patterns and because they are highly regulated during transformation. The group of five tropomyosins that has been previously identified (Matsumura et al. 1983a) is an excellent example of a family of cytoskeletal proteins whose synthesis is altered by transformation (Matsumura et al. 1983b). Our results are in agreement with these studies; we have found no exception to the rules that TM-1 is repressed in Ad5-transformed cells and that TM-2 and TM-3 are repressed in KiMSV-transformed cells.

In REF52 cells, we have found two cytoskeletal proteins, a smooth-muscle form of actin and the newly identified tropomyosin-6 (TM-6), that are detectable in

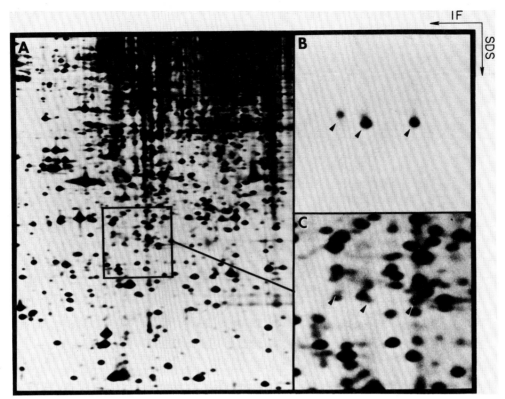

Figure 7 p21 proteins from REF52 cells. Proteins of dividing REF52-Ag6 cells (a line derived from REF52-WT6A by selection for growth in soft agar) were labeled for 24 hr with [^{35}S]methionine and separated on gels containing pH 3.5–10 ampholytes in the first dimension and 12.5% acrylamide in the second dimension. To detect minor proteins, 980,000 dpm were loaded onto the gel, and the gel was exposed to film for 8 days. The area indicated in A is enlarged in C. Proteins immunoprecipitated from these cells by the monoclonal antibody Y13-259 are shown in B. These proteins are detectable among the proteins of the total cell lysate (▶).

dividing cells and become prominent among the proteins synthesized at confluence. The rate of synthesis of TM-6 increases relative to total protein synthesis. The rate of synthesis of the smooth-muscle actin increases dramatically relative to the other actins but is lower relative to total protein synthesis in quiescent cells, probably because actins in general are synthesized at lower rates in quiescent cells. We do not yet have enough evidence to classify REF52 as a smooth-muscle cell, but it is likely that these proteins are characteristic of the differentiated state of the embryonic cell that gave rise to the cell line. The synthesis of these two growth-regulated proteins is highly repressed in each of the three transformed lines studied. In adenovirus and KiMSV-transformed cells, these proteins have never been detected even in high-density cultures. In SV40-transformed cells, they are not detected in low-density cultures but are induced to low but detectable levels when the cultures become crowded (data not shown). A transformation-sensitive actin in the same position as the smooth-muscle actin reported here has been found in cultures of chick embryo fibroblasts (Witt et al. 1983).

The differential expression of the actins and tropomyosins is not understood, but given that the smooth-muscle actin and TM-6 are regulated differently from the other actins and tropomyosins in normal cells and that numerous other REF52 proteins are repressed (or induced) more strongly by adenovirus than by SV40 (data not shown), it is likely that these two cytoskeletal proteins are regulated as part of a larger, possibly differentiation-specific, regulatory program that is affected by each of the transforming viruses. It would be interesting to know if the repression of these two proteins is essential to achieve the transformed state or whether their repression is simply an indirect consequence of the altered regulatory program.

We have made efforts to identify some proteins of lesser abundance that have been shown to be altered in transformed cells. The protein cyclin (Bravo et al. 1981), which we have shown to be identical with PCNA (Miyachi et al. 1978), is one of the most consistently induced proteins of transformed cells. It is a nuclear protein preferentially synthesized in S phase, but its rate of synthesis does not strictly correlate with the cell-cycle time. The level of PCNA in adenovirally transformed cells is substantially higher than in SV40-transformed cells, yet the cell-cycle times are similar (Table 1). Even in quiescent REF52 cells, detectable levels of PCNA are still present (Fig. 3). These observations argue against the function of PCNA as a protein needed only for DNA replication.

The p53 protein has been studied intensively because of its association with SV40 T antigen and adenoviral

E1B-57K protein. We find a complex pattern of p53 spots in immunoprecipitates from SV40-transformed cells and a somewhat simpler pattern of spots in immunoprecipitates from adenovirally transformed cells. The complexity of the p53 pattern is probably not due to multiple genes (Oren and Levine 1983) and is likely to result from posttranslational modifications. The most basic members of the p53 spot complex could be identified on our two-dimensional separations of total cell lysates. Protein patterns representing adenovirally transformed cells showed elevation of these two spots, whereas SV40-transformed cells showed the same levels of incorporation as control REF52 cells. These results suggest that the adenovirus E1B-57K protein and the SV40 T antigen interact differently with the cellular p53 protein.

The products of a cellular proto-oncogene have been identified by using monoclonal antibodies to the p21 ras proteins. In REF52 the three forms detected are probably products of the c-Ha-ras gene, based on known monoclonal antibody specificities. These p21 forms were detected in each of the transformed lines studied. In the KiMSV transformant, a series of more basic p21 spots, probably the products of the v-Ki-ras gene, was also detected using monoclonal antibodies with broad specificity. Although the functions of the normal and the activated ras gene products are unknown, the examination of their synthesis and modification during normal growth responses might give clues to their function.

The identification of the p21 and p53 proteins on two-dimensional gels shows that minor proteins involved in primary events of transformation can be detected by two-dimensional gel electrophoresis of total cell lysates. The ability to detect these minor proteins suggests that changes in their synthesis and modification can be followed and correlated with changes in more abundant proteins. As the analysis of data from two-dimensional gels becomes more routine, and as more of the known proteins become identified on the two-dimensional gel patterns, we can hope to obtain a more complete understanding of the protein changes that occur as cells become transformed through the action of known transforming genes.

Acknowledgments

We thank Lynn Cascio, Jane Emanuele, Patricia Smith, and Phil Renna for excellent technical assistance. We thank Bill Topp and Don McClure for cell lines, and our many colleagues who have supplied antibodies. We thank Fumio Matsumura and Jim Lin for helpful discussions, and we thank Dr. J.D. Watson for encouragement and support. This work was supported by grants from the National Institutes of Health (CA-13106) and from the National Science Foundation (PCM-7922769).

References

Bonner, W.M. and R.A. Laskey. 1974. A film detection method for tritium-labeled proteins and nucleic acids in polyacrylamide gels. *Eur. J. Biochem.* **46**: 83.

Bravo, R. and J.E. Celis. 1980. Gene expression in normal and virally transformed mouse 3T3B and hamster BHK21 cells. *Exp. Cell Res.* **127**: 249.

Bravo, R., S.J. Fey, J. Bellatin, P.M. Larsen, J. Arevalo, and J.E. Celis. 1981. Identification of a nuclear and of a cytoplasmic polypeptide whose relative proportions are sensitive to changes in the rate of cell proliferation. *Exp. Cell Res.* **136**: 311.

Brzeski, H. and T. Ege. 1980. Changes in polypeptide pattern in ASV-transformed rat cells are correlated with the degree of morphological transformation. *Cell* **22**: 513.

Crawford, L. and E. Harlow. 1982. Uniform nomenclature for monoclonal antibodies directed against virus-coded proteins from simian virus 40 and polyoma virus. *J. Virol.* **41**: 709.

Fransen, L., F. Van Roy, and W. Fiers. 1983. Changes in gene expression and protein phosphorylation in murine cells, transformed or abortively infected with wild type and mutant simian virus 40. *J. Biol. Chem.* **258**: 5276.

Furth, M.E., L.J. Davis, B. Fleurdelys, and E.M. Scolnick. 1982. Monoclonal antibodies to the p21 products of the transforming gene of Harvey murine sarcoma virus and of the cellular ras gene family. *J. Virol.* **43**: 294.

Garrels, J.I. 1979. Two-dimensional gel electrophoresis and computer analysis of proteins synthesized by clonal cell lines. *J. Biol. Chem.* **254**: 7961.

———. 1983. Quantitative two-dimensional gel electrophoresis of proteins. *Methods Enzymol.* **100**: 411.

Garrels, J.I., J.T. Farrar, and C.B. Burwell IV. 1984. The QUEST system for computer-analyzed two-dimensional gel electrophoresis of proteins. In *Two-dimensional gel electrophoresis of proteins: Methods and applications* (ed. J.E. Celis and R. Bravo). Academic Press, New York. (In press.)

Harlow, E., L.V. Crawford, D.C. Pim, and N.M. Williamson. 1981. Monoclonal antibodies specific for simian virus 40 tumor antigens. *J. Virol.* **39**: 861.

Hendricks, M. and H. Weintraub. 1981. Tropomyosin is decreased in transformed cells. *Proc. Natl. Acad. Sci.* **78**: 5633.

Lane, D.P. and L.V. Crawford. 1979. T antigen is bound to a host protein in SV40-transformed cells. *Nature* **256**: 495.

Leavitt, J. and R. Moyzis. 1978. Changes in gene expression accompanying neoplastic transformation of Syrian hamster cells. *J. Biol. Chem.* **253**: 2497.

Leonardi, C.L., R.H. Warren, and R.W. Rubin. 1982. Lack of tropomyosin correlates with the absence of stress fibers in transformed rat kidney cells. *Biochim. Biophys. Acta* **720**: 154.

Lin, J.J.-C. 1981. Monoclonal antibodies against myofibrillar components of rat skeletal muscle decorate the intermediate filaments of cultured cells. *Proc. Natl. Acad. Sci.* **78**: 2335.

Linzer, D.I.H. and A.J. Levine. 1979. Characterization of a 54K dalton cellular SV40 tumor antigen present in SV40-transformed cells and uninfected embryonal carcinoma cells. *Cell* **17**: 43.

Matsumura, F., S. Yamashiro-Matsumura, and J.J.-C. Lin. 1983a. Isolation and characterization of tropomyosin-containing microfilaments from cultured cells. *J. Biol. Chem.* **258**: 6636.

Matsumura, F., J.J.-C. Lin, S. Yamashiro-Matsumura, G.P. Thomas, and W.C. Topp. 1983b. Differential expression of tropomyosin forms in the microfilaments isolated from normal and transformed rat cultured cells. *J. Biol. Chem.* **258**: 13954.

McClure, D., M.J. Hightower, and W.C. Topp. 1982. Effect of SV40 transformation on the growth-factor requirements of the rat embryo cell line REF52 in serum-free medium. *Cold Spring Harbor Conf. Cell Proliferation* **9**: 345.

Miyachi, K., M.J. Fritzler, and E.M. Tan. 1978. Autoantibody to a nuclear antigen in proliferating cells. *J. Immunol.* **121**: 2228.

Oren, M. and A.J. Levine. 1983. Molecular cloning of a cDNA specific for the murine p53 cellular tumor antigen. *Proc. Natl. Acad. Sci.* **80**: 56.

Radke, K. and G.S. Martin. 1979. Transformation by Rous sarcoma virus: Effects of src gene expression on the synthesis and phosphorylation of cellular polypeptides. *Proc. Natl. Acad. Sci.* **76:** 5212.

Sarnow, P., Y.S. Ho, J. Williams, and A.J. Levine. 1982. Adenovirus E1B-58K tumor antigen and SV40 large tumor antigen are physically associated with the same 54kd cellular protein in transformed cells. *Cell* **28:** 387.

Strand, M. and J.T. August. 1977. Polypeptides of cells transformed by RNA or DNA tumor viruses. *Proc. Natl. Acad. Sci.* **74:** 2729.

Takasaki, Y., J. Deng, and E.M. Tan. 1981. A nuclear antigen associated with cell proliferation and blast transformation. *J. Exp. Med.* **154:** 1899.

Vandekerckhove, J. and K. Weber. 1978. At least six different actins are expressed in a higher mammal: An analysis based on the amino acid sequence of the amino-terminal tryptic peptide. *J. Mol. Biol.* **126:** 783.

Welch, W.J., J.I. Garrels, G.P. Thomas, J.J.-C. Lin, and J.R. Feramisco. 1983. Biochemical characterization of the mammalian stress proteins and identification of two stress proteins as glucose- and Ca^{2+} ionophore-regulated proteins. *J. Biol. Chem.* **258:** 7102.

Witt, D.P., D.J. Brown, and J.A. Gordon. 1983. Transformation-sensitive isoactin in passaged chick embryo fibroblasts transformed by Rous sarcoma virus. *J. Cell Biol.* **96:** 1766.

Changes Induced by Epidermal Growth Factor in the Polypeptide Synthesis of A431 Cells

R. Bravo
European Molecular Biology Laboratory, 6900 Heidelberg, Federal Republic of Germany

Epidermal growth factor (EGF), a polypeptide originally isolated from mouse submaxillary gland (Carpenter and Cohen 1979a), is a potent mitogen for a number of cell types in culture (Gospodarowicz et al. 1978).

A431 cells, a human epidermal carcinoma cell line that expresses an unusually large number of high-affinity cell-membrane receptors for EGF (Fabricant et al. 1977; Haigler et al. 1978), have been used in recent years as a model for the study of early events after the interaction of the growth factor with its specific cell-surface receptor. These cells have been used to study the rapid effects of EGF on cell morphology (Chinkers et al. 1979, 1981; Schlessinger and Geiger 1981) and to demonstrate the internalization of EGF by whole cells (Haigler et al. 1978, 1979; Cohen et al. 1979; McKanna et al. 1979; Willingham et al. 1983). These cells also have been used for identifying and isolating the EGF receptor (Wrann and Fox 1979; Haigler and Carpenter 1980) as well as for studying the effect of EGF on protein phosphorylation in membrane preparations (Carpenter et al. 1978; Carpenter and Cohen 1979b; Cohen et al. 1980; Ushiro and Cohen 1980; Buss et al. 1982; Lipshitz et al. 1983) or in whole cells (Erikson et al. 1981; Gill and Lazar 1981; Hunter and Cooper 1981). Although EGF increases the levels of phosphotyrosine in A431 cells (Gill and Lazar 1981; Hunter and Cooper 1981), it inhibits the replication of these cells (Gill and Lazar 1981; Barnes 1982). Several studies have been carried out in order to determine whether EGF-stimulated protein kinase activity plays a role in the inhibition of A431 cell proliferation by EGF (Buss et al. 1982; Gill et al. 1982; Lipshitz et al. 1983).

In spite of all the great efforts to understand the early events after the interaction of EGF with its specific cell-surface receptor in A431 cells, not much has been reported on the late effects that possibly occur at the level of protein synthesis in these cells.

We have recently started a detailed study of the effect of EGF in A431 cells at the protein synthesis level, aiming to gain a better knowledge of the biological response of these cells to the growth factor. The initial results of these studies are reported here.

Experimental Procedures

Cells and growth conditions
Human epidermal carcinoma cells (A431) were kindly provided by Dr. G. Todaro. Cells were routinely grown in Dulbecco's modified Eagle's medium (DMEM) containing 10% fetal calf serum (FCS) and antibiotics (penicillin 100 units/ml; streptomycin 50 μg/ml).

Effect of EGF on A431 cell growth
Cells were plated in 35-mm dishes in DMEM containing 10% FCS, at a density of 1.5×10^4 cells/dish and left to attach overnight. Medium was replaced by fresh DMEM containing 10% FCS and different concentrations of EGF. The medium was changed after 3 days and cells were counted on day 6. All plates were done in triplicate.

Labeling of cells with [^{35}S]methionine
Cells were grown in 0.25-ml flat-bottomed microtiter plates (NUNC) for at least 24 hours before labeling. To label the cells, the normal medium was replaced by 0.1 ml of DMEM containing 1 mg/liter cold methionine supplemented with 10% dialyzed FCS in the presence of 100 μCi [^{35}S]methionine (Amersham SJ204) (Bravo and Celis 1980; Bravo et al. 1982).

For sample preparation, the labeling medium was aspirated carefully, and immediately 20–40 μl of lysis buffer was added. These aliquots were kept at −70°C.

Triton cytoskeletons and Triton supernatants
The [^{35}S]methionine-labeled, EGF-treated cells were rinsed in Hank's buffer, and 0.1 ml of 0.5% Triton X-100 in PIPES cytoskeleton buffer (a Ca^{++}-free Hank's solution containing 2 mM $MgCl_2$, 2 mM EGTA, 5 mM PIPES, pH 6.1) was added. Treatment with Triton X-100 was carried out for 60–90 seconds at room temperature. Then the solution was carefully removed, and the extracted cells (Triton cytoskeleton) were rinsed with Hank's buffer before adding 20–40 μl of lysis buffer. The Triton X-100 solution containing the extracted proteins (Triton supernatant) was immediately frozen and freeze-dried. The sample was resuspended in 20–40 μl of lysis buffer.

Two-dimensional gel electrophoresis
The procedures used are those described by O'Farrell (1975) and Bravo et al. (1982) with further modifications.

The first-dimension separations (IEF) were performed in 230×1.2-mm 4% (w/v) polyacrylamide gels containing 2% ampholytes (1.6% pH 5–7, 0.4% pH 3.5–10) at 1200 V for 20 hours. After the run, the gels were incubated for 5 minutes in equilibration buffer and stored at −70°C. The second-dimension electrophoretic separations were carried out by laying the IEF gels onto a 15% polyacrylamide gel (25 × 25 cm) and run at room temperature for 16 hours at 13 mA.

The gels were processed for fluorography as described by Laskey and Mills (1975). Approximately 10^6 trichloraceticacid-precipitable cts/min were routinely applied on a gel. For quantitation, the spots were cut out and processed as described by Bravo et al. (1982).

Results

Effect of EGF on the growth of A431 cells

We have observed that EGF at 100 ng/ml inhibits colony formation of A431 cells and that this is not due to an alteration in the plating efficiency (R. Bravo and H. Macdonald-Bravo, unpubl.). To understand more about the effect of EGF on A431 cell proliferation, a growth response curve to EGF was carried out.

As shown in Figure 1A, there is a clear inhibition of A431 cell growth by EGF when used at concentrations higher than 100 pg/ml. Cells are slightly stimulated to grow when 10–100 pg/ml of EGF are used. A typical growth curve of A431 cells in the presence of 100 ng/ml EGF is presented in Figure 1B. Similar results have been obtained with cells grown in 0.5% FCS and in a conditioned medium.

Studies using [^{3}H]thymidine incorporation have also clearly demonstrated that the incorporation into an asynchronous A431 cell population is greatly reduced after 24 hours of EGF treatment (not shown; R. Bravo and P. Blundell, unpubl.).

Polypeptides sensitive to EGF in A431 cells

The two-dimensional polypeptide maps of control A431 cells and EGF-treated cells labeled for 20 hours with [^{35}S]methionine are shown in Figure 2. Visual inspection of many films exposed for different intervals did not reveal new, major polypeptides in A431 cells after EGF treatment. However, significant changes in the abundance of certain polypeptides that are present in both untreated and EGF-treated A431 cells were observed. In some cases these proteins were synthesized in very small amounts in either the control or treated cells, but they could be detected after prolonged exposure of the gels. Only those polypeptides whose relative proportions varied consistently more than 60% have been considered as variable and are indicated in Figure 2. The positions of actin and α- and β-tubulins are indicated for reference.

The percentages of some of the major EGF-sensitive polypeptides and their molecular weights are presented in Table 1.

Since it is known that EGF induces great changes in the morphology of A431 cells (Chinkers et al. 1979, 1981; Schlessinger and Geiger 1981), several drugs that alter the cytoskeleton were tested to see whether they were able to produce changes in the polypeptide synthesis similar to those effected by EGF. Neither taxol, colchicine, nocodazole, or cytochalasin B were able to mimic the effect of EGF. Furthermore, as shown in Figure 3, B and C, the presence of these compounds did

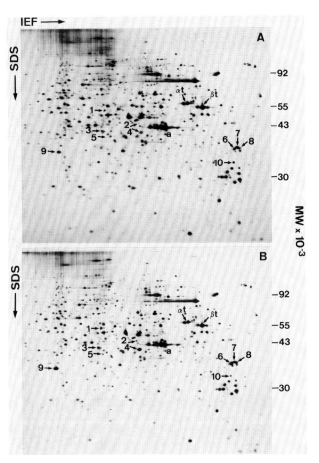

Figure 2 EGF-sensitive polypeptides in A431 cells. Two-dimensional gel electrophoresis of total [^{35}S]methionine-labeled polypeptides from untreated asynchronous cells (A) and EGF-treated cells (B). EGF (100 ng/ml) was added simultaneously with the [^{35}S]methionine. The gels have been exposed so as to keep the intensity of the spots within the linear range of the film response. Spots indicated with numbers are the EGF-sensitive polypeptides. (αt) α-Tubulin; (βt) β-tubulin.

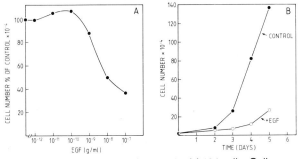

Figure 1 Effect of EGF on the growth of A431 cells. Cells were grown as described in Experimental Procedures. (A) Number of cells present in dishes for each concentration of EGF compared with control dishes without EGF; number of cells in control dishes was taken as 100% (B) EGF was used at 100 ng/ml. All results are the averages of triplicate dishes.

Table 1 Percentages and Molecular Weights of Some Major EGF-Sensitive Polypeptides in A431 Cells

Polypeptide (IEF)	m.w. (× 10³)	Percentage of total label[a]	
		control	+EGF[b]
4	41	0.07	0.70
6	35	0.03	0.20
7	36	0.10	<0.01
8	35.5	<0.01	0.20
9	34.5	0.20	0.40

[a]Cells were labeled with a mixture of 16 ¹⁴C-labeled amino acids as described by Bravo et al. (1981b). Percentages were calculated with respect to the total number of counts applied to the gel. The data have been corrected for the efficiency of counting.

[b]Cells were treated for 24 hr with 100 ng/ml EGF before being labeled. EGF was present during the labeling.

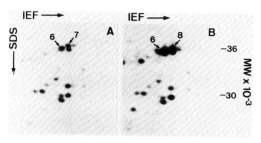

Figure 4 In vitro translation products of total RNA from A431 cells: (A) control cells; (B) EGF-treated cells. Cells were grown for 24 hr in the presence of EGF before RNA extraction. In each case 4 μg of total RNA were translated in a rabbit reticulocyte system.

not alter the cellular response to EGF. Only actinomycin D inhibited the EGF-induced changes in polypeptide synthesis (see Fig. 3A).

Since the result obtained with actinomycin D suggested that EGF required an active transcription of the genome to induce changes in the protein synthesis, total RNA from untreated and EGF-treated A431 cells was prepared. The in vitro products of these RNAs are shown in Figure 4. It is clear from the figure that EGF greatly increases the amount of mRNA for polypeptides IEF 6 and 8. These increases are also observed for the other EGF-sensitive polypeptides but not so dramatically.

Some of the effects induced by EGF, such as the increase in the levels of phosphotyrosine in proteins and changes in cellular morphology, seem to be transient. The response decreases in intensity during a period of time in spite of the continuous presence of EGF in the medium (Chinkers et al.1981; Hunter and Cooper 1981; Schlessinger and Geiger 1981). To test whether the effect of EGF in the polypeptide pattern of A431 cells was also a transient phenomenon, cells were grown in the presence of the growth factor for several days before being labeled with [³⁵S]methionine. A typical result of these experiments is illustrated in Figure 5. It is clear from this polypeptide pattern that the intensity of the cellular response to EGF remains constant even for 1 week or more while the growth factor is present (cf. Figs. 2 and 5).

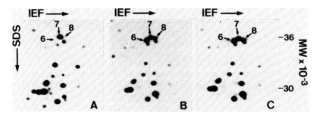

Figure 3 Effect of different drugs on the EGF response. (A) Actinomycin D, 5 μg/ml; (B) taxol, 0.1 μg/ml; (C) cytochalasin B, 10 μg/ml. In all cases cells were incubated for 2 hr in the presence of the drug before adding EGF (100 ng/ml). Cells were labeled with [³⁵S]methionine for 6 hr. Only a part of the two-dimensional gel is shown.

Triton skeleton and Triton supernatants

As a first attempt to assign a possible cellular location for the EGF-sensitive polypeptides, the Triton skeleton and Triton-soluble polypeptides of EGF-treated cells were analyzed by two-dimensional gel electrophoresis. As shown in Figure 6, the EGF-sensitive polypeptides are not components of the Triton skeleton (Fig. 6B), but they are extracted more than 95% with the Triton X-100 treatment (Fig. 6A). The arrowheads indicate the positions of the cytokeratins that are not extracted by Triton X-100.

The results obtained suggest that the EGF-sensitive polypeptides could be components of some of the membrane systems of the cell. To analyze this further, EGF-treated cells were extracted with Triton X-114, which has been demonstrated to selectively extract integral membrane proteins and can be used to identify them from a crude cellular extract (Bordier 1981). Figure 7B

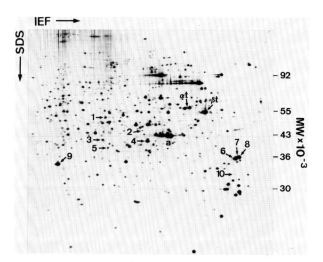

Figure 5 Two-dimensional gel electrophoresis of total [³⁵S]methionine-labeled polypeptides from EGF-treated A431 cells. Cells were grown in the presence of EGF (100 ng/ml) for 7 days before labeling.

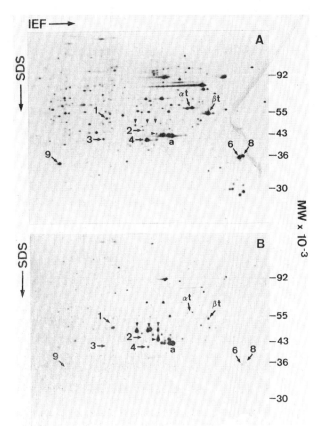

Figure 6 Two-dimensional gel electrophoresis of [^{35}S]-methionine-labeled polypeptides from EGF-treated A431 cells: (A) Triton X-100 supernatants; (B) Triton X-100 cytoskeletons. Cells were incubated with EGF for 24 hr before labeling. Only major EGF-sensitive polypeptides are indicated. Arrowheads show the positions of the cytokeratins.

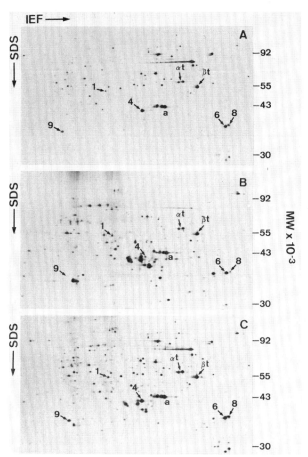

Figure 7 Triton-soluble polypeptides from EGF-treated A431 cells: (A) Triton X-100 supernatant; (B) Triton X-114 supernatant; (C) Triton X-100 + Triton X-114 supernatants. Only some of the major EGF-sensitive polypeptides are indicated.

shows a two-dimensional gel map of the proteins extracted by Triton X-114 from a total cellular lysate. The figure shows that many of these polypeptides are not detected in the Triton X-100 supernatant (Fig. 7A). Interestingly enough, some of the EGF-sensitive polypeptides, mainly polypeptides IEF 6 and 8, are selectively extracted in Triton X-114 in contrast to polypeptides IEF 4 and 9, which are barely detected in Figure 7B.

Discussion

The studies presented in this report reveal that EGF induces important changes in the relative proportion of a few of the total [^{35}S]methionine-labeled polypeptides of A431 cells as detected by two-dimensional gel electrophoresis. These studies however, have not revealed new polypeptides in the EGF-treated cells, at least at the level of detection currently achieved by this technique.

Only 10 polypeptides of a total of over 1000 polypeptides resolved were found to be EGF-sensitive, meaning that less than 1% of the polypeptides are affected by EGF. Of these, only the identity of polypeptide IEF 7 and 10 is known. IEF 7 corresponds to HeLa cell polypeptide IEF 49 (cyclin; Bravo and Celis 1982), which has been demonstrated to be a karyoplastic protein that varies during the cell cycle and is sensitive to changes in cell proliferation (Bravo and Celis 1980; Bravo et al. 1981a,c). The synthesis of this polypeptide in A431 cells decreases dramatically after a few hours of EGF action. This result is in line with those previously reported (Bravo et al. 1981c) as growth of these cells is inhibited by EGF. IEF 10, whose synthesis also decreases after EGF treatment, corresponds to HeLa cell polypeptide IEF 55 (Bravo and Celis 1982), which has been demonstrated to be a tropomyosin-related polypeptide by means of one-dimensional peptide mapping (Bravo et al. 1981c). Recently, we have observed that IEF 10 does not contain proline, a characteristic of the tropomyosins. Given the striking changes in morphology that occur in A431 cells treated with EGF, it is likely that IEF 10 may play a role in modulating cell morphology.

Some of the evidence presented in this report demonstrates that none of the EGF-sensitive polypeptides whose synthesis is increased by the growth factor is a component of the cellular cytoskeleton. Furthermore, the results obtained with Triton X-114 extraction (Bordier

1981) strongly suggest that polypeptides IEF 6 and 8 could possibly be integral membrane proteins.

The mechanism by which EGF is able to stimulate or inhibit the synthesis of specific polypeptides in A431 cells is far from being understood, but it becomes clear from our studies that EGF affects the transcription of certain mRNAs, altering therefore the relative proportion of the proteins encoded by these mRNAs.

The possible relation that the EGF-sensitive polypeptides could have with the inhibition of growth in A431 cells needs further investigation. For this, we have isolated clones that are resistant to the growth-inhibitory effects of EGF, and we are currently studying the effect of the growth factor in the protein synthesis of these clones.

Work is now in progress to analyze the polypeptide composition of different cellular fractions and to prepare antibodies against those polypeptides of interest in order to get an insight into the cellular locations of the EGF-sensitive polypeptides.

Acknowledgments

I would like to thank H. Macdonald-Bravo and P. Blundell for their excellent technical assistance. I am also grateful to Dr. S.L. Jorcano for helping with the RNA preparation.

References

Barnes, D.W. 1982. Epidermal growth factor inhibits growth of A431 human epidermoid carcinoma in serum-free cell culture. *J. Cell Biol.* **93:** 1.

Bordier, C. 1981. Phase separation of integral membrane proteins in Triton X-114 solution. *J. Biol. Chem.* **256:** 1604.

Bravo, R. and J.E. Celis. 1980. A search for differential polypeptide synthesis throughout the cell cycle of HeLa cells. *J. Cell Biol.* **48:** 795.

———. 1982. Updated catalogue of HeLa cell proteins: Percentages and characteristics of the major cell polypeptides labelled with a mixture of 16 [^{14}C]-labelled amino acids. *Clin. Chem.* **28:** 766.

Bravo, R., A. Celis, D. Moses, and J.E. Celis. 1981a. Distribution of HeLa cell polypeptides in cytoplasts and karyoplasts. *Cell Biol. Int. Rep.* **5:** 479.

Bravo, R., S.J. Fey, J.V. Small, P. Mose Larsen, and J.E. Celis. 1981b. Coexistence of three major isoactins in a single sarcoma 180 cell. *Cell* **25:** 195.

Bravo, R., J.V. Small, S.J. Fey, P. Mose Larsen, and J.E. Celis. 1982. Architecture and polypeptide composition of HeLa cell cytoskeletons. Modification of cytoarchitectural proteins during mitosis. *J. Mol. Biol.* **154:** 121.

Bravo, R., S.J. Fey, J. Bellatin, P. Mose Larsen, J. Arevalo, and J.E. Celis. 1981c. Identification of a nuclear and of a cytoplasmic polypeptide whose relative proportions are sensitive to changes in the rate of cell proliferation. *Exp. Cell Res.* **136:** 311.

Buss, J.E., J.E. Kudlow, C.S. Lazar, and G.N. Gill. 1982. Altered epidermal growth factor (EGF)–stimulated protein kinase activity in variant A431 cells with altered growth response to EGF. *Proc. Natl. Acad. Sci.* **79:** 2574.

Carpenter, G. and S. Cohen. 1979a. Epidermal growth factor. *Annu. Rev. Biochem.* **48:** 193.

———. 1979b. Rapid enhancement of protein phosphorylation in A431 cell membrane preparations by epidermal growth factor. *J. Biol. Chem.* **254:** 4884.

Carpenter, G., L. King, and S. Cohen. 1978. Epidermal growth factor stimulates phosphorylation in membrane preparations in vitro. *Nature* **276:** 409.

Chinkers, M., J.A. McKanna, and S. Cohen. 1979. Rapid induction of morphological changes in human carcinoma cells A431 by epidermal growth factor. *J. Cell Biol.* **83:** 260.

———. 1981. Rapid rounding of human epidermoid carcinoma cells A431 induced by epidermal growth factor. *J. Cell Biol.* **88:** 422.

Cohen, S., G. Carpenter, and L. King. 1980. Epidermal growth factor-receptor-protein kinase interaction. *J. Biol. Chem.* **255:** 4834.

Cohen, S., H.T. Haigler, G. Carpenter, L. King, Jr., and J.A. McKanna. 1979. Epidermal growth factor: Visualization of the binding and internalization of EGF in cultured cells and enhancement of phosphorylation by EGF in membrane preparations in vitro. *Cold Spring Harbor Conf. Cell Proliferation* **6:** 131.

Erikson, E., D.J. Shealy, and R.L. Erikson. 1981. Evidence that viral transforming gene products and epidermal growth factor stimulate phosphorylation of the same cellular protein with similar specificity. *J. Biol. Chem.* **256:** 11381.

Fabricant, R.N., J.E. Delarco, G.J. Todaro. 1977. Nerve growth factor receptors on human melanoma cells in culture. *Proc. Natl. Acad. Sci.* **74:** 565.

Gill, G.N. and C.S. Lazar. 1981. Increased phosphotyrosine content and inhibition of proliferation in EGF-treated A431 cells. *Nature* **293:** 305.

Gill, G.N., J.E. Buss, C.S. Lazar, A. Lipshitz, and J.A. Cooper. 1982. Role of epidermal growth factor (EGF)–stimulated protein kinase in control of proliferation of A431 cells. *J. Cell. Biochem.* **19:** 249.

Gospodarowicz, D., G. Greenburg, H. Bialecki, and B.R. Zetter. 1978. Factors involved in the modulation of cell proliferation in vivo and in vitro: The role of fibroblast and epidermal growth factors in the proliferative response of mammalian cells. *In Vitro* **14:** 85.

Haigler, H.T. and G. Carpenter. 1980. Production and characterization of antibody affecting epidermal growth factor: Receptor interactions. *Biochim. Biophys. Acta* **598:** 314.

Haigler, H.T., J.A. McKanna, and S. Cohen. 1979. Direct visualization of the binding and internalization of a ferritin conjugate of epidermal growth factor in human carcinoma cells A431. *J. Cell Biol.* **81:** 382.

Haigler, H., J.F. Ash, S.J. Singer, and S. Cohen. 1978. Visualization by fluorescence of the binding and internalization of epidermal growth factor in human carcinoma cells A431. *Proc. Natl. Acad. Sci.* **75:** 3317.

Hunter, T. and J.A. Cooper. 1981. Epidermal growth factor induces rapid tyrosine phosphorylation of proteins in A431 human tumor cells. *Cell* **24:** 741.

Laskey, R.A. and A.D. Mills. 1975. Quantitative film detection of [^{3}H] and [^{14}C] in polyacrylamide gels by fluorography. *Eur. J. Biochem.* **56:** 335.

Lipshitz, A., C.S. Lazar, J.E. Buss, and G.N. Gil. 1983. Analysis of morphology and receptor metabolism in clonal variant A431 cells with differing growth response to epidermal growth factor. *J. Cell. Physiol.* **115:** 235.

McKanna, J.A., H.T. Haigler, and S. Cohen. 1979. Hormone receptor topology and dynamics: A morphological analysis using ferritin-labelled epidermal growth factor. *Proc. Natl. Acad. Sci.* **76:** 5689.

O'Farrell, P.H. 1975. High resolution two-dimensional electrophoresis of proteins. *J. Biol. Chem.* **250:** 4007.

Schlessinger, J. and B. Geiger. 1981. Epidermal growth factor induces redistribution of actin and α-actinin in human epidermal carcinoma cells. *Exp. Cell Res.* **134:** 273.

Ushiro, H. and S. Cohen. 1980. Identification of phosphotyrosine as a product of epidermal growth factor–activated protein kinase in A431 cell membranes. *J. Biol. Chem.* **255:** 8363.

Willingham, M.C., H.T. Haigler, D.J.P. Fitzgerald, M.G. Gallo, A.V. Rutherford, and I.H. Pastan. 1983. The morphologic pathway of binding and internalization of epidermal growth factor in cultured cells. *Exp. Cell Res.* **146:** 163.

Wrann, M.M. and C.F. Fox. 1979. Identification of epidermal growth factor receptors in a hyperproducing human epidermoid carcinoma cell line. *J. Biol. Chem.* **254:** 8083.

Intermediate Filaments—from Wool α-Keratins to Neurofilaments: A Structural Overview

K. Weber and N. Geisler

Max-Planck Institute for Biophysical Chemistry, D-3400 Goettingen, Federal Republic of Germany

Intermediate, or 10-nm, filaments (IFs) are present in almost all vertebrate cells. Together with actin-based microfilaments and the tubulin-containing microtubules, they form the three fibrous systems of the cytoskeleton. Although known to electron microscopists for some time, their abundance and wide distribution began to be appreciated only when suitable antibodies to them were used in immunofluorescence microscopy. Immunological data were also instrumental in the recognition of the five distinct IF subclasses: epithelial (cyto)keratins, neuronal neurofilaments, astrocyte-specific glial fibrillary acidic protein (GFAP), myogenic desmin, and "mesenchymal" vimentin. The emerging, strict cell- and tissue-specific expression patterns parallel to histologically distinct cell types and known differentiation pathways have made these proteins excellent markers both in embryogenesis and surgical pathology. The early realization that (cyto)keratins form a particularly complex subfamily possibly containing some 20 distinct polypeptides, which in majority follow strict expression patterns in morphologically distinct epithelia, has posed additional questions as to the molecular basis of IF protein divergence. Since divergence cannot be understood until the common structural formula is delineated, the problem of IF structure soon became very important.

Here we summarize how amino acid sequence data derived originally by protein chemistry and more recently by DNA technology have put the field of IF proteins on a firm molecular basis and have allowed the formulation of reasonable models of IF structure that account for both the common and variable properties of these molecules. Instrumental in coming to these conclusions were three events: (1) the development of extended amino acid sequences and the recognition of their informational content, which was expected to be very high, since coiled coils were predicted and actually found; (2) the proof by sequence analysis that IF proteins and the classical α-keratins are closely related has allowed several important predictions of mutual benefit for both fields; (3) as in myosin, the different domains of IF proteins defined by proteolytic and biochemical experiments can now be firmly related with parallelly developed sequences and their structural interpretation.

Development of a general topographical model: A structurally conserved coiled-coil domain flanked by hypervariable domains

We started sequence work on nonepithelial IF proteins for three reasons: to put the immunological classification on a firm molecular footing, to reveal a possible relationship between the five IF classes, and also to begin to understand IF structure and organization. The first study centered on the m.w. 15,000 (15K) carboxyterminal fragments of vimentin and desmin (Geisler and Weber 1981). The two proteins were characterized as being highly related but clearly distinct molecules with tissue specificity overriding a small species divergence. We also recognized a seven-residue repeat pattern (a,b,c,d,e,f,g) with a and d being hydrophobic residues, which started 56 residues from the carboxyl end and continued toward the very amino terminus of the fragments. Since such a pattern is the molecular formula of α-helices able to form coiled coils, the amino acid sequences were directly connected with the α-type X-ray pattern of intermediate filaments (Geisler and Weber 1981). Having shown that the two other major nonepithelial IF proteins (neurofilament 68K protein and GFAP) followed the same organizational principles (Geisler et al. 1982b), we concentrated on desmin and developed, by proteolytic studies, a three-domain structure parallel to further sequence data (Fig. 1; Geisler et al. 1982a). Three contiguous but distinct arrays account for the whole molecule. The highly α-helical middle domain (residues 70–415) had a rodlike morphology when analyzed by electron microscopy. Since a coiled-coil domain of

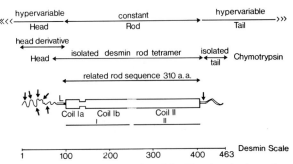

Figure 1 Schematic presentation of the IF molecule given along the desmin sequence line. The three proteolytically defined domains (head, rod, and tail) are indicated. Note that the desmin rod, isolated by chymotrypsin treatment, extends at each end the related rod sequence of about 310 residues found in all IF molecules. The extension at the aminoterminal end (L) is so far only seen in the nonepithelial IF proteins. (↓) Points of cleavage by chymotrypsin. Head and tail are non-α-helical and hypervariable in sequence and length (see Fig. 2 and Table 1). the rod is highly α-helical, and its sequence defines three arrays (boxes) able to form parallel interpolypeptide coiled coils (coils Ia, Ib, and II). The dots indicate the possibility to extend coil Ib further (see text). In this case the sum of coil Ia and Ib (I) would be equal in length to coil II (i.e., ~140 residues or 210 Å).

about 140 residues was documented by the heptade convention (coil II), the high α-helical content of the rod (85%) allowed us to predict that a similar array (coil I) was present prior to a short spacer. The rod domain is flanked by two non-α-helical domains, the aminoterminal headpiece and the carboxyterminal tailpiece. Following up an earlier observation that sheep wool α-keratins and human epidermal keratins show immunological cross-reactivity (Weber et al. 1980), we compared the published partial sequence data on two distinct wool α-keratins, 8c-1 and 7c (Crewther et al. 1980; Sparrow and Inglis 1980), with the desmin molecule. We could easily delineate a common rod domain related in sequence and length for all three proteins, and we discussed the terminal non-α-helical arrays as hypervariable regions (Weber and Geisler 1982). The completion of the desmin sequence documented coils Ia and Ib (Figs. 1 and 2) and allowed for a full alignment of all α-keratin data (Geisler and Weber 1982). Meanwhile, several of the predicted alignments have been verified by further extension of the α-keratin data (Dowling et al. 1983). With the characterization of α-keratins as IF proteins, the desmin model accounted for both epithelial and nonepithelial proteins. When the first and almost complete DNA sequence of a human epidermal keratin sequence became available (Hanukoglu and Fuchs 1982), we showed its alignment parallel to desmin and its striking relation with α-keratin 8c-1 along the common rod domain (Geisler and Weber 1982). In an independent approach aimed at understanding human epidermal keratin divergence, E.V. Fuchs and co-workers used modern DNA technology and established two prototype keratins (Fuchs et al. 1981; Hanukoglu and Fuchs 1982; Kim et al. 1983). Their first report on one of the prototypes provided a sequence structurally related generally to the rod of the desmin-type molecules and noted the strong homology with α-keratin 8 in a comparison with a rather short fragment of this protein (Hanukoglu and Fuchs 1982). As predicted from all the combined information, the second human epidermal keratin prototype showed a pronounced relation to the second α-keratin prototype, i.e., wool α-keratin 7c (Hanukoglu and Fuchs 1983).

Consolidating information came from several additional studies. Quax-Jeuken et al. (1983) extended our previous 180 residue sequence of vimentin (Geisler and Weber 1981; Geisler et al. 1982b) to a nearly complete sequence obtained from DNA data. We were able to greatly expand the protein data on both the neurofilament 68K protein (Geisler et al. 1983) and GFAP (Geisler and Weber 1983), and Steinert et al. (1983) reported the complete sequence of a 59K mouse epidermal keratin deduced from DNA studies. Although none of these reports have actually changed our view of the informational content of IF protein sequences, they all have contributed greatly by consolidating the structural principle and have also helped in understanding divergence. Clearly, without sequence information, it would not have been possible to unravel IF relation on the molecular level and to begin to try to understand IF structure.

The conserved rod domain dictates the structural organization of IF

The sequences of 9 different IF proteins (Fig. 2) identify the common part of the rod domain as about 310 amino acids in length (desmin residues 97–407). A preceding leader-type sequence of 17 residues (desmin positions 80–96) of α-helical character but poor coiled-coil forming potential is found in the four nonepithelial proteins. Although we do not know if it also exists in type-II keratins (α-keratin 5/7–related), it is absent in all three known type-I keratins (α-keratin 8–related). Along the rod there are three distinct coiled-coil domains, which we have called Ia (~35 residues), Ib (95 residues), and II (~140 residues) (Geisler and Weber 1982). Coils I and II are of approximately equal size and account for a calculated length of 195–210 Å, a value possibly related to the 210-Å filament periodicity seen in shadowed filament specimens (Henderson et al. 1982; Milam and Erickson 1982). We note, however, that coil Ib could easily be extended by close to 15 residues (Steinert et al. 1983), leading to a rather short, non-α-helical sequence separating this coil from coil II (see legend to Fig. 2 and below).

Predicted coiled coils are easily aligned and do not allow for any deletions or additions when the terminal residues are fixed by maximal homology in all known

Figure 2 (*see facing page*) Sequence relation between different IF proteins. Alignment is based on previous arguments (Geisler and Weber 1982; see also text). For primary sequence data, see the following references: Geisler et al. (1982a, 1983), Geisler and Weber (1982, 1983), Hanukoglu and Fuchs (1982, 1983), Dowling et al. (1983), Quax-Jeuken et al. (1983), Steinert et al. (1983). Abbreviations for individual proteins are: (ME_1) mouse epidermal keratin 59K, (HE_1) human epidermal keratin 50K, (HE_2) human epidermal keratin 56K, (8 and 7) sheep wool hard α-keratin components 8c-1 and 7c, (D) chicken desmin, (V) hamster vimentin, (G) porcine GFAP, and (NF) porcine 68K neurofilament protein. Horizontal lines indicate not-yet-established sequences. Deletions (dashes) allow for some better alignments, mainly in the non-α-helical regions. Individual members of a prototype sequence are arranged next to each, and identical residues along the prototypes are given by bold letters. The three types are: the four nonepithelial proteins (D, V, G, NF), the type-I keratin (HE_1, ME_1, 8), and the type-II keratin (HE_2, 7). Note that keratins arise by coassembly of types I and II, whereas nonepidermal proteins can form homopolymeric filaments (see text). Identical residues present within the coiled-coil arrays of all proteins have arrowheads pointing up. Note the consensus sequences early in coil Ia and at the end of coil II. The three structural domains (head, rod, and tail) are indicated, as are the hydrophobic a and d positions (dots) in the consecutive heptades. Presumptive coiled-coil arrays (Ia, Ib, and II) are marked by lines above and below the sequence blocks. For some irregularities in coil II and the possible extension of coil Ib, see text and Fig. 1. In the latter case, dashes past coil Ib have to be removed and the array would stop around the proline residue found in most of the proteins. Arrowheads pointing down mark the length of the isolated desmin rod (Geisler et al. 1982a; see Fig. 1). A leader-type sequence (underlined) occurring early in the rod is found only in nonepidermal proteins. The X in NF is either R or K. Underlined residues in the carboxyterminal sequence of GFAP may have ambiguities (Geisler and Weber 1983).

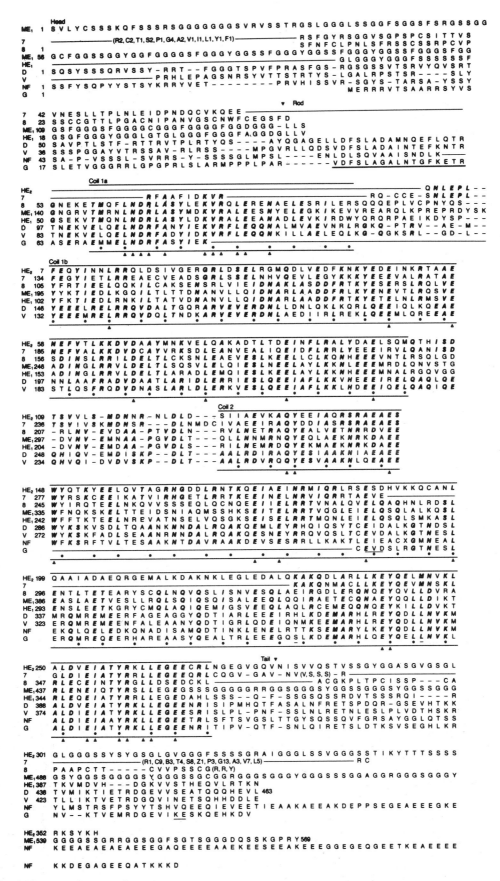

Figure 2 (See facing page for legend.)

proteins and the heptade convention is rigorously applied. All coils are separated by rather short spacers, many but not all of which reveal proline residues. These spacers show rather limited length and sequence variability. No evidence for the major inserts predicted earlier by indirect arguments (Steinert et al. 1980) is seen beyond that indicated only by compositional data on wool 7c-1 (Sparrow and Inglis [1980], as discussed in Geisler and Weber [1982]). Even if this were verified by sequence analysis, it does not occur in the human keratin II (Hanukoglu and Fuchs 1982) and a further but distinct keratin II isolated from human placenta (see below) where the current sequence gap in coil II of α-keratin 7c is fully covered by consecutive heptades.

Three prototype sequences (Fig. 2) are found along the common part of the rods (Geisler and Weber 1982, 1983; Hanukoglu and Fuchs 1983). One is provided by the four nonepithelial proteins (desmin, vimentin, GFAP, and neurofilament 68K protein), where sequence identity values are between maximally 70% (vimentin and desmin) and minimally 55% (desmin and NF68K) (Geisler and Weber 1982, 1983). The second encompasses the keratin I group and is found in α-keratin 8c-1, human keratin 50K, and a mouse epidermal 59K keratin. Here sequence identity is at least 60%, with the α-keratin being the most remote molecule. The third type accounts for the keratin II group and combines wool α-keratin 7c and the human 56K protein. Although evidence for a coil Ia domain is so far only provided in one case for type-II keratins (Geisler and Weber 1982), the overall sequence identity is around 55% (Hanukoglu and Fuchs 1983; for a more extended comparison, see Fig. 2).

The three prototype sequences, when compared with each other, reveal in each case only about 30% sequence identity. Nevertheless, all nonepithelial proteins are able to form normal 10-nm filaments from a single protein (for review, see Geisler and Weber 1983), whereas the keratins need at least two components (Steinert et al. 1982), and no epithelial cell with only a single keratin polypeptide is known (e.g., Moll et al. 1982). Although we cannot rigorously exclude a direct participation of the two terminal domains in this obligatory heteropolymer system, it seems more likely that keratin structure involves some "complementation" between rod sequences I and II to achieve a self-assembly-competent interaction system comparable to that provided in the nonepithelial proteins by a single type of rod domain. Individual keratins seem not to be able to form proteolytically stable derivatives (Steinert et al. 1982), and therefore this complementation could already occur at the level of the coiled coil. Alternatively, it is surely present when coiled-coil Ib fragments interact to form a tetramer (Gruen and Woods 1983).

Along the coiled coils we notice the required monotony of consecutive heptades in their display of hydrophobic a and d residues. In the majority of cases where a nonhydrophobic residue or even a charged residue is encountered in such a position, it is found for almost all the distinct proteins. Other examples breaking the monotony are the possibility of an a-to-d reversal (or skip residue) in coil II around desmin residue 343 (Dowling et al. 1983; Geisler et al. 1983; Steinert et al. 1983), the allowance for a very short non-coiled-coil region or skip residue around the centrally located tryptophan-desmin residue 286 (Dowling et al. 1983; Steinert et al. 1983; Weber et al. 1983), or even a possible division into a 35-residue coil IIa followed after 10 residues by a 95-residue coil IIb (Hanukoglu and Fuchs 1983). Which of these solutions will be verified when more-detailed data become available remains to be seen; however, in any case it should be applicable for all proteins.

Filament structure is thought to arise from many interacting coiled coils (Fraser and MacRae 1983), and the various protein sequences reveal a pronounced 28-residue (42-Å) repeat pattern (Parry et al. 1977; Geisler et al. 1982a; McLachlan and Stewart 1982; Steinert et al. 1983) also known for myosin but not for the simple system of tropomyosin. One such segment of three consecutive 28-residue units has been analyzed in detail (McLachlan and Stewart 1982), and it is generally assumed that the ordered arrangement of basic and acidic residues is important in packing neighboring coiled coils.

The alignment of the rod sequences predicted for desmin and the α-keratins (Weber and Geisler 1982) has been confirmed with each additional sequence report. There is indeed only one unambiguous alignment, since any shift leads to a drastic loss of homology. The reason for this unique solution cannot be appreciated when only secondary structural prediction rules concerning α-helices and the location of putative β-turns is performed, as has been done in the studies of Hanukoglu and Fuchs (1982, 1983). The basis for a unique solution lies simply in the fact that IF structure is based on interacting coiled coils, and thus we and others have executed coiled-coil conventions. Normal α-helices could clearly be shorter or longer or even be shifted against each other since they are not restricted by the heptade rule of coiled coil–forming α-helices. But even simple, noninteracting coiled coils do not have to show extreme length restriction, as seen by the different sizes of muscle and nonmuscle tropomyosin. Interacting coiled coils, however, must be extremely finely tuned in sequence due to the 28-residue repeat pattern with its inherent 28/3 and 28/10 repeats in charged amino acid residues, which is necessary to pack neighboring coiled coils.

Two further aspects are important. Sequence identity values along the rod are unevenly distributed (Fig. 2). They clearly peak at the beginning of coil Ia and at the end of coil II, providing consensus sequences, which certainly will be of particular importance for IF structure (Geisler and Weber 1982). Thus, it is not surprising that the epitope of a general IF monoclonal antibody that also reacts with the mixture of α-keratins has been mapped to the last 20 residues in coil II (Geisler et al. 1983). Relatively high sequence identity is also apparent at the end of coil Ib and the beginning of coil II. As shown in Figure 2, within the remainder of the coils there are many regions of low sequence identity, even among the closest members of each prototype. Thus, there is no reason why these arrays cannot give rise to immunogenic sites accounting for highly specific monoclonal and polyclonal antibodies able to distinguish the different proteins without resorting to the curious hypervariable regions as the putative sites of the epitopes (E. Debus et al., in prep.).

Heads and tails are hypervariable regions but are involved in filament stability

Head- and tailpieces are the hypervariable regions of IF proteins (Geisler and Weber 1982, 1983; see Table 1). A complete characterization is available only for the four nonepithelial proteins. Here the headpieces reveal a non-α-helical structure in line with their multiple β-bends and several proline residues. The presence of 10 arginine residues and many hydroxyamino acids, the virtual lack of acidic groups, and the consequent high proteolytic sensitivity explain various degradation patterns seen in the literature (Geisler et al. 1982a). In spite of a generally related non-α-helical structure, even the extensive introduction of gaps allows only for very few and extremely short sequence identities rarely above the length of a tri- or tetrapeptide. The size of the headpieces from nonepithelial proteins varies from about 45 residues in GFAP to about 70 residues in the three other proteins. Tailpieces of desmin, vimentin, and GFAP are about 50 residues long and clearly are closely related. By contrast, NF68K shows about 154 residues easily separable into two distinct subdomains (Geisler et al. 1983). The first 48 residues have some resemblance to headpiece structure, indicating the possibility of a head-to-tail type assembly, whereas the final subdomain of 106 residues is highly acidic due to about 50 glutamic acid residues. This feature is unique to the neurofilament protein and probably indicates some scaffold type of extension. The most remarkable length variation is seen with the two higher neurofilament proteins (160K and 200K), where the rod domain is followed by tailpieces with molecular weights of ~100K and ~160K, respectively. We have discussed elsewhere how the tailpiece of at least the 200K component could act as an interfilament crossbridge (Geisler et al. 1983; Weber et al. 1983).

A distinctly different chemistry is seen in the wool α-keratins, where cysteine-rich regions are encountered particularly in the tailpieces (Crewther et al. 1980; Sparrow and Inglis 1980; for discussion, see Weber and Geisler 1982). This feature, typical of a final frozen state relying on disulfide bridges, seems restricted to α-keratins, since the epidermal keratins reveal a new motif. Here, repetitive sequences of several glycine residues flanked by large hydrophobic residues have been documented or predicted for either both domains or the headpiece only (Hanukoglu and Fuchs 1982, 1983; Steinert et al. 1983). Although such sequences raise the problem of understanding "polyglycine"-derived structures, they could again indicate a head- and tail-type assembly mechanism. Further variations on the theme are given by two observations. First, the human epidermal 50K keratin of type I lacks the "polyglycine" tailpiece and shows a sequence related to desmin, vimentin, and GFAP (Hanukoglu and Fuchs 1982; Geisler and Weber 1983). Second, a porcine, 52K, intestinal epithelium–derived keratin has an amino acid composition indicating a greatly reduced glycine content (K. Weber and N. Geisler, unpubl.), as does a 50K protein from murine endodermal cells (Oshima 1981). Therefore we can predict that keratins without extended polyglycine tracts exist in simple epithelia (Table 1).

Given this bewildering divergence of sequence and length in the terminal domains, it is important to state that at least for desmin the isolated rod (residues 70–415) is unable to form filaments under physiological conditions since it behaves as a soluble protein. Thus, at least for nonepithelial IF proteins, there has to be a further structural contribution by either one or both terminal domains (Geisler et al. 1982a; Geisler and Weber 1982). We have argued before for a direct participation of headpiece regions, but we also anticipate an involvement of the tailpiece. It is not clear whether parts or the entire length of the terminal domains are required in assembly. Given the widely different chemistry of these domains, one expects that individual differences in physical-chemical properties of IF proteins are strongly influenced by them.

How can we explain human keratin divergence?

The documented sequence relation between two human epidermal keratins and their corresponding counterparts

Table 1 Sequence Types in the Hypervariable Regions

Sequence type	Head	Tail
"Normal" sequence	none (so far)	D, V, GFAP human epidermal keratin 50K
Arginine, proline, hydroxyamino acids	D, V, GFAP, NF68K, NF160K	NF68K (domain a)
"Glycine repeats"	mouse epidermal keratin 59K human epidermal keratin 50K human epidermal keratin 56K[a]	human epidermal keratin 56K mouse epidermal keratin 59K
Absence of "glycine repeats"	keratins from simple epithelia[a]	keratins from simple epithelia[a]
Cysteine-rich	wool α-keratins	wool α-keratins
Glutamic acid–rich		domain b of NF68K
Long extensions		high-molecular-weight triplet proteins (NF160K, 200K)

[a]Not yet documented by sequence data; for sequence data, see Fig. 2.

in wool α-keratins offers several avenues for understanding human keratin divergence. The eight α-keratins of sheep wool follow only two prototype rod sequences, i.e., 8 (keratin type I) and 5/7 (keratin type II). Although the two prototypes differ by about 70% in amino acid sequence, the four individual members of each type may show approximately 80% sequence identity, and evidence for some of this small divergence can be seen in various reports (Crewther et al. 1978; Dowling et al. 1979; for some references, see Gruen and Woods 1983). The known interaction between type-I and -II keratin molecules, explored particularly for their coil Ib fragments (Gruen and Woods 1983), readily explains that self-assembly of epidermal keratins in vitro is based on heteropolymer formation (Steinert et al. 1982) and that there is no epithelial cell that expresses only one "cytokeratin" (Moll et al. 1982). Elegant hybridization studies have identified only two subclasses of epithelial keratin mRNAs strictly parallel to the two known prototype molecules (Kim et al. 1983). We think that individual differences in hybridization behavior could be accounted for by a differential degree of relation between the various molecules of one subclass, which along their rod regions could easily be even more closely related than desmin and vimentin. Finally, the expected strong restrictions to build a filament using two somehow "complementary" molecules makes it highly unlikely that there are more than the two prototype rods already known.

Although further sequence data would be very helpful, we know already about a second member of the type-II human keratins. A protein isolated from human placenta and several carcinomas, tissue polypeptide antigen (TPA), has previously been studied. From partial sequence data (Redelius et al. 1980), B. Lüning has now proposed that TPA resembles an IF protein, most likely a keratin (B. Lüning, in prep.). Inspection of the published data (Redelius et al. 1980) allows the firm conclusion that TPA is a type-II keratin. Two longer sequences accounting for half of coil II show 67% sequence identity with the corresponding regions of the 56K human epidermal keratin. The third sequence covers the consensus sequence seen in coil Ia so far for all IF proteins. A more detailed account of this keratin X will be given elsewhere (B. Lüning and K. Weber, in prep.).

Further divergence of keratins surely arises due to the hypervariable terminal domains (Table 1). Regions extremely rich in glycine can exist on both ends of type-I and -II molecules. Since one human epidermal keratin shows such an array only at the head and has a desmin-related tailpiece, we expect that the opposite may also be found. In addition, we have already proposed that keratins from certain simple epithelia have a normal glycine content approaching that of the nonepithelial proteins (see above). Finally, we have to consider the human hard α-keratins from nail and hair, thus increasing the human keratin catalog of 19 components (Moll et al. 1982). The former reveal the unusual high cysteine content typical also of wool α-keratins but not of epidermal keratins. Thus, the five nail components, which seem to follow Mendelian rules (Marshall 1980), will probably closely resemble the wool α-keratins, where most cysteines are clustered into the hypervariable domains (Fig. 2), providing for a high degree of disulfide linkage. We note that since neither the α-keratins nor the desmin group of proteins need the unusually high urea concentration necessary to "melt" certain epidermal keratin mixtures (Franke et al. 1983), this requirement may simply reflect interacting "polyglycine" tracts.

From coiled coils to the 10-nm filament

The isolated desmin rod (residues 70–415) as well as the mixture of the coil Ib fragments of α-keratin have a tetrameric organization, indicating a dimer of a normal double-stranded coiled coil. The various arguments have been reviewed recently (Geisler and Weber 1982; Gruen and Woods 1983), and we see no ready explanation to reconcile this view with the triple coiled-coil prediction of Steinert et al. (1980, 1982). We believe our data show that as in any other double-stranded coiled coil, the α-helical segments are parallel and in register and that in the desmin tetramer the two dimers are most likely arranged against each other in an antiparallel orientation. To account for a particularly strong interaction between coil Ib tetramers (Gruen and Woods 1983) and various proteolytic data on IF proteins (e.g., see Steinert et al. 1980), one is tempted to envision a rather long tetrameric unit in which the two type-I coils are antiparallely oriented with the two coil-II domains protruding without partners on each side (N. Geisler and K. Weber, in prep.). Although, from various electron micrographs, such a protofilamentous unit is somewhat longer than expected, it can be used to build a framework of stacked coiled coils that seems to account for most of the published X-ray information on α-keratins (Fraser and MacRae 1983). In their presentation, there are seven or eight such tetramers in a filament diameter, depending on whether there is an inner core of an additional tetramer. Thus, the number of molecules present per cross-sectioned area is 28 or 32. Since in such a structure one expects some limitations in twisting higher-order coiled-coil derivatives around each other, the question of subfilament arrangements or protofibrils is raised (N. Geisler and K. Weber, in prep.). Approaching the architecture by morphological criteria, using partially unraveled filaments, leads to ropelike models. For instance, the presence of four protofibrils each containing two protofilaments built as tetrameric structures has been discussed (Aebi et al. 1983). However, each of these avenues still requires much more additional information as to the various interaction patterns to arrive at a three-dimensional model useful to the cell biologist interested in intracellular assembly and disassembly. Given the established wealth of sequence data and the possibility of envisioning a consecutive set of 28 residue segments with an ordered charge pattern, we are currently involved in their analysis in order to design experiments to identify nearest-neighbor segments. We hope that this approach will lead to a detailed structural model of IF organization.

References

Aebi, U., W.E. Fowler, P. Rew, and T.T. Sun. 1983. The fibrillar substructure of keratin filaments unravelled. *J. Cell Biol.* **97:** 1131.

Crewther, W.G., L.M. Dowling, and A.S. Inglis. 1980. Amino acid sequence data from a microfibrillar protein of α-keratin. In *Proceedings from the 6th Quinquennial International Wool Textile Research Conference*, Pretoria, vol. 2, p. 79.

Crewther, W.G., K.H. Gough, A.S. Inglis, and N.M. McKern. 1978. Sequence homologies in helical segments from α-keratin. *Text. Res. J.* **48:** 160.

Dowling, L.M., D.A.D. Parry, and L.G. Sparrow. 1983. Structural homology between hard α-keratin and the intermediate filament proteins desmin and vimentin. *Biosci. Rep.* **3:** 73.

Dowling, L.M., K.H. Gough, A.S. Inglis, and L.G. Sparrow. 1979. Comparison of some microfibrillar proteins from wool. *Aust. J. Biol. Sci.* **32:** 437.

Franke, W.W., D.L. Schiller, M. Hatzfeld, and S. Winter. 1983. Protein complexes of intermediate-sized filaments. *Proc. Natl. Acad. Sci.* **80:** 7113.

Fraser, R.D.B. and T.P. MacRae. 1983. The structure of the α-keratin myofibril. *Biosci. Rep.* **3:** 517.

Fuchs, E.V., S.M. Coppock, H. Green, and D.W. Cleveland. 1981. Two distinct classes of keratin genes and their evolutionary significance. *Cell* **27:** 75.

Geisler, N. and K. Weber. 1981. Comparison of the proteins of two immunologically distinct intermediate-sized filaments by amino acid sequence analysis: Desmin and vimentin. *Proc. Natl. Acad. Sci.* **78:** 4120.

———. 1982. The amino acid sequence of chicken muscle desmin provides a common structural model for intermediate filament proteins. *EMBO J.* **1:** 1649.

———. 1983. Amino acid sequence data on glial fibrillary acidic protein (GFA); implications for the subdivision of intermediate filaments into epithelial and non-epithelial members. *EMBO J.* **2:** 2059.

Geisler, N., E. Kaufmann, and K. Weber. 1982a. Protein-chemical characterization of three structurally distinct domains along the protofilament unit of desmin 10 nm filaments. *Cell* **30:** 277.

Geisler, N., U. Plessmann, and K. Weber. 1982b. Related amino acid sequences in neurofilament and non-neuronal intermediate filaments. *Nature* **296:** 448.

Geisler, N., E. Kaufmann, S. Fischer, U. Plessmann, and K. Weber. 1983. Neurofilament architecture combines structural principles of intermediate filaments with carboxyterminal extensions increasing in size between triplet proteins. *EMBO J.* **2:** 1295.

Gruen, L.C. and E.F. Woods. 1983. Structural studies on the microfibrillar proteins of wool. *Biochem. J.* **209:** 587.

Hanukoglu, I. and E. Fuchs. 1982. The cDNA sequence of a human epidermal keratin: Divergence of sequence but conservation of structure among intermediate filament proteins. *Cell* **31:** 243.

———. 1983. The cDNA sequence of a type II cytoskeletal keratin reveals constant and variable structural domains among keratins. *Cell* **33:** 915.

Henderson, D., N. Geisler, and K. Weber. 1982. A periodic ultrastructure in intermediate filaments. *J. Mol. Biol.* **155:** 173.

Kim, K.H., J.G. Rheinwald, and E. Fuchs. 1983. Tissue specificity of epithelial keratins: Differential expression of mRNAs from two multigene families. *Mol. Cell. Biol.* **3:** 495.

Marshall, R.C. 1980. Genetic variation in the proteins of human nail. *J. Invest. Dermatol.* **75:** 264.

McLachlan, A.D. and M. Stewart. 1982. Periodic charge distribution in the intermediate filament proteins desmin and vimentin. *J. Mol. Biol.* **162:** 293.

Milam, L. and H.P. Erickson. 1982. Visualization of a 21 nm axial periodicity in shadowed keratin filaments and neurofilaments. *J. Cell Biol.* **94:** 592.

Moll, R., W.W. Franke, D.L. Schiller, B. Geiger, and R. Krepler. 1982. The catalogue of human cytokeratins. *Cell* **31:** 11.

Oshima, R.G. 1981. Identification and immunoprecipitation of cytoskeletal proteins from murine extra embryonic endodermal cells. *J. Biol. Chem.* **256:** 8124.

Parry, D.A.D., W.G. Crewther, R.D.B. Fraser, and T.P. MacRae. 1977. Structure of α-keratin: Structural implication of the amino acid sequences of the type I and type II chain segments. *J. Mol. Biol.* **113:** 449.

Quax-Jeuken, Y.E.F.M., W.J. Quax, and H. Bloemendal. 1983. Primary and secondary structure of hamster vimentin predicted from the nucleotide sequence. *Proc. Natl. Acad. Sci.* **80:** 3548.

Redelius, P., B. Lüning, and B. Björklund. 1980. Chemical studies of tissue polypeptide antigen. *Acta Chem. Scand. B.* **34:** 265.

Sparrow, L.G. and A.S. Inglis. 1980. Characterization of the cyanogen bromide peptides of component 7c, a major microfibrillar protein from wool. In *Proceedings from the 6th Quinquennial International Wool Textile Research Conference*, Pretoria, vol. 2, p. 237.

Steinert, P.M., W.W. Idler, and R.D. Goldman. 1980. Intermediate filaments of baby hamster kidney (BHK-21) cells and bovine epidermal keratinocytes have similar ultrastructures and subunit structures. *Proc. Natl. Acad. Sci.* **77:** 4534.

Steinert, P.M., W.W. Idler, M. Aynardi-Whitman, R. Zackroff, and R.D. Goldman. 1982. Heterogeneity of intermediate filaments assembled in vitro. *Cold Spring Harbor Symp. Quant. Biol.* **46:** 465.

Steinert, P.M., R.H. Rice, D.R. Roop, B.L. Trus, and A.C. Steven. 1983. Complete amino acid sequence of a mouse epidermal keratin subunit and implications for the structure of intermediate filaments. *Nature* **302:** 794.

Weber, K. and N. Geisler. 1982. The structural relation between intermediate filament proteins in living cells and the α-keratins of sheep wool. *EMBO J.* **1:** 1155.

Weber, K., M. Osborn, and W.W. Franke. 1980. Antibodies to merokeratin from sheep wool decorate cytokeratin filaments in non-keratinizing epithelial cells. *Eur. J. Cell Biol.* **23:** 110.

Weber, K., G. Shaw, M. Osborn, E. Debus, and N. Geisler. 1983. Neurofilaments, a subclass of intermediate filaments; Structure and expression. *Cold Spring Harbor Symp. Quant. Biol.* **48:** 717.

Differential Expression of Two Classes of Keratins in Normal and Malignant Epithelial Cells and Their Evolutionary Conservation

E. Fuchs, M.P. Grace, K.H. Kim, and D. Marchuk
Department of Biochemistry, The University of Chicago, Chicago, Illinois 60637

The keratins are a family of 10–20 proteins (m.w. 40,000–67,000) produced by most vertebrate epithelial cells (Baden et al. 1973; Steinert and Idler 1975; Brysk et al. 1977; Culbertson and Freedberg 1977; Sun and Green 1978; Fuchs and Marchuk 1983). These polypeptides have the capacity to assemble into 8-nm filaments that, together with the 6-nm actin microfilaments and the 23-nm microtubules, comprise the cytoskeletal network of the epithelial cells. The keratin subunits have been classified as intermediate filament (IF) proteins, which include four additional groups of proteins that form 8–10-nm cytoskeletal filaments in other vertebrate cells: desmins (muscle cells), vimentins (cells of mesenchymal origin), glial filament proteins (astrocytes), and neurofilament proteins (neurons) (for a review, see Lazarides 1982). Although all IF subunits assemble into 8-nm filaments, the resulting properties of the filaments differ widely, depending on the class of IF subunit. Thus, in contrast to the actins and tubulins, which are ubiquitous in all eukaryotic cells, the IF proteins appear to have evolved to meet specialized structural requirements of individual vertebrate cell types.

Within the five classes of IF proteins, the keratins represent the largest and most diverse family. Of the 10–20 different keratins, however, only a small subset (2–5) are expressed by an epithelial cell at any one time. To determine whether individual keratins might perform different structural and functional roles, we and others have investigated in detail the complexity of the keratin family and their differential expression. We now know that there are different keratins, not only in different tissues, but also during different stages of differentiation and development. In addition, changes in the expression of keratins can occur upon malignant transformation of an epithelium, once again illustrating the dynamic flexibility in keratin gene expression.

Our studies have shown that all of the keratins can be divided into at least two distinct groups, each of which is encoded by its own small multigene family. Members of both groups of keratins are coordinately expressed in all epithelial cells, from human to fish and possibly even to lower eukaryotes, suggesting that both of these types of keratins play a fundamental role in filament assembly. The tissue- and differentiation-specific expression of the different members within each of these two groups most likely indicates a subtle tailoring of the resulting filaments to suit the individual needs of each epithelial cell. The results of these findings are summarized in this paper.

Expression of keratins in normal epithelia and in malignant neoplasms derived from normal epithelia

In epidermis, the keratins are especially abundant, comprising 30–85% of the total protein of these cells, depending on their relative stages of differentiation. In humans, the smaller keratins (M_r 40K–58K) are expressed in all epidermal cells, including the basal cells, whereas the larger keratins (M_r 60K–67K) are found only in the terminally differentiating cells (Fig. 1, lanes 1 and 2) (Dale et al. 1976; Fuchs and Green 1980). In most other epithelial cells, only the smaller keratins are expressed, but different patterns are produced that are characteristic of each specific cell type (Doran et al. 1980; Fuchs and Green 1980; Milstone and McGuire 1981; Tseng et al. 1982; Moll et al. 1982). Figure 1, lanes 3–6, shows the keratin patterns produced by human mesothelial, conjunctival, tongue, and tracheal epithelial cells, respectively. These keratins were identified by their cross-reactivity with antiserum against human epidermal keratins.

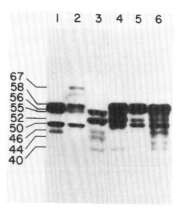

Figure 1 Keratins were extracted from different tissue biopsies of normal human epithelia. Proteins were resolved by electrophoresis through an 8.5% polyacrylamide gel, and the resolved proteins were transferred electrophoretically to nitrocellulose paper (Burnette 1981). The paper was then hybridized in a serum albumin-saline solution containing a 1:100 dilution of antiserum specific for the total mixture of human epidermal keratins. After thorough washing, the keratins were then indirectly radiolabeled by incubating the blot in a solution of 10^5 cpm/ml ^{125}I-labeled *S. aureus* protein A (Burnette 1981). Bands were visualized by autoradiography. (Lane 1) Cultured epidermal cells; (lane 2) epidermis; (lane 3) mesothelia; (lane 4) conjunctiva; (lane 5) tongue; (lane 6) trachea. Numbers at left indicate m.w. $\times 10^3$.

Epithelial-derived malignancies (carcinomas) continue to produce keratins (Batlifora et al. 1980; Schlegel et al. 1980; Krepler et al. 1983; Moll et al. 1983; Osborn and Weber 1983). However, a comparison of the carcinoma and its normal cellular counterpart has revealed that the pattern of keratins produced by these abnormal cells is sometimes quite different (Wu and Rheinwald 1981; Moll et al. 1983). This recently identified phenotypic difference between normal and malignant epithelial cells is extremely important from both biochemical and medical perspectives. Thus, a precise description of keratins in malignant cells might be useful not only in the diagnosis and possible therapy of a particular cancer, but also in elucidating the functional significance of the multiplicity of keratins.

Some carcinomas, such as adenocarcinoma of the colon, have been found to express keratin patterns nearly identical with those produced from their specific cells of origin (Moll et al. 1983). In contrast, some tumors of glandular epithelia, including breast cancer and adenocarcinoma of the lung, have been found to express only a subset of the keratins produced in the corresponding normal tissue (Moll et al. 1983). Even more unusual have been some of the tumors from different stratified squamous epithelia, which were found to produce keratins that are significantly different from their normal tissue counterparts (Wu and Rheinwald 1981; Moll et al. 1983). Wu and Rheinwald (1981) were the first investigators to demonstrate that squamous cell carcinomas of the skin frequently showed the absence of a major 67K epidermal keratin and the presence of a 40K keratin not normally seen in keratinized, stratified squamous epithelium of the skin (see Fig. 2, lanes 1 and 2).

In our own studies, we investigated the keratin patterns of (1) esophageal stratified squamous epithelium and squamous cell carcinoma of the esophagus and (2) bronchial ciliated, pseudostratified columnar epithelium and squamous cell carcinoma of the lung. Water-insoluble proteins were extracted from tissues and resolved by gel electrophoresis. Keratins were identified in an immunoblot analysis by their cross-reactivity with antiserum against human epidermal keratins (Fig. 2). The normal esophageal epithelium of six different patients contained three major keratins of M_r 58K, 56K, and 52K (Fig. 2, lane 3). In contrast, samples of esophageal squamous cell carcinomas from three of these same patients contained major keratins of M_r 58K, 56K, and 50K and a minor keratin of 46K (Fig. 2, lanes 4–6). In at least one case (Fig. 2, lane 4), a faint band of 44K could also be detected. Occasionally (Fig. 2, lanes 5–6), small amounts of 52K keratin were also found. Whether this band reflected the presence of a small number of normal cells in the tumor tissue was not determined.

While the keratin pattern of different samples of normal human bronchial epithelium was constant (Fig. 2, lane 7), the pattern produced from different lung cancers varied and was usually different from the normal tissue counterpart (Fig. 2, lanes 8–10). Normal bronchial epithelium, which consists primarily of ciliated, columnar epithelial cells, consistently gave a pattern of major ker-

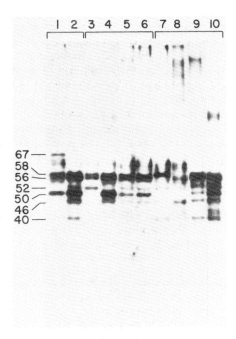

Figure 2 Keratins were extracted from different tissue biopsies of normal human epithelia of esophagus and lung and of malignant neoplasms derived from these tissues. Proteins were resolved by electrophoresis through an 8.5% polyacrylamide gel, and the resolved proteins were transferred electrophoretically to nitrocellulose paper (Burnette 1981). The paper was hybridized with antikeratin antibodies according to the procedure outlined in the legend to Fig. 1. (Lane 1) Epidermis; (lane 2) SCC-15, a cell line similar to SCC-13, a carcinoma of the skin (Wu and Rheinwald 1981); (lane 3) esophageal epithelium; (lanes 4–6) squamous cell carcinoma of the esophagus; (lane 7) bronchial epithelium; (lane 8) large cell carcinoma of the lung; (lanes 9–10) squamous cell carcinoma of the lung.

atin bands at 58K, 52K, 46K, and 40K (Fig. 2, lane 7). In the particular sample shown in lane 7, the antikeratin antisera used did not cross-react strongly with the 46K and 40K keratins, although these bands were major keratins in this sample as judged by Coomassie blue staining of the proteins (not shown). In contrast, two squamous cell carcinomas of the lung produced major keratins of M_r 56K, 52K, 50K, 46K, and 40K (Fig. 2, lanes 9–10). A large cell carcinoma of the lung produced a keratin pattern that was even more different from that of the normal bronchial epithelium, with major keratins of M_r 56K and 46K, and others of M_r 54K, 52K, and 42K (Fig. 2, lane 8).

Although additional studies will certainly be necessary to determine whether any rules can be formulated for the abnormal expression of keratins in malignant neoplasms from the esophagus and lung, it is clear from our results that even within a particular type of carcinoma, the pattern of keratins for different tumors is not always the same. This was initially surprising since previously a fixed pattern of keratins had been observed for a number of squamous cell carcinomas of the epidermis and oral epithelium (Wu and Rheinwald 1981). Even in some malignant tumors of the epidermis, however, local variations in the degree of cornification of the carcinoma have recently been shown to result at least in quantitative, if not qualitative, changes in the pattern of keratins synthesized (Moll et al. 1983). Our finding that the pattern of keratins produced by carcinomas of the esophagus and the lung were strikingly different not only from each other, but also from squamous cell carcinomas of the skin, suggests even more strongly that the variability in keratin expression upon malignant transformation is greater than was initially realized. It is important to note, however, that the altered pattern of keratins produced upon malignant transformation, although variable, does not appear to be arbitrary, and in no case have tumors been found to produce the 67K keratin characteristic of differentiating epidermis. It seems that the altered expression of keratins in each carcinoma follows a general set of guidelines, within which there is some flexibility.

Keratins of normal epithelia and malignant neoplasms are encoded by different mRNAs

To determine whether changes in the expression of keratins involve changes in the posttranslational modification and processing of these proteins or, alternatively, changes in the synthesis of new keratin mRNAs, we and others have isolated keratin mRNAs from different epithelial cells (Kim et al. 1983; Magin et al. 1983) and from epidermal cells at different stages of differentiation (Fuchs and Green 1979, 1980; Schweizer and Goerttler 1980; Roop et al. 1983). It is clear from these studies that there are multiple mRNAs for the keratins and that most of the changes in keratin patterns are due to the appearance of newly synthesized keratin mRNAs in the cytoplasm of the cells.

When mRNAs were isolated from the squamous cell carcinoma line SCC-15 (Rheinwald and Beckett 1981) and translated in vitro in a reticulocyte lysate translation system, it was discovered that even the altered pattern of keratins produced upon malignant transformation of cells of the epidermis and oral epithelium is due primarily to a change in the expression of different keratin mRNAs (Fig. 3A; Kim et al. 1983). Immunoprecipitation of the translation products from poly(A)$^+$ RNA isolated from normal epidermal cells showed the presence of major keratins of M_r 58K, 56K, 50K, and 46K (Fig. 3A, lane 3) but not 40K (Fig. 3A, lane 4), whereas poly(A)$^+$ RNA from SCC-15 cells contained an mRNA for the 40K keratin (Fig. 3A, lane 8) in addition to the others (Fig. 3A, lane 7). Thus, whereas malignant transformation of cells has been shown to result in numerous changes in glycosylation (for a review, see Hynes 1976) and phosphorylation (Erikson et al. 1979) of proteins, such processes do not appear to be responsible at least for the gross changes in keratin patterns.

Keratins can be divided into two distinct groups that have evolved from lower eukaryotes to form the intermediate filaments of mammalian epithelia

The multiple keratin mRNAs can be grouped into at least two major classes as judged by their ability to hybridize with different types of cloned keratin cDNA sequences (Fuchs et al. 1981; Kim et al. 1983; Roop et al. 1983).

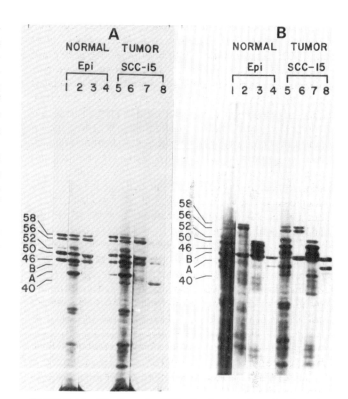

Figure 3 Identification of mRNAs for the keratins of cells cultured from normal stratified squamous epithelia and squamous cell carcinoma of the epidermis (A) and their dissection into two distinct families (B). (A) Keratins and poly(A)$^+$ RNA were extracted from epidermal cells (Epi) and squamous cell carcinoma of the oral epithelium (SCC-15), shown by Wu and Rheinwald (1981) to make keratins identical to squamous cell carcinoma of the epidermis. The RNAs were translated in vitro, and translation products were precipitated with antiserum. All samples were resolved by electrophoresis through an 8.5% polyacrylamide gel. (Lanes 1,5) Keratins from [^{35}S]methionine-labeled cellular extracts; (lanes 2,6) total [^{35}S]methionine-labeled translation products of poly(A)$^+$ RNA; the 45K band (B) is an mRNA-independent endogenous product from the reticulocyte system; (lanes 3,7) same translation products as lanes 2 and 6 after precipitation with antiserum to keratins from cultured human epidermal cells; (lanes 4,8) same translation products as lanes 2 and 6 after precipitation with antiserum to the 40K keratin of SCC-15 cells. Sizes at left are m.w. × 10^3. A stands for actin. (B) Poly(A)$^+$ RNA was isolated from epidermal cells (Epi) and squamous cell carcinoma cells (SCC-15). (Lanes 1,5) Translation products of unfractionated poly(A)$^+$ RNA; (lanes 2,6) translation products of mRNAs that specifically hybridized to DNA from pKA-1, a hybrid plasmid containing a 56K epidermal keratin cDNA insert; (lanes 3,7) translation products of mRNAs that specifically hybridized to DNA from pKB-2, a hybrid plasmid containing a 50K epidermal keratin cDNA insert; (lanes 4,8) translation products of mRNAs that specifically hybridized to plasmid DNA from pA-1, a hybrid plasmid containing human actin cDNA. B identifies an mRNA-independent artifact of the reticulocyte system; A represents actin. For further details, see Kim et al. (1983).

Two classes of cDNAs corresponding to the mRNAs of cultured human epidermal keratins have been identified: One class removes from total human epidermal mRNA a fraction that translates into keratins of M_r 58K and 56K (Fig. 3B, lane 2), and the other class removes from total human epidermal mRNA a fraction that translates into keratins of M_r 50K and 46K (Fig. 3B, lane 3). Keratin mRNAs from different types of cultured human epithelial cells also hybridize with one or the other, but not both, of these two different cloned cDNAs from cultured human epidermal cells (Kim et al. 1983). In all epithelia thus far investigated, at least one member of each of the two classes of keratin mRNAs always seems to be expressed, suggesting the functional importance that must be linked to their coordinate expression. Figure 3B (lanes 5–8) shows that even the keratin mRNAs from the squamous cell carcinoma line SCC-15 all hybridize with one or the other of these two cDNAs (Kim et al. 1983). Thus, although malignant transformation may result in an altered pattern of keratin biosynthesis, it does not seem to change the apparent requirement for the expression of both classes of keratin mRNAs in human epithelial cells.

The two classes of keratins have been named type I and type II (Hanukoglu and Fuchs 1983). The type-I keratins are acidic (pk_a 4.5–5.3) and are mostly of size 40K–55K, whereas the type-II keratins are more basic (pk_a 5.5–7.5) and typically of M_r 53K–70K. At least one type-I and one type-II keratin have always been found to be expressed in all epithelial cell types studied thus far. Recently, this has been demonstrated using monoclonal antibodies to a type-I and a type-II keratin from human epidermal cells (Nelson and Sun 1983). The mRNAs encoding these two groups of keratins have been shown to hybridize with varying degrees of stringency to each of two human epidermal keratin cDNAs, indicating that the individual members within each group are of similar, but not identical sequence (Kim et al. 1983).

To explore the possible functional significance of these two classes of keratin sequences, we have investigated their evolutionary conservation (Fuchs and Marchuk 1983). Using the two cloned cDNAs for the human type-I and type-II keratin mRNAs, we examined the genomes of other vertebrate species for the presence of sequences similar to these two human keratin mRNAs. DNAs from all vertebrates, including human, rat, chick, turtle, and catfish, were found to contain nonoverlapping sets of sequences that hybridized with each type of human keratin cDNA (Fig. 4, lanes 1–5). Under reduced stringency conditions, even specific sequences in yeast were found to hybridize with ^{32}P-labeled probes prepared from these cDNAs (Fig. 4, lane 6). At intermediate stringencies, only a single band hybridized with each probe (lane 6, arrows). Although cloning and sequencing of these hybridizing yeast fragments will be necessary before their precise relation to human keratin genes can be elucidated, it is possible that the evolutionary conservation of these two types of keratin sequences extends even to the genomes of lower eukaryotes. Thus, appreciable evolutionary pressure has been exerted to maintain at least some portions of both of these types of keratin sequences relatively unaltered.

To determine whether the two types of keratin genes in lower vertebrates are both expressed, we investigated the water-insoluble proteins of the epidermis of different species. All vertebrate skins were found to express a

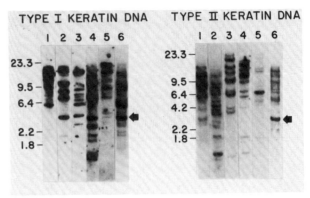

Figure 4 Presence of sequences homologous to human keratin mRNAs in other eukaryotic genomes. DNAs (2–10 μg) were digested with a fivefold excess of restriction endonuclease EcoRI, which does not cleave within the type-I or type-II human keratin cDNAs. Fragments were subjected to 0.8% agarose gel electrophoresis in duplicate and transferred to nitrocellulose paper (Southern 1975). The paper was then cut and each half was hybridized except as noted below under stringent conditions (0.75 M NaCl, 50% formamide, 41°C) to ^{32}P-labeled probe copied from the purified cloned cDNA inserts corresponding to the type-I and type-II keratin sequences (Fig. 1). DNAs were from human (1), rat (2), chick (3), turtle (4), catfish (5), and yeast (6). The yeast was hybridized under lowered conditions of stringency (0.9 M NaCl, 15% formamide, 50°C). The arrow to the right of lane 6 in each diagram indicates a single hybridizing band remaining in yeast DNA when the stringency was raised to 0.9 M NaCl, 30% formamide, 41°C.

group of proteins of M_r 40K–70K, which were similar to human keratins as judged by their solubility properties and one-dimensional peptide mapping (Fuchs and Marchuk 1983). When antibodies raised against electrophoretically pure 50K (type-I) and 56K (type-II) human epidermal keratins were used to investigate the immunologic relatedness of these other vertebrate proteins, it was found that they cross-reacted specifically with all putative epidermal keratins (Fig. 5). Most vertebrate IF polypeptides of M_r 40K–55K cross-reacted with antiserum directed against the type-I human keratin, whereas most IF polypeptides of M_r 56K–70K cross-reacted with antiserum directed against the type-II human keratins (Fig. 5). Occasionally, a large keratin was seen to cross-react with antiserum directed against the type-I human keratin (not shown). These few cases most likely are acidic type-I keratins, which are unusually large in size. At least one case of a large, acidic type-I keratin of M_r 59K has already been reported (Steinert et al. 1983). The finding that both types of keratins are expressed in most, if not all, normal and malignant human epithelial cells and that these two distinct types of keratins are coordinately conserved throughout vertebrate evolution leaves no doubt as to their importance in the formation of the cytoskeletal backbone of these cells.

Discussion

Why is there a difference between the keratin patterns of normal epithelia and malignant neoplasms derived from these tissues?

One possible explanation for the difference in keratin patterns between normal and malignant tissues is that

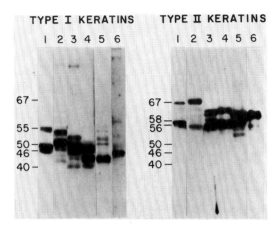

Figure 5 Keratins of type I and type II are both present in all vertebrate epidermis. Antibodies were raised against electrophoretically purified 50K (type-I) and 56K (type-II) human keratins. These antisera were used in immunoblot analysis to detect the presence of immunoreactive forms of both classes of keratins in other vertebrate epidermal keratinocytes. Different vertebrate samples of IF proteins were from the epidermis of human (lane 1), mouse (lane 2), rabbit (lane 3), bovine (lane 4), chicken (lane 5), and frog (lane 6).

the malignant tissue may represent a homogeneous population of cells at a particular stage of differentiation, whereas the normal tissue may consist of cells at multiple stages of differentiation (Moll et al. 1983). Alternatively, some heterogeneity in the degree of differentiation of different cells within the tumor would also cause a shift in the percentage of cells at different stages of differentiation. Either of these changes would result in variations in the total keratin pattern of the tissue. Although an imbalance in the population of cells at different stages of differentiation might explain the altered keratin patterns for some malignant neoplasms, it does not appear to account for the change in epidermis, since the 40K keratin expressed by several epidermal squamous cell carcinomas is neither a normal constituent of basal cells nor of epidermal cells at any subsequent stage of terminal differentiation.

It is also possible that an apparent change in the expression of keratins during malignant transformation might be an artifact due to the transformation of a minor epithelial cell type in a complex epithelial tissue. Although we cannot yet rule out this possibility, we have investigated one complex epithelial tissue, namely bronchial epithelium, and found that most of the multiple keratins produced by the normal bronchial tissue were also made by several different tumors, including both large cell and squamous cell carcinomas. Although the relative amounts of these keratins varied among the different tumors, even malignant cells cultured from the tumor showed a complex keratin pattern (M.P. Grace and E. Fuchs, unpubl.), suggesting that the complexity of the bronchial keratin pattern does not result from a combination of multiple, simpler patterns from different cell types within the tissue. Two-dimensional gel analyses of the keratins of several different lung cancers (Moll et al. 1983) have provided additional statistics on the keratin patterns of squamous cell carcinomas of the lung, and the studies from the two separate laboratories are generally in agreement. Thus, although some differences in keratin patterns between normal and malignant tissues may reflect differences in the ratios of different epithelial cell types within the two tissues, some changes in keratin patterns must be generated by other means.

There are several other reasons as to why the keratin patterns of the carcinomas and normal tissues might be different. It has been suggested that the appearance of a new keratin upon malignant transformation might be characteristic for a specific carcinoma and might represent either a fetal keratin (Moll et al. 1983) or a mutational alteration that unmasks its gene (Wu and Rheinwald 1981). Although these explanations are certainly plausible, we now know that at least for epidermis, the 40K human keratin present in epidermal squamous cell carcinomas is a normal constituent of a number of adult internal epithelia (Tseng et al. 1982). Moreover, the "abnormal" pattern of keratins produced by these epidermal squamous cell carcinomas can be mimicked by normal epidermal cells in culture by adding vitamin A to the medium at physiological concentrations (Fuchs and Green 1981). This finding suggests the intriguing possibility that malignant transformation of at least one epithelial cell type results in an altered response of the cells to the differentiation-specific factor, vitamin A. Clearly, more-detailed knowledge of the many factors that influence the expression of different keratins will be necessary in order to explain the changes in keratin patterns resulting from malignant transformation. Nonetheless, studies from our laboratory and others suggest that several different mechanisms may play a role in this complex process.

What is the functional significance of the two classes of keratins and how do the type-I and type-II keratins differ?

The finding of two distinct classes of keratins that are coordinately expressed throughout vertebrate evolution suggests that similar to the α- and β-tubulins of microtubules, the type-I and type-II keratins are both essential for the formation of the 8-nm keratin filament. Early in vitro filament-assembly studies indicated that homopolymers composed of a single keratin polypeptide are not possible, whereas as few as two different keratins are sometimes sufficient for filament formation (Steinert et al. 1976). Since type-I and type-II keratins are both expressed in all epithelial cells, in all stages of differentiation and development, it seems likely that these two classes of keratins represent the building blocks of the 8-nm filament.

Like the α- and β-tubulins (Cleveland et al. 1980; Cowan et al. 1981), the type-I and type-II keratins are each encoded by small multigene families, the individual members of which appear to be differentially expressed (Fuchs et al. 1981; Fuchs and Marchuk 1983; Kim et al. 1983; Roop et al. 1983). In contrast to the tubulins, which show only microheterogeneities in their multiple forms, both the type-I and type-II keratins demonstrate considerable variation in size. Both a 50K human and a 59K mouse epidermal keratin were recently shown to be approximately 85% homologous according to their amino acid sequences predicted from sequencing their corresponding cDNAs (Hanukoglu and Fuchs 1982; Steinert et al. 1983). It seems that substantial variations in the sizes of the keratins have been tolerated, even though major portions of their sequences have clearly been conserved throughout vertebrate evolution.

The amino acid sequences recently deduced from the cDNA sequences of representative members of each of the two groups of keratins have now elucidated the major differences between the two groups of keratins. Despite their similar amino acid compositions, being rich in glutamic acid, glycine, and serine, both a 50K keratin from human (Hanukoglu and Fuchs 1982) and also a similar type-I keratin from mouse (Steinert et al. 1983) have little sequence homology (~30%) with a 56K type-II human keratin (Hanukoglu and Fuchs 1983). Within each of the two classes of keratins, however, the individual members are quite similar to one another, particularly in the central, large α-helical domain of each polypeptide. The two types of wool microfibrillar keratins, for instance, share 50–60% homology with the two types of cytoskeletal keratins (Crewther et al. 1980; Hanukoglu and Fuchs 1982, 1983; Dowling et al. 1983). The major source of diversity among the individual members of each keratin type lies in their amino and carboxyl termini. Thus, the keratins of epidermis have termini that are rich in glycine, serine, and phenylalanine, whereas the keratins of wool have termini that are rich in cysteine. The termini are also markedly variable in their lengths, thus accounting for the large variation in the size of the keratins.

The nonhelical terminal domains of the keratin filament subunits most likely interact with one another not only end to end to extend the length of the filament, but also side by side to produce at least some of the interprotofibrillar interactions (Skerrow et al. 1973; Geisler et al. 1982; Steinert et al. 1983). If this is true, then the resulting properties of the 8–10 nm filaments will differ widely, depending on the subunit selected. Thus, for example, it is easy to see that the ends of the microfibrillar wool and hair keratins can be cross-linked through disulfide bonding to form the highly stable and insoluble filaments characteristic of wool fibers. The variability among the nonhelical termini of the multiple cytoskeletal keratins is likely to be more subtle, since most of these polypeptides have been shown to be rich in glycine and serine. Additional sequencing of tissue-specific and differentiation-specific keratins will need to be made before their precise relation is established. Nevertheless, the muliplicity of sequences within the type-I and type-II keratin families appears to be a reflection of refining the properties of the 8-nm filament to suit the particular structural needs of each epithelial cell.

We do not yet know precisely how the use of different keratin polypeptides in the assembly of the 8-nm filaments might alter the resulting properties of these fibrils. To date, there are only a few examples where we can even associate the synthesis of a new keratin with specific morphological changes in an epithelial cell. One such example is the synthesis of a 67K keratin that seems to be the first indication of a human epidermal cell's commitment to terminal differentiation and stratum corneum formation. This keratin has been shown to be specifically associated with the 8-nm filaments that are attached to the desmosomal plaques in differentiating epidermis (Drochmans et al. 1978). How this keratin might enable the ends of the 8-nm filament to interact with certain desmosomal components must await the determination of the sequence of this polypeptide.

Assigning functional roles to each of the individual members of the type-I and type-II keratin families will require the combined knowledge not only of the sequences of these keratins, but also of the ultrastructure of the cytoskeleton in specific epithelial cells in which a particular keratin is expresed. The finding that the pattern of keratin expression is altered upon malignant transformation should help us to associate specific changes in keratin synthesis with alterations in the cytoskeletal architecture. The alterations in keratin patterns should be especially useful since for each type of epithelial malignancy, there seems to be strict limitations as to what the new pattern will be.

Thus, even though the keratins are differentially expressed under many diverse conditions, their pattern of synthesis is always restricted, suggesting that the expression of keratins is by no means arbitrary but reflects true, though as yet unidentified, differences in the structural requirements of each epithelial cell. The field of keratins now appears to be at the threshold of elucidating the molecular significance of the multiplicity of keratin sequences.

Acknowledgments

We would like to thank Dr. Israel Hanukoglu, Ms. Angela Tyner, Ms. Faina Schwartz, and Ms. Naoko Tanese in our laboratory for their important research contributions that support the interpretations of the data presented in this paper. In addition, we extend our gratitude to Dr. James Rheinwald (Harvard Medical School) for his valuable advice and contributions to the mesothelial cell work. Finally, we thank Ms. Lisa Fuller for her expeditious typing of the manuscript. E.F. is a Searle Scholar and the recipient of a Career Development Award from the National Institutes of Health. M.P.G. is a Postdoctoral Fellow funded by the Damon Runyon–Walter Winchell Cancer Fund; K.H.K. is a Postdoctoral Fellow funded by the National Cancer Institute; and D.M. is a Predoctoral Trainee funded by the National Institutes of Health. This work has been supported by a grant from the National Institutes of Health.

References

Baden, H.P., L.A. Goldsmith, and B. Fleming. 1973. The polypeptide composition of epidermal prekeratin. *Biochim. Biophys. Acta* **317**: 303.

Batlifora, H., T.-T. Sun, R.M. Bahu, and S. Rao. 1980. The use of antikeratin antiserum as a diagnostic tool: Thymoma versus lymphoma. *Hum. Pathol.* **11:** 635.

Brysk, M.M., R.M. Gray, and I.A. Bernstein. 1977. Tonofilament protein from newborn rat epidermis. *J. Biol. Chem.* **252:** 2127.

Burnette, W.N. 1981. "Western blotting": Electrophoretic transfer of proteins from sodium dodecyl sulfate–polyacrylamide gels to unmodified nitrocellulose and radiographic detection with antibody and radioiodinated protein A. *Anal. Biochem.* **223:** 195.

Cleveland, D.W., M.A. Lopata, R.J. MacDonald, N.J. Cowan, W.J. Rutter, and M.W. Kirschner. 1980. Number and evolutionary conservation of α- and β-tubulin and cytoplasmic β- and γ-actin genes using specific cloned cDNA probes. *Cell* **20:** 95.

Cowan, N.J., C.D. Wilde, L.T. Chow, and F.C. Wefald. 1981. Structural variation among human β-tubulin genes. *Proc. Natl. Acad. Sci.* **78:** 4877.

Crewther, W.G., L.M. Dowling, and A.S. Inglis. 1980. Amino acid sequence data from a microfibrillar protein of α-keratin. In *Proceedings from the 6th Quinquennial International Wool Textile Research Conference*, Pretoria, vol. 2, p. 79.

Culbertson, V.B. and I.M. Freedberg. 1977. Isolation and characterization of the α-helical proteins from newborn rat. *Biochim. Biophys. Acta* **490:** 178.

Dale, B.A., I.B. Stern, M. Rabin, and L.-Y. Huang. 1976. The identification of fibrous proteins in fetal rat epidermis by electrophoretic and immunologic techniques. *J. Invest. Dermatol.* **66:** 223.

Doran, T.I., A. Vidrich, and T.-T. Sun. 1980. Intrinsic and extrinsic regulation of the differentiation of skin, corneal and esophageal epithelial cells. *Cell* **22:** 17.

Dowling, L.M., D.A.D. Parry, and L.G. Sparrow. 1983. Structural homology between hard α-keratin and the intermediate filament proteins desmin and vimentin. *Biosci. Rep.* **3:** 73.

Drochmans, P., C. Freudenstein, J.-C. Wanson, L. Laurent, T.W. Keenan, J. Stadler, R. LeLoup, and W.W. Franke. 1978. Structure and biochemical composition of desmosomes and tonofilaments isolated from calf muzzle epidermis. *J. Cell Biol.* **79:** 427.

Erikson, R.L., M.S. Collett, E. Erikson, and A.F. Purchio. 1979. Evidence that the avian sarcoma virus transforming gene product is a cyclic AMP-independent protein kinase. *Proc. Natl. Acad. Sci.* **76:** 6260.

Fuchs, E. and H. Green. 1979. Multiple keratins of cultured human epidermal cells are translated from different mRNA molecules. *Cell* **17:** 573.

———. 1980. Changes in keratin gene expression during terminal differentiation of the keratinocyte. *Cell* **19:** 1033.

———. 1981. Regulation of terminal differentiation of cultured human keratinocytes by vitamin A. *Cell* **25:** 617.

Fuchs, E. and D. Marchuk. 1983. Type I and type II keratins have evolved from lower eukaryotes to form the epidermal intermediate filaments in vertebrate skin. *Proc. Natl. Acad. Sci.* **80:** 5857.

Fuchs, E.V., S.M. Coppock, H. Green, and D.W. Cleveland. 1981. Two distinct classes of keratin genes and their evolutionary significance. *Cell* **27:** 75.

Geisler, N., E. Kaufman, and K. Weber. 1982. Protein chemical characterization of three structurally distinct domains along the protofilament unit of desmin 10-nm filaments. *Cell* **30:** 277.

Hanukoglu, I. and E. Fuchs. 1982. The cDNA sequence of a human epidermal keratin: Divergence of sequence but conservation of structure among intermediate filament proteins. *Cell* **31:** 243.

———. 1983. The cDNA sequence of a type II cytoskeletal keratin reveals constant and variable structural domains among keratins. *Cell* **33:** 915.

Hynes, R.O. 1976. Cell surface proteins and malignant transformation. *Biochim. Biophys. Acta* **458:** 73.

Kim, K.H., J.G. Rheinwald, and E. Fuchs. 1983. Tissue specificity of epithelial keratins: Differential expression of mRNAs from two multigene families. *Mol. Cell. Biol.* **3:** 495.

Krepler, R., H. Denk, E. Weirich, E. Schmid, and W.W. Franke. 1983. Keratin-like proteins in normal and neoplastic cells of human and rat mammary gland as revealed by immunofluorescence microscopy. *Differentiation* **20:** 242.

Lazarides, E. 1982. Intermediate filaments: A chemically heterogeneous developmentally regulated class of proteins. *Annu. Rev. Biochem.* **51:** 219.

Magin, T.M., J.L. Jorcano, and W.W. Franke. 1983. Translational products of mRNAs coding for non-epidermal cytokeratins. *EMBO J.* **2:** 1387.

Milstone, L.M. and J. McGuire. 1981. Different polypeptides form the intermediate filaments in bovine hoof and esophageal epithelium and in aortic endothelium. *J. Cell Biol.* **88:** 312.

Moll, R., R. Krepler, and W.W. Franke. 1983. Complex cytokeratin polypeptide patterns observed in certain human carcinomas. *Differentiation* **23:** 256.

Moll, R., W.W. Franke, D.L. Schiller, B. Geiger, and R. Krepler. 1982. The catalog of human cytokeratin polypeptides: Patterns of expression of specific cytokeratins in normal epithelia, tumors and cultured cells. *Cell* **31:** 11.

Nelson, W.G. and T.-T. Sun. 1983. The 50- and 58-Kdalton keratin classes as molecular markers for stratified squamous epithelia: Cell culture studies. *J. Cell Biol.* **97:** 244.

Osborn, M. and K. Weber. 1983. Tumor diagnosis by intermediate filament typing: A novel tool for surgical pathology. *Lab. Invest.* **48:** 372.

Rheinwald, J.G. and M.A. Beckett. 1981. Tumorigenic keratinocyte lines requiring anchorage and fibroblast support cultured from human squamous cell carcinoma. *Cancer Res.* **41:** 1647.

Roop, D.R., P. Hawley-Nelson, C.K. Cheng, and S.H. Yuspa. 1983. Keratin gene expression in mouse epidermis and cultured epidermal cells. *Proc. Natl. Acad. Sci.* **80:** 716.

Schlegel, R., S. Banks-Schlegel, and G.S. Pinkus. 1980. Immunohistochemical localization of keratin in normal human tissues. *Lab. Invest.* **42:** 91.

Schweizer, J. and K. Goerttler. 1980. Synthesis in vitro of keratin polypeptides directed by mRNA isolated from newborn and adult mouse epidermis. *Eur. J. Biochem.* **112:** 243.

Skerrow, D., G. Matoltsy, and M. Matoltsy. 1973. Isolation and characterization of the helical regions of epidermal prekeratin. *J. Biol. Chem.* **248:** 4820.

Southern, E.M. 1975. Detection of specific sequences among DNA fragments separated by gel electrophoresis. *J. Mol. Biol.* **98:** 503.

Steinert, P.M. and W.W. Idler. 1975. The polypeptide composition of bovine epidermal α-keratin. *Biochem. J.* **151:** 603.

Steinert, P.M., W.W. Idler, and S.B. Zimmerman. 1976. Self-assembly of bovine epidermal keratin filaments in vitro. *J. Mol. Biol.* **108:** 547.

Steinert, P.M., R.H. Rice, D.R. Roop, B.L. Trus, and A.C. Steven. 1983. Complete amino acid sequence of a mouse epidermal keratin subunit and implications for the structure of intermediate filaments. *Nature* **302:** 794.

Sun, T.-T. and H. Green. 1978. Keratin filaments of cultured human epidermal cells. *J. Biol. Chem.* **253:** 2053.

Tseng, S.C.G., M.J. Jarvinen, W.G. Nelson, J.-W. Huang, J. Woodcock-Mitchell, and T.-T. Sun. 1982. Correlation of specific keratins with different types of epithelial differentiation: Monoclonal antibody studies. *Cell* **30:** 361.

Wu, Y.J. and J.G. Rheinwald. 1981. A new small (40kd) keratin filament protein made by some cultured human squamous cell carcinomas. *Cell* **25:** 627.

Classification, Expression, and Possible Mechanisms of Evolution of Mammalian Epithelial Keratins: A Unifying Model

T.-T. Sun,* R. Eichner,† A. Schermer,* D. Cooper,* W.G. Nelson,‡ and R.A. Weiss§

*Departments of Dermatology and Pharmacology, New York University School of Medicine, New York, New York 10016; †Departments of Dermatology and Cell Biology and Anatomy, and ‡Department of Urology, The Johns Hopkins University School of Medicine, Baltimore, Maryland 21205; §Dermatology Branch, National Cancer Institute, National Institutes of Health, Bethesda, Maryland 20205

The keratins are a family of water-insoluble proteins (40K–70K) that form intermediate-sized (10-nm) filaments in almost all epithelial cell types (Sun and Green 1977, 1978a; Franke et al. 1978a,b, 1979; Sun et al. 1979). By high-resolution, two-dimensional gel electrophoresis, more than 17 keratin species have been identified in various human epithelia (Sun and Green 1978b; Franke et al. 1981c; Moll et al. 1982c; Wu et al. 1982). Usually a subset consisting of 2–10 keratins are expressed in any given epithelium; the detailed composition of the subset is highly heterogeneous, however, and can vary depending on cell type, period of embryonic development, cellular growth environment, disease state, and stage of histological differentiation (for references, see Green et al. 1982; Moll et al. 1982c; Woodcock-Mitchell et al. 1982). To better understand the biological significance of such keratin heterogeneity, we have produced several monoclonal antikeratin antibodies and used them to (1) localize the four major epidermal keratins (50K, 56.5K, 58K, and 65K–67K) in different epidermal layers (Woodcock-Mitchell et al. 1982); (2) compare the keratins expressed by a single epithelial cell type, the human epidermal keratinocyte, under several growth conditions (Eichner et al. 1984); (3) follow the expression of keratins in the epidermis as a function of embryonic development and various disease states, including neoplasms (Tseng et al. 1982a; Weiss et al. 1983, 1984; Nelson et al. 1984); (4) correlate specific keratin species with different types of epithelial differentiation (Tseng et al. 1982c; Nelson and Sun 1983); (5) study the expression of keratins in various nonkeratinized tissues during vitamin A deficiency in an in vivo rabbit model system (Tseng et al. 1982b); and finally (6) compare the keratins of several mammalian species (D. Cooper et al., unpubl.). A consistent and relatively simple picture has emerged from our studies. In this paper, we will review results from these studies that shed light on the possible functional significance of specific keratin molecules. We will also present new evidence to establish that *all* known human epithelial keratins can be divided into two mutually exclusive subfamilies. From these findings, we have constructed a unifying model that defines and emphasizes the concepts of "keratin subfamilies" and "keratin pairs." The implications of this model on the structure, expression, and possible mechanisms of evolution of mammalian epithelial keratins will be discussed.

Experimental Procedures

Mouse monoclonal antibodies AE1 and AE3 were prepared against SDS-denatured human epidermal (callus) keratins by the hybridoma technique and were shown to be keratin-specific by both immunofluorescence and immunoblot techniques (Woodcock-Mitchell et al. 1982; Tseng et al. 1982c; for a recent discussion of the definition of keratins, see Sun et al. 1983b). Human epidermal cells derived from newborn foreskin were grown in the presence of lethally irradiated 3T3 cells (Rheinwald and Green 1975). Cultured human mesothelial cells were kindly provided by J. Rheinwald of Harvard Medical School (Wu and Rheinwald 1981; Wu et al. 1982). The extraction, electrophoretic separation, and immunoblot analysis of keratins were done according to our previously described procedures (Woodcock-Mitchell et al. 1982).

Results and Discussion

Keratin subfamilies A and B

Water-insoluble cytoskeletal proteins from five representative human epithelial tissues and cells were analyzed by two-dimensional polyacrylamide gel electrophoresis. A total of 17–18 keratins, which account for all known keratin species, could be identified, with different subsets being expressed in abdominal epidermis (Fig. 1a), corneal epithelium (Fig. 2a), esophageal epithelium (Fig. 2b), as well as in cultured epidermal cells (Fig. 1b) and mesothelial cells (Fig. 3a; Moll et al. 1982c; Woodcock-Mitchell et al. 1982; Wu et al. 1982; Eichner et al. 1984). These human keratins represent distinct entities since, first, the fact that specific antibodies can be raised against several small keratins (Wu and Rheinwald 1981; Debus et al. 1982) strongly suggests that they cannot be degradative products of the larger keratins, and second, almost all of these keratins have been shown to be genuine translational products encoded by their own mRNAs (Fuchs and Green 1979; Kim et al. 1983; Magin et al. 1983).

For immunological characterization, keratins from unstained gels were transferred electrophoretically to nitrocellulose sheets and reacted with our AE1 and AE3 monoclonal antikeratin antibodies by the peroxidase-anti-peroxidase technique. Consistent with our earlier findings (Tseng et al. 1982c; Woodcock-Mitchell et al. 1982; Sun et al. 1983b; Eichner et al. 1984), the results

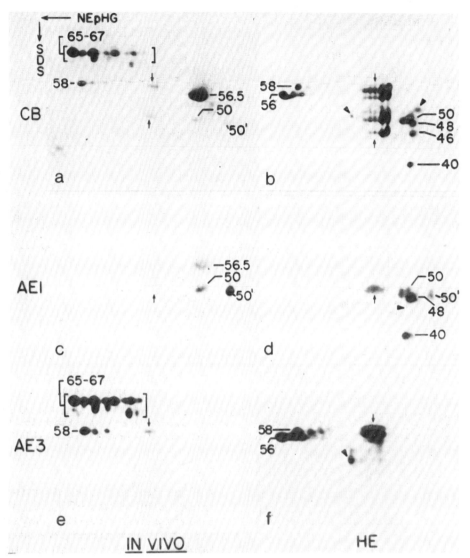

Figure 1 Two-dimensional gel electrophoretic and immunoblot analysis of the keratins of human abdominal epidermis (in vivo; a,c,e) and cultured human foreskin epidermal cells (HE; b,d,f). (CB) Coomassie blue–stained gels; (AE1 and AE3) immunoblot analysis of parallel gels using AE1 and AE3 monoclonal antikeratin antibodies, respectively; note that in these two samples AE1 recognizes the acidic 40K, 48K, 50K, 50′K, and 56.5K keratins, whereas AE3 recognizes the basic 56K, 58K, and 65–67K keratins. (From Eichner et al. 1984; reprinted with permission from the Rockefeller University Press.)

indicated that all known human epithelial keratins can be divided into two subfamilies according to their charge properties and their immunoreactivities with the AE1 and AE3 monoclonal antibodies.

The 65K–67K (nos. 1 and 2 according to Moll et al. 1982c), 64K (no. 3), 59K (no. 4), 58K (no. 5), 56K (no. 6), 54K (no. 7), and 52K (no. 8) keratins are all relatively basic (pI>6.0) and share an AE3 antigenic determinant (Figs. 1e,f; 2e,f; 3c). These results are in accordance with earlier data from Franke and co-workers who demonstrated by two-dimensional peptide mapping that all of these proteins are structurally related (subfamily of "relatively large and basic cytokeratins"; Schiller et al. 1982). The same group further demonstrated that a monoclonal antikeratin antibody, $K_G 8.13$, reacted strongly with the 65K–67K, 58K, 56K, 54K, and 52K keratins (Gigi et al. 1982). Data from positive hybrid-selection experiments also indicated that some of these keratins are related (58K, 56K, 54K, and 52K) (Fuchs et al. 1981; Kim et al. 1983). Therefore, results from several laboratories using diverse techniques are consistent and suggest that these keratins are closely related and form a subfamily (referred to as subfamily B ["basic"]; see Bladon et al. 1982; Schiller et al. 1982; Eichner et al. 1984).

In contrast, the 56.5K (no. 10), 55K (no. 12), 54′K (no. 13), 50K (no. 14), 50′K (no. 15), 48K (no. 16), 46K (no. 17), 45K (no. 18), and 40K (no. 19) keratins are all relatively acidic in charge (pI <5.7), and many of them (56.5K, 54′K, 50K, 50′K, 48K, and 40K) share an AE1 antigenic determinant (Figs. 1c,d; 2c,d; 3b). Moll et al. (1982b) have shown by peptide mapping that the 50K,

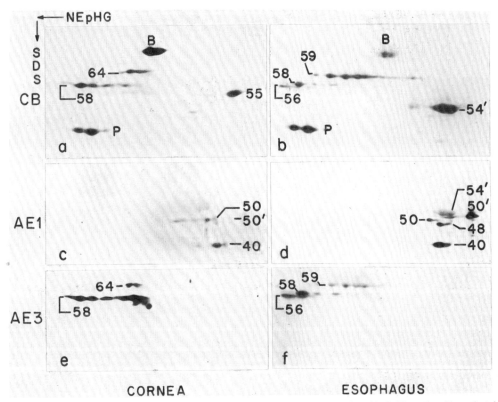

Figure 2 Two-dimensional gel analysis of human corneal (a,c,e) and esophageal (b,d,f) epithelial keratins. Note that AE1 recognizes in the two epithelia the acidic 40K, 48K, 50K, 50'K, and 54'K keratins, whereas AE3 recognizes the basic 56K, 58K, 59K, and 64K keratins. B and P denote bovine serum albumin and 3-phosphoglycerate kinase (added as standard proteins), respectively.

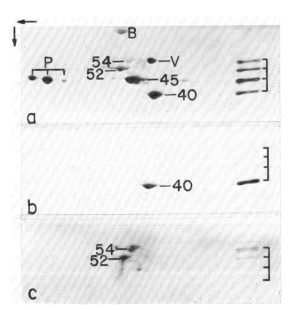

Figure 3 Two-dimensional gel analysis of keratins expressed by cultured mesothelial cells. (a) Coomassie blue staining; (b) AE1 staining; (c) AE3 staining. A sample of human mesothelial keratins was generously provided by J. Rheinwald (Wu and Rheinwald 1981). AE1 recognizes the 40K keratin, whereas AE3 reacts strongly with the 52K keratin as well as the 54K keratin (the major 45K keratin is AE1- and AE3-negative).

48K, and 46K keratins are structurally related. cDNA data also suggest that the 50K, 46K, 45K, and 40K keratins share similar sequences (Fuchs et al. 1981; Kim et al. 1983). Taken together, the data indicate that these keratins form another subfamily (subfamily A ["acidic"]; see Bladon et al. 1982; Eichner et al. 1984) that is immunologically and biochemically distinct from subfamily B. These results are summarized in Figure 4.

Tissue distribution and functional significance of individual keratins: Concept of "keratin pairs"

Except for the 40K component, which is present in almost all nonepidermal epithelia (Moll et al. 1982c; Tseng et al. 1982c), all other A subfamily components have a corresponding member in the B subfamily, forming "keratin pairs." Within each pair, the two keratins have identical "size ranks" in their respective subfamilies and, in general, follow similar rules for expression (Table 1). For example, the 45K and 52K keratins of the A and B subfamilies, respectively, are the smallest members of the two subfamilies (except for the 40K component). Both are found in significant quantities only in simple epithelia (Franke et al. 1981c; Moll et al. 1982c; Tseng et al. 1982c; Wu et al. 1982) and thus may be regarded as markers for such tissues (see below).

The next larger keratins, i.e., the 46K and 54K keratins of the A and B subfamilies, respectively, are also present in simple epithelia. However, these keratins are also

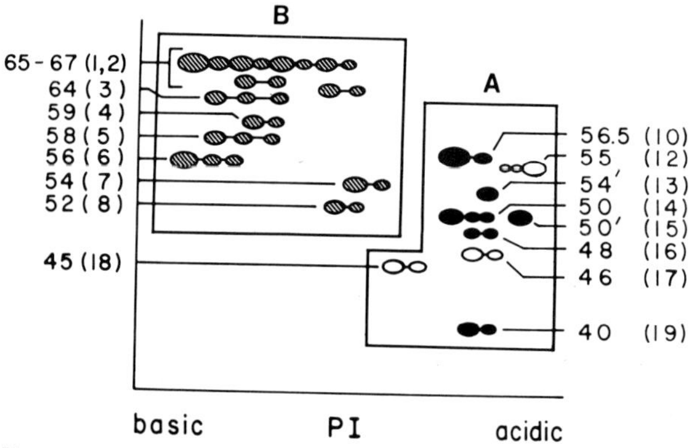

Figure 4 Classification of human epithelial keratins into A and B subfamilies. Keratins (Figs. 1–3) are represented by circles according to their molecular weights (ordinate) and relative charge properties (pI; abscissa). This scheme is in excellent agreement with that of Moll et al. (1982c), whose designations are shown in parentheses. There are, however, some doubts as to whether components no. 9 (64K) and no. 11 (56K) of Moll et al. (1982c) represent intact, genuine keratin species, and therefore they are not included here. (●) AE1-positive keratins; (◉) AE3-positive keratin; (○) keratins that react with neither antibody.

thelial cells (Sun and Green 1977; Doran et al. 1980; Wu et al. 1982; Nelson and Sun 1983). They are usually not detectable, however, in normal abdominal epidermis or ichthyosis vulgaris, which is a nonhyperproliferative epidermal disease (Weiss et al. 1984). These results suggest that the 48K and 56K keratins represent molecular markers for hyperproliferative keratinocytes (Weiss et al. 1983, 1984).

The 50K (A) and 58K (B) components are two of the four major keratins (50K, 56.5K, 58K, and 65K–67K) of normal human epidermis. Immunolocalization and cell fractionation experiments have shown that they are synthesized early during epidermal differentiation by the basal cells (Woodcock-Mitchell et al. 1982; Skerrow and Skerrow 1983; also see Fuchs and Green 1980). Since these two keratins are present in various quantities in all stratified squamous epithelia, including cultured cells and neoplasms, but not in any simple epithelia, they serve as useful markers for cells of a stratified epithelial origin (Tseng et al. 1982c; Nelson and Sun 1983; Nelson et al. 1984; also see Franke et al. 1981c; Moll et al. 1982a,c; Wu et al. 1982).

The 54′K (A) and 59K (B) keratins are the major tonofilament proteins of the nonkeratinized, stratified squamous epithelia of internal organs, such as the esophagus and cervix (Franke et al. 1981c; Moll et al. 1982c). The 55K (A) and 64K (B) keratins are among the most prominent cytoskeletal components in corneal epithelium and are cornea-specific (Doran et al. 1980; Franke et al. 1981c; Moll et al. 1982c; Tseng et al. 1982c). The largest members of the two subfamilies, the 56.5K (A) and 65K–67K (B) keratins, are normally found only in keratinized epidermis (Moll et al 1982c; Tseng et al. 1982c). The expression of these two keratins appears to be tightly coupled with the process of keratinization (Tseng et al. 1982b). For instance, they are not expressed by cultured human or rabbit epidermal cells that form nonkeratinized colonies; their synthesis can be induced coordinately, however, under conditions permissive for keratinization (Doran et al. 1980; Fuchs and Green 1981; Eichner et al. 1984). Immunolocalization and cell fractionation data have further illustrated that, in the epidermis, these two keratins are present only in supra-

found in some epidermal-derived glands, as well as in some cultured keratinocytes[1] (Franke et al. 1981c; Moll et al. 1982c; Wu et al. 1982).

The 48K and 56K keratins of subfamilies A and B, respectively, are characteristic of hyperproliferative keratinocytes, including those of the sole epidermis, hair follicle, and a number of hyperproliferative epidermal diseases, including psoriasis, verrucae, actinic keratoses, basal cell carcinoma, and squamous cell carcinoma (Moll et al. 1982a,b,c; Woodcock-Mitchell et al. 1982; Nelson et al. 1984; Weiss et al. 1984). In addition, these two keratins are also expressed in large quantities by cultured skin, corneal, conjunctival, and esophageal epi-

---

[1]The keratinocyte is the major cell type of stratified squamous epithelia. It is characterized by a high keratin content (> 30% of total cellular protein vs. < 5% in simple epithelial cells) and by its ability to make a cornified envelope.

Table 1 Classification and Functional Significance of Human Epithelial Keratins

Keratin subfamily							
A (acidic)			B (basic)			d.m.w. $\times 10^{-3}$ (B−A)	
m.w. $\times 10^{-3}$	pI	antibody	m.w. $\times 10^{-3}$	pI	antibody		Markers for
56.5 (10)	5.3	AE1	65–67 (1,2)	6–8	AE3	8	keratinization
55 (12)	4.9	—	64 (3)	7.5	AE3	9	corneal epithelium
54′, AE1 (13)	5.1	AE1	59 (4)	7.3	AE3	5	esophageal, exocervical, tongue epithelia
50/50′ (14/15)	5.3/4.9	AE1	58 (5)	7.4	AE3	8	all stratified epithelia
48 (16)	5.1	AE1	56 (6)	7.8	AE3	8	hyperproliferative keratinocytes
46 (17)	5.1	—	54 (7)	6.0	AE3	8	simple and some stratified epithelia
45 (18)	5.7	—	52 (8)	6.1	AE3	7	mainly simple epithelia
40 (19)	5.2	AE1					

Keratins are listed under subfamilies A and B (see Figs. 4 and 5; also see text). For the original description of A and B keratin subfamilies, see Eichner et al. (1984). The numbers in parentheses are keratin designations according to Moll et al. (1982c).

basal, terminally differentiated cells (Woodcock-Mitchell et al. 1982; Skerrow and Skerrow 1983; Sun et al. 1983a; also see Fuchs and Green 1980; Viac et al. 1980). Taken together, these results strongly suggest that the 56.5K and 65K–67K keratins may be regarded as molecular markers for phenotypic keratinization.

Within each keratin pair the B member is always larger than the A member by approximately 8K: A model

Figure 5 shows a schematic arrangement of all keratins of the two subfamilies according to their molecular weights. In subfamily A, the 40K, 45K, 46K, 48K, 50K, and 56.5K keratins are arranged vertically, in ascending order, and are connected with a vertical line to indicate a possible evolutionary relationship (Tseng et al. 1982c; see below). Both the 54'K keratin of the internal stratified epithelia and the 55K cornea-specific keratin are connected to the 50K keratin; they are placed sideways, however, to indicate that these two cell-type-specific keratins, like the 56.5K keratin, probably represent markers for terminal differentiation. The keratins of subfamily B are similarly arranged according to a molecular-weight scale that is 8K higher but otherwise identical to that used for subfamily A. This arrangement provides a striking demonstration of the following: First, keratins of the the two subfamilies are closely related in terms of relative size distribution; second, keratins with identical size ranks in subfamilies A and B always form a pair and follow similar rules of expression; and third, within each keratin pair, the member of the B subfamily is always larger than that of the A subfamily by approximately 8K (Fig. 5 and Table 1). The implications of this model are discussed below.

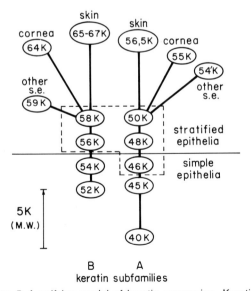

Figure 5 A unifying model of keratin expression. Keratins of subfamilies A and B are arranged according to their molecular weights (bar, 5K). The 8K difference between the members of each keratin pair most likely resides in the nonhelical regions (Hanukoglu and Fuchs 1983); the functional significance of this 8K peptide is currently unclear. (s.e.) Stratified epithelia.

Keratin structure

Steinert et al. (1976) and Lee and Baden (1976) have demonstrated that two different keratin subunits are required for the proper in vitro reconstitution of tonofilaments. Recent cross-linking data further suggest that intermediate filaments are composed of protofilament units built as a dimer of coiled coils (Geisler and Weber 1982). The findings that all known epithelial keratins can be divided into two subfamilies and that at least one member from each subfamily is expressed in all epithelia examined, including in vivo normal tissues (Gigi et al. 1982; Schiller et al. 1982; Tseng et al. 1982c), carcinomas (Moll et al. 1982a; Nelson et al. 1984), and cultured cells (Fuchs et al. 1981; Kim et al. 1983; Nelson and Sun 1983), lend support to the dimer model, perhaps with keratins within the "pairs" playing complementary roles in keratin assembly and/or function.

Evolution of keratin heterogeneity

The embryonic development of human epidermis provides a clue concerning the possible evolutionary sequence of various epithelial structures. In the early embryo, the epidermis is a simple epithelium that only later becomes stratified and finally keratinized (Fig. 6). If this ontogenic sequence recapitulates what may have happened during evolution, one can speculate that simple epithelia may have evolved first as a primitive, surface-covering device. Stratified epithelia may have occurred later through hyperproliferation and piling up of the simple epithelia to provide improved protection for the underlying organs. Finally, the keratinized epidermis may have evolved in certain terrestrial species in order to provide a tough external integument as well as to prevent water loss (Fig. 6). This scheme implies that the smallest keratins, which are characteristic of simple epithelia and appear first during embryogenesis (Jackson et al. 1980; Banks-Schlegel 1982; Franke et al. 1982; Tseng et al. 1982a; Moll et al. 1983), are relatively primitive. Genes coding for these keratins may have given rise sequentially to those for larger keratins, perhaps through gene duplication and subsequent size increases and diversification (e.g., see Craik et al. 1983). This hypothesis explains why, within each keratin subfamily, members with increasing sizes are associated with progressively more complex structures—from simple epithelia, to hyperproliferative keratinocytes, to normal keratinocytes, and finally, to several advanced states of keratinocyte differentiation (e.g., epidermal keratinization; Figs. 5 and 6).

Two possible explanations as to why larger keratins may have evolved to be associated with the more com-

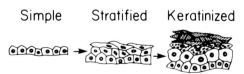

Figure 6 A schematic drawing showing the embryonic development as well as the postulated evolutionary history of human epidermis.

plex epithelia may be considered. First, the additional sequences could produce some changes in keratin structure resulting in different functional properties, including physical stability; tendency to interact with neighboring filaments to form stronger bundles; interaction with membranes, other cytoskeletal elements, or intermediate filament–associated molecules; resistance to proteolytic turnover; and ability to form intra- or intermolecular disulfide bonds. Second, different regulatory mechanisms could have arisen with the newly duplicated genes, resulting in, for instance, different rates or levels of synthesis. This could explain why keratins of the "stratified epithelial type" are usually made in much higher quantities per cell than those of the "simple epithelial type" (Nelson and Sun 1983).

Coordinated size increases in two keratin subfamilies

If this pattern of keratin evolution is correct, it is interesting to note that the two keratin subfamilies have evolved through an almost identical set of size increases of 1K–2K, 2K, 2K, and 7K, leading to the formation of 46K–54K, 48K–56K, 50K–58K, and 56.5K–65K keratin pairs (Fig. 5). Such highly coordinated size increases in the two keratin subfamilies suggest the existence of structural constraints, perhaps through protein-protein interactions between corresponding members of the two keratin subfamilies (keratin pairs) during dimer formation.

The positions and sequences of the additional peptides that keratins may have gained during evolution are unknown. These additional peptides could be located primarily in either the helical or the nonhelical portions of the molecules. The former possibility is attractive since the coiled-coil interaction may impose severe structural constraints on corresponding members of the two subfamilies, thus explaining the coordinated size increases. Alternatively, the additional sequences could be located mainly in the nonhelical domains; this possibility is actually supported by several lines of evidence, listed below.

(1) The amino acid sequence of a mouse 59K epidermal keratin has recently been published (Steinert et al. 1983). This keratin shows extensive sequence homologies with the 50K human keratin (Hanukoglu and Fuchs 1982; Steinert et al. 1983), and is expressed only by in vivo keratinized epidermis (Roop et al. 1983). Such results strongly suggest that this mouse keratin is equivalent to the 56.5K keratin in humans. A comparison of the amino acid sequences of the mouse 59K (human 56.5K equivalent) and human 50K keratins (see Fig. 5 for their possible evolutionary relationship) clearly indicates that the two keratins share a similar central, α-helical portion; the 59K keratin is longer than the 50K keratin, however, by about 100 amino acid residues located mainly in the nonhelical, carboxyterminal region (Steinert et al. 1983). This finding suggests that between the 50K and 56.5K keratins, the size increase is primarily due to the addition of nonhelical sequences.

(2) Steinert and Idler (1975) have shown previously that, as a general rule, smaller keratins tend to have a higher helical content than the larger keratins, suggesting the "dilution" of the helical content in larger keratins by nonhelical sequences.

(3) In the glycine-rich, nonhelical portions of keratin molecules, there exist several tandem peptide repeats of 5–20 amino acid residues (Hanukoglu and Fuchs 1983; Steinert et al. 1983). These sequences are of unknown functional significance but could be related to some of the repeated size increases of keratins.

(4) The helical portions of intermediate filaments appear to be relatively conserved, whereas the nonhelical portions are highly variable and are likely to play structural and/or functional roles that are filament-type or subunit specific (Geisler and Weber 1982; Hanukoglu and Fuchs 1982; Steinert et al. 1983). Whether and how the nonhelical domains of the corresponding keratins from subfamilies A and B may interact, exerting pressure for coordinated size increases during evolution, is currently unknown.

Specific keratins as markers for tumor diagnosis: Some general rules

According to their tissue-distribution patterns, human epithelial keratins can be divided into two major groups: those that are mainly expressed by simple epithelia and those expressed by stratified squamous epithelia (Fig. 5). The expression of these two groups of keratins appears to follow some general rules:

1. Depending on cell type, most simple epithelia express two to four keratins from a set of five "simple epithelial keratins" consisting of the 40K (subfamily A), 45K (A), 46K (A), 52K (B), and 54K (B) components (Fig. 5, below the line; Franke et al. 1981a,b,c; Moll et al. 1982c; Wu et al. 1982). This pattern appears to be relatively stable and is usually retained in cell culture as well as in neoplasms (Denk et al. 1982; Moll et al. 1982c).

2. The keratins expressed by stratified squamous epithelia are, within each subfamily, larger than those of the simple epithelia. These include the 48K, 50K, 54'K, 55K, and 56.5K keratins of subfamily A and the 56K, 58K, 59K, 64K, and 65K–67K keratins of subfamily B (Fig. 5, above the line). It should be noted that the 46K keratin (subfamily A), although mainly associated with simple epithelia, can also be found in some glands and in hyperproliferative keratinocytes (Sun and Green 1977, 1978b; Moll et al. 1982c; Eichner et al. 1984).

3. The 50K/50'K (A) and 58K (B) keratins are present in various quantities in all stratified epithelia and thus provide the most reliable markers for identifying cultured cells or neoplasms with a stratified epithelial origin (Tseng et al. 1982c; Nelson and Sun 1983; Nelson et al. 1984; see also Moll et al. 1982c, 1983).

4. The expression of the largest three keratin pairs appears to be differentiation specific. Thus, the 56.5K (A) and 65K–67K (B) keratins are expressed primarily in keratinized epidermis; the 55K (A) and 64K (B) keratins are corneal epithelium specific; and the 54K (A) and 59K (B) keratins are found in many nonker-

atinized, stratified epithelia of internal organs, including the esophagus (Table 1). In cultured cells as well as neoplasms, the expression of these keratins is frequently diminished, perhaps due to a loss of their normal differentiated state.

5. During hyperproliferation, all stratified squamous epithelial cells express a common set of five keratins (the 46K, 48K, and 50K keratins of subfamily A, and the 56K and 58K keratins of subfamily B; Fig. 5, box). Therefore, these five keratins are frequently found as the major intermediate filament subunits in squamous cell carcinomas (of esophagus, epiglottis, tongue, and epidermis; Moll et al. 1982a; Nelson et al. 1984), as well as in a wide variety of epidermal hyperproliferative diseases (Kubilus et al. 1980; Weiss et al. 1984). In this regard, it is interesting to note that we have proposed earlier that various in vitro–cultured, stratified epithelia may revert to more "primitive," and thus less distinctive, programs of differentiation (Doran et al. 1980). This explains why cultured human skin, corneal, and conjunctival epithelial cells lose many of their in vivo morphological and biochemical (keratin) differences and become indistinguishable from one another (Sun and Green 1977). A similar mechanism may also apply to neoplasms of various stratified epithelia.

Generality of the model and concluding remarks

Although a great majority of epithelial tissues can be classified as being either simple or stratified, a few do not fall into these two categories. These include the bladder (transitional) epithelium, the tracheal (pseudostratified columnar) epithelium, and some glandular epithelia. In general, keratins expressed by these unusual epithelia represent a combination of "stratified" and "simple" types (Moll et al. 1982c; Tseng et al. 1982c; Wu et al. 1982).

Although our model is primarily based on data derived from studies of human epithelial keratins, similar results have been obtained in monkey, rabbit (Tseng et al. 1982c), cow (D. Cooper, unpubl.), and mouse (R. Eichner, unpubl.), suggesting that keratins are highly conserved during mammalian evolution. Thus, the major points depicted in our model (Fig. 5) concerning the expression and evolution of keratin components may also be applicable to other mammalian species.

In conclusion, we have constructed a unifying model that can account for almost all known variations in keratin expression as a function of epithelial cell type, embryonic development, cellular growth environment, and diseases, including neoplasms. This model defines and emphasizes the concepts of keratin subfamilies and keratin pairs. Although the model will undoubtedly be modified and expanded as new data become available, we hope that even in its present form it will stimulate the formulation of new working hypotheses concerning keratin structure, the possible significance of keratin subfamilies and keratin pairs, the mechanisms of the highly coordinated size increases (2K, 2K, 2K, and 7K) in the two keratin subfamilies, and the functional roles of the differentiation-specific keratins.

Acknowledgments

This investigation was aided in part by grants from the National Institutes of Health (EY-02472, EY-04722, and AM-25140) and the Estee Lauder and Gillette Companies. We thank J. Rheinwald for providing a sample of mesothelial keratins, S.C.G. Tseng, J. Woodcock-Mitchell, and M. Jarvinen for earlier contributions, T. Hronis for helpful discussions, and M. Lynch for critical reading of the manuscript. T.-T.S. and R.E. were the recipients of an NIH Research Career Development Award (EY-0125) and a Dermatology Foundation Fellowship from Avon Products, Inc., respectively.

References

Banks-Schlegel, S.P. 1982. Keratin alterations during embryonic epidermal differentiation: A presage of adult epidermal maturation. *J. Cell Biol.* **93**: 551.

Bladon, P.T., P.E. Bowden, W.J. Cunliffe, and E.J. Wood. 1982. Prekeratin biosynthesis in human scalp epidermis. *Biochem. J.* **208**: 179.

Craik, C.S., W.J. Rutter, and R. Fletterick. 1983. Splice junctions: Association with variation in protein structure. *Science* **220**: 1125.

Debus, E., K. Weber, and M. Osborn. 1982. Monoclonal cytokeratin antibodies that distinguish simple from stratified squamous epithelia: Characterization on human tissues. *EMBO J.* **1**: 1641.

Denk, H., R. Krepler, E. Lackinger, U. Artlieb, and W.W. Franke. 1982. Biochemical and immunohistochemical analysis of the intermediate filament cytoskeleton in human hepatocellular carcinomas and hepatic neoplastic nodules of mice. *Lab. Invest.* **46**: 584.

Doran, T.I., A. Vidrich, and T.-T. Sun. 1980. Intrinsic and extrinsic regulation of the differentiation of skin, corneal and esophageal epithelial cells. *Cell* **22**: 17.

Eichner, R., P. Bonitz, and T.-T. Sun. 1984. Classification of epidermal keratins according to their immunoreactivity, isoelectric point, and mode of expression. *J. Cell Biol.* (in press).

Franke, W.W., H. Denk, R. Kait, and E. Schmid. 1981a. Biochemical and immunological identification of cytokeratin proteins present in hepatocytes of mammalian liver tissue. *Exp. Cell Res.* **131**: 299.

Franke, W.W., E. Schmid, M. Osborn, and K. Weber. 1978a. Different intermediate-sized filaments distinguished by immunofluorescence microscopy. *Proc. Natl. Acad. Sci.* **75**: 5034.

Franke, W.W., C. Grund, C. Kuhn, B.W. Jackson, and K. Illmensee. 1982. Formation of cytoskeletal elements during mouse embryogenesis. III. Primary mesenchymal cells and the first appearance of vimentin filaments. *Differentiation* **23**: 43.

Franke, W.W., K. Weber, M. Osborn, E. Schmid, and C. Freudenstein. 1978b. Antibody to prekeratin: Decoration of tonofilament-like arrays in various cells of epithelial character. *Exp. Cell Res.* **116**: 429.

Franke, W.W., B. Appelhans, E. Schmid, C. Freudenstein, M. Osborn, and K. Weber. 1979. Identification and characterization of epithelial cells in mammalian tissues by immunofluoresence microscopy using antibodies to prekeratin. *Differentiation* **15**: 7.

Franke, W.W., S. Winter, C. Grund, E. Schmid, D.L. Schiller, and E.D. Jarasch. 1981b. Isolation and characterization of desmosome-associated tonofilaments from rat intestinal brush border. *J. Cell Biol.* **90**: 116

Franke, W.W., D.L. Schiller, R. Moll, S. Winter, E. Schmid, I. Engelbrecht, H. Denk, R. Krepler, and E. Platzer. 1981c. Diversity of cytokeratins: Differentiation-specific expression of cytokeratin polypeptides in epithelial cells and tissues. *J. Mol. Biol.* **153**: 933.

Fuchs, E. and H. Green. 1979. Multiple keratins of cultured human epidermal cells are translated from different mRNA molecules. *Cell* **17**: 573.
———. 1980. Changes in keratin gene expression during terminal differentiation of the keratinocyte. *Cell* **19**: 1033.
———. 1981. Regulation of terminal differentiation of cultured human keratinocytes by vitamin A. *Cell* **25**: 617.
Fuchs, E., S.M. Coppock, H. Green, and D.W. Cleveland. 1981. Two distinct classes of keratin genes and their evolutionary significance. *Cell* **27**: 75.
Geisler, N. and K. Weber. 1982. The amino acid sequences of chicken muscle desmin provides a common structural model for intermediate filament proteins. *EMBO J.* **1**: 1649.
Gigi, O., B. Geiger, Z. Eshhar, R. Moll, E. Schmid, S. Winter, D.L. Schiller, and W.W. Franke. 1982. Detection of a cytokeratin determinant common to diverse epithelial cells by a broadly cross-reacting monoclonal antibody. *EMBO J.* **1**: 1429.
Green, H., E. Fuchs, and F. Watt. 1982. Differentiated structural components of the keratinocytes. *Cold Spring Harbor Symp. Quant. Biol.* **46**: 293.
Hanukoglu, I. and E. Fuchs. 1982. The cDNA sequence of a human epidermal keratin: Divergence of sequence but conservation of structure among intermediate filament proteins. *Cell* **31**: 243.
———. 1983. The cDNA sequence of a type II cytoskeletal keratin reveals constant and variable structural domains among keratins. *Cell* **33**: 915.
Jackson, B.W., C. Grund, E. Schmid, K. Burki, W.W. Franke, and K. Illmensee. 1980. Formation of cytoskeletal elements during mouse embryogenesis. I. Intermediate filaments of the cytokeratin type and desmosomes in preimplantation embryos. *Differentiation* **17**: 161.
Kim, K.H., J. Rheinwald, and E.V. Fuchs. 1983. Tissue specificity of epithelial keratins: Differential expression of mRNAs from two multigene families. *Mol. Cell. Biol.* **3**: 495.
Kubilus, J., H.P. Baden, and N. McGilvray. 1980. Filamentous proteins of basal cell epithelioma: Characteristics *in vivo* and *in vitro*. *J. Natl. Cancer Inst.* **65**: 869.
Lee, L.D. and H.P. Baden. 1976. Organization of the polypeptide chains in mammalian keratin. *Nature* **264**: 377.
Magin, T.M., J.L. Jorcano, and W.W. Franke. 1983. Translational products of mRNAs coding for non-epidermal cytokeratins. *EMBO J.* **2**: 1387.
Moll, R., R. Krepler, and W.W. Franke. 1982a. Complex cytokeratin polypeptide patterns observed in certain human carcinomas. *Differentiation* **23**: 256.
Moll, R., I. Moll, and W. Wiest. 1983. Changes in the pattern of cytokeratin polypeptides in epidermis and hair follicles during skin development in human fetuses. *Differentiation* **23**: 170.
Moll, R., W.W. Franke, B. Volc-Platzer, and R. Krepler. 1982b. Different keratin polypeptides in epidermis and other epithelia of human skin: A specific cytokeratin of molecular weight 46,000 in epithelia of the pilosebaceous tract and basal cell epitheliomas. *J. Cell Biol.* **95**: 285.
Moll, R., W.W. Franke, D.L. Schiller, B. Geiger, and R. Krepler. 1982c. The catalog of human cytokeratins: Patterns of expression in normal epithelia, tumors and cultured cells. *Cell* **31**: 11.
Nelson, W.G. and T.-T. Sun. 1983. The 50- and 58-kdalton keratin classes as molecular markers for stratified squamous epithelia: Cell culture studies. *J. Cell Biol.* **97**: 244.
Nelson, W.G., H. Battifora, H. Santana, and T.-T. Sun. 1984. Specific keratins as markers for neoplasms with a stratified squamous origin. *Cancer Res.* (in press).
Rheinwald, J.G. and H. Green. 1975. Serial cultivation of strains of human epidermal keratinocytes: The formation of keratinizing colonies from single cells. *Cell* **6**: 331.
Roop, D.R., P. Hawley-Nelson, C.K. Cheng, and S.H. Yuspa. 1983. Keratin gene expression in mouse epidermis and cultured epidermal cells. *Proc. Natl. Acad. Sci.* **80**: 716.

Schiller, D.L., W.W. Franke, and B. Geiger. 1982. A subfamily of relatively large and basic cytokeratin polypeptides as defined by peptide mapping is represented by one or several polypeptides in epithelial cells. *EMBO J.* **1**: 761.
Skerrow, D. and C.J. Skerrow. 1983. Tonofilament differentiation in human epidermis: Isolation and polypeptide chain composition of keratinocyte subpopulation. *Exp. Cell Res.* **143**: 27.
Steinert, P.M. and W.W. Idler. 1975. The polypeptide composition of bovine epidermal alpha-keratin. *Biochem. J.* **151**: 603.
Steinert, P.M., W.W. Idler, and S.B. Zimmerman. 1976. Self-assembly of bovine epidermal keratin filaments in vitro. *J. Mol. Biol.* **108**: 547.
Steinert, P.M., R.H. Rice, D.R. Roop, L.T. Benes, and A.C. Steven. 1983. Complete amino acid sequence of a mouse epidermal keratin subunit and implications for the structure of intermediate filaments. *Nature* **302**: 794.
Sun, T.-T. and H. Green. 1977. Cultured epithelial cells of cornea, conjunctiva and skin: Absence of marked intrinsic divergence of their differentiated states. *Nature* **269**: 489.
———. 1978a. Immunofluorescent staining of keratin fibers in cultured cells. *Cell* **14**: 469.
———. 1978b. Keratin filaments of cultured human epidermal cells: Formation of intermolecular disulfide bonds during terminal differentiation. *J. Biol. Chem.* **253**: 2053.
Sun, T.-T., C. Shih, and H. Green. 1979. Keratin cytoskeletons in epithelial cells of internal organs. *Proc. Natl. Acad. Sci.* **76**: 2813.
Sun, T.-T., R. Eichner, W.G. Nelson, A. Vidrich, and J. Woodcock-Mitchell. 1983a. Keratin expression during normal epidermal differentiation. In *Normal and abnormal epidermal differentiation* (ed. M. Seiji and I.A. Bernstein), p. 277. University of Tokyo Press, Tokyo.
Sun, T.-T., R. Eichner, W.G. Nelson, S.C.G. Tseng, R.A. Weiss, M. Jarvinen, and J. Woodcock-Mitchell. 1983b. Keratin classes: Molecular markers for different types of epithelial differentiation. *J. Invest. Dermatol.* **81**: 109s.
Tseng, S.C.G., J.-W. Huang, and T.-T. Sun. 1982a. Developmental changes in keratin antigens of skin and corneal epithelia. *Invest. Ophthalmol. Visual Sci.* **73**. (ARVO Abstr.)
Tseng, S.C.G., D. Hatchell, J.-W. Huang, and T.-T. Sun. 1982b. Specific keratins as molecular markers of keratinization. *J. Cell Biol.* **95**: 231a.
Tseng, S.C.G., M. Jarvinen, W.G. Nelson, J.-W. Huang, J. Woodcock-Mitchell, and T.-T. Sun. 1982c. Correlation of specific keratins with different types of epithelial differentiation: Monoclonal antibody studies. *Cell* **30**: 361.
Viac, J., M.J. Staguet, J. Thivolet, and C. Goujon. 1980. Experimental production of antibodies against stratum corneum keratin polypeptides. *Arch. Dermatol. Res.* **267**: 179.
Weiss, R.A., R. Eichner, and T.-T. Sun. 1984. Monoclonal antibody analysis of keratin expression in epidermal diseases: A 48kd and a 56kd keratin as molecular markers for keratinocyte hyperproliferation. *J. Cell Biol.* (in press).
Weiss, R.A., G.Y.A. Guillet, I.M. Freedberg, E.R. Farmer, E.A. Small, M.M. Weiss, and T.-T. Sun. 1983. The use of monoclonal antibody to keratin in human epidermal disease: Alterations in immunohistochemical staining pattern. *J. Invest. Dermatol.* **81**: 224.
Woodcock-Mitchell, J., R. Eichner, W.G. Nelson, and T.-T. Sun. 1982. Immunolocalization of keratin polypeptides in human epidermis using monoclonal antibodies. *J. Cell Biol.* **95**: 580.
Wu, Y.-J. and J.G. Rheinwald. 1981. A new small (40 kd) keratin filament protein made by some cultured human squamous cell carcinomas. *Cell* **25**: 627.
Wu, Y.-J., L.M. Parker, N.E. Binder, M.A. Beckett, J.H. Sinard, C.T. Griffiths, and J.G. Rheinwald. 1982. The mesothelial keratins: A new family of cytoskeletal proteins identified in cultured mesothelial cells and nonkeratinizing epithelia. *Cell* **31**: 693.

Cytokeratins: Complex Formation, Biosynthesis, and Interactions with Desmosomes

W.W. Franke, D.L. Schiller, M. Hatzfeld, T.M. Magin, J.L. Jorcano,*
S. Mittnacht,* E. Schmid, J.A. Cohlberg, and R.A. Quinlan

Division of Membrane Biology and Biochemistry, Institute of Cell and Tumor Biology, German Cancer Research Center, and *Center of Molecular Biology, University of Heidelberg, D-6900 Heidelberg, Federal Republic of Germany

Cytokeratin filaments form a subclass of intermediate-sized (7–11-nm) filaments (IF) that consists of proteins related to, but not identical with, the "low-sulfur" keratins of wool, hair, and epidermis. They occur in a diversity of epithelial cells, including epithelial tissues and cultured epithelial cells (Franke et al. 1978a,b, 1979b,c; Sun and Green 1978; Schmid et al. 1979; Sun et al. 1979; Moll et al. 1982b; for loss of expression in certain cultures, see Venetianer et al. 1983), as well as in epithelia-derived tumors and cultured carcinoma cells (Franke et al. 1978a,b, 1979a; Sun et al. 1979; Bannasch et al. 1980; Schlegel et al. 1980; Gabbiani et al. 1981; Moll et al. 1982b, 1983; for a review, see Osborn and Weber 1983).

Comparisons of patterns of expression of cytokeratin polypeptides in different tissues have shown that different sets of cytokeratin polypeptides are expressed in different epithelia, the specific cytokeratin polypeptide pattern being characteristic of a given type of epithelium (Doran et al. 1980; Winter et al. 1980; Franke et al. 1981a,b,c,d, 1982d; Fuchs and Green 1981; Wu and Rheinwald 1981; Moll et al. 1982b). To date, 19 different cytokeratin polypeptides have been distinguished in human cells (Fig. 1; cf. Moll et al. 1982b); 15 of these have also been recognized in human transformed cells and tumors (Fig. 1; Moll et al. 1982b, 1983). Similar patterns of diversity exist within the cytokeratin polypeptides of rodents and the cow (Franke et al. 1981a,b,c,e, 1982d; Schiller et al. 1982).

On the basis of peptide mapping (Schiller et al. 1982), nucleic acid hybridization (Fuchs et al. 1981; Kim et al. 1983), and analyses of determinants of monoclonal antibodies (Tseng et al. 1982; see also Gigi et al. 1982; Sun et al., this volume), the cytokeratin polypeptides have been subdivided into two families, one comprising the relatively large and basic members (nos. 1–8 of the human cytokeratin catalog of Moll et al. 1982b) and the other the more acidic cytokeratin polypeptides.

Cytokeratin polypeptide complexes

Studies of IF reconstitution in vitro have shown that purified polypeptides of vimentin, desmin, glial filament protein, and the M_r 68,000 neurofilament polypeptide can form homopolymer IFs (Small and Sobieszek 1977; Rueger et al. 1979; Geisler and Weber 1980, 1981b; Huiatt et al. 1980; Renner et al. 1981; Steinert et al. 1981a,b; Liem et al. 1982). Moreover, hybrid IFs containing heteropolymers of vimentin with desmin or glial filament protein have been demonstrated by reconstitution in vitro

(Steinert et al. 1981a,b, 1982), immunoelectron microscopy (Sharp et al. 1982), and chemical cross-linking of heterodimers in IFs (Quinlan and Franke 1982, 1983). However, experiments to reconstitute IFs from purified, single cytokeratin polypeptides have failed, and specific combinations of two or more cytokeratin polypeptides appear to be required for assembly of regular IFs (Lee and Baden 1976; Steinert et al. 1976; Milstone 1981). This suggests that (1) the diverse cytokeratin polypeptides combine with each other in very specific ways and (2) the subunits of polymerization may not be monomers of the individual cytokeratin polypeptides but complexes containing two or more different polypeptide molecules. We have recently studied such subunit complexes in some detail.

Native or reconstituted cytokeratin filaments, or total cytoskeletal residues obtained after treatment with high-salt buffers and detergent, can be solubilized in low-salt buffer solutions of nondenaturing concentrations of

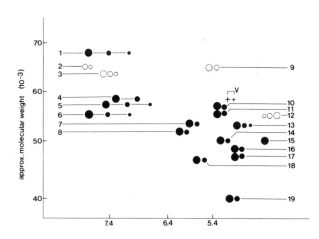

Figure 1 Diagram of human cytokeratin polypeptides arranged according to their isoelectric positions after separation by two-dimensional gel electrophoresis (for details, see Moll et al. 1982b). Of the 19 polypeptides described in various human tissues, 15 cytokeratins (●) have so far also been identified in human tumors (different sets of polypeptides are expressed in different types of tumors). The sizes of the circles identifying one polypeptide indicate the relative intensities of the non-phosphorylated (left circle in each group) and the various phosphorylated molecules in a representative tissue or cell type. The degree of phosphorylation, however, varies between different cell types and different physiological conditions of the same cell type. (V) Position of vimentin and phosphorylated vimentin.

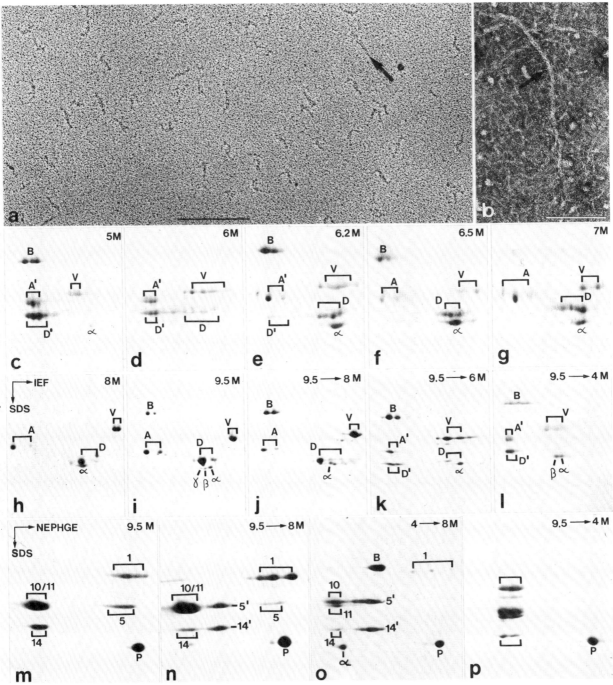

Figure 2 Complexes of cytokeratin polypeptides solubilized in low-salt buffer (10 mM Tris·HCl, pH 8.0) containing 25 mM or 700 mM 2-mercaptoethanol and 4 M urea are examined by electron microscopy (a,b) and by two-dimensional gel electrophoresis at different urea concentrations (c–p). (a) Cytokeratin molecules from cytoskeletons of cultured rat hepatoma cells (line MH,C,), which contain only two cytokeratin polypeptides (A and D in a 1:1 ratio; cf. Franke et al. 1981b, 1983a), have been solubilized in 4 M urea and visualized by electron microscopy using the rotary metal shadow-cast technique (for details, see Franke et al. 1982a). The molecules appear as flexible rods of ~48 nm in length, sometimes revealing two threadlike subfilaments (→). Bar, 0.2 μm. (b) Partial IF reconstitution from complexes as shown in a after dialysis against 10 mM Tris-buffer (pH 7.0) containing 10 mM 2-mercaptoethanol for 10 hr is visualized by electron microscopy using negative staining with uranyl acetate (for details of preparation, see Franke et al. 1982a). Note coexistence of rodlike elements of 2–3-nm diameter and ~50-nm length, longer (proto-)filaments of 2–3-nm diameter, and typical IF (→). Bar, 0.1 μm. (c–l) Two-dimensional gel electrophoresis of cytokeratins from cultured bovine kidney cells of the MDBK line (IEF in first dimension as indicated in h; downward arrow denotes direction of second-dimension electrophoresis in the presence of SDS; acidic proteins are to the right; for conditions of electrophoresis, see Moll et al. 1982b; Franke et al. 1983a). Urea concentrations and specific treatments are indicated in the upper right corners. B (BSA) and α (α-actin) denote coelectrophoresed reference proteins. (β,γ)Endogenous, β, γ-actin; (V) endogenous vimentin (cf. Franke et al. 1979b, 1982b,d). (c) After isoelectric focusing of total cytoskeletal proteins in 5 M urea, the two cytokeratin polypeptides A and D appear at isoelectric positions A′ and D′ due to their association in a complex.

urea, usually 4 M. For example, cytokeratin material solubilized by treatment with 4 M urea from cytoskeletons of hepatocytes or cultured rat hepatoma cells of the MH_1C_1 line (for immunofluorescence microscopy and cytoskeletal composition, see also Franke et al. 1981b) consists predominantly of molecules of M_r 95,000 (putative heterodimers) and M_r 206,000 (putative "double dimers"), as determined by sedimentation equilibrium centrifugation at 19.7°C and gel electrophoresis of cross-linked molecules (see below). In the electron microscope, the majority of these molecules appear as rodlike structures of a mean length of 48 ± 4 nm, as determined after rotary shadowing (Fig. 2a; values corrected for metal deposition), and a width of approximately 2 nm, as seen in shallow negative-staining preparations (Fig. 2b). Occasionally, these rods reveal a composition of two parallel threads, which possibly represent the individual polypeptides (Fig. 2a, arrow). The rodlike structures of these intact cytokeratin molecules are similar in size and shape to the protease-resistant α-helical core particles prepared from epidermal keratin from cow and mouse (Steinert 1978; Steinert et al. 1983) and chicken gizzard desmin (Geisler et al. 1982). On dialysis to buffers of 10 mM Tris·HCl (pH 7.2) or of higher ionic strength, these rodlike structures can reassemble gradually into protofilaments and finally into normal-looking IFs (Fig. 2b).

On electrophoretic gels, this solubilized cytokeratin material appears in the form of two polypeptides of M_r 49,000 and 55,000 (D and A; cf. Franke et al. 1981a,b,c,d; equivalent to components nos. 8 and 18 of Moll et al. 1982b) in an approximately 1:1 ratio, as estimated from Coomassie blue staining. These two polypeptides are isoelectric at 4 M urea but are separated from each other at higher urea concentrations (Franke et al. 1983a). Similar observations have been made for the A:D complex present in bovine liver and cultured MDBK cells (Fig. 2c–i). On dialysis of cytokeratin denatured in 9.5 M or 10 M urea to lower urea concentrations (6 M–4 M), these two cytokeratin polypeptides not only refold as individual polypeptides, but also reassociate into a complex isoelectric at the same pH as the native A:D complex (Fig. 2j–l). Upon further removal of urea, these reconstituted complexes assemble into protofilamentous and annular structures and, at appropriate ionic strength, finally into IFs (Fig. 2b; for details of intermediate structures during IF reconstitution, see also Franke et al. 1982b). The strand separation and complex reassociation kinetics are defined by the "midpoint" of strand separation and reassociation in urea (U_m; Franke et al. 1983a), which is a specific property of a given cytokeratin complex subunit. The stability characteristics of different cytokeratin complexes exhibit a remarkably broad range of variability from $U_m = 5.9$ M to U_m values of approximately 9 M urea. An example of the latter type of unusually stable cytokeratin complexes is represented by the complexes of cytokeratins nos. 5 and 14 of human epidermis (Fig. 2m–p) and cytokeratins nos. 6 and 14 of cultured human keratinocytes (not shown), which require more than 8 M urea for strand separation. A diagrammatic presentation of the strand separation and reassociation curves of different cytokeratin complexes is shown in Figure 3 (for details, see also Franke et al. 1983a).

The specificity of cytokeratin complex formation is also demonstrated by the exclusion of vimentin (Fig. 2c–l), which coexists with cytokeratins in a number of cultured epithelial and carcinoma cells, usually displaying fibril arrays different from those of the cytokeratin IFs (Franke et al. 1978a, 1979a,b, 1982d; Osborn et al. 1980; Virtanen et al. 1981). This shows that vimentin and both families of cytokeratins differ in important recognition signals, despite considerable amino acid sequence homology in their α-helical central core regions (Geisler and Weber 1981a; Geisler et al. 1982, 1983; Hanukoglu and Fuchs 1982, 1983; Quax-Jeuken et al. 1983; Steinert et al. 1983; see also Steinert et al. 1980). In all cytokeratin complexes examined, we have found combinations of at least one polypeptide of the "basic" family (apparently equivalent to the "type-II" keratins described for wool keratins; Crewther et al. 1978; Gough et al. 1978) with, at least, one polypeptide of the more acidic family (some of which are related to "type-I" keratins of wool). We propose that these complexes are the actual molecular subunits of cytokeratin IF assembly. All examples so far studied in detail are hetero-

Figure 2 (*continued*) (*d*) Dialysis and focusing in 6 M urea show a similar situation as in *c* but also some displacement of polypeptide D (from D' to D). (*e*) In 6.2 M urea, more of polypeptide D is displaced to position D. (*f*) In 6.5 M urea, most of D is displaced from the complex-bound form D' to position D. In 7 M (*g*), 8 M (*h*), and 9.5 M (*i*) urea, polypeptides A and D are completely separated. (*j*) When cytokeratins are first denatured in 9.5 M urea, then dialyzed against buffer containing 8 M urea, and finally focused in 8 M urea, polypeptides A and D are still separated. (*k*) Similar to *j*, but after further dialysis and focusing in 6 M urea, half of D is already detected in the position of the complex D'. (*l*) In 4 M urea, all cytokeratin cofocuses and the re-formed complex is indistinguishable from that directly solubilized in 4 M urea (cf. with *c*); note that vimentin (V) is always separated from the cytokeratin polypeptides, demonstrating the exclusion from cytokeratin complex formation. (*m*–*p*) Two-dimensional gel electrophoretic analysis of human epidermal keratins (for preparation of samples, see Moll et al. 1982a; nonequilibrium pH gradient electrophoresis, NEPHGE, in first dimension; second-dimension electrophoresis is as in *c*–*l*) after dialysis at different concentrations of urea containing 25 mM or 700 mM 2-mercaptoethanol (10 mM Tris·HCl, pH 8.0). (P) Phosphoglycerokinase as a reference protein (isoelectric at pH 7.5); other symbols are as in *c*–*l*. (*m*) At 9.5 M urea, the major epidermal cytokeratin polypeptides (nos. 1, 5, 10, 11, and 14 of the catalog of Moll et al. 1982b) are separated. (*n*) When human epidermal keratins are solubilized in buffer containing 9.5 M urea and are then dialyzed against the same buffer containing 8 M urea, most of cytokeratins nos. 5 and 14 appear in isoelectric positions (5' and 14') due to their reassociation into complexes; by contrast, most of components nos. 1 and 10 + 11 are still separated and appear in a similar position as in 9.5 M urea. (*o*) When epidermal keratin is solubilized in buffer containing 4 M urea and then dialyzed against buffer containing 8 M urea, a similar situation as in *n* is found. (*p*) At 4 M urea, all epidermal keratin polypeptides appear in complexes that have nearly coincident isoelectric pH values (brackets).

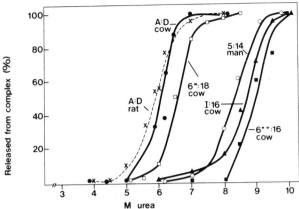

Figure 3 Diagram presenting some melting and reassociation curves of different cytokeratin complexes: (X) Complex of A and D from cultured rat hepatoma MH_1C_1 cells; (●) complex of A and D from cultured bovine MDBK cells; (□) complex of bovine cytokeratins 6* and 18 of esophagus tissue; (○) complex of cytokeratins 5 and 14 of human epidermis; (▲) complex of cytokeratins I (nos. 1–3) and 16 in bovine muzzle epidermis; (■) complex of bovine cytokeratins 6** and 16 BMGE + H cells (Schmid et al. 1983). For determinations, see Franke et al. (1983a).

dimers that, in turn, may associate to pairs ("double dimers"), in agreement with our finding of 1:1 dimers, $(1:1) + 1$ trimers, and $2 \times (1:1)$ tetramers by chemical cross-linking with dimethylsuberimidate (not shown). Similar "double-heterodimer" ("tetramer") complexes have been proposed from studies on proteolytic fragments of wool keratin (Gruen and Woods 1983). It will be important to determine which cytokeratin polypeptides can combine with each other and which sequences and conformations in the partners of a specific cytokeratin complex govern this unusual complex formation taking place even at high urea concentrations.

Our analyses of the melting and reassociation behavior of cytokeratins of clonal cell lines clearly demonstrate that different cytokeratin complexes can occur in the same cell (Fig. 4a–h), that the different cytokeratin complexes present in one cell may differ in isoelectric pH values and U_m (Fig. 4e, f), and that in the same cell certain cytokeratin polypeptides can form complexes with different partners (e.g., cytokeratin no. 8 can combine with cytokeratin no. 18 as well as with cytokeratin no. 19; Fig. 4a–d). We do not know whether such different complexes occur in the same filament or whether the IFs of such epithelial cells are heterogeneous in composition. Our mRNA microinjection experiments (see below) at least indicate that different cytokeratins and hence cytokeratin complexes can occur in the same filament. Whether the differences in affinity and melting behavior between different cytokeratin complexes reflect different forms of coiled-coil interactions and whether these are related to biological functions of the cytokeratin IFs in the diverse tissues remain to be elucidated.

Biosynthesis of cytokeratins

Studies using in vitro translation of total or poly(A)$^+$ RNAs have shown that desmin, vimentin, and neurofil-

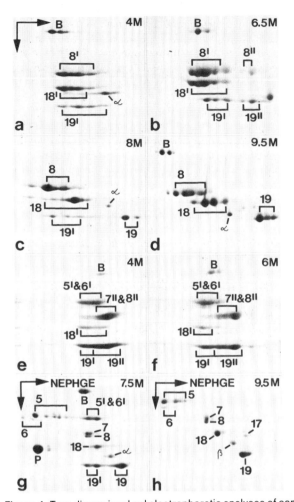

Figure 4 Two-dimensional gel electrophoretic analyses of complexes of human cytokeratin polypeptides from cultured mammary gland carcinoma cells of the MCF-7 line (a–d) and cultured pharyngeal carcinoma (PC) cells of the line Detroit 562 (ATCC; e–h), using different concentrations of urea. (IEF is used in the first dimension, except for g and h where NEPHGE has been used.) Symbols are as in Fig. 2; (β) endogenous β-actin. Cytokeratins are designated with their numbers from the human cytokeratin polypeptide catalog of Moll et al. (1982b), and conditions of solubilization and dialysis are as described by Franke et al. (1983a). (a) At 4 M urea, all three cytokeratin polypeptides of MCF-7 cells appear in isoelectric positions due to their inclusion in complexes (8′, 18′, and 19′). (b) At 6.5 M urea, differences are obvious between the complexes of cytokeratins 8 and 18, most of which are still in complex position, and those of cytokeratins 8 and 19, a considerable proportion of which appear in positions 8″ and 19″. (c) At 8 M urea, most of the complexes between cytokeratins 8 and 18 and 8 and 19 have separated, but some residual complex material is still recognized. (d) Complete separation of the three MF-7 cytokeratins at 9.5 M urea. (e) At 4 M urea, all cytokeratin material of PC cells is recognized to be in the form of complexes, but there are at least two different types of complexes: cytokeratins 5, 6, and 18 as well as a good proportion of cytokeratin 19 appear at a less acidic pH value (5′ & 6′, 18′, and 19′) than complexes containing cytokeratins 7 and 8, which appear together with most of cytokeratin 19 (7″, 8″, and 19″). (f) At 6 M urea, the two different types of complexes are still recognized. (g) At 7.5 M urea, partial separation of the constituents of the different complexes is recognized. (h) At 9.5 M urea, all cytokeratin polypeptides are separated; note that the minor cytokeratin 17 present in these cells (cf. Moll et al. 1983) is not identified at the low protein loadings of the gels shown in e–g.

ament proteins are coded by specific mRNAs and that the translation products obtained in vitro are identical, in two-dimensional gel electrophoresis, to the corresponding polypeptides isolated from tissues or cultured cells (Czosnek et al. 1980; Franke et al. 1980a; Schmid et al. 1980; O'Connor et al. 1981; Dodemont et al. 1982; McTavish et al. 1983). This indicates that these IF proteins of the nonepithelial type are synthesized in their final form and are not derived from considerably longer precursor molecules or subject to extensive posttranslational modifications other than distinct phosphorylations and acetylations (for review, see Lazarides 1982). This identity of translational products and mature polypeptides has also been substantiated by the comparison of the amino acid sequence of vimentin (Geisler and Weber 1981a) with the nucleotide sequence of the corresponding genomic and cDNA clones (Quax et al. 1983; Quax-Jeuken et al. 1983).

Similar results have been obtained for cytokeratins identified after in vitro translation of RNAs from epidermis of diverse species (Fuchs and Green 1979; Schweizer and Goerttler 1980; Gibbs and Freedberg 1982; Schiller et al. 1982; Roop et al. 1983). Interestingly, however, two of the keratin polypeptides of human epidermis have not been identified among the translational products of mRNA isolated from this tissue (Fuchs and Green 1979, 1980), and it has been suggested that these two polypeptides are proteolytically produced from larger precursor polypeptides during terminal differentiation in upper layers of human epidermis. In contrast, we have been able to identify all major and minor cytokeratin polypeptides of bovine muzzle epidermis among the translation products of the poly(A)$^+$ RNA isolated from this tissue, using both the reticulocyte lysate in vitro system and microinjection into Xenopus oocytes (Fig. 5a–d; cf. Schiller et al. 1982; Kreis et al. 1983; Magin et al. 1983). Electrophoretic fractionation of poly(A)$^+$ RNA has shown further that the different cytokeratin polypeptides are products of mRNAs of different molecular weights (e.g., Fig. 5b; Fuchs and Green 1980; Franke et al. 1983b). In those cases in which mRNAs from nonepidermal cells have been examined by translation in vitro, cytokeratin polypeptides found in the cytoskeletons of the specific cells have been identified among the translational products (Kim et al. 1983; Magin et al. 1983). These results indicate that the diverse cytokeratin polypeptides expressed in different epithelia of a species are not the result of posttranslational processing and modifications of a common precursor but are products of different mRNAs synthesized in the specific cell type.

Amino acid and nucleotide analyses have revealed a certain degree of sequence homology in the α-helical central core portion between members of the two cytokeratin families as well as between the cytokeratins and other IF proteins, besides striking differences in the amino- and carboxyterminal portions (Geisler and Weber 1981a; Geisler et al. 1982, 1983; Hanukoglu and Fuchs 1982, 1983; Weber and Geisler 1982; Dowling et al. 1983; Steinert et al. 1983). Interestingly, in the limited sequence comparisons performed so far, relatively little homology ($\sim 30\%$) has been found between the two keratin families (Crewther et al. 1978; Gough et al. 1978; Hanukoglu and Fuchs 1983; see also Fuchs et al., this volume).

Using cDNA clones to various mRNAs encoding epidermal keratin polypeptides, Roop et al. (1983) have reported that mRNA sequences of different types of murine keratins cross-hybridize, including members of the two different families, whereas no cross-hybridization has been noted in the 3' noncoding portions. Fuchs et al. (1981) have demonstrated that, under their hybridization conditions, cross-hybridization is significant only between members of the same family; i.e., genes coding for polypeptides of the basic (type-II) cytokeratin family do not cross-hybridize with those of the acidic (type-I) family (see also Kim et al. 1983). These results suggest that there are extended sequence homologies within the specific cytokeratin families. However, using various cDNA clones of about 1000 bp containing sequences coding for the 3' ends of all major mRNAs of bovine epidermal cytokeratins in hybridization-selection experiments under very stringent conditions, we have specifically hybridized and separated each of the diverse cytokeratin mRNAs present in this tissue (Fig. 5e–g). These differences of hybridization indicate that even within the same cytokeratin polypeptide family there is limited homology and that different cytokeratin polypeptides of the same family do not contain extended 3'-end regions identical in sequence. These hybridization data also suggest that the different cytokeratin polypeptides are coded by different genes (cf. Fuchs et al. 1981).

It has been shown that vimentin obtained by translation in vitro using the reticulocyte lysate system as well as a small proportion of vimentin pulse-labeled with [^{35}S]methionine in RVF-SM cells (cf. Franke et al. 1980a) exists in a soluble form, which by gel filtration appears with molecular sizes (M_r 140,000–190,000) larger than the monomeric polypeptide (M_r 50,000–60,000). This suggests that vimentin is present as oligomers and/or combined with other molecules (Schmid et al. 1980). Oxidative cross-linking of the soluble vimentin (for conditions, see Quinlan and Franke 1983) has revealed the existence of vimentin dimers in such prepartions, but other forms of molecular associations have not yet been excluded (R.A. Quinlan et al., unpubl.). Lazarides and colleagues (Blikstad and Lazarides 1983; Moon and Lazarides 1983) have recently described, in erythroid cells from chicken embryos, a pool of soluble vimentin that is enriched in newly synthesized molecules, but the molecular form of this soluble vimentin has not been further characterized. Currently, we do not know the state of the newly synthesized cytokeratin molecules in the living cell. Possibly there exists a relatively small pool of unpolymerized ("soluble") cytokeratins, as suggested for rat liver tissue by Sahyoun et al. (1982), but it is not clear whether the unpolymerized molecules occur as individual polypeptides or in the form of hetero-oligomer complexes of the type described above. In view of the high tendency of cytokeratins to assemble to IFs, it is rea-

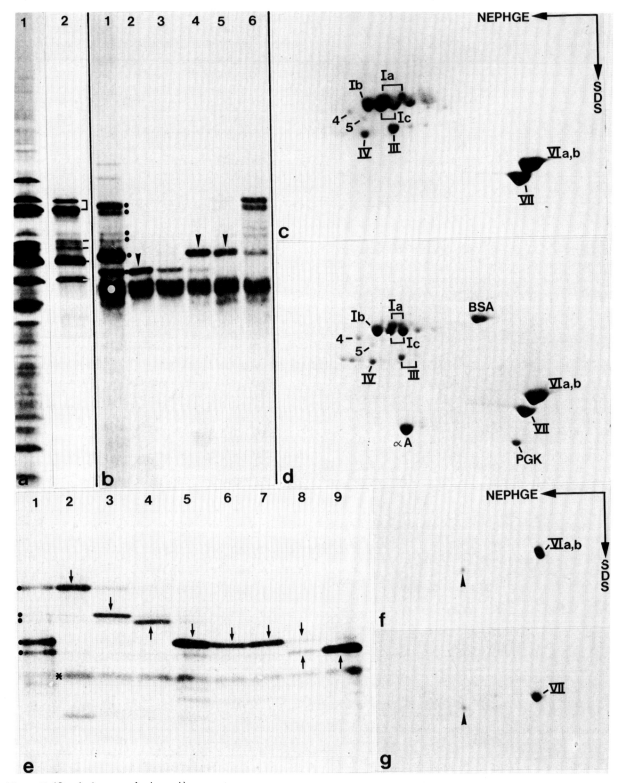

Figure 5 (See facing page for legend.)

sonable to consider the possible existence of special "cytokeratin-binding" molecules that might stabilize the unpolymerized form(s) and thus contribute to regulatory mechanisms of IF formation and rearrangement (for rearrangements of cytokeratin IFs, see also Franke et al. 1982b; Geiger et al., this volume).

Desmosome attachment and filament assembly

In most epithelial cells, numerous bundles of IFs of the cytokeratin type (tonofilaments) are found attached to the dense cytoplasmic plaques of a special type of intercellular junction, the desmosome (for ultrastructural aspects, see Campbell and Campbell 1971; Drochmans et al. 1978; Henderson and Weber 1981; Kelly and Kuda 1981). Recently, certain proteins have been identified by immunolocalization techniques as components of the desmosomal plaque, i.e., the structure that mediates the attachment of these IFs to a specific plasma membrane domain (Franke et al. 1981d, 1982c, 1983b,c; Cohen et al. 1983; Cowin and Garrod 1983; Mueller and Franke 1983). The major components of the desmosomal plaque of bovine epidermal cells are desmoplakins I (M_r 250,000) and II (M_r 215,000), together with two smaller polypeptides of M_r 83,000 and M_r 75,000 (for details, see Franke et al. 1981d, 1982c, 1983b,c; Cowin and Garrod 1983; Mueller and Franke 1983). Proteins of similar molecular weights that are immunologically related have been detected in desmosomal plaques of other tissues and species. Combined use of antibodies to such desmosomal plaque proteins and antibodies to cytokeratins in double-label immunofluorescence microscopy has allowed the demonstration of the specific arrays and interconnections of these two different structures in a certain type of cell or tissue (e.g., Fig. 6). In many cultured epithelial cells that express both cytokeratin and vimentin, IF cytokeratin filaments are associated with desmosomes, whereas vimentin IFs are often located in other parts of the cell (e.g., Franke et al. 1982d; Schmid et al. 1983). However, attachment to desmosomal plaques is not exclusive for cytokeratin IFs. In myocardiac tissue and Purkinje fiber cells of heart as well as in cultured myocardiac cells, desmin IFs are attached to desmosomes (Franke et al. 1982c; Kartenbeck et al. 1983; Tokuyasu et al. 1983), and in certain arachnoidal cells of the brain and in meningiomas derived therefrom, vimentin IFs are the only filaments anchored to the desmosomes (J. Kartenbeck et al., in prep.). This indicates that specific attachment to desmosomal plaques is a feature of different subclasses of IF and may be regulated by a common feature of primary or

Figure 5 Characterization of mRNAs coding for cytokeratins of bovine muzzle epidermis. (a) Immunoprecipitation of cytokeratins synthesized by translation in vivo. 0.5 μg of poly(A)$^+$ mRNA, isolated as described (Kreis et al. 1983; Magin et al. 1983), was translated in vitro using a commercially available reticulocyte lysate and L-[^{35}S]methionine as radioactive label. One half of the assay was immunoprecipitated with guinea pig antibodies directed against gel-electrophoretically-purified, cow muzzle keratins (cf. Franke et al. 1980b), following the protocol of Bravo et al. (1983). The immunoprecipitate (lane 2) and the other half of the total in vitro translation assay (lane 1) were analyzed by SDS-polyacrylamide (9%) gel electrophoresis and autoradiofluorographed. Note the specific enrichment of the major cytokeratins Ia–c (bracket), III, IV, VI, and VII (horizontal bars; cf. Schiller et al. 1982; Magin et al. 1983) in the immunoprecipitate. (b) For fractionation of bovine muzzle epidermis RNA, 100 μg of total RNA from cow muzzle epidermis were fractionated by preparative agarose (1.5%)–6 M urea gel electrophoresis as described in Franke et al. (1983b); one fifth of the RNA isolated from each fraction was translated in vitro, and the translation products were analyzed by electrophoresis as described above. (Lane 1) Translation products of total poly(A)$^+$ mRNA; (lanes 2–6) translation products obtained from RNA fractions of increasing apparent molecular weights. (●) Positions of the major keratin polypeptides of this tissue; (○) position of the artificially labeled protein endogenous to the lysate used; (▼) separation of cytokeratin VII (lane 2) and VI (lanes 4 and 5). For better separation of larger keratin polypeptides, see Franke et al. (1983b). (c and d) Specific recovery of keratin polypeptides synthesized in vitro by copolymerization. 0.5 μg of cow muzzle poly(A)$^+$ mRNA were translated in vitro (see above). The translation assay mixtures were diluted 10 times with distilled water, and the in vitro-synthesized keratins were allowed to copolymerize with 30 μg of cow muzzle keratin in partly protofilamentous form added in 1 mM Tris·HCl (pH 7.6) as described (Kreis et al. 1983; Magin et al. 1983). At the end of the incubation, the precipitated filaments were pelleted and analyzed by two-dimensional gel electrophoresis; the symbols and reference proteins added are as in Fig. 2. (c) Fluorography showing the [^{35}S]methionine-labeled keratins obtained during translation in vitro; Arabic and Roman numerals denote the different bovine epidermal keratin polypeptides (Schiller et al. 1982). (d) Coomassie blue staining of the same gel as in c; note the purity of the translation products harvested in this way and the electrophoretic identity of the keratin polypeptides in c and d. (e–g) Identification of cDNA clones coding for bovine epidermal keratins by hybridization-selection-translation. Double-stranded cDNA was synthesized from poly(A)$^+$ mRNA of cow muzzle epidermis and cloned in the PstI site of plasmid pUC8 by the dC-dG-tailing method (Roewekamp and Firtel 1980). Five hundred recombinants were screened using ^{32}P-labeled cow muzzle cDNA. Plasmid DNA (20 μg) isolated from strongly hybridizing colonies that contained cDNA inserts of about 1000 bp or more was bound to nitrocellulose filters and hybridized for 3 hr at 50°C against poly(A)$^+$ mRNA in the presence of 65% formamide in 0.4 M NaCl. At the end of the hybridization, the filters were washed 10 times at 63°C with 0.1 × SSC containing 0.2% SDS and 3 times with 0.1 × SSC at the same temperature (SSC is 0.15 M NaCl, 0.015 M Na-citrate). The hybridized RNA was eluted by eluting each filter individually in 0.3 ml of boiling water and precipitated with ethanol and was used in an in vitro translation assay (see above). The products synthesized in vitro were analyzed by one- or two-dimensional gel electrophoresis. (e) Fluorography of an SDS-polyacrylamide (9%) gel showing the in vitro translation products of RNAs selected by different clones (lanes 2–9). (Lane 1) In vitro translation products of total cow muzzle poly(A)$^+$ mRNA used for reference. (●) Positions of the cow muzzle keratins; (*) artificially labeled, endogenous component of the lysate; (→) clones specific for cytokeratin I (lane 2), III (lane 3), IV (lane 4), VI (three different clones have been used to select the mRNA analyzed in lanes 5–7), and VII (lane 9). The hybridization of the clone analyzed in lane 8, which shows some cross-hybridization between mRNAs coding for polypeptides VI and VII, is presently not explained. (f and g) Fluorographies presenting further identification of the keratin polypeptides defined by two of the clones shown in e, using two-dimensional gel electrophoresis (symbols are as in c). (➤) Position of an endogenous, labeled product used here for reference. Note the high specificity of the cDNA clones for either cytokeratin polypeptides VIb (f) or VII (g).

higher order structure. Therefore, our observations of a preferential association of cytokeratin IFs in cells that also contain IFs of the vimentin type may best be explained by a higher affinity of cytokeratins to desmosomal plaques.

It should also be emphasized that desmosomal plaques are not required for growth, proliferation, and monolayer formation of epithelial cells in culture nor for cytokeratin expression, as shown by certain cell lines that are devoid of desmosomes but not of cytokeratins (Zerban and Franke 1978; Venetianer et al. 1983). Moreover, microinjection experiments have shown that cytokeratin mRNAs can be translated in nonepithelial cells lacking desmosomes and that the newly synthesized cytokeratins gradually assemble into IF structures in the absence of desmosomes (Kreis et al. 1983; Geiger et al., this volume; for absence of desmosomes in bovine lens-forming cells, see Ramaekers et al. 1980 and Mueller and Franke 1983). Independence of cytokeratin expression and IF bundle formation from desmosomal structures and desmosomal proteins is also illustrated by the PtK_2 cell line derived from rat kangaroo kidney epithelium, which lacks desmosomes (Zerban and Franke 1978; Franke et al. 1982c) but forms extensive bundles of cytokeratin IFs. After injection into PtK_2 cells of mRNAs coding for epidermal keratins of cow muzzle, a tissue known for its abundance of desmosomes (Matoltsy 1975; Drochmans et al. 1978), both the endogenous "kidney-type" and the heterologous, introduced "epidermal-type" cytokeratins are integrated into the same fibril meshwork pattern, indicating that both types of cytokeratins assemble into normal IF arrays in the absence of desmosomes (Fig. 7).

To study the distribution of newly synthesized cytokeratin IFs in cells that, in contrast to PtK_2 cells, are functionally and morphologically polarized and do contain desmosomes, we have injected epidermal mRNAs coding for all bovine muzzle keratins into cultured bovine kidney epithelial cells of the MDBK line (for desmosome morphology and literature, see Franke et al. 1981d; Geiger et al. 1983), which normally express only two cytokeratin polypeptides (A and D, Franke et al. 1982b; see also Fig. 2c–l). As shown in Figure 8 the injected cells synthesize epidermal-type cytokeratins, as visualized by antibodies that specifically react only with the epidermal keratins. These "foreign" keratins are assembled into IF bundles that extend throughout the whole cytoplasm and approach the desmosomes. Double-label immunofluorescence microscopy again has shown that the newly synthesized "foreign" epidermal polypeptides

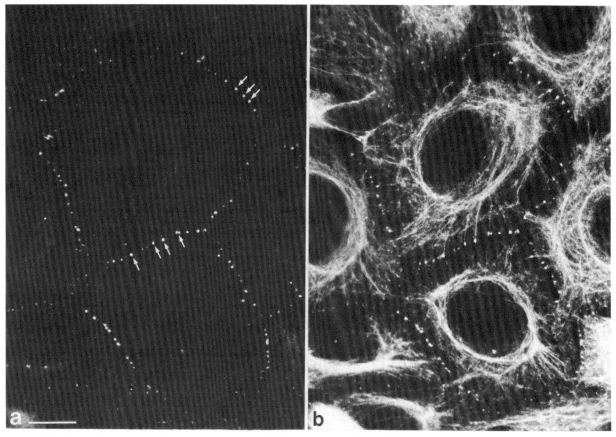

Figure 6 Double immunofluorescence microscopy of cultured bovine kidney epithelial (MDBK) cells using antibodies to desmosomal protein (a; antigen "DP1"; Franke et al. 1981d; Geiger et al. 1983) and antibodies to cytokeratins plus antibodies to desmosomal proteins (b). The arrowheads in a point to desmosomes, which are clearly associated with cytokeratin IF bundles (cf. b). Bar, 10 μm.

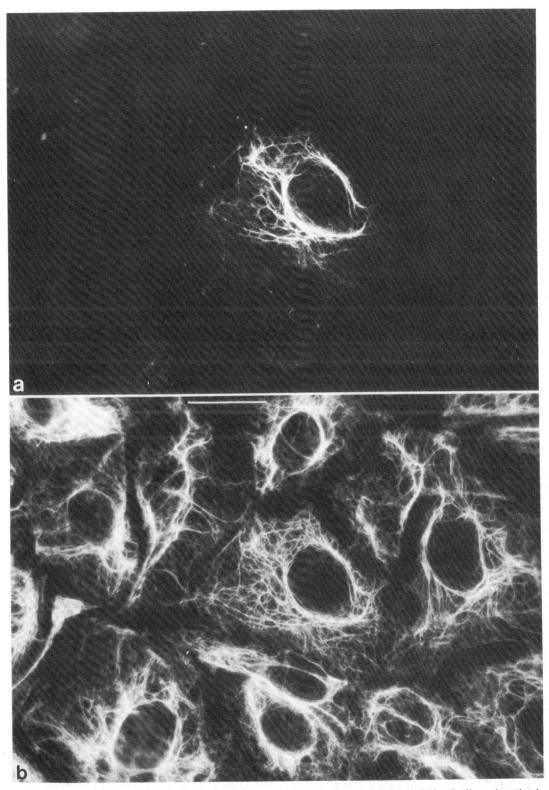

Figure 7 Double-label immunofluorescence microscopy showing localization of endogenous and "foreign" cytokeratins in cultured rat kangaroo kidney epithelial (PtK$_2$) cells, one of which has been injected with poly(A)$^+$ RNA isolated from bovine muzzle epidermis 15 hr before fixation (for conditions, see Kreis et al. 1983). (a) Rabbit antibodies of serum K999 raised against bovine epidermal keratins (kindly provided by Dr. S. Neumann, E. Merck, Darmstadt, FRG) react only with the epidermal keratin polypeptides synthesized in the injected cell (fluoresceine-labeled second antibodies as described by Kreis et al. 1983). (b) Murine monoclonal antibody K$_s$18.8 reacts with the endogenous cytokeratins of PtK$_2$ cells (rhodamine-labeled second antibodies as described by Gigi et al. 1982). Note assembly of epidermal keratins of cow into fibril arrays also containing rat kangaroo kidney–type cytokeratins of the PtK$_2$ cells, which lack desmosomes. Bar, 20 μm.

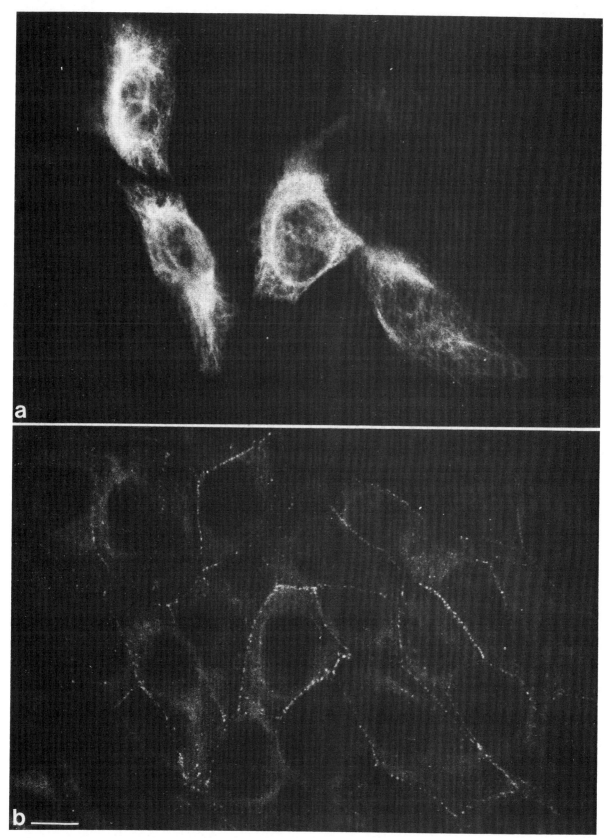

Figure 8 Double-label immunofluorescence microscopy showing synthesis of epidermal cytokeratins after injection of poly(A)$^+$ RNA from bovine muzzle epidermis into bovine kidney epithelial cells of the MDBK line. Four cells in a dense cell monolayer were injected 24 hr prior to fixation and processing for microsopy. (*a*) Injected cells synthesize epidermal keratins (visualized by staining with rabbit antibodies K999; see above), which are integrated into extended cytoplasmic fibrils. (*b*) Staining with guinea pig antibodies against desmoplakins I and II reveals typical dotted lines of desmosomal units along cell-to-cell boundaries. (For details of antibodies and fluoresceine-labeled second antibodies, see Franke et al. 1982c; Kreis et al. 1983; Mueller and Franke 1983.) Bar, 15 μm.

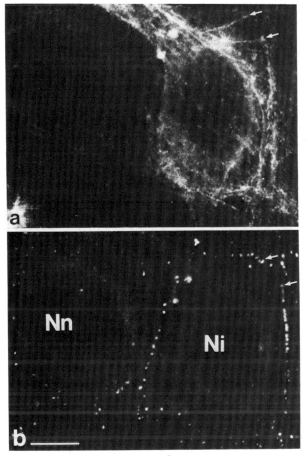

Figure 9 Higher magnification of a similar double-label experiment as described in the previous figure, showing that the epidermal keratins synthesized from the injected poly(A)$^+$ RNA, which are specifically visualized by rabbit antibodies in a, are incorporated into tonofilaments attached to desmosomes (→), which are visualized by the reaction with guinea pig antibodies to desmoplakins (b). (Nn) Nucleus of noninjected cell; (Ni) nucleus of injected cell. Bar, 10 μm.

integrate into the bundles of the endogenous "kidney-type" cytokeratin IFs but not into the endogenous vimentin IFs (data not shown). Detailed double-label immunofluorescence microscopy using antibodies reacting exclusively with the epidermal keratins synthesized after injection of mRNA in combination with antibodies to desmoplakins I and II further allows demonstration of the integration of the "foreign" cytokeratin molecules into cytokeratin IF bundles ("tonofibrils") anchoring at desmosomal plaques (Fig. 9). This suggests that tonofilaments can incorporate newly synthesized cytokeratin molecules and that different types of cytokeratin complexes can coassemble and contribute to the formation of the same tonofilament bundle, as also demonstrable by electron microscopic immunolocalization (results not shown).

We do not yet know which components of the desmosomal plaque interact with the specific IF proteins and what forces are involved in this attachment. The fact, however, that the desmosome-IF complex as a whole is so stable under various kinds of treatments, including low- and high-salt buffers as well as nondenaturing detergents, indicates that the interaction is a very strong one. Tight and stable association of IFs and desmosomes is also demonstrated by observations made during experimental dissociations of epithelial or myocardiac cells using Ca^{++}-chelating agents or proteases (Overton 1968; Kartenbeck et al. 1982, 1983). When desmosomes are split, the resulting half-desmosome equivalents are taken up into the cell by an endocytotic process, resulting in cytoplasmic vesicles that bear a "cap" of desmosomal plaque material with bundles of IF still attached. Clearly, mechanisms must exist that control desmosome formation as well as IF association at newly forming desmosomes or hemidesmosomes during embryogenesis, cell growth, and tissue regeneration as well as during tumor growth and formation of metastases. Hopefully, the recent indentification of the major proteins involved in this special membrane-cytoskeletal framework will provide an adequate tool to ask the right experimental questions and to elucidate the biological functions of these components and the regulation of their interaction.

Acknowledgments

We thank Drs. J. Kartenbeck (German Cancer Research Center, Heidelberg, FRG), R. Moll (Department of Pathology, University of Mainz, FRG), T.E. Kreis (European Molecular Biology Laboratory, Heidelberg), and B. Geiger (Weizmann Institute of Science, Rehovot, Israel) for valuable discussions. The work has been supported in part by grants from the Deutsche Forschungsgemeinschaft (DFG) and the Federal Ministry for Research and Technology (Bonn, FRG).

References

Bannasch, P., H. Zerban, E. Schmid, and W.W. Franke. 1980. Liver tumors distinguished by immunofluorescence microscopy with antibodies to proteins of intermediate-sized filaments. *Proc. Natl. Acad. Sci.* **77:** 4948.

Blikstad, I. and E. Lazarides. 1983. Vimentin filaments are assembled from a soluble precursor in avian erythroid cells. *J. Cell Biol.* **96:** 1803.

Bravo, R., S.J. Fey, P.M. Larsen, N. Coppard, and J.E. Celis. 1983. Proteins IEF (isoelectric focusing) 31 and IEF 46 are keratin-type components of the intermediate-sized filaments: Keratins of various human cultured epithelial cells. *J. Cell Biol.* **96:** 416.

Campbell, R.D. and J.H. Campbell. 1971. Origin and continuity of desmosomes. In *Origin and continuity of cell organelles* (ed. J. Reinert and H. Ursprung), p. 261. Springer-Verlag, Berlin.

Cohen, S.M., G. Gorbsky, and M.S. Steinberg. 1983. Immunochemical characterization of related families of glycoproteins in desmosomes. *J. Biol. Chem.* **258:** 2621.

Cowin, P. and D.R. Garrod. 1983. Antibodies to epithelial desmosomes show wide tissue and species cross-reactivity. *Nature* **302:** 148.

Crewther, W.G., A.S. Inglis, and N.M. McKern. 1978. Amino acid sequences of α-helical segments from S-carboxymethylkerateine-A. *Biochem. J.* **173:** 365.

Czosnek, H., D. Soifer, and H.M. Wisniewski. 1980. Studies on the biosynthesis of neurofilament proteins. *J. Cell Biol.* **85:** 726.

Dodemont, H.J., P. Soriano, W.J. Quax, F. Ramaekers, J.A. Lenstra, M.A.M. Groenen, G. Bernardi, and H. Bloemendal. 1982. The genes coding for the cytoskeletal proteins actin and vimentin in warm-blooded vertebrates. *EMBO J.* **1:** 167.

Doran, T.I., A. Vidrich, and T.-T. Sun. 1980. Intrinsic and extrinsic regulation of the differentiation of skin, corneal and esophageal epithelial cells. *Cell* **22:** 17.

Dowling, L.M., D.A.D. Parry, and L.G. Sparrow. 1983. Structural homology between hard α-keratin and the intermediate filament proteins desmin and vimentin. *Biosci. Rep.* **3:** 73.

Drochmans, P., C. Freudenstein, J.-C. Wanson, L. Laurent, T.W. Keenan, J. Stadler, R. Leloup, and W.W. Franke. 1978. Structure and biochemical composition of desmosomes and tonofilaments isolated from calf muzzle epidermis. *J. Cell Biol.* **79:** 427.

Franke, W.W., E. Schmid, M. Osborn, and K. Weber. 1978a. Different intermediate-sized filaments distinguished by immunofluorescence miscroscopy. *Proc. Natl. Acad. Sci.* **75:** 5034.

Franke, W.W., K. Weber, M. Osborn, E. Schmid, and C. Freudenstein. 1978b. Antibody to prekeratin. Decoration of tonofilament-like arrays in various cells of epithelial character. *Exp. Cell Res.* **116:** 429.

Franke, W.W., E. Schmid, K. Weber, and M. Osborn. 1979a. HeLa cells contain intermediate-sized filaments of the prekeratin type. *Exp. Cell Res.* **118:** 95.

Franke, W.W., E. Schmid, S. Winter, M. Osborn, and K. Weber. 1979b. Widespread occurrence of intermediate-sized filaments of the vimentin-type in cultured cells from diverse vertebrates. *Exp. Cell Res.* **123:** 25.

Franke, W.W., B. Appelhans, E. Schmid, C. Freudenstein, M. Osborn, and K. Weber. 1979c. Identification and characterization of epithelial cells in mammalian tissues by immunofluorescence microscopy using antibodies to prekeratin. *Differentiation* **15:** 7.

Franke, W.W., E. Schmid, J. Vandekerckhove, and K. Weber. 1980a. A permanently proliferating rat vascular smooth muscle cell with maintained expression of smooth muscle characteristics, including actin of the vascular smooth muscle type. *J. Cell Biol.* **87:** 594.

Franke, W.W., E. Schmid, C. Freudenstein, B. Appelhans, M. Osborn, K. Weber, and T.W. Keenan. 1980b. Intermediate-sized filaments of the prekeratin type in myoepithelial cells. *J. Cell Biol.* **84:** 633.

Franke, W.W., H. Denk, R. Kalt, and E. Schmid. 1981a. Biochemical and immunological identification of cytokeratin proteins in hepatocytes of mammalian liver tissue. *Exp. Cell Res.* **131:** 299.

Franke, W.W., D. Mayer, E. Schmid, H. Denk, and E. Borenfreund. 1981b. Differences of expression of cytoskeletal proteins in cultured rat hepatocytes and hepatoma cells. *Exp. Cell Res.* **134:** 345.

Franke, W.W., S. Winter, C. Grund, E. Schmid, D.L. Schiller, and E.-D. Jarasch. 1981c. Isolation and characterization of desmosome-associated tonofilaments from rat intestinal brush border. *J. Cell Biol.* **90:** 116.

Franke, W.W., E. Schmid, C. Grund, H. Müller, I. Engelbrecht, R. Moll, J. Stadler, and E.-D. Jarasch. 1981d. Antibodies to high molecular weight polypeptides of desmosomes: Specific localization of a class of junctional proteins in cells and tissues. *Differentiation* **20:** 217.

Franke, W.W., D.L. Schiller, R. Moll, S. Winter, E. Schmid, I. Engelbrecht, H. Denk, R. Krepler, and B. Platzer. 1981e. Diversity of cytokeratins: Differentiation specific expression of cytokeratin polypeptides in epithelial cells and tissues. *J. Mol. Biol.* **153:** 933.

Franke, W.W., D.L. Schiller, and C. Grund. 1982a. Protofilamentous and annular structures as intermediates during reconstitution of cytokeratin filaments in vitro. *Biol. Cell* **46:** 257.

Franke, W.W., E. Schmid, C. Grund, and B. Geiger. 1982b. Intermediate filament proteins in nonfilamentous structures: Transient disintegration and inclusion of subunit proteins in granular aggregates. *Cell* **30:** 103.

Franke, W.W., R. Moll, D.L. Schiller, E. Schmid, J. Kartenbeck, and H. Müller. 1982c. Desmoplakins of epithelial and myocardial desmosomes are immunologically and biochemically related. *Differentiation* **23:** 115.

Franke, W.W., E. Schmid, D.L. Schiller, S. Winter, E.-D. Jarasch, R. Moll, H. Denk, B.W. Jackson, and K. Illmensee. 1982d. Differentiation-related expression of proteins of intermediate-size filaments in tissues and cultured cells. *Cold Spring Harbor Symp. Quant. Biol.* **46:** 431.

Franke, W.W., D.L. Schiller, M. Hatzfeld, and S. Winter. 1983a. Protein complexes of intermediate-sized filaments: Melting of cytokeratin complexes in urea reveals different polypeptide separation characteristics. *Proc. Natl. Acad. Sci.* **80:** 7113.

Franke, W.W., H. Müller, S. Mittnacht, H.-P. Kapprell, and J.L. Jorcano. 1983b. Significance of two desmosome plaque-associated polypeptides of molecular weights 75,000 and 83,000. *EMBO J.* **2:** 2211.

Franke, W.W., R. Moll, H. Müller, E. Schmid, C. Kuhn, R. Krepler, U. Artlieb, and H. Denk. 1983c. Immunocytochemical identification of epithelium-derived human tumors using antibodies to desmosomal plaque proteins. *Proc. Natl. Acad. Sci.* **80:** 543.

Fuchs, E. and H. Green. 1979. Multiple keratins of cultured human epidermal cells are translated from different mRNA molecules. *Cell* **17:** 573.

———. 1980. Changes in keratin gene expression during terminal differentiation of the keratinocyte. *Cell* **19:** 1033.

———. 1981. Regulation of terminal differentiation of cultured human keratinocytes by vitamin A. *Cell* **25:** 617.

Fuchs, E.V., S.M. Coppock, H. Green, and D.W. Cleveland. 1981. Two distinct classes of keratin genes and their evolutionary significance. *Cell* **27:** 75.

Gabbiani, G., Y. Kapanci, P. Barazzone, and W.W. Franke. 1981. Immunochemical identification of intermediate-sized filaments in human neoplastic cells. *Am. J. Pathol.* **104:** 206.

Geiger, B., E. Schmid, and W.W. Franke. 1983. Spatial distribution of proteins specific for desmosomes and adhaerens junctions in epithelial cells demonstrated by double immunofluorescence microscopy. *Differentiation* **23:** 189.

Geisler, N. and K. Weber. 1980. Purification of smooth-muscle desmin and a protein-chemical comparison of desmins from chicken gizzard and hog stomach. *Eur. J. Biochem.* **111:** 425.

———. 1981a. Comparison of the proteins of two immunologically distinct intermediate-sized filaments by amino acid sequence analysis: Desmin and vimentin. *Proc. Natl. Acad. Sci.* **78:** 4120.

———. 1981b. Self-assembly in vitro of the 68,000 molecular weight component of the mammalian neurofilament triplet proteins into intermediate-sized filaments. *J. Mol. Biol.* **151:** 565.

Geisler, N., E. Kaufmann, and K. Weber. 1982. Protein-chemical characterization of three structurally distinct domains along the protofilament unit of desmin 10 nm filaments. *Cell* **30:** 277.

Geisler, N., E. Kaufmann, S. Fischer, U. Plessmann, and K. Weber. 1983. Neurofilament architecture combines structural principles of intermediate filaments with carboxy-terminal extensions increasing in size between triplet proteins. *EMBO J.* **2:** 1295.

Gibbs, P.E.M. and I.M. Freedberg. 1982. Epidermal keratin messenger RNAs. A heterogeneous family. *Biochim. Biophys. Acta* **696:** 124.

Gigi, O., B. Geiger, Z. Eshhar, R. Moll, E. Schmid, S. Winter, D.L. Schiller, and W.W. Franke. 1982. Detection of a cytokeratin determinant common to diverse epithelial cells by a broadly cross-reacting monoclonal antibody. *EMBO J.* **1:** 1429.

Gough, K.H., A.S. Inglis, and W.G. Crewther. 1978. Amino acids sequences of α-helical segments from S-carboxymethylkerateine-A. *Biochem. J.* **173:** 373.

Gruen, L.C. and E.F. Woods. 1983. Structural studies on the microfibrillar proteins of wool. *Biochem. J.* **209:** 587.

Hanukoglu, I. and E. Fuchs. 1982. The cDNA sequence of a human epidermal keratin: Divergence of sequence but conservation of structure among intermediate filament proteins. *Cell* **31:** 243.

———. 1983. The cDNA sequence of a type II cytoskeletal keratin reveals constant and variable structural domains among keratins. *Cell* **33:** 915.

Henderson, D. and K. Weber. 1981. Immuno-electron microscopical identification of the two types of intermediate filaments in established epithelial cells. *Exp. Cell Res.* **132:** 297.

Huiatt, T.W., R.M. Robson, N. Arakawa, and M.H. Stromer. 1980. Desmin from avian smooth muscle. *J. Biol. Chem.* **255:** 6981.

Kartenbeck, J., W.W. Franke, J.G. Moser, and U. Stoffels. 1983. Specific attachment of desmin filaments to desmosomal plaques in cardiac myocytes. *EMBO J.* **2:** 735.

Kartenbeck, J., E. Schmid, W.W. Franke, and B. Geiger. 1982. Different modes of internalization of proteins associated with adhaerens junctions and desmosomes: Experimental separation of lateral contacts induces endocytosis of desmosomal plaque material. *EMBO J.* **1:** 725.

Kelly, D.E. and A.M. Kuda. 1981. Traversing filaments in desmosomal and hemidesmosomal attachments: Freeze-fracture approaches toward their characterization. *Anat. Rec.* **199:** 1.

Kim, K.H., J.G. Rheinwald, and E.V. Fuchs. 1983. Tissue specificity of epithelial keratins: Differential expressions of mRNAs from two multigene families. *Mol. Cell. Biol.* **3:** 495.

Kreis, T.E., B. Geiger, E. Schmid, J.L. Jorcano, and W.W. Franke. 1983. De novo synthesis and specific assembly of keratin filaments in nonepithelial cells after microinjection of mRNA for epidermal keratin. *Cell* **32:** 1125.

Lazarides, E. 1982. Intermediate filaments: A chemically heterogeneous, developmentally regulated class of proteins. *Annu. Rev. Biochem.* **51:** 219.

Lee, L.D. and H.P. Baden. 1976. Organisation of the polypeptide chains in mammalian keratin. *Nature* **264:** 377.

Liem, R.K.H., C.H. Keith, J.F. Leterrier, E. Trenkner, and M.L. Shelanski. 1982. Chemistry and biology of neuronal and glial intermediate filaments. *Cold Spring Harbor Symp. Quant. Biol.* **46:** 341.

Magin, T.M., J.L. Jorcano, and W.W. Franke. 1983. Translational products of mRNAs coding for non-epidermal cytokeratins. *EMBO J.* **2:** 1387.

Matoltsy, A.G. 1975. Desmosomes, filaments and keratohyaline granules: Their role in the stabilization and keratinization of the epidermis. *J. Invest. Dermatol.* **65:** 127.

McTavish, C.F., W.J. Nelson, and P. Traub. 1983. Synthesis of vimentin in a reticulocyte cell-free system programmed by poly(A)-rich RNA from several cell lines and rat liver. *Eur. J. Biochem.* **130:** 211.

Milstone, L.M. 1981. Isolation and characterization of two polypeptides that form intermediate filaments in bovine esophageal epithelium. *J. Cell Biol.* **88:** 317.

Moll, R., R. Krepler, and W.W. Franke. 1983. Complex cytokeratin polypeptide patterns observed in certain human carcinomas. *Differentiation* **23:** 256.

Moll, R., W.W. Franke, B. Volc-Platzer, and R. Krepler. 1982a. Different keratin polypeptides in epidermis and other epithelia of human skin: A specific cytokeratin of molecular weight 46,000 in epithelia of the pilosebaceous tract and basal cell epitheliomas. *J. Cell Biol.* **95:** 285.

Moll, R., W.W. Franke, D.L. Schiller, B. Geiger, and R. Krepler. 1982b. The catalog of human cytokeratins: Patterns of expression in normal epithelia, tumors and cultured cells. *Cell* **31:** 11.

Moon, R.T. and E. Lazarides. 1983. Synthesis and post-translational assembly of intermediate filaments in avian erythroid cells: Vimentin assembly limits the rate of synemin assembly. *Proc. Natl. Acad. Sci.* **80:** 5495.

Mueller, H. and W.W. Franke. 1983. Biochemical and immunological characterization of desmoplakins I and II, the major polypeptides of the desmosomal plaque. *J. Mol. Biol.* **163:** 647.

O'Connor, C.M., D.J. Asai, C.N. Flytzanis, and E. Lazarides. 1981. In vitro translation of the intermediate filament proteins desmin and vimentin. *Mol. Cell. Biol.* **1:** 303.

Osborn, M. and K. Weber. 1983. Tumor diagnosis by intermediate filament typing: A novel tool for surgical pathology. *Lab. Invest.* **48:** 372.

Osborn, M., W.W. Franke, and K. Weber. 1980. Direct demonstration of the presence of two immunologically distinct intermediate-sized filament systems in the same cell by double immunofluorescence microscopy. *Exp. Cell Res.* **125:** 37.

Overton, J. 1968. The fate of desmosomes in trypsinized tissue. *J. Exp. Zool.* **168:** 203.

Quax, W., W.V. Egberts, W. Hendriks, Y. Quax-Jeuken, and H. Bloemendal. 1983. The vimentin gene: Complete structure and evolutionary aspects. *Cell* **35:** 215.

Quax-Jeuken, Y.E.F., W.J. Quax, and H. Bloemendal. 1983. Primary and secondary structure of hamster vimentin predicted from the nucleotide sequence. *Proc. Natl. Acad. Sci.* **80:** 3548.

Quinlan, R.A. and W.W. Franke. 1982. Heteropolymer filaments of vimentin and desmin in vascular smooth muscle tissue and cultured baby hamster kidney cells demonstrated by chemical crosslinking. *Proc. Natl. Acad. Sci.* **79:** 3452.

———. 1983. Molecular interactions in intermediate-sized filaments revealed by chemical cross-linking. *Eur. J. Biochem.* **132:** 477.

Ramaekers, F.C.S., M. Osborn, E. Schmid, K. Weber, H. Bloemendal, and W.W. Franke. 1980. Identification of the cytoskeletal proteins in lens-forming cells, a special epithelioid cell type. *Exp. Cell Res.* **127:** 309.

Renner, W., W.W. Franke, E. Schmid, N. Geisler, K. Weber, and E. Mandelkow. 1981. Reconstitution of intermediate-sized filaments from denatured monomeric vimentin. *J. Mol. Biol.* **149:** 285.

Roewekamp, W. and R.A. Firtel. 1980. Isolation of developmentally regulated genes from *Dictyostelium*. *Dev. Biol.* **79:** 409.

Roop, D.R., P. Hawley-Nelson, C.K. Cheng, and S.H. Yuspa. 1983. Keratin gene expression in mouse epidermis and cultured epidermal cells. *Proc. Natl. Acad. Sci.* **80:** 716.

Rueger, D.C., J.S. Huston, D. Dahl, and A. Bignami. 1979. Formation of 100 Å filaments from purified glial fibrillary acidic protein in vitro. *J. Mol. Biol.* **135:** 53.

Sahyoun, N., P. Stenbuck, H. LeVine III, D. Bronson, B. Moncharmont, C. Henderson, and P. Cuatrecasas. 1982. Formation and identification of cytoskeletal components from liver cytosolic precursors. *Proc. Natl. Acad. Sci.* **79:** 7341.

Schiller, D.L., W.W. Franke, and B. Geiger. 1982. A subfamily of relatively large and basic cytokeratin polypeptides as defined by peptide mapping is represented by one or several polypeptides in epithelial cells. *EMBO J.* **1:** 761.

Schlegel, R., S. Banks-Schlegel, J.A. McLeod, and G.S. Pinkus. 1980. Immunoperoxidase localization of keratin in human neoplasms. *Am. J. Pathol.* **101:** 41.

Schmid, E., D. Ghosal, and W.W. Franke. 1980. Biosynthesis of an intermediate filament protein, vimentin, in vivo and by translation in vitro. *Eur. J. Cell Biol.* **22:** 374.

Schmid, E., D.L. Schiller, C. Grund, J. Stadler, and W.W. Franke. 1983. Tissue type-specific expression of intermediate filament proteins in a cultured epithelial cell line from bovine mammary gland. *J. Cell Biol.* **96:** 37.

Schmid, E., S. Tapscott, G.S. Bennett, J. Croop, S.A. Fellini, H. Holtzer, and W.W. Franke. 1979. Differential location of different types of intermediate-sized filaments in various tissues of the chicken embryo. *Differentiation* **15:** 27.

Schweitzer, J. and K. Goerttler. 1980. In vitro synthesis of keratin polypeptides directed by mRNA isolated from newborn and adult mouse epidermis. *Eur. J. Biochem.* **132:** 477.

Sharp, G., M. Osborn, and K. Weber. 1982. Occurrence of two different intermediate filament proteins in the same filament in situ within a human glioma cell line. *Exp. Cell Res.* **141:** 385.

Small, J.V. and A. Sobieszek. 1977. Studies on the function and composition of the 10-nm (100-Å) filaments of vertebrate smooth muscle. *J. Cell Sci.* **23:** 243.

Steinert, P.M. 1978. Structure of the three-chain unit of the bovine epidermal keratin filament. *J. Mol. Biol.* **123:** 49.

Steinert, P.M., W.W. Idler, and R.D. Goldman. 1980. Intermediate filaments of baby hamster kidney (BHK-21) cells and bovine epidermal keratinocytes have similar ultrastructures and subunit domain structures. *Proc. Natl. Acad. Sci.* **77:** 4534.

Steinert, P.M., W.W. Idler, and S.B. Zimmerman. 1976. Self-assembly of bovine epidermal keratin filaments in vitro. *J. Mol. Biol.* **108:** 547.

Steinert, P., R. Zackroff, M. Aynardi-Whitman, and R.D. Goldman. 1982. Isolation and characterization of intermediate filaments. *Methods Cell Biol.* **24:** 399.

Steinert, P., W. Idler, M. Aynardi-Whitman, R. Zackroff, and R.D. Goldman. 1981a. Heterogeneity of intermediate filaments assembled in vitro. *Cold Spring Harbor Symp. Quant. Biol.* **46:** 465.

Steinert, P.M., W.W. Idler, F. Cabral, M.M. Gottesman, and R.D. Goldman. 1981b. In vitro assembly of homopolymer and copolymer filaments from intermediate filament subunits of muscle and fibroblastic cells. *Proc. Natl. Acad. Sci.* **78:** 3692.

Steinert, P.M., R.H. Rice, D.R. Roop, B.L. Trus, and A.C. Steven. 1983. Complete amino acid sequence of a mouse epidermal keratin subunit and implications for the structure of intermediate filaments. *Nature* **302:** 794.

Sun, T.-T. and H. Green. 1978. Immunofluorescent staining of keratin fibers in cultured cells. *Cell* **14:** 469.

Sun, T.-T., C. Shih, and H. Green. 1979. Keratin cytoskeletons in epithelial cells of internal organs. *Proc. Natl. Acad. Sci.* **76:** 2813.

Tokuyasu, K.T., A.H. Dutton, and S.J. Singer. 1983. Immunoelectron microscopic studies of desmin (skeletin) localization and intermediate filament organization in chicken cardiac muscle. *J. Cell Biol.* **96:** 1736.

Tseng, S.C.G., M.J. Jarvinen, W.G. Nelson, J.-W. Huang, J. Woodcock-Mitchell, and T.-T. Sun. 1982. Correlation of specific keratins with different types of epithelial differentiation epithelial tissues and cells. *Cell* **30:** 361.

Venetianer, A., D.L. Schiller, T. Magin, and W.W. Franke. 1983. Cessation of cytokeratin expression in a rat hepatoma cell line lacking differentiated functions. *Nature* **305:** 730.

Virtanen, I., V.-P. Lehto, E. Lehtonen, T. Vartio, S. Stenman, P. Kurki, O. Wager, J.V. Small, D. Dahl, and R.A. Badley. 1981. Expression of intermediate filaments in cultured cells *J. Cell Sci.* **50:** 45.

Weber, K. and N. Geisler. 1982. The structural relation between intermediate filament proteins in living cells and the α-keratins of sheep wool. *EMBO J.* **1:** 1155.

Winter, S., E.-D. Jarasch, E. Schmid, W.W. Franke, and H. Denk. 1980. Differences in polypeptide composition of cytokeratin filaments, including tonofilaments, from different epithelial tissues and cells. *Eur. J. Cell Biol.* **22:** 371.

Wu, Y.-J. and J.G. Rheinwald. 1981. A new small (40 kd) keratin filament protein made by some cultured human squamous cell carcinomas. *Cell* **25:** 627.

Zerban, H. and W.W. Franke. 1978. Modified desmosomes in cultured epithelial cells. *Cytobiologie* **18:** 360.

Conventional and Monoclonal Antibodies to Intermediate Filament Proteins in Human Tumor Diagnosis

M. Osborn, M. Altmannsberger,* E. Debus, and K. Weber

Max-Planck-Institute for Biophysical Chemistry, Goettingen, Federal Republic of Germany and
*Department of Pathology, University of Goettingen, Goettingen, Federal Republic of Germany

Pathological diagnosis is based to a large extent on the morphological appearance of the specimens. Usually the specimen is formaldehyde-fixed, treated with a stain such as hemotoxylin-eosin or May-Grünwald-Giemsa, and then examined under the light microscope. Although perhaps 90–95% of tumors are relatively easy to diagnose by such methods, the remaining 5–10% are often classified by describing them as pleomorphic, anaplastic, round-cell, spindle-shaped, or organoid, which are terms that give no hint as to their histogenetic origin. Further subdivision of this type of tumor usually depends on the experience of the pathologist in diagnosing tumors of a given type. The purpose of this article is to provide an overview of a method that may yield histogenetic information and therefore be of help in the classification of some of these mostly undifferentiated tumors. This approach has emerged from studies on the intermediate, or 10-nm, filaments, a component of the cytoskeleton of most animal cells.

Intermediate filaments (IFs) were positively identified in 1969 and can be easily distinguished from microfilaments by both electron microscopy and by immunofluorescence techniques. However, they are still often erroneously identified as microfilaments when detected in electron micrographs of, for instance, tumor cells. Immunological, biochemical, and subsequent protein sequence studies have shown that cells in tissues can be subdivided into six major classes due to their IF content. These are shown in Table 1. Thus both keratinizing and nonkeratinizing epithelia contain (cyto)keratin; most but not all neurons contain neurofilaments; astrocytes and a few other cells of glial origin contain glial fibrillary acidic protein (GFAP); myogenic derivatives (with the exception of certain vascular smooth-muscle cells) contain desmin; and cells of "mesenchymal origin" as well as a few other nonepithelial cell types contain vimentin. In addition, there is a sixth class of cells in which IFs cannot be found. The names (cyto)keratin, neurofilament, GFAP, desmin, and vimentin are thus the trivial names used to describe the major proteins that constitute the IFs present in the different cell types. The striking feature about Table 1 is that the subdivisions follow histological principles; i.e., they approximate well the major subdivisions made in textbooks of histology. That IFs should influence the structural appearance of the cytoplasm is not unexpected, given their abundance and insolubility. That the subdivisions should, however, agree well with those used in normal pathology was unexpected and suggested that the histogenic origin of cells in abnormal tissues and in tumors might be determined by typing the IFs present in such cells. Here we summarize studies on human tumor material that have shown that IF typing can distinguish the major classes of human tumors, either in biopsy specimens or in cytological samples obtained by fine-needle aspiration biopsies. Such studies have used both conventionally prepared and monoclonal antibodies specific for each of the IF polypeptides.

Isolation and characterization of monoclonal antibodies specific for each IF type

In certain instances monoclonal antibodies can offer advantages over their conventional counterparts. In prin-

Table 1 Types of Intermediate Filaments

Type	Protein	Examples in situ
Epithelial	(cyto)keratin multiple polypeptides (40K–68K)	keratinizing and nonkeratinizing epithelia
Neuronal	neurofilament (68K, 145K, and 220K)	most but probably not all neurons in central and peripheral nerves
Glial	GFAP (55K)	astrocytes, Bergmann glia (some also V^+)
Muscle	desmin (53K)	sarcomeric muscle; smooth muscle; vascular smooth-muscle cells can be D^+V^+, D^+V^-, or D^-V^+
Mesenchymal	vimentin (57K)	fibroblasts, chondrocytes, macrophages, endothelial cells, etc.
No IFs		certain cells of the early embryo; certain neurons

(V) Vimentin; (D) desmin.

ciple, they provide a readily exchangeable reagent of unlimited quantity where the epitope has to be defined only once. In addition, at least for those IF classes that contain more than one polypeptide, i.e., the neurofilaments and the cytokeratins, further subdivision of these classes may be easier with monoclonal rather than with conventional antibodies. For these reasons, in 1981 we started to develop a set of monoclonal antibodies, each member of which would recognize one and only one IF type. Our current collection includes monoclonal antibodies specific for each of the three neurofilament polypeptides (i.e., 68K, 160K, and 200K) as well as for GFAP, desmin, and vimentin (Debus et al. 1982a, 1983a,b; Osborn et al. 1984). In addition, we have isolated several cytokeratin monoclonal antibodies that specifically recognize a single cytokeratin (Debus et al. 1982b) (no. 18 in the nomenclature of Moll et al. 1982).

The strategy used to isolate and characterize these monoclonals has depended on the availability of purified IF polypeptides used in parallel protein chemical studies and on data accumulated from studies using conventionally prepared antibodies in which the IF content of cells in culture, or of cells in tissues, has been determined. In general, mice were immunized with purified IF proteins, although in the case of the cytokeratin monoclonals HeLa cell cytoskeletons were used. After fusion with mice myeloma lines, hybridomas were assayed both by immunochemical techniques such as ELISA or western blot analysis and by immunofluorescence techniques, both on cell lines and on tissues. Where possible, human cell lines and tissues were used since species-specific changes have been reported for IFs (e.g., Geisler and Weber 1981). In general, we tried to select for monoclonals against IF polypeptides that (1) recognize only a single polypeptide when tested in the immunoblotting procedure, (2) recognize IFs in immunofluorescence microscopy (e.g., on cell lines), (3) recognize IFs in as wide a range of species as possible, including man, and (4) recognize IFs in tissues after formaldehyde fixation. Although not all monoclonal antibodies fulfill criterion 4, the set of IF monoclonals shown in Figure 1 has now been well characterized. The specificities of these monoclonals appear in most instances to resemble those of IF antibodies prepared in a conventional manner, and thus these monoclonals appear to be useful reagents in their own right in human pathology (see below).

The specificity of each monoclonal antibody has been tested both in western blot analysis and by immunofluorescence on sections of normal human and animal material. Some representative results are illustrated in Figures 1–3. They show that monoclonals recognize in each case a single polypeptide chain of the correct molecular weight (Fig. 1)—i.e., for vimentin, 57K; for desmin, 53K; for GFAP, 50K; and for neurofilaments, either 68K, 160K, or 200K. The specificity of the cytokeratin monoclonals CK1–CK4 for cytokeratin 18 (45K, pI 5.7) is documented in Debus et al. (1984). In immunofluorescence microscopy, each monoclonal antibody recognizes cells of the type expected from previous studies with the conventionally produced antibodies (see Table 1). Thus, the vimentin monoclonals identify cells such as

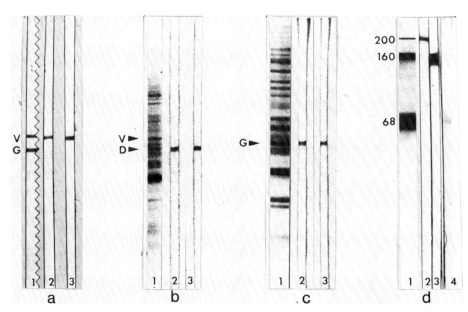

Figure 1 Western blots of monoclonal antibodies to different IF proteins. (a) Immunoblots of glioma cells (V$^+$GFAP$^+$); (lane 1) vimentin monoclonal V-1 and GFAP monoclonal G-K-7; (lane 2) vimentin monoclonal V-2; (lane 3) vimentin monoclonal V-3. (b) (Lane 1) Coomassie blue–stained gel of RD cell extract (V$^+$D$^+$); (lanes 2 and 3) immunoblot with desmin monoclonals DE-A-7 (lane 2) and DE-C-3 (lane 3). (c) (Lane 1) Glioma cell extract (V$^+$GFAP$^+$) stained with Coomassie blue; (lanes 2 and 3) immunoblot with GFAP monoclonals G-A-5 (lane 2) and G-C-6 (lane 3). (d) (Lane 1) Porcine neurofilament preparation enriched in 68K and 160K polypeptides; (lanes 2–4) immunoblots with neurofilament monoclonal antibodies NE14 (lane 2), which recognizes only the 200K polypeptide, NN18 (lane 3), which recognizes only the 160K polypeptide, and NR4 (lane 4), which recognizes only the 68K polypeptide. Note that all the monoclonal antibodies recognize only a single band in the immunoblot.

Figure 2 (a,b) Frozen sections of human skeletal muscle stained with vimentin monoclonal V-9 (a) or desmin monoclonal DE-1-1 (b). Note that only cells lying between the muscle fibers are stained with the vimentin monoclonal, while the desmin monoclonal stains Z-lines within the muscle fiber. (c,d) Frozen sections of motor cortex of human brain stained either with a GFAP monoclonal antibody G-G-4 (c) or with a neurofilament monoclonal NR4 specific for the 68K polypeptide (d). Note the typical morphology of the astrocytes in c, and the density of staining of neural processes in d. Magnifications: (a,b) 250×; (c,d) 400×.

fibroblasts and chondrocytes (Fig. 2a), the desmin monoclonals identify skeletal, visceral, and certain vascular smooth-muscle cells (Fig. 2b), the GFAP monoclonals identify astrocytes and Bergmann glia (Fig. 2c), and the neurofilament monoclonals identify neuronal cells. The cytokeratin monoclonals CK1–CK4 recognize simple epithelia as well as transitional epithelia, such as that of bladder, but do not recognize squamous epithelia.

Thus the CK1–CK4 series, in contrast to the broad specificity cytokeratin antibodies, allows a subdivision of epithelia.

Identification of epitopes

Progress has been made in identifying the epitopes of three of the monoclonal antibodies that recognize IF polypeptides. Such identification relies on the availability of defined fragments of one or the other IF protein, and on a comparison of the available sequence information for the different IF polypeptides (for summary, see Weber et al. 1984; Weber and Geisler, this volume). Thus the NF1 antibody, which is specific for the NF200K polypeptide (Debus et al. 1982a), recognizes the 150K carboxyterminal region, not the aminoterminal rodlike region (Geisler et al. 1983). Desmin monoclonals appear to recognize an epitope between residues 325–372. A different desmin monoclonal that recognizes not only desmin, but also other IF proteins, can be assigned to the region between residues 70–280 (for arguments, see Debus et al. 1983a).

IF typing in histology

Since the discovery that IF proteins are histogenetic markers of cell origin (Franke et al. 1982; Holtzer et al. 1982; Osborn et al. 1982b), a variety of normal and pathological material has been typed as to IF content. In general, in adult animals and in humans many differentiated cells in situ contain only one IF type (see Table 1). Cells that contain two IF types include (1) astrocytes, which may contain GFAP and vimentin, (2) vascular smooth-muscle cells, which may contain desmin and vimentin, and (3) horizontal cells of the retina, which contain neurofilaments and vimentin. Other examples of coexpression occur during embryogenesis. Thus, cells of parietal but not visceral endoderm have cytokeratin and vimentin (Lane et al. 1983), and developing chick neurons coexpress neurofilaments and vimentin (Holtzer et al. 1982). Notice in all the cases where two IF types are expressed that vimentin is one member of the pair. Of particular interest with respect to possible functions of IFs is the finding that certain cells, including those in the early embryo, lack IFs. The first IFs to be observed appear to be cytokeratin filaments in the trophoectoderm of the blastocyst.

What happens when pathological material, particularly human tumor material, is examined? Is the IF type typical of the cell of origin kept? Or are additional IF types acquired either in the primary or in metastases? Several thousand tumors have by now been examined in several different laboratories, including those of Gabbiani, Denk, Ramaekers, Virtanen, Franke, Moll, and Nagle. Our data obtained from frozen sections and from material that has been fixed with ethanol and embedded in paraffin is summarized in Tables 2, 3, 4, and 5 and has been published in part elsewhere (Altmannsberger et al. 1981a,b, 1982a,b, 1984a and in prep.; Osborn et al. 1982a). The IF antibodies used were raised in rabbits and affinity purified on their original antigens; more recently we have begun to use monoclonal antibodies either directly from the hybridoma supernatant fluids or after dilution of ascitic fluid. Table 2 lists those tumors shown to be cytokeratin-positive with a broad-specificity cytokeratin antibody raised in rabbits. Examination of a series of both breast and gastrointestinal tumors initially established that broad-specificity cytokeratin antibodies were excellent markers for carcinomas, identifying all carcinomas in material that was properly fixed. Interestingly, signet ring carcinomas of the stomach could be particularly easily identified by using such techniques (Altmannsberger et al. 1982a). It seemed possible to define two subgroups of nephroblastomas. In the first, blastema cells appeared to contain both cytokeratin and vimentin; the second group contained only vimentin. Table 2, as well as other studies (e.g., Nagle et al. 1983; Ramaekers et al. 1982), shows that a variety of other tumors of epithelial origin are also stained positively by broad-specificity cytokeratin antibodies but not by vimentin antibodies. Vimentin antibodies did, however, positively stain tumor cells in all nonmuscle sarcomas. These tumors were positive with vimentin but negative for other IF types (e.g., Fig. 3). Use of (cyto)keratin and vimentin antibodies to discriminate carcinomas from nonmuscle sarcomas has also been described by others, including Gabbiani et al. (1981).

One of our particular interests has been in the use of desmin antibodies to diagnose muscle sarcomas and, in particular, rhabdomyosarcomas (Altmannsberger et al. 1982a) (Table 4). Currently we have examined 30 specimens sent to us as possible rhabdomyosarcomas. Of these, 18 were desmin-positive. In approximately 70% of these cases, these tumors can be diagnosed as rhabdomyosarcoma by using the usual pathological criteria. The remaining 30%, although desmin-positive, showed a morphology typical of round-cell tumors. Evaluation of the 12 tumors in the second group that did not stain with desmin showed that 10 of these could be classified into other tumor groups by conventional pathological techniques. The 2 remaining desmin-negative cases were reevaluated by an expert because of the finding that they were negative with the desmin antibody. Both were then reclassified into other tumor groups because of this reevaluation. The use of desmin as a marker for rhabdomyosarcoma would be consistent with the finding that desmin expression is an early event in skeletal muscle differentiation. Thus, it is perhaps not surprising that we found that both the large rhabdomyoblasts and also the smaller cells, which are critical for the evaluation of rhabdomyosarcoma, express desmin. However, although the use of desmin as a marker for rhabdomyosarcoma has been confirmed in another study (Miettinen et al. 1982), other laboratories using different desmin antisera have not always been able to identify rhabdomyosarcomas reliably. Gabbiani et al. (1981) reported one case in which only the larger rhabdomyoblasts were positive, and Kahn et al. (1982) report that only 50% of their rhabdomyosarcomas were desmin-positive, a result possibly attributable to the different fixation methods used (M.J. Phillips, pers. comm.). The development of the desmin monoclonal antibodies, which also stain rhabdomyosarcomas of human and rat origin positively (Fig. 3b), should allow us to resolve these differences and to see whether our belief that desmin is currently the best available immunological marker for rhabdomyosarcomas can be further substantiated.

GFAP-positive tumors include astrocytomas and ependymoblastomas (Bignami et al. 1980). Neurofilament antibodies specific for one or more of the neuro-

Table 2 Cytokeratin-positive Tumors

Diagnosis	No. cases	CK	Vimentin	Desmin	GFAP	NF	Fixation
Squamous cell carcinoma	4	+	−	−	−	−	E
Well- and moderately differentiated adenocarcinoma stomach/large bowel	15	+	−	−	−	−	E
Poorly differentiated and undifferentiated adenocarcinoma stomach/large bowel	9	+	−	−	−	−	E
Ductal breast cancer	22	+	−	−	−	−	E
Intraductal breast cancer	5	+	−	−	−	−	E
Embryonal carcinoma testis	2	+	−	−	−	−	E
Teratocarcinoma	1	+	−	−	−	−	F
Yolk sac tumor	1	+	−	−	−	−	F
Mature teratoma	1	+	+[a]	+[b]	−	−	F
Papillary carcinoma thyroid	2	+	−	−	−	−	F
Follicular carcinoma thyroid	1	+	−	−	−	−	F
Hürthle adenoma thyroid	1	+	−	−	−	−	F
Nasopharyngeal carcinoma	6	+	−	−	−	−	F
Nephroblastoma:							
tubuli and blastema	5	+[c]	+[d]	(+)[b]	−	−[e]	F
blastema	2	−	+	−	−	−	E
Bone metastasizing renal tumor of childhood	1	−	+	−	−	−	

Data taken in part from Altmannsberger et al. (1981b, 1982a, 1984b). (F)Frozen section; (E) ethanol-fixed, paraffin-embedded.
[a]Mesenchymal part of tumor.
[b]Muscle differentiation of tumor in teratoma and in 2 nephroblastomas.
[c]Tubuli and blastema, cytokeratin-positive.
[d]Blastema, vimentin-positive; tubulin, vimentin-negative.
[e]A few neurofilament-positive processes were detected in 3 cases.

filament triplet proteins clearly identify some tumors originating from sympathetic neurons, e.g., pheochromocytoma (Fig. 3d), ganglioneuroblastoma, and at least some neuroblastomas (Osborn et al. 1982a; Osborn and Weber 1983) (Table 5). Whether those neuroblastomas that currently appear not to contain IFs are those that are least differentiated when conventional staging criteria are used is under study.

It should be stressed that IF typing of human tumor material has been developed independently in several laboratories, using different antibodies (e.g., Gabbiani et al. 1981; Miettinen et al. 1982; Ramaekers et al. 1982). Usually, all tumors of a given type display the same IF pattern. The agreement of results of the testing of similar tumors by different laboratories is striking (for review, see Osborn and Weber 1983). There are, however, still some minor disagreements, and these usually concern either (1) rare tumors (e.g., alveolar soft-part sarcoma) or (2) tumors with which it may be possible to use IF typing to further subdivide the group; e.g., bronchial carcinoids have been reported as neurofilament-positive (Altmannsberger et al. 1984a; Lehto et al.

Table 3 Vimentin-positive Tumors

	No. cases	CK	Vimentin	Desmin	GFAP	NF
Malignant histiocytosis	1	−	+	−	−	−
Non-Hodgkin's lymphoma	5	−	+	−	−	−
Giant lymph node hyperplasia (Castleman's disease)	1	−	+	−	−	−
Dermatofibrosarcoma protuberans	1	−	+	−	−	−
Monophasic synovial sarcoma	2	−	+	−	−	−
Malignant fibrous histiocytoma	3	−	+	−[a]	−	−
Pleomorphic liposarcoma	2	−	+	−	−	−
Malignant Schwannoma	2	−	+	−	−	−
Angiosarcoma	2	−	+	−	−	−
Infantile fibrosarcoma	1	−	+	−	−	−
Aggresive fibromatosis	2	−	+	−[a]	−	−
Intramuscular lipoma	1	−	+	−	−	−
Malignant fibrous mesothelioma	2	−	+	−	−	−
Kaposi's sarcoma[b]	1	−	+			
Ewing's sarcoma	4	−	+	−	−	−
Osteosarcoma	2	−	+	−	−	−
Chondrosarcoma	1	−	+	−	−	−

Data taken in part from Altmannsberger et al. (1982b).
[a]In one case, focal smooth-muscle metaplasia.
[b]Frozen sections; all other samples were ethanol-fixed.

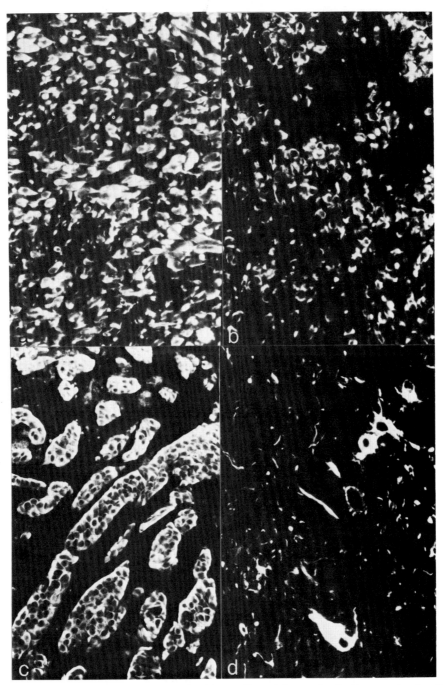

Figure 3 Human tumor material fixed in alcohol and embedded in paraffin and then stained with monoclonal antibodies. (a) Infantile fibrosarcoma stained with the vimentin monoclonal V-1; (b) rhabdomyosarcoma stained with the desmin monoclonal DE-I-1; (c) breast carcinoma stained with the cytokeratin monoclonal CK1; (d) pheochromocytoma stained with NF-1, a monoclonal specific for the 200K neurofilament polypeptide. Magnifications: (a) 400×; (b,d) 250×; (c) 160×.

1983), whereas carcinoids in another site appear to be (cyto)keratin-positive (Kahn et al. 1983). Perhaps the most important current disagreement concerns oat cell carcinomas. Six cases have been reported as neurofilament-positive and keratin-negative by Lehto et al. (1983), although our own data on 16 cases suggest that some of these tumors are clearly keratin-positive. The existence of such discrepancies underlines the necessity for well-characterized antibodies and for definition of the epitope where possible. The known amino acid sequences of IF proteins suggest that "surprises" such as common recognition of neurofilaments and only some of the 19 cytokeratins may occur with some antibodies. Thus, further careful study is required both of the antibodies and of additional cases before these tumors can be added to the list of those for which additional information as to histogenetic origin may be obtained by IF typing.

Table 4 Desmin-positive Tumors

	No. cases	CK	Vimentin	Desmin	GFAP	NF
Leiomyoma	2	−	+[a]	+	−	−
Leiomyosarcoma	2	−	+[a]	+	−	−
Embryonal rhabdomyosarcoma	15	−	+[a]	+	−	−
Alveolar rhabdomyosarcoma	2	−	+[a]	+	−	−
Sarcoma botryoides	1	−	+[a]	+	−	−

Data taken in part from Altmannsberger et al. (1982b). All samples were ethanol-fixed and paraffin-embedded.
[a]Tumor cells express desmin and some tumor cells coexpress vimentin.

Primaries and solid metastases appear to retain the IF type typical of the cell of origin. Ramaekers et al. (1983) have suggested that certain carcinomas growing as ascites may coexpress (cyto)keratin and vimentin.

Further subdivision of carcinomas, using monoclonal antibodies

Cytokeratins are the most complex class of IF proteins. At least 19 different cytokeratins have been identified on two-dimensional gels, and it has been shown that expression of those cytokeratins depends on the epithelium under study, as well as on the degree of cellular differentiation (Fuchs and Green 1978; Moll et al. 1982; Wu et al. 1983). An alternative method of studying cytokeratin complexity in different epithelia is to isolate and characterize cytokeratin antibodies that are specific either to a single cytokeratin or to subgroups of cytokeratins (Debus et al. 1982a; Lane 1982; Tseng et al. 1982). Thus, monoclonal antibodies CK1–CK4, which recognize cytokeratin 18, stain simple epithelia, as well as transitional epithelia of the bladder, but do not recognize stratified squamous epithelia from skin or esophagus (Debus et al. 1982b). CK1–CK4 have recently been used to subclassify carcinomas. Tumors originating from simple epithelia are strongly stained by CK1–CK4. Tumors such as squamous cell carcinoma of the skin, tongue, or esophagus derived from stratified squamous epithelia are not stained, although in two cases (squamous cell carcinoma of the epiglottis and of the cervix uteri) the least differentiated tumor cells are stained whereas the most differentiated are not (Debus et al. 1984). In contrast, the broad-specificity cytokeratin monoclonal antibody isolated by Gigi et al. (1982) stained all the carcinomas described above (Debus et al. 1984). This study, as well as studies using monoclonals to other cytokeratins in other laboratories (e.g., Gatter et al. 1982), thus suggests that cytokeratin typing may be an easy way to further subdivide carcinomas. Ideally, if it proves possible to isolate a sufficient number of cytokeratin monoclonals with different cell or tissue specificities, it may be possible to decide—for instance, from a lymph node metastasis—not only that the primary tumor is a carcinoma, but also in which epithelium the primary is located.

IF typing in cytology

Currently it is generally accepted that the difficulties in classifying tumor cells in cytology are greater than in histology because the positional information normally present in histological specimens is obviously lacking. This perhaps accounts for the fact that fine-needle aspiration biopsy "even now enjoys only recent and still selectively enthusiastic acceptance within the United States" (Frable 1983). However, the economic savings as well as advantages for the patient, in that the method needs no anesthesia, is fast and usually painless, and has little risk of causing bleeding or infection, also merit consideration.

We thus decided to see whether IF typing could also be applied to biopsies obtained by fine-needle aspiration or to imprints. A summary of our data is shown in Table 6, and an example is illustrated in Figure 4. Examination of Table 6 makes it again clear that the major human tumor groups can be separated using IF typing and that the results parallel those obtained by histology. Thus, carcinomas are cytokeratin-positive, nonmuscle sarcomas are vimentin-positive, muscle sarcomas are desmin-positive, and certain tumors originating from the sympathetic nervous system, as well as bronchus carcinoid, appear to respond positively to antisera to neurofilaments. The diagnoses given in Table 6 are based on conventional stains, usually of both cytological and histological samples. In most cases IF typing was also performed on a biopsy specimen (for further details, see Altmannsberger et al. 1984a).

The results show that tumor cells retain the IF-type characteristic of their cells of origin and do not acquire additional IF types. Certain problems of differential di-

Table 5 Neurofilament- and GFAP-positive Tumors

	No. cases	CK	Vimentin	Desmin	GFAP	NF
Neuroblastoma	12	−	−	−	−	6+,6−
Ganglioneuroblastoma	3	−	−	−	−	+
Pheochromocytoma	3	−	−	−	−	+
Malignant ependymoblastoma	1	−	−	−	+	−
Alveolar soft-part sarcoma	1	−	−	−	−	−

For further details, see Osborn et al. (1982a) and Osborn and Weber (1983).

Table 6 IF Typing in Aspiration Biopsy Cytology

Diagnosis	No. cases	CK	Vimentin	Desmin	GFAP	NF
Ductal breast carcinoma	3	+	−	−	−	−
Squamous cell carcinoma	1	+	−	−	−	−
Thymoma	1	+	−	−	−	−
Retroperitoneal teratocarcinoma	1	+	−	−	−	−
Papillary carcinoma of thyroid	10	+	(+)[a]	−	−	−
Follicular carcinoma of thyroid	3	+	(+)[a]	−	−	−
Hürthle cell tumor	1	+	(+)[a]	−	−	−
Pulmonary blastoma (biphasic):	1					
malignant epithelial part		+	−			
malignant mesenchymal part		−	+			
Ewing's sarcoma	2	−	+	−	−	−
Lymphoblastic lymphoma	2	−	+	−	−	−
Centroblastic lymphoma	1	−	+	−	−	−
Centrocytic/centroblastic lymphoma	1	−	+	−	−	−
Immunoblastic lymphoma	1	−	+	−	−	−
Hodgkin's disease	1	−	+	−	−	−
Plasmocytoma	1	−	+	−	−	−
Rhabdomyosarcoma (embryonal)	6	−	(+)[b]	+	−	−
Rhabdomyosarcoma (alveolar)	1	−	(+)[b]	+	−	−
Leiomyosarcoma	2	−	(+)[b]	+	−	−
Neuroblastoma (Hughes III)	1	−	−	−	−	+
Pheochromocytoma	1	−	−	−	−	+
Bronchus carcinoid	2	−	−	−	−	+

Data taken in part from Altmannsberger et al. 1984a.
[a]Tumor cells express keratin and most tumor cells coexpress vimentin.
[b]Tumor cells express desmin and some tumor cells coexpress vimentin.

agnosis can clearly be clarified by such methods (for further discussion, see Altmannsberger et al. 1984a). Examples include (1) mediastinal tumors, where it is important to distinguish malignant lymphoma from teratoma or thymoma, (2) retroperitoneal tumors, where it is important to know the tumor type before the operation, and (3) the small round-cell tumors in children (see below). Obviously IF typing of cytological specimens can be performed with monoclonal as well as with conventional antibodies. The combined data leave no doubt that aspiration cytology in combination with IF typing is not only a useful method for diagnosis and for preoperative planning of surgery, but can also increase the accuracy of the diagnosis.

Small round-cell tumors of children

This group of tumors include neuroblastoma, lymphoma, Ewing's sarcoma, Wilm's tumor, and rhabdomyosarcoma. Differential diagnosis of these tumors can be very difficult because, as the name implies, these tumors often share similar, relatively undifferentiated morphologies. However, as is clear from Tables 2–6, IF typing provides a relatively easy way to subgroup these tumors. Thus, rhabdomyosarcoma is desmin-positive, lymphoma and Ewing's sarcoma contain only vimentin, neuroblastoma appears to contain either neurofilaments or no IFs, and the blastema cells in Wilm's tumor contain either (cyto)keratin and vimentin or only vimentin. This discrimination is nontrivial since the clinical regimens to treat these tumors are different, as is the prognosis for several of these tumors.

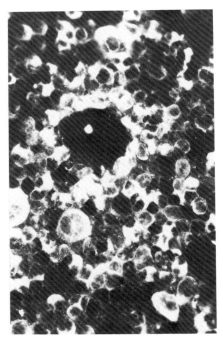

Figure 4 Touch imprint of morbus Hodgkin's cells stained with antibodies to vimentin. Note the positive staining of Hodgkin's cells and also of lymphoid cells. Magnification, 400×.

Conclusions

1. Tumor cells retain the IF-type characteristic of their cells of origin and thus far, at least in primaries and in solid metastases, appear not to acquire additional IF types.
2. In some cases where diagnosis is difficult by conventional techniques, IF typing obviously provides additional infor-

mation. Examples include (1) carcinoma vs. lymphoma testing, (2) examination of small round-cell tumors of children, and (3) the assessment of cytological specimens. In some instances, IF typing has already led to a revision or reconsideration of the original light-microscopic diagnosis.

3. Results with monoclonal antibodies thus far fully support the conclusions made with their conventional counterparts and suggest that these antibodies are also useful reagents in surgical pathology. It should be possible to identify the epitopes recognized by these antibodies. In addition, use of such antibodies may allow further subdivisions, particularly of carcinomas.

Acknowledgments

We thank Sabine Schiller and Susanne Isenberg for technical help and Professor Alfred Schauer for discussion. This work has been supported by the Max-Planck-Gesellschaft and the Deutsche Forschungsgemeinschaft.

References

Altmannsberger, M., M. Osborn, A. Schauer, and K. Weber. 1981a. Antibodies to different intermediate filament proteins: Cell type-specific markers on paraffin-embedded human tissues. *Lab. Invest.* **45:** 427.

Altmannsberger, M., M. Osborn, M. Droese, K. Weber, and A. Schauer. 1984a. Diagnostic value of intermediate filament antibodies in clinical cytology. *Klin. Wochenschr.* **62:** 114.

Altmannsberger, M., M. Osborn, A. Hölscher, A. Schauer, and K. Weber. 1981b. The distribution of keratin type intermediate filaments in human breast cancer: An immunohistological study. *Virchows Arch. B Cell Pathol.* **37:** 277.

Altmannsberger, M., M. Osborn, D. Schäfer, A. Schauer, and K. Weber. 1984b. Distinction of nephroblastomas from other childhood tumors using antibodies to intermediate filaments. *Virchows Arch. B Cell Pathol.* (in press).

Altmannsberger, M., K. Weber, A. Hölscher, A. Schauer, and M. Osborn. 1982a. Antibodies to intermediate filaments as diagnostic tools: Human gastrointestinal carcinomas express keratin. *Lab. Invest.* **46:** 520.

Altmannsberger, M., M. Osborn, J. Treuner, A. Hölscher, K. Weber, and A. Schauer. 1982b. Diagnosis of human childhood rhabdomyosarcoma by antibodies to desmin, the structural protein of muscle-specific intermediate filaments. *Virchows Arch. B Cell Pathol.* **39:** 203.

Bignami, A., D. Dahl, and D.C. Rueger. 1980. Glial fibrillary acidic protein (GFA) in normal neural cells and in pathological conditions. *Adv. Cell Neurobiol.* **1:** 285.

Debus, E., K. Weber, and M. Osborn. 1982a. Monoclonal cytokeratin antibodies that distinguish simple from stratified human epithelia: Characterization on human tissues. *EMBO J.* **1:** 1641.

———. 1983a. Monoclonal antibodies to desmin, the muscle-specific intermediate filament protein. *EMBO J.* **2:** 2305.

———. 1983b. Monoclonal antibodies specific for glial fibrillary acidic (GFA) protein and for each of the individual neurofilament triplet polypeptides. *Differentiation* **25:** 193.

Debus, E., G. Flügge, K. Weber, and M. Osborn. 1982b. A monoclonal antibody specific for the 200K polypeptide of the neurofilament triplet. *EMBO J.* **1:** 41.

Debus, E., R. Moll, W.W. Franke, K. Weber, and M. Osborn. 1984. Immunohistochemical distinction of human carcinomas by cytokeratin typing with monoclonal antibodies. *Am. J. Pathol.* **114:** 121.

Frable, W.J. 1983. Fine needle aspiration biopsy: A review. *Hum. Pathol.* **14:** 9.

Franke, W.W., E. Schmid, D.L. Schiller, S. Winter, E. Jarasch, R. Moll, H. Denk, B.W. Jackson, and K. Illmensee. 1982. Differentiation patterns of expression of proteins of intermediate-sized filaments in tissues and cultured cells. *Cold Spring Harbor Symp. Quant. Biol.* **46:** 413.

Fuchs, E. and H. Green. 1978. The expression of keratin genes in epidermis and cultured epithelial cells. *Cell* **15:** 887.

Gabbiani, G., Y. Kapanci, P. Barazzone, and W.W. Franke. 1981. Immunochemical identification of intermediate-sized filaments in human neoplastic cells. *Am. J. Pathol.* **104:** 206.

Gatter, K.C., Z. Abdulaziz, M. Mota, J.R.G. Nash, K. Pulford, H. Stein, J. Taylor-Papadimitriou, C. Woodhouse, and D.Y. Mason. 1982. Use of monoclonal antibodies for the histopathological diagnosis of human malignancy. *J. Clin. Pathol.* **35:** 1253.

Geisler, N. and K. Weber. 1981. Comparison of the proteins of two immunologically distinct intermediate sized filaments by amino acid sequence analysis: Desmin and vimentin. *Proc. Natl. Acad. Sci.* **78:** 4120.

———. 1982. The amino acid sequence of chicken muscle desmin provides a common structural model for intermediate filament proteins, including the wool α-keratins. *EMBO J.* **1:** 1649.

Geisler, N., E. Kaufmann, S. Fischer, U. Plessman, and K. Weber. 1983. Neurofilament architecture combines structural principles of intermediate filaments with carboxyterminal extensions increasing in size between triplet proteins. *EMBO J.* **2:** 1295.

Gigi, O., B. Geiger, Z. Eshkar, R. Moll, E. Schmid, S. Winter, D.L. Schiller, and W.W. Franke. 1982. Detection of a cytokeratin determinant common to diverse epithelial cells by a broadly crossreacting monoclonal antibody. *EMBO J.* **1:** 1429.

Holtzer, H., G.S. Bennett, S.J. Tapscott, J.M. Croop, and Y. Toyama. 1982. Intermediate-sized filaments: Changes in synthesis and distribution in cells in myogenic and non-myogenic lineages. *Cold Spring Harbor Symp. Quant. Biol.* **46:** 317.

Kahn, H.J., A. Garrido, S.-N. Huang, and R. Baumal. 1983. Demonstration of cytokeratin (CK) filaments in carcinoid tumors. *Lab. Invest.* **49:** 509.

Kahn, H.J., H. Yeger, O. Kassim, A.M. Jorgenson, D. McLennan, R. Baumal, and M.J. Phillips. 1982. Immunohistochemical and electron microscopic assessment of rhabdomyosarcomas: Increased accuracy of diagnosis over routine histological methods. *Lab. Invest.* **46:** 7P (Abstr.).

Lane, E.B. 1982. Monoclonal antibodies provide specific intramolecular markers for the study of epithelial tonofilament organization. *J. Cell Biol.* **92:** 665.

Lane, E.B., B.L.M. Hogan, M. Kurkinen, and J.I. Garrels. 1983. Coexpression of vimentin and cytokeratins in parietal endoderm cells of the early mouse embryo. *Nature* **303:** 701.

Lehto, V.-P., S. Stenman, M. Miettinen, D. Dahl, and I. Virtanen. 1983. Expression of a neural type of intermediate filament as a distinguishing feature between oat cell carcinomas and other lung cancers. *Am. J. Pathol.* **110:** 113.

Miettinen, M., V.-P. Lehto, R.A. Badley, and I. Virtanen. 1982. Expression of intermediate filaments of soft tissue sarcomas. *Int. J. Cancer* **29:** 541.

Moll, R., W.W. Franke, D.L. Schiller, B. Geiger, and R. Krepler. 1982. The catalogue of human cytokeratin polypeptides: Patterns of expression of cytokeratins in normal epithelia, tumors and cultured cells. *Cell* **31:** 11.

Nagle, R.B., K.M. McDaniel, V.A. Clark, and C.M. Payne. 1983. The use of antikeratin antibodies in the diagnosis of human neoplasms. *Am. J. Clin. Pathol.* **79:** 458.

Osborn, M. and K. Weber. 1983. Tumor diagnosis by intermediate filament typing. *Lab Invest.* **48:** 372.

Osborn, M., E. Debus, and K. Weber. 1984. Monoclonal antibodies specific for vimentin. *Eur. J. Cell Biol.* (in press).

Osborn, M., M. Altmannsberger, G. Shaw, A. Schauer, and K. Weber. 1982a. Various sympathetic derived human tumors differ in neurofilament expression. *Virchows Arch. B Cell Pathol.* **40:** 141.

Osborn, M., N. Geisler, G. Shaw, G. Sharp, and K. Weber. 1982b. Intermediate filaments. *Cold Spring Harbor Symp. Quant. Biol.* **46:** 413.

Ramaekers, F.C.S., D. Haag, A. Kant, O. Moesker, P.H.K. Jap, and G.P. Vooijs. 1983. Co-expression of keratin and vimentin-type intermediate filaments in human metastatic carcinoma cells. *Proc. Natl. Acad. Sci.* **80:** 2618.

Ramaekers, F.C.S., J.J.G. Puts, A. Kant, O. Moesker, P.H.K. Jap, and G.P. Vooijs. 1982. Use of antibodies to intermediate filaments in the characterization of human tumors. *Cold Spring Harbor Symp. Quant. Biol.* **46:** 331.

Tseng, S.C.G., M.J. Jarvinen, W.G. Nelson, J.W. Huang, J. Woodcock-Mitchell, and T.T. Sun. 1982. Correlation of specific keratins with different types of epithelial differentiation: Monoclonal antibody studies. *Cell* **30:** 361.

Weber, K., G. Shaw, M. Osborn, E. Debus, and N. Geisler. 1984. Neurofilaments, a subclass of intermediate filaments: Structure and expression. *Cold Spring Harbor Symp. Quant. Biol.* **48:** 717.

Wu, Y.J., L.M. Parker, N.E. Binder, M.A. Beckett, J.H. Sinard, C.T. Griffiths, and J.G. Rheinwald. 1983. The mesothelial keratins: A new family of cytoskeletal proteins identified in cultured mesothelial cells and nonkeratinizing epithelia. *Cell* **31:** 693.

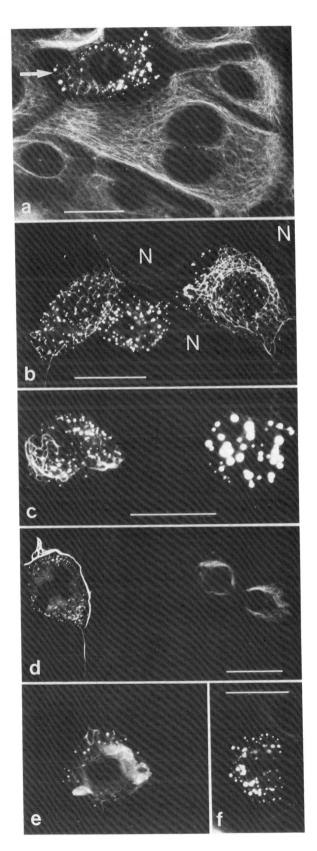

Figure 2 Immunofluorescent micrographs of MDBK cells in mitoses, using different guinea pig antibodies against cytokeratin, reacting either with (i) both the cytokeratin IFs of interphase cells and the altered cytokeratin arrays, including spheroidal aggregates of mitotic cells (a) or with (ii) the perimitotic cytokeratin configurations alone (b–f). (a) Fluorescent fibrils are seen in interphase cells but not in a mitotic cell (→), which contains fluorescent "dots" (antibodies of type i above). (b) With antibodies of type ii, interphase cells are unstained but mitotic cells of various stages show positive fluorescence in fibrillar (cell at right) or in dotted patterns. The number of mitotic cells in this culture has been increased by treatment with colcemid (for details, see Franke et al. 1982a). (N) Nuclei. (c) Same antibody as used in b, showing the variability in sizes of the mitotic cytokeratin aggregates ("dots") in two adjacent mitoses. (d) Two different mitotic stages stained with the same antibodies as in b, showing positive staining of both residual fibrils and small dots in a metaphase cell (left) and persistence of positive staining on IF bundles present in early postmitotic cells (two daughter cells are shown at right), as opposed to absence of staining in the cytokeratin fibrils present in all the adjacent interphase cells. (e,f) Same antibodies as in b, showing two different aspects of the fibril-to-ball transition in rounded-off anaphase cells. The mitotic cell shown in e still reveals some fibril aggregates, together with dots, whereas the mitotic cell shown in f contains exclusively fluorescent dots. This indicates that the extent of fibril-to-ball transformation can differ in different cells. Bars, 25 μm.

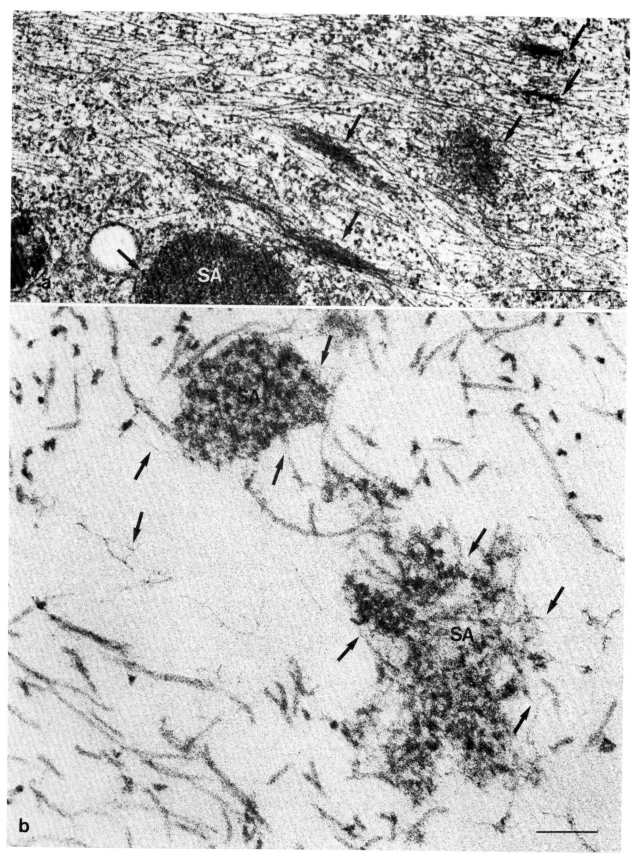

Figure 3 (See facing page for legend.)

rial, which often are associated with residual IF bundles and seem to form on such IF bundles (Fig. 3a). Regardless of the angle of sectioning or section thickness, typical IFs of 7-11-nm diameter are not detected in central positions of these spheres, but tangles of fine (2-4-nm) filaments are occasionally detected within these mitotic balls. Occasionally, we have found short fragments of IFs that seem to unravel and fray at their ends into several distinct filaments of 2-4-nm diameter, suggestive of a protofilamentous nature. From such observations we tend to conclude that the structural rearrangement of cytokeratin involves unwinding of IFs into the constituting protofilaments or protofilamentlike structures, which then aggregate into a different spheroidal mass, probably mediated by factors active only during mitosis (Franke et al. 1982a).

Transient mitotic rearrangment of cytokeratin IFs into spheroidal masses is not exclusive to certain cultured cells but can also take place in mitoses of certain cells growing in the body. Figure 4, a-e, presents the example of cells of a primary tumor of a human anorectal carcinoma, showing regions of IF bundle-ball associations and also the involvement of desmosome-attached tonofibrils in this transition process (Fig. 4d; for cultured cells, see Fig. 5c,d of Franke et al. 1982a).

Reorganization of cytokeratin during mitosis: Changes in the organization of IFs as expressed by selective masking and unmasking of specific antigenic determinants

For the experiments involving immunofluorescent localization of cytokeratin in mitotic cells, we have used a large library of antibodies, including some obtained conventionally (prepared in rabbits or in guinea pigs) but also some monoclonal antibodies (prepared in mice). In these experiments we have often found antibodies that selectively react with certain cells of the same culture but not with others. Similar observations have been reported by Lane and Klymkowsky (1982). Initially we have attributed this variability to differences of expression of cytokeratin polypeptides in certain subpopulations in these cell cultures (for differences of cytokeratin patterns in cell lines, see Franke et al. 1981a,b; Wu and Rheinwald 1981; Moll et al. 1982b; Schiller et al. 1982). However, in many cases we have found that the absence of immunofluorescent labeling with the antibodies results from masking of the epitope (antigenic site) rather than from the absence of the protein. A very common observation is that mentioned above, i.e., staining of cytokeratin in mitotic balls but not in IFs of mid-interphase cells. This finding has not been too surprising in view of the "IF-unraveling" hypothesis presented above since epitopes on the surfaces of cytokeratin molecules facing the interior of the IFs would not be reactive unless the filament unfolds into its protofilament constituents.

More surprising, however, are the results with cells that retain their meshwork of cytokeratin and vimentin IFs during mitosis, such as PtK_2 cells (Franke et al. 1978a; Aubin et al. 1980; Horwitz et al. 1981; for simultaneous synthesis of vimentin and cytokeratins, see Franke et al. 1978b; Osborn et al. 1980). Screening these cells with different monoclonal antibodies (mAb) has shown that a hybridoma clone $K_G 8.13$ (Gigi et al. 1982), which is not reactive with IFs of PtK_2 during interphase, is highly reactive with IFs of mitotic cells of the same line (Fig. 5a,b). It should be emphasized that, unlike the mitotic changes described above, the cytokeratin in mitotic PtK_2 cells is still associated with typical 7-11-nm filaments as defined by electron microscopy and by their resistance toward extraction (Franke et al. 1978a). This could be explained either by mitosis-specific changes in the cytokeratin composition or by structural rearrangements within the filaments. The first possibility appears rather unlikely since mitotic cells contain the same two polypeptides as interphase cells, in largely the same proportions, and also do not show considerable differences of phosphorylation, as demonstrated by two-dimensional gel electrophoresis (B. Geiger et al., unpubl.).

To substantiate the masking-unmasking hypothesis, we have exposed coverslip-attached PtK_2 cells to different conditions of fixation and labeling, aiming at unmasking the epitope recognized by $K_G 8.13$. Figure 6, a-d, illustrates the results of several such treatments. As shown, exposure to detergent and organic solvents at room temperature, or short treatment with trypsin or with low (2 M) urea concentrations, provides effective unmasking. These and additional experiments (cf. Franke et al. 1983b) have provided some insight into the dynamic rearrangement of cytokeratin IFs in living cells and have allowed the following conclusions:

1. Living epithelial cells contain physiological cell-cycle–dependent factors that can induce changes in the organization of cytokeratin IFs.
2. The initial step in this process seems to affect the fine disposition and interaction of cytokeratin polypeptides or their protofilaments, followed by complete or partial unwinding of the IFs. Interestingly, in early stages of mitosis (early prophase) in MDBK cells, the assembled IFs are already reactive with such antibodies prior to the formation of the mitotic balls, which is in line with this conclusion.

Figure 3 Electron micrographs of sections through the cytoplasm of a mitotic MDBK cell (a) and cytoskeletons from MDBK cell cultures enriched in mitoses, isolated by extraction with high-salt buffer and Triton X-100 (b) (for details, see Franke et al. 1982a). (a) Note the coexistence of typical intermediate-sized filaments, spheroidal cytokeratin aggregates (SA; one is denoted by arrow at lower left), and numerous densities probably representing localized regions of early stages of fibril-to-aggregate transitions (arrows at right). (b) Details of the close association of IFs and spheroidal aggregates (SA) of cytokeratin material are revealed at higher magnification in isolated cytoskeletons. Note numerous filaments of 2-4-nm diameter, suggestive of unraveling of IFs into protofilamentous subunits (some are denoted by arrows). Bars: (a) 0.5 μm, (b) 0.1 μm.

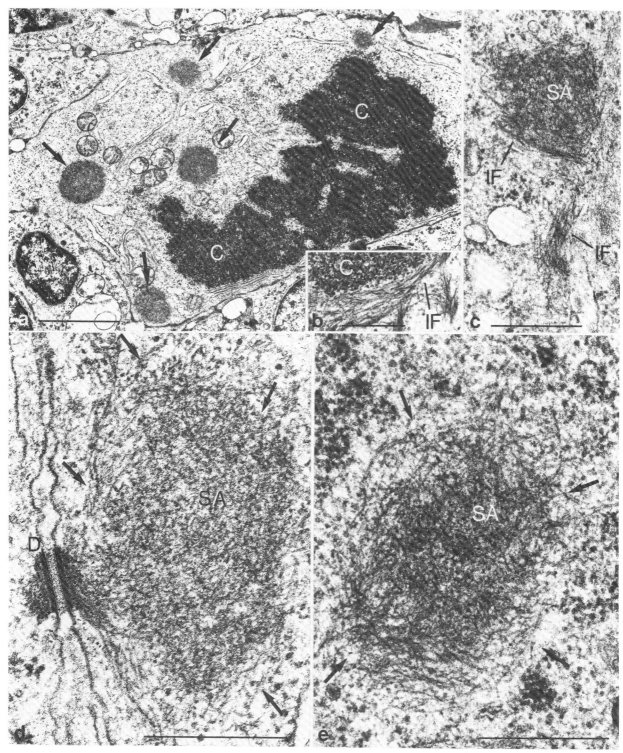

Figure 4 Electron micrographs showing spheroidal aggregates (SA) of cytokeratin material in mitotic cells of a solid human carcinoma (highly differentiated, transmurally growing, anorectal primary tumor of a 58-year-old woman). Fixation and other procedures are as described for another anorectal tumor by Moll et al. (1982a). (*a*) Survey micrograph showing occurrence of numerous spheroidal aggregates (→) in a mitotic cell. (C) Chromosomes. (*b*) Persistence of some bundles of IFs in a mitotic tumor cell, which contains spheroidal aggregates in other regions. Note close vicinity of IFs and chromosomes (C). (*c*) Note close vicinity of IFs and SA, sometimes suggesting direct transitions. (*d*) Desmosome(D)-attached SA in a mitotic cell, showing the association of the mitotically transformed cytokeratin material with residual tonofilaments and the desmosomal plaque, respectively. Filamentous structures are also revealed in the periphery of the SA (→). (*f*) SA body showing composition by granular and by nonordered filamentous structures, most of them thinner than IFs (→). Bars: (*a*) 2 μm, (*b–e*) 0.5 μm.

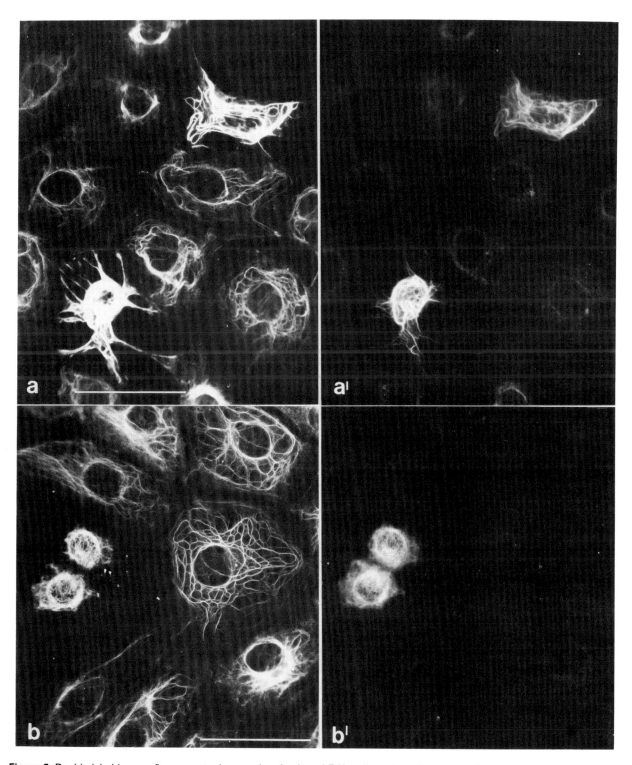

Figure 5 Double-label immunofluorescent micrographs of cultured PtK$_2$ cells, using guinea pig antibodies against cytokeratin (a,b) and murine monoclonal antibody K$_G$8.13 (a',b'; details have been presented by Franke et al. 1983b). Note that antibody K$_G$8.13 reacts only with cytokeratin IFs of mitotic cells (a') and early postmitotic interphase (b') cells but is negative on IFs of other interphase stages. Bars, 50 μm.

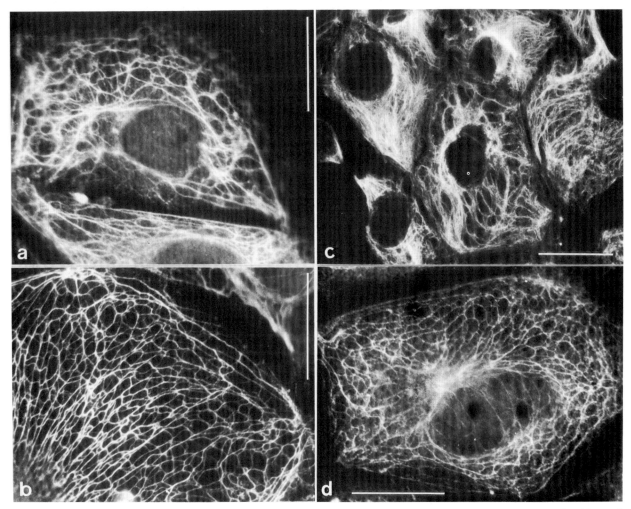

Figure 6 Immunofluorescent micrographs of PtK$_2$ cells with monoclonal antibody K$_G$8.13 after brief treatment of the cells with trypsin (a,b), Triton X-100 followed by methanol and acetone (c), or 2 M urea (d), prior to application of antibodies. Details of the various procedures have been presented elsewhere (Franke et al. 1983b). Bars: (a,c,d) 30 μm, (b) 20 μm.

3. These changes are reversible, but the typical interphase organization of IFs is reestablished, not immediately in telophase, but often into the early stages of interphase.
4. The molecular nature of these changes in IF structure, which are expressed by unmasking, is not clear. They may involve the selective removal of peripheral "masking molecules" attached to the surface of the IFs during interphase or reorientation of cytokeratin polypeptides within the filaments. It is conceivable that changes of either type could lead to subsequent alterations in filament stability.

The marked differences between the rearrangement found in PtK$_2$ cells on the one hand and in MDBK and HeLa cells on the other hand may only be a quantitative difference leading to limited and localized changes in IF organization in the former and more extensive reorganization in the latter cell types. In both cases the observed rearrangements seem to be of functional significance in as much as they are associated with the radical changes of cell shape and organelle distribution that take place during cell division.

Our observations also clearly demonstrate that antigenic determinants present on cytokeratin polypeptides may not always be exposed and available for immunolabeling. This seems to be a general phenomenon that we have encountered during screening of different monoclonal antibodies on sections through frozen tissues and/or on cultured cells (O.G. et al., unpubl.). Recent results of T.-T. Sun and colleagues (Woodcock-Mitchell et al. 1982) on the selective exposure of different cytokeratin epitopes in different regions of the epidermis also point to the same phenomenon. In view of the widespread use of cytokeratin antibodies for diagnostic purposes (see Schlegel et al. 1980; Gabbiani et al. 1981; Osborn et al. 1982), it seems to be important that negative results in immunocytochemical labeling of cells or tissues should be carefully interpreted and confirmed by other approaches (electrophoretic separation, immunoblotting analysis, etc.).

Moreover, dynamic changes of epitope exposure do not seem to be restricted to IFs of the cytokeratin type since changes of epitope exposure have recently also been observed in desmin IFs during myogenesis (Lazarides et al. 1982) and in vimentin IFs in relation to cell growth and density in vitro (Dulbecco et al. 1983).

De novo synthesis of cytokeratin polypeptides and their assembly into IFs in living cells

Another phase during which cytokeratin must be expected to go through a dynamic structural transition is IF assembly, i.e., from the time of synthesis through the formation of oligomeric polymerizing subunits up to the assembly of IFs or IF bundles. One approach for the characterization of filament assembly processes, which was successfully used for the study of microfilament and microtubule dynamics, is the microinjection of fluorescently labeled proteins into living cells (for review, see Kreis and Birchmeier 1982). Unfortunately, this approach is not applicable to IFs due to their very low solubility in buffers of physiological or near-physiological ionic strength. We have therefore decided to study the assembly of IFs of the cytokeratin type through the injection of bovine epidermal mRNA into living nonepithelial and epithelial cells. The technical details of these experiments have been described elsewhere (Kreis et al. 1983).

Epidermal poly(A)$^+$ RNA extracted from bovine snout contains mRNA species coding for all the major and minor epidermal cytokeratins found in the tissue, as demonstrated by translation in vitro (Schiller et al. 1982; Kreis et al. 1983; Magin et al. 1983). Initially, solutions of poly(A)$^+$ mRNA from this tissue have been injected with glass capillaries into cultured nonepithelial cells such as bovine lens-forming cells (LFC; Ramaekers et al. 1980), SV40-transformed human fibroblasts ("SV40 fibroblasts"; cf. Franke et al. 1979a), rat mammary carcinoma–derived fibroblasts (RMCD cells; Rathke et al. 1975), and rat vascular smooth-muscle–derived cells (RVF-SMC; Franke et al. 1980). All these cells contain IF only of the vimentin type. At different intervals after injection, the cells have been fixed and immunolabeled for cytokeratin. Relatively early after injection (e.g., 1 hr and less), positively labeled material is detected throughout the cytoplasm of the injected cells (Kreis et al. 1983). Initially, only small cytokeratin-containing granules or "flakes" are detected in the cytoplasm (Fig. 7a). Subsequently, with increasing amounts of cytokeratin translated from the injected mRNAs, longer arrays of filaments are formed that, after 12 hours or more, often develop into a dense and elaborate network of fibrils similar to those found in many cultured epithelial cells (Fig. 7b). Electron microscopic immunolocalization has shown that the keratin molecules synthesized from the injected mRNA indeed do form IFs (Kreis et al. 1983). The fibril patterns of the newly synthesized and assembled cytokeratin IFs in these cells are different from the patterns of the endogenous vimentin IFs, microtubules, and actin microfilaments (cf. Kreis et al. 1983). Similar results have been obtained with different mesenchymal cells, regardless of whether the mRNA and the recipient cells have been from the same species (bovine) or have been used in heterologous combinations (e.g., human or rat cells; Fig. 8a,b).

These observations provide some insight into the specificity of cytokeratin assembly in a cytoplasmic environment and its exclusive interaction with other cytokeratin molecules to form a normal-looking fibrillar cytoskeleton. This is, however, an "artificial" system since here cytokeratins are assembled in a cell that does not express its genes for cytokeratins and possibly is also devoid of other proteins involved in cytokeratin IF regulation, whereas in epithelial cells newly formed cytokeratin encounters the preexisting network of cytokeratin filaments. To study the assembly of cytokeratin in epithelial cells, we have injected the same bovine epidermal keratin mRNA into cultured epithelial cells derived from kidney, which express different (i.e., nonepidermal) types of cytokeratins of the same species (MDBK cells; for cytokeratin pattern, see Franke et al. 1981b, 1982a,b) or of a different species (PtK$_2$ cells from rat kangaroo; cf. Franke et al. 1981b, 1982b). At different intervals after injection, the cells have been fixed and double-labeled for cytokeratins, using two types of antibodies coupled to different fluorochromes. One type of antibody reacts with epidermal keratins only, whereas the other recognizes exclusively nonepidermal cytokeratins. The results shown in Figure 9 indicate that where the host cells contain endogenous cytokeratin IFs, the newly formed cytokeratin polypeptides of a different type encoded by the injected epidermal mRNA readily assemble into the endogenous fibril network. Initially, the newly synthesized polypeptides are nonuniformly distributed over the endogenous fibril meshwork, but it takes long incubations until nearly complete coincidence is achieved. Sometimes, short fibril regions can be seen that are stained only by the antibodies to the epidermal-type cytokeratins but not by those reacting with the endogenous ones (Fig. 10a). Whether this is due, however, to the selective assembly of certain fibril portions by the "new" type of keratin only or to the masking of "old" IFs by their interaction with new cytokeratin molecules cannot be decided. By immunoelectron microscopy we have also shown that the newly synthesized epidermal-type cytokeratin is associated with typical IF structures that are resistant to extractions with detergents and high-salt buffers (data not shown). This interaction of the newly synthesized cytokeratin with preexisting cytokeratin IFs regardless of the species differences and tissue diversity of expression of polypeptides of the cytokeratin family is in marked contrast to the independence of these IFs from vimentin and other cytoplasmic filaments in the same cells (Fig. 10b; for differences of vimentin and cytokeratin IF distribution in

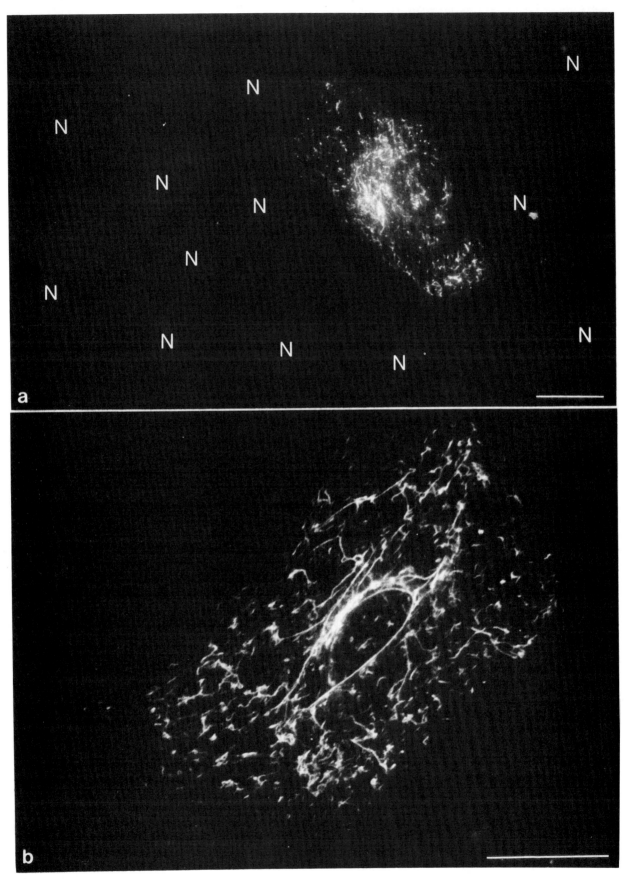

Figure 7 (See facing page for legend.)

Several conclusions can be drawn from these microinjection experiments.

1. Cytokeratin polypeptides synthesized in vivo in a "foreign" cytoplasmic environment (i.e., a nonepithelial cell) can assemble into a cytoplasmic IF fibril network with major structural features of a typical epithelial cytoskeleton. It is yet to be determined whether the overall pattern of assembly is exclusively determined by the major structural elements (i.e., cytokeratins) or whether minor "regulatory" proteins that are also present in nonepithelial cells or are encoded by the injected mRNA are of importance.
2. In line with cytochemical evidence (cf. Osborn et al. 1980) and the well-documented requirement for at least two different cytokeratin polypeptides for complex formation and IF assembly (Steinert et al. 1976, 1982; Franke et al. 1983a), cytokeratins do not intermix with vimentin into the same IF.
3. The efficient association of newly formed polypeptides with the resident cytokeratin fibril network of PtK$_2$ cells suggests that the latter can serve as an organizing template for newly formed cytokeratin molecules.

Concluding Remarks

The studies described here illuminate some of the dynamic properties of the cytokeratin-containing IF cytoskeleton. They indicate that, in contrast to the impressive rigidity and chemical stability of these IFs in vitro, they may be readily rearranged in living cells. The molecular nature of the regulatory mechanisms that determine assembly and rearrangements are unclear now. The results presented here, however, strongly suggest that mechanisms altering and modulating IF structures exist and are of major importance for the organization of the cytoplasm and for dynamic activities of the cell.

Acknowledgments

We thank Drs. R. Moll and R. Quinlan (Heidelberg) for valuable discussions and Drs. E. Debus and M. Osborn (Max-Planck-Institute for Biophysical Chemistry, Goettingen, FRG) for providing the monoclonal antibody CK4. The work has been supported in part by the Deutsche Forschungsgemeinschaft and a special German-Israeli research grant provided to B.G. and W.W.F. T.E.K. has been a recipient of an EMBO long-term fellowship.

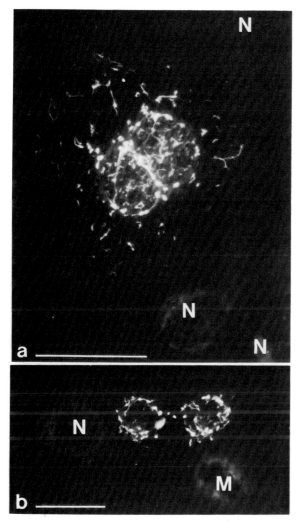

Figure 8 Immunofluorescent micrographs of human SV40 virus–transformed fibroblasts (a) and rat smooth-muscle–derived RVF-SM cells (b), which normally only express IFs of the vimentin type, after microinjection of keratin mRNA from bovine snout epidermis, incubation for 5 hr (a) and 15 hr (b), followed by treatment with rabbit (a) or guinea pig (b) antibodies against cytokeratin (for procedures of immunofluorescence microscopy, see Franke et al. 1982a; Kreis et al. 1983). Note the expression of short fibrils of cytokeratin in the injected cells, as opposed to the absence of fluorescence in the noninjected cells. (N) Nucleus; (M) mitotic cell. Bars, 25 μm.

noninjected PtK$_2$ cells, see Osborn et al. 1980). Essentially similar results have been obtained with injected MDBK cells (Franke et al., this volume).

Figure 7 Immunofluorescent micrograph of calf lens-forming cells (CLF cells), which contain vimentin IFs but normally do not synthesize cytokeratins, after injection of individual cells with keratin mRNA from bovine snout epidermis (for details, see Kreis et al. 1983) and processing with guinea pig antibodies against cytokeratin. (a) Cytokeratin material synthesized de novo from injected keratin mRNA appears in small dots and flakes throughout the cytoplasm of the injected cell, in contrast to the absence of staining in the surrounding noninjected cells. (N) Nucleus. (b) Higher magnification of a cell injected with cytokeratin mRNA, fixed three days after injection, and stained with antibodies to cytokeratin. Note formation of IF bundle structures. Adjacent, noninjected cells are negative. Bars, 25 μm.

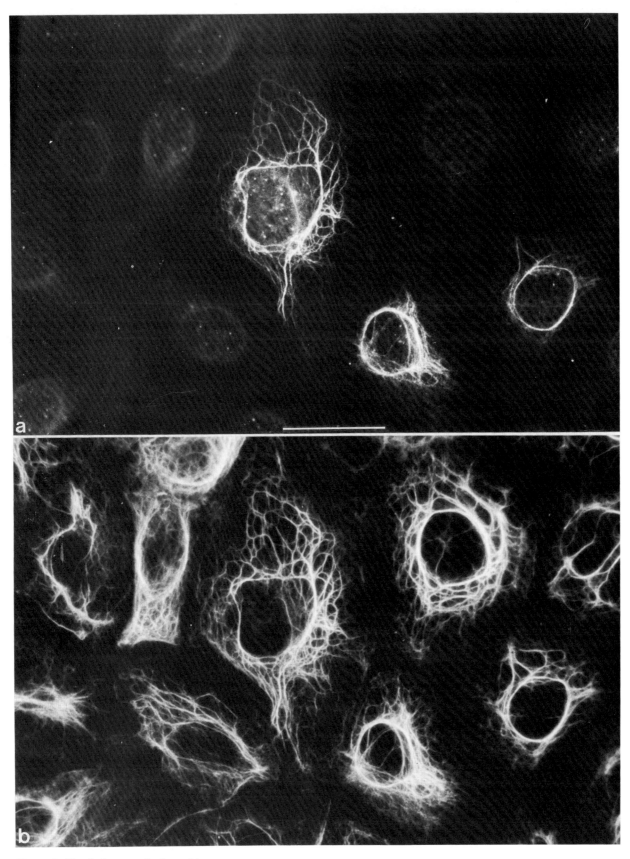

Figure 9 (See facing page for legend.)

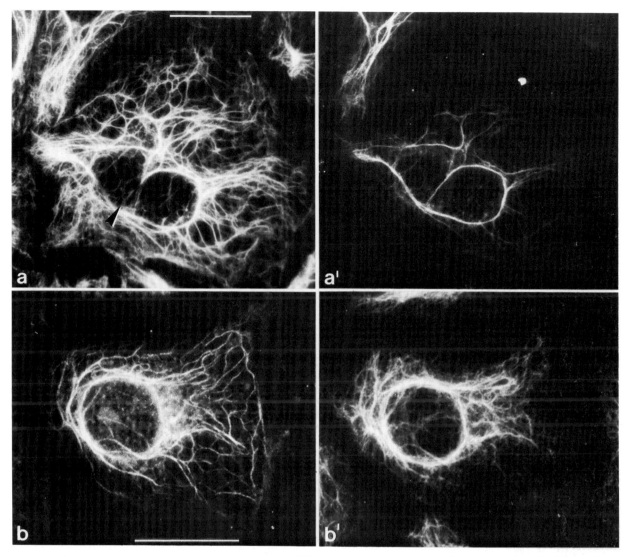

Figure 10 Double-label immunofluorescent micrographs of PtK$_2$ cells 6 hr after injection with bovine epidermal mRNA coding for keratins and processing, using monoclonal murine antibody specifically staining the endogenous PtK$_2$ cytokeratin IFs (antibody K$_s$18.8) (a), rabbit antibodies specific for the bovine epidermal cytokeratins (a',b), and guinea pig antibodies specific for vimentin (b'). (a) The antibody stains the endogenous cytokeratin IF meshwork. Many of the fibrils include newly synthesized epidermal-type keratins, except for certain IF bundles that are only poorly stained with this antibody (one is denoted by the arrow). (a') Epidermal-type keratins synthesized from injected mRNA have integrated into some bundles of the endogenous IF pattern but not into all. Note also strong staining of the bundle marked by the arrow in a. (b) Bovine epidermal keratins synthesized de novo after microinjected epidermal mRNAs have been incorporated into most of the endogenous kangaroo kidney-type cytokeratin fibrils of the PtK$_2$ cell but not into the vimentin filaments shown in b'. Bars, 20 μm.

Figure 9 Double-label immunofluorescent micrographs of cultured PtK$_2$ cells, using rabbit cytokeratin antibodies specific for epidermal keratin polypeptides (a) and monoclonal murine antibody CK4 against cytokeratin D (b; equivalent to human cytokeratin no. 18 of the catalog of Moll et al. 1982b), described by Debus et al. (1982). (a) Three cells have been injected with keratin mRNA from bovine snout epidermis, and exclusively these three cells show positive fibrillar staining with the specific rabbit antibodies used (indirect immunofluorescence using fluorescein-coupled antibodies against rabbit Ig as second antibodies), whereas the endogenous IFs present in the noninjected cells are not recognized. (b) Same cells seen by staining with the monoclonal antibody, showing cytokeratin IFs in both the injected and noninjected cells (visualized by rhodamine-labeled goat antibodies against murine Ig as second antibodies). Note that fibrils containing the newly synthesized bovine epidermal keratins in the injected cells (a) are also stained by the antibodies reacting with the kidney-type cytokeratins presented in PtK$_2$ cells (cytokeratin D is not present in bovine epidermis). Bar, 25 μm.

References

Anderton, B.H. 1981. Intermediate filaments: a family of homologous structures. *J. Muscle Res. and Cell Motil.* **2:** 141.

Aubin, J.E., M. Osborn, W.W. Franke, and K. Weber. 1980. Intermediate filaments of the vimentin-type and the cytokeratin-type are distributed differently during mitosis. *Exp. Cell Res.* **129:** 149.

Bennett, G.S., S.A. Fellini, J.M. Croop, J.J. Otto, J. Bryan, and H. Holtzer. 1978. Differences among 100-Å filament subunits from different cell types. *Proc. Natl. Acad. Sci.* **75:** 4364.

Debus, E., K. Weber, and M. Osborn. 1982. Monoclonal cytokeratin antibodies that distinguish simple from stratified squamous epithelia: Characterization on human tissues. *EMBO J.* **1:** 1641.

Dulbecco, R., R. Allen, S. Okada, and M. Bowman. 1983. Functional changes of intermediate filaments in fibroblastic cells revealed by a monoclonal antibody. *Proc. Natl. Acad. Sci.* **80:** 1915.

Franke, W.W., C. Grund, M. Osborn, and K. Weber. 1978a. The intermediate-sized filaments in rat kangaroo PtK$_2$ cells. I. Morphology in situ. *Cytobiologie* **17:** 365.

Franke, W.W., E. Schmid, C. Grund, and B. Geiger. 1982a. Intermediate filament proteins in nonfilamentous structures: Transient disintegration and inclusion of subunit proteins in granular aggregates. *Cell* **30:** 103.

Franke, W.W., D.L. Schiller, M. Hatzfeld, and S. Winter. 1983a. Protein complexes of intermediate-sized filaments: Melting of cytokeratin complexes in urea reveals different polypeptide separation characteristics. *Proc. Natl. Acad. Sci.* **80:** 7113.

Franke, W.W., E. Schmid, M. Osborn, and K. Weber. 1978b. Different intermediate-sized filaments distinguished by immunofluorescence microscopy. *Proc. Natl. Acad. Sci.* **75:** 5034.

Franke, W.W., E. Schmid, J. Vandekerckhove, and K. Weber. 1980. A permanently proliferating rat vascular smooth muscle cell with maintained expression of smooth muscle characteristics, including action of the vascular smooth muscle type. *J. Cell Biol.* **87:** 594.

Franke, W.W., D. Mayer, E. Schmid, H. Denk, and E. Borenfreund. 1981a. Differences of expression of cytoskeletal proteins in cultured rat hepatocytes and hepatoma cells. *Exp. Cell Res.* **134:** 345.

Franke, W.W., E. Schmid, S. Winter, M. Osborn, and K. Weber. 1979a. Widespread occurrence of intermediate-sized filaments of the vimentin-type in cultured cells from diverse vertebrates. *Exp. Cell Res.* **123:** 25.

Franke, W.W., K. Weber, M. Osborn, E. Schmid, and C. Freudenstein. 1978c. Antibody to prekeratin. Decoration of tonofilament-like arrays in various cells of epithelial character. *Exp. Cell Res.* **116:** 429.

Franke, W.W., B. Appelhans, E. Schmid, C. Freudenstein, M. Osborn, and K. Weber. 1979b. Identification and characterization of epithelial cells in mammalian tissues by immunofluorescence microscopy using antibodies to prekeratin. *Differentiation* **15:** 7.

Franke, W.W., E. Schmid, J. Wellsteed, C. Grund, O. Gigi, and B. Geiger. 1983b. Change of cytokeratin filament organization during the cell cycle: Selective masking of an immunologic determinant in interphase PtK$_2$ cells. *J. Cell Biol.* **97:** 1255.

Franke, W.W., D.L. Schiller, R. Moll, S. Winter, E. Schmid, I. Engelbrecht, H. Denk, R. Krepler, and B. Platzer. 1981b. Diversity of cytokeratins: Differentiation specific expression of cytokeratin polypeptides in epithelial cells and tissues. *J. Mol. Biol.* **153:** 933.

Franke, W.W., E. Schmid, D.L. Schiller, S. Winter, E.-D. Jarasch, R. Moll, H. Denk, B.W. Jackson, and K. Illmensee. 1982b. Differentiation-related expression of proteins of intermediate-size filaments in tissues and cultured cells. *Cold Spring Harbor Symp. Quant. Biol.* **46:** 431.

Gabbiani, G., Y. Kapanci, P. Barazzone, and W.W. Franke. 1981. Immunochemical identification of intermediate-sized filaments in human neoplastic cells. *Am. J. Pathol.* **104:** 206.

Gigi, O., B. Geiger, Z. Eshhar, R. Moll, E. Schmid, S. Winter, D.L. Schiller, and W.W. Franke. 1982. Detection of a cytokeratin determinant common to diverse epithelial cells by a broadly cross-reacting monoclonal antibody. *EMBO J.* **1:** 1429.

Horwitz, B., H. Kupfer, Z. Eshhar, and B. Geiger. 1981. Reorganization of arrays of prekeratin filaments during mitosis. *Exp. Cell Res.* **134:** 281.

Huiatt, T.W., R.M. Robson, N. Arakawa, and M.H. Stromer. 1980. Desmin from avian smooth muscle. *J. Biol. Chem.* **255:** 6981.

Kreis, T.E. and W. Birchmeier. 1982. Microinjection of fluorescently labeled proteins into living cells with emphasis on cytoskeletal proteins. *Int. Rev. Cytol.* **75:** 209.

Kreis, T.E., B. Geiger, E. Schmid, J.L. Jorcano, and W.W. Franke. 1983. De novo synthesis and specific assembly of keratin filaments in nonepithelial cells after microinjection of mRNA for epidermal keratin. *Cell* **32:** 1125.

Lane, E.B. and M.W. Klymkowsky. 1982. Epithelial tonofilaments: Investigating their form and function using monoclonal antibodies. *Cold Spring Harbor Symp. Quant. Biol.* **46:** 387.

Lane, E.B., S.L. Goodman, and L.K. Trejdosiewicz. 1982. Disruption of the keratin filament network during epithelial cell division. *EMBO J.* **1:** 1365.

Lazarides, E. 1982. Intermediate filaments: A chemically heterogeneous, developmentally regulated class of proteins. *Annu. Rev. Biochem.* **51:** 219.

Lazarides, E., B.L. Granger, D.L. Gard, C.M. O'Connor, J. Breckler, M. Price, and S.I. Danto. 1982. Desmin- and vimentin-containing filaments and their role in the assembly of the z disk in muscle cells. *Cold Spring Harbor Symp. Quant. Biol.* **46:** 351.

Magin, T.M., J.L. Jorcano, and W.W. Franke. 1983. Translational products of mRNAs coding for non-epidermal cytokeratins. *EMBO J.* **2:** 1387.

Moll, R., D.B. von Bassewitz, U. Schulz, and W.W. Franke. 1982a. An unusual type of cytokeratin filament in cells of a human cloacogenic carcinoma derived from the anorectal transition zone. *Differentiation* **22:** 25.

Moll, R., W.W. Franke, D.L. Schiller, B. Geiger, and R. Krepler. 1982b. The catalog of human cytokeratins: Patterns of expression in normal epithelia, tumors and cultured cells. *Cell* **31:** 11.

Osborn, M., W.W. Franke, and K. Weber. 1980. Direct demonstration of the presence of two immunologically distinct intermediate-sized filament systems in the same cell by double immunofluorescence microscopy. *Exp. Cell Res.* **125:** 37.

Osborn, M., N. Geisler, G. Shaw, and K. Weber. 1982. Intermediate filaments. *Cold Spring Harbor Symp. Quant. Biol.* **46:** 413.

Ramaekers, F.C.S., M. Osborn, E. Schmid, K. Weber, H. Bloemendal, and W.W. Franke. 1980. Identification of the cytoskeletal proteins in lens-forming cells, a special epithelioid cell type. *Exp. Cell Res.* **127:** 309.

Rathke, P.C., E. Schmid, and W.W. Franke. 1975. The action of the cytochalasins at the subcellular level. I. Effects and binding of cytochalasin B in cells of a line derived from a rat mammary adenocarcinoma and in rat erythrocytes. *Cytobiologie* **10:** 366.

Renner, W., W.W. Franke, E. Schmid, N. Geisler, K. Weber, and E. Mandelkow. 1981. Reconstitution of intermediate-sized filaments from denatured monomeric vimentin. *J. Mol. Biol.* **149:** 285.

Rueger, D.C. J.S. Huston, D. Dahl, and A. Bignami. 1979. Formation of 100 Å filaments from purified glial fibrillary acidic protein in vitro. *J. Mol. Biol.* **135:** 53.

Schiller, D.L., W.W. Franke, and B. Geiger. 1982. A subfamily of relatively large and basic cytokeratin polypeptides as defined by peptide mapping is represented by one or several polypeptides in epithelial cells. *EMBO J.* **1:** 761.

Schlegel, R., S. Banks-Schlegel, J.A. McLeod, and G.S. Pinkus. 1980. Immunoperoxidase localization of keratin in human neoplasms. *Am. J. Pathol.* **101**: 41.

Schmid, E., S. Tapscott, G.S. Bennett, J. Croop, S.A. Fellini, H. Holtzer, and W.W. Franke. 1979. Differential location of different types of intermediate-sized filaments in various tissues of the chicken embryo. *Differentiation* **15**: 27.

Steinert, P.M., W.W. Idler, and S.B. Zimmerman. 1976. Self-assembly of bovine epidermal keratin filaments in vitro. *J. Mol. Biol.* **108**: 547.

Steinert, P., R. Zackroff, M. Aynardi-Whitman, and R.D. Goldman. 1982. Isolation and characterization of intermediate filaments. *Methods Cell Biol.* **24**: 399.

Sun, T.-T. and H. Green. 1978. Immunofluorescent staining of keratin fibers in cultured cells. *Cell* **14**: 469.

Sun, T.-T., C. Shih, and H. Green. 1979. Keratin cytoskeletons in epithlial cells of internal organs. *Proc. Natl. Acad. Sci.* **76**: 2813.

Woodcock-Mitchell, J., R. Eichner, W.G. Nelson, and T.-T. Sun. 1982. Immunolocalization of keratin polypeptides in human epidermis using monoclonal antibodies. *J. Cell Biol.* **95**: 580.

Wu, Y.-J. and J.G. Rheinwald. 1981. A new small (40 kd) keratin filament protein made by some cultured human squamous cell carcinomas. *Cell* **25**: 627.

Expression of Specific Keratin Subsets and Vimentin in Normal Human Epithelial Cells: A Function of Cell Type and Conditions of A Growth during Serial Culture

J.G. Rheinwald, T.M. O'Connell, N.D. Connell, S.M. Rybak, B.L. Allen-Hoffmann, P.J. LaRocca, Y.-J. Wu, and S.M. Rehwoldt

Division of Cell Growth and Regulation, Dana-Farber Cancer Institute, and Department of Physiology and Biophysics, Harvard Medical School, Boston, Massachusetts 02115

The mechanisms by which lining and surface epithelia regulate their growth and perform their differentiated functions are poorly understood. These epithelia are particularly interesting in the context of this volume because many become malignant in humans at high frequency (e.g., bronchial, colonic, exocervical, and oral epithelia and the epidermis). The diverse types of epithelia that function as protective barriers against deleterious environments, either external (e.g., epidermis, esophageal, exocervical, and tracheal epithelia) or internal (e.g., urothelium), or that serve as a nonadherent surface between organs (e.g., mesothelium) all share the common feature that their function is performed by their structure. Thus, the molecules responsible for their differentiated function are likely to be abundant, relatively stable macromolecules rather than specific enzymes or hormones, as characterize the "nonstructural" epithelia (such as the liver, pancreas, and pituitary), which have yielded to the more classical enzymological and endocrinological methods of analysis. Rather recent advances in electrophoretic separation of proteins and immunochemical techniques have resulted in a rapid accumulation of information about the identities of many structural proteins (see Cold Spring Harbor Symp. Quant. Biol., vol. 46, 1982). Nevertheless, the functions of most of these proteins and the regulation of their expression are as yet unknown.

This paper describes our use of cell culture to identify and study cell type–specific patterns of expression and regulation of intermediate filament proteins in epithelia. We discuss our research in the following areas:

1. the diversity of keratins and the cell type–specific expression of keratin subsets stably maintained by normal cells in culture;
2. the specific keratin subsets, colony morphologies, and growth requirements that identify three different epithelial cell types in the urinary tract and expelled in urine;
3. the growth-related regulation of keratin and vimentin synthesis in mesothelial cells;
4. the cell type–specific patterns of vimentin induction and suppression in normal epithelia.

Methods

Normal human cell strains and the 3T3 line
Epidermal keratinocyte strain N and dermal fibroblast strain A1F were cultured from newborn foreskins. Urothelial strains HBl-6 and HBl-10 were cultured from 22-week-old fetal bladders, and strain HB1-8 was from an 18-year-old adult bladder. Kidney epithelial strain HKi-10 was cultured from the renal cortex of a 22-week-old fetus. Epithelial cell population TU was cultured from the urine of a normal 24-year-old adult. The 3T3 fibroblast line used as a feeder layer to support the growth of epithelial cells (Rheinwald and Green 1975a,b) was derived from a randomly bred Swiss fetal mouse (Todaro and Green 1963).

Culture conditions
Keratinocytes, urothelial cells, and kidney epithelial cells were cocultivated with mitomycin-treated 3T3 mouse fibroblasts in DMEM/F12 (3:1 v/v) plus 5% fetal calf serum (FCS) plus 0.4 µg/ml hydrocortisone (HC) plus 5 ng/ml epidermal growth factor (EGF) plus 10^{-10} M cholera toxin plus 1.8×10^{-4} M adenine plus 5 µg/ml insulin (see Wu et al. 1982 and references therein). For growth of these cells in the absence of 3T3 feeder cells, Ca^{++}-free M199/F12 (3:1 v/v) plus 5–20% FCS (dialyzed against Ca^{++}-free Earle's salts) was conditioned by incubation for 7–18 hours with confluent 3T3 cell cultures. The conditioned medium was filter-sterilized, and the above supplements were then added.

Mesothelial cells were cultured without a 3T3 feeder layer in M199/MCDB202 (K.C. Biologicals, Inc.) (1:1 v/v) plus 15% FCS plus 0.4 µg/ml HC plus 10 ng/ml EGF for rapid growth, or without EGF to slow growth and favor keratin synthesis (Connell and Rheinwald 1983). Human fibroblasts were cultured in M199 plus 15% FCS.

Antibodies used in indirect immunofluorescence microscopy
Anti-$40K_M$ keratin antiserum was raised in rabbits and is specific for this keratin (Wu and Rheinwald 1981; Wu et al. 1982). Anti–stratum corneum keratins antiserum was raised in rabbits and cross-reacts to various degrees with all the "mesothelial" or "simple epithelial" keratins (Wu and Rheinwald 1981; Wu et al. 1982). An-

tivimentin antiserum raised in guinea pigs was kindly provided by W. Franke (German Cancer Research Center, Heidelberg). Mouse monoclonal antibody OC125, specific for a surface antigen of ovarian carcinoma cells (Bast et al. 1981), was obtained from R. Bast of this Institute. Methods used in fluorescein- and rhodamine-coupled indirect immunofluorescence are described in Connell and Rheinwald (1983).

Biochemical methods
Protein labeling, extraction, and electrophoresis were done as described in Wu et al. (1982) and Connell and Rheinwald (1983). Proteins were labeled by incubating cultures for 4 hours with 100 µCi of [^{35}S]methionine (sp. act. ~800 Ci/mmole; New England Nuclear), in otherwise methionine-free medium. When 3T3 feeder cells were used, they were removed selectively before extraction by gently spraying them off the dish with pipette streams of medium, a procedure that leaves the epithelial cells adherent to the dish. Keratins and vimentin were extracted as the major components of the Triton/0.6 M KCl-insoluble, SDS-soluble or 9.5 M urea–soluble fraction of cultured cells. Cells were lysed in the presence of phenylmethylsulfonylfluoride to inhibit proteolysis, and at 4°C to depolymerize microtubules so that the tubulins did not fractionate with the intermediate filament proteins. Two-dimensional gel electrophoresis was done by the method of O'Farrell (1975) for isoelectric focusing in the first dimension or by the method of O'Farrell et al. (1977) for nonequilibrium pH gradient electrophoresis in the first dimension in order to detect basic keratins.

Poly(A)$^+$ RNA preparation, in vitro translation, and Northern blot analysis
Preconfluent mesothelial cultures growing rapidly in the presence of EGF and confluent mesothelial cultures quiescent in the absence of EGF in roller bottles were chilled to 4°C, rinsed with cold RSB (10 mM NaCl, 10 mM Tris·HCl (pH 7.3), 1.5 mM MgCl$_2$ in H$_2$0), and then lysed by the addition of 15 ml of cold RSB plus 10% NP-40, a procedure that left the nuclei and much of the cytoskeleton adherent to the vessel surface. RNA in the supernatant was purified by phenol and chloroform/isoamyl alcohol (99:1 v/v) extractions, and was ethanol-precipitated at −20°C (Ginzburg et al. 1983). Poly(A)$^+$ RNA was isolated by oligo(dT)-cellulose chromatography (Aviv and Leder 1972). Growing and quiescent poly(A)$^+$ RNA (0.5 µg of each) was translated in a rabbit reticulocyte lysate system by the method of Pelham and Jackson (1976). Aliquots containing 3 × 10^5 TCA-precipitable cpm of the [^{35}S]methionine-labeled translation products were separated by two-dimensional, nonequilibrium pH gradient electrophoresis.

RNA was separated on the basis of size by electrophoresis through a 1.2% agarose/formaldehyde gel and was transferred to nitrocellulose paper by the method of Thomas (1980). pBR322 plasmids containing large portions of the coding sequences for epidermal keratin 50K$_E$ (plasmid pKB2) and for epidermal keratin 56K$_E$ (plasmid pKA1), kindly provided by E. Fuchs (Univ. of Chicago) (Fuchs et al. 1981; Kim et al. 1983), were labeled with ^{32}P by nick translation (Rigby et al. 1977). The RNA filter blots were prehybridized overnight with denatured salmon sperm DNA at 42°C and then incubated for 3 days at 42°C with 2 × 10^6 cpm of ^{32}P-labeled plasmid DNA. The filter blots were then rinsed at 50°C, dried, and applied with an intensifying screen to X-ray film for 5 days at −70°C.

Keratin nomenclature
Identification of the keratins by size and isoelectric point and their classification as "mesothelial" or "epidermal" type on the basis of their expression in cultured mesothelial cells or epidermal keratinocytes are described in Wu et al. (1982). The major "mesothelial" keratins (40K$_M$, 44K$_M$, 52K$_M$, and 55K$_M$) are the same proteins as the "simple epithelial" keratins numbers 19, 18, 8, and 7, respectively, according to the nomenclature of Moll et al. (1982b). Keratins 46K$_E$, 50K$_E$, 52K$_E$, 56K$_E$, and 58K$_E$, first identified and studied in cultured epidermal keratinocytes by Sun and Green (1978) and Fuchs and Green (1978, 1981), are keratins 17, 14, 13, 6, and 5, respectively, according to the nomenclature of Moll et al. (1982b).

Results and Discussion

Growth characteristics of different epithelial cell types in culture

We have recently found (Wu et al. 1982) that keratinocyte growth is greatly improved by several additions to the original culture system, which included a 3T3 feeder layer, hydrocortisone, EGF, and cholera toxin (Rheinwald and Green 1975a,b, 1977; Green 1978). These are the substitution of a DMEM/F12 mixture for DMEM, an increase in the concentration of adenine (as employed by Peehl and Ham [1980] in their completely defined, short-term keratinocyte culture medium), the addition of insulin, and the reduction of the FCS supplement to 5%. This medium yields substantially higher colony-forming efficiencies and longer replicative life spans for keratinocytes and is permissive for clonal growth and serial cultivation of the more fastidious urothelial cells (T.M. O'Connell et al., unpubl.). Keratinocytes and urothelial cells grow optimally in this medium with an FCS supplement of 5% and adequately with as little as 0.2–0.5% serum. They can also be serially propagated in the absence of intact 3T3 feeder cells (albeit at 3-fold to 10-fold lower colony-forming efficiencies), provided that the medium is "conditioned" by 3T3 cells and that the Ca^{++} concentration of the medium is reduced to 0.1 mM or less (following the observations of Hennings et al. [1980] and Peehl and Ham [1980] for short-term keratinocyte growth in the absence of 3T3 feeder cells). In the absence of a 3T3 feeder layer, urothelial cells and keratinocytes prefer higher serum concentrations (5–20%) and an F12/M199 mixture as the defined component.

Kidney cortex epithelial cells also grow well in the feeder layer system, with two morphologically distinct cell types apparent. Type-I cells survive a maximum of three serial subcultures. Type-II cells grow faster, exhibit a higher colony-forming ability at each passage, and can

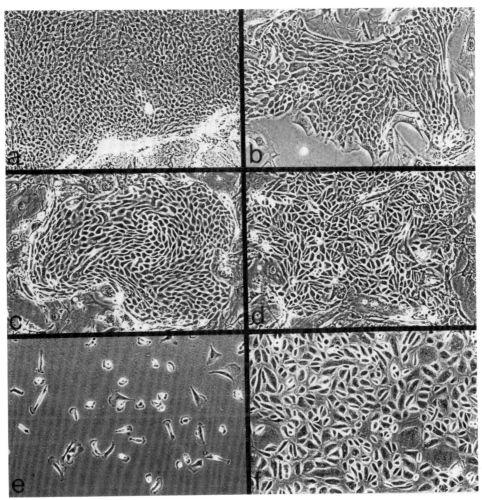

Figure 1 Colony morphologies of normal, human epithelial cells. (a) Epidermal keratinocyte strain N; (b) bladder urothelial strain HBl-10; (c) kidney epithelial strain HKi-10, type-I colony; (d) kidney epithelial strain HKi-10, type-II colony. (a–d were cultured with 3T3 feeder cells, which can be seen around the colony peripheries.) (e) Peritoneal mesothelial strain LP-9, preconfluent and growing rapidly in the presence of EGF; (f) LP-9, confluent and quiescent in the absence of EGF. (e and f reprinted, with permission, from Connell and Rheinwald 1983. Copyright 1983 by M.I.T.)

be serially passaged five to ten times. Neither kidney epithelial cell type is 3T3-dependent or inhibited by Ca^{++}.

Normal mesothelial cells are not fibroblast feeder–dependent, but they do possess a stringent requirement for EGF and hydrocortisone (Connell and Rheinwald 1983). They grow optimally with a mixture of M199 and MCDB202 as the defined medium component, require higher concentrations of serum than the other epithelial cell types described here, and are not sensitive to Ca^{++}.

The morphologies of keratinocytes, urothelial cells, types-I and -II kidney epithelial cells, and mesothelial cells are very distinctive (Fig. 1). Keratinocytes form tightly packed, stratified colonies. Urothelial cells form moderately adherent, unstratified colonies. Type-I kidney epithelial cells are plump and form tightly adherent, unstratified colonies. Type-II kidney epithelial cells are flatter and sometimes intermingle with the feeder cells at the colony periphery. Mesothelial cells grow in a dispersed fashion and adopt a stubby, somewhat fibroblastoid morphology when growing rapidly with EGF, continuing to divide after reaching confluence to form multiple layers of elongated cells, similar to the behavior of fibroblasts. In the absence of EGF, mesothelial cells grow very slowly and form an epithelioid monolayer similar to, although less regular than, their appearance in vivo. Each of these normal, human epithelial cell types stably retains its particular morphological and growth-requirement characteristics during its finite replicative life span of 25 to more than 100 cell generations, depending upon cell type and age of donor.

Nonepidermal keratins and cell type–specific patterns of keratin synthesis

Mesothelial cells synthesize a completely different set of Triton/high salt–insoluble, "cytoskeletal" proteins from that of epidermal keratinocytes (cf. Figs. 2a and 3f and see Fig. 2f). The mesothelial cytoskeleton consists of vimentin plus four keratins (40KM, 44KM, 52KM, and 55KM) with isoelectric points intermediate between the

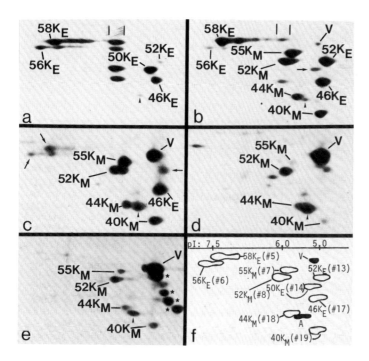

Figure 2 Intermediate filament proteins synthesized by different epithelial cell types in culture. Autoradiographs of [^{35}S] methionine-labeled, Triton/high salt–insoluble proteins separated by nonequilibrium pH gradient electrophoresis in the horizontal dimension (left, basic; right, acidic) and by SDS-polyacrylamide gel electrophoresis in the vertical dimension (top, higher molecular weight; bottom, lower molecular weight). (a) Epidermal keratinocyte strain N. (b) Bladder urothelial strain HBI-6; arrow points to a small amount of $50K_E$. (c) Type-I cells of kidney epithelial strain HKi-10; arrows point to small amounts of $58K_E$, $56K_E$, and $52K_E$. The vimentin in this culture is from incompletely removed 3T3 feeder cells. (d) Type-II cells of kidney epithelial strain HKi-10, cultured in the absence of 3T3 feeder cells. (e) Urine-derived epithelial cell population TU, cultured in the absence of 3T3 feeder cells; asterisks indicate proteolytic digestion intermediates of vimentin. (d and e were greatly overexposed to reveal the keratins, which were present at low levels.) (f) Map of keratins synthesized by normal epithelial cells in culture, labeled according to the nomenclatures both of Wu et al. (1982) and of Moll et al. (1982b). Spots beneath pairs of vertical lines are keratins that migrated as heterodimers (one basic + one acidic keratin) during pH gradient electrophoretic separation in 9.5 M urea and consequently were at an intermediate position (see Franke et al., this volume). V indicates vimentin, and arrowheads point to actin.

larger, basic ($56K_E$ and $58K_E$) and the smaller, acidic ($46K_E$ and $50K_E$) epidermal keratins (Wu et al. 1982). The mesothelial keratins are specifically immunoprecipitated, with different degrees of efficiency, by an antiserum raised against total epidermal stratum corneum keratins. The two smaller ones ($40K_M$ and $44K_M$) are more closely related to the smaller epidermal keratins ($46K_E$ and $50K_E$) on the basis of their immunoprecipitation with an antiserum raised against the $46K_E$ epidermal keratin. One-dimensional peptide maps of individual epidermal and mesothelial keratins did not support the hypothesis that any of the mesothelial keratins are cleavage products of larger keratins (Wu et al. 1982). In more recent experiments in collaboration with K.H. Kim and E. Fuchs at the University of Chicago, we have found that the two smaller mesothelial keratins ($40K_M$ and $44K_M$) are translated from mRNAs that hybridize under low stringency conditions with a cloned cDNA sequence of $50K_E$ keratin mRNA, while the two larger mesothelial keratins ($52K_M$ and $55K_M$) are trans-

Table 1 Keratins Expressed by Different Epithelia In Vivo and in Culture

		$40K_M$ (#19)	$44K_M$ (#18)	$46K_M$	$52K_M$ (#8)	$55K_M$ (#7)	$46K_E/47K_E$ (#17, #16)	$50K_E$ (#14)	$50K_E$ (#15)	$52K_E$ (#13)	$56K_E$ (#6)	$58K_E$ (#5)	$56.5K_E$ (#10, #11)	$64-67K_E$ (#2, #1)
Epidermal keratinocyte	V[a]						(+)	++		±	(+)	++	++	++
	C[b]						++	++		±	++	++		
Mesothelial cell	V[c]	++	++		++	(+)								
	C[d,e]	++	++	(+)	++	+								
Urothelial cell	V[f]	++	+		++	++	±			++	±			
	C[d,g]	++	+		++	++	++	±		++	±	++		
Kidney epithelial cell, type I	C[g]	+	++		++	++	+			±	±	±		
Kidney epithelial cell, type II	C[g]	+	++		++	+								
Urine-derived epithelial cell	C[g]	+	+		++	+								
Mammary epithelial cell	V[f]	++			+	+	++	++	±		++			
	C[d]	(+)			+	+	++	++	+		++	++		

(V) In vivo; (C) in serially cultured cell populations; (++) most prominent keratins; (+) keratins present at ~¼ the level of ++ keratins; (±) keratins present at ≤1/20 the level of ++ keratins; ((+)) keratins present in variable amounts or absent in some samples.
[a]Moll et al. (1982a); Woodcock-Mitchell et al. (1982); Fuchs and Green (1978).
[b]Sun and Green (1978); Fuchs and Green (1978).
[c]P.J. LaRocca and J.G. Rheinwald (in prep.).
[d]Wu et al. (1982).
[e]Connell and Rheinwald (1983).
[f]Moll et al. (1983).
[g]T.M. O'Connell and J.G. Rheinwald (in prep.).

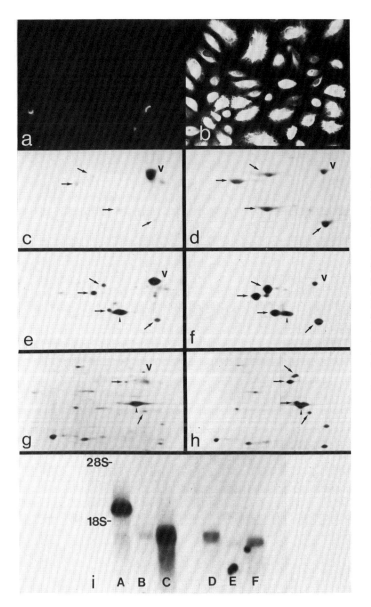

Figure 3 Growth-related modulation of keratin content, synthesis, and mRNA levels in mesothelial cells. (a,c,e,g) Preconfluent mesothelial cultures, growing rapidly in the presence of EGF; (b,d,f,h) confluent mesothelial cultures, quiescent in the absence of EGF. (a,b) Antikeratin indirect immunofluorescence; 14 cells are shown in a, 11 of which had no detectable keratin. (c,d) Coomassie blue–stained, Triton/high salt–insoluble proteins extracted from equal amounts of total protein of growing and quiescent mesothelial cultures. (e,f) Autoradiographs of Triton/high salt–insoluble, [^{35}S]methionine-labeled proteins (isoelectric focusing in the first dimension). (g,h) Autoradiographs of total [^{35}S]methionine-labeled proteins synthesized in rabbit reticulocyte lysates, directed by equal amounts of poly(A)$^+$ RNA from growing and quiescent mesothelial cultures (nonequilibrium pH gradient electrophoresis in the first dimension); arrows point to keratins 55K_M, 52K_M, 44K_M, and 40K_M; V indicates vimentin, and arrowheads point to actin in c–h. (i) Northern blots of epithelial cell RNA separated by agarose gel electrophoresis and hybridized with ^{32}P-labeled pKA1 (lanes A–C) and pKB2 (lanes D–F) plasmids, bearing keratin cDNA sequences. (Lanes A and D) 10 μg of total RNA of epidermal keratinocyte strain N; (lanes B and E) 5 μg of poly(A)$^+$ RNA from rapidly growing, strain LP-9 mesothelial cells; (lanes C and F) 5 μg of poly(A)$^+$ RNA from quiescent, strain LP-9 mesothelial cells. (pKA1 hybridizes with 56K_E and 58K_E mRNA in epidermal keratinocytes and with 52K_M and 55K_M RNA in mesothelial cells; pKB2 hybridizes with 46K_E and 50K_E mRNA in epidermal keratinocytes and with 40K_M and 44K_M mRNA in mesothelial cells; see Fuchs et al. [1981] and Kim et al. [1983].) (a–f reprinted, with permission, from Connell and Rheinwald 1983. Copyright 1983 by M.I.T.)

lated from mRNAs that hybridize under low stringency conditions with a cloned cDNA sequence of 56K_E keratin mRNA (Kim et al. 1983; Fuchs et al., this volume).

Our two-dimensional gel analysis of the Triton/high salt–insoluble proteins synthesized by a variety of human epithelial cell types in culture has disclosed that many epithelial cell types express a qualitatively and/or quantitatively unique combination of keratins from the "epidermal" and "mesothelial" families (Wu et al. 1982). The independent identification of the nonepidermal keratins and the tissue-specific expression of keratins in vivo by W.W. Franke, R. Moll, and collaborators (Franke et al. 1981a; Moll et al. 1982b, 1983) and by T.T. Sun and collaborators (Doran et al. 1980; Tseng et al. 1982; Sun et al. 1983) are reviewed by Franke et al. and Sun et al. in this volume. An immediate clinical application of this research to the problem of cancer would be the use of antisera specific for individual keratins (e.g., anti-40K_M [Wu and Rheinwald 1981] or anti-44K_M [Debus et al. 1982] as immunohistopathological tools for earlier or more certain diagnosis of cancer, since some carcinoma cells express an additional keratin or keratins not expressed by the normal epithelial cell type of origin (Wu and Rheinwald 1981; Wu et al. 1982; Moll et al. 1983; Rheinwald et al. 1983).

Comparing our cell culture results with the native tissue analyses published by others (Table 1), it is clear that there are marked, intrinsic differences among cell types with respect to the pattern of keratins they express in culture in the absence of specific regional influences, but that the in vivo and in culture patterns are sometimes substantially different. Except for the involvement of vitamin A in suppressing synthesis of the highest-molecular-weight, "keratinization-specific" keratins (64K–67K) in epidermal keratinocytes growing in culture (Fuchs and Green 1981), nothing is known about the mechanisms by which keratin gene expression is regulated in epithelial cells. Cultured populations of various

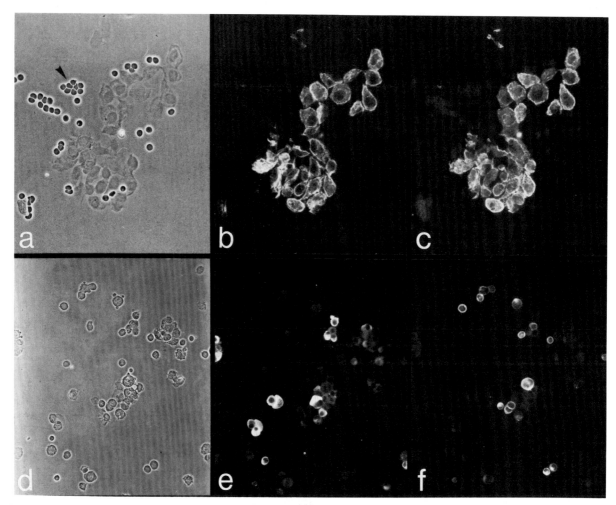

Figure 4 Keratin and vimentin coexpression in normal mesothelial cells in situ, in ascites fluid, and after 1 day of culture. (a–c) Adult rat pleural mesothelium, dislodged with a rubber policeman and applied to a slide by cytocentrifugation; (d–i) cells concentrated by cytocentrifugation from ascites fluid of a patient with ovarian neoplasm; (j–l) cells from same ascites fluid after 1 day in primary culture. (a,d,g,j) Phase-contrast microscopy; (b,e,h,k) fluorescein-coupled indirect immunofluorescence with antivimentin antiserum;

normal epithelial cell types promise to be excellent material for studies directed at elucidating these mechanisms.

Identification of different epithelial cell types in the urinary tract and in urine by their growth requirements and patterns of keratin synthesis in culture

It has been known for several years that cells of epithelioid morphology can be cultured from normal human urine (Hoehn et al. 1975; Felix and Littlefield 1979), but their tissue of origin has remained a mystery. Our gel electrophoresis, immunofluorescence microscopy, and growth requirement experiments prove that the cells cultured from urine are not urothelial cells but possess some characteristics of kidney epithelial cells (T.M. O'Connell and J.G. Rheinwald, in prep.). Urothelial cells from the urinary bladder, ureter, and renal pelvis all require the 3T3 feeder layer for clonal growth and serial culture and synthesize the "urothelial-specific" set of keratins in culture (see Fig. 2b and Table 1). Explant cultures of kidney cortex (the outer region of the kidney, excluding the central collecting region or "pelvis") consist of type-I and -II epithelial cells. Kidney type-I cells and urothelial cells have a somewhat similar keratin pattern, although the former synthesize only trace amounts of $58K_E$ and $52K_E$. Kidney type-I cells differ from type-II cells by the much greater keratin content and the presence of $46K_E$ keratin in the former and by the high levels of vimentin in the latter (Figs. 2c,d and 6g–j). Cultures of cells collected from urine consist predominantly of cells possessing low to moderate levels of keratins $55K_M$, $52K_M$, $44K_M$, and $40K_M$ plus high levels of vimentin (Fig. 2e). Epithelial cells derived from urine can be cultured without a 3T3 feeder layer or conditioned medium and in the presence of greater than 1 mM Ca^{++} (Hoehn et al. 1975; Felix and Littlefield 1979), unlike urothelial cells, and they are similar to kidney epithelial cells morphologically.

Hoehn et al. (1975) noticed the strong morphological similarity between the cells cultured from fetal urine and those cultured from amniotic fluid. Ochs et al. (1983) recently analyzed the intermediate filament composition of the several morphological types of cells that grow in

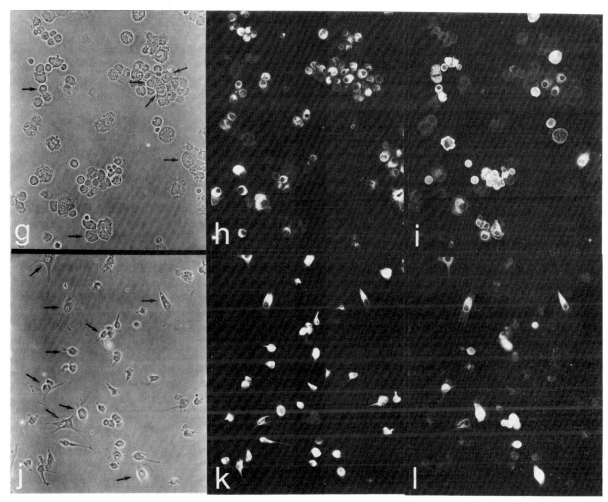

Figure 4 (*continued*) (*c,i,l*) rhodamine-coupled indirect immunofluorescence with antikeratin antiserum; (*f*) rhodamine-coupled indirect immunofluorescence with mouse monoclonal antibody OC125 against ovarian carcinoma cell surface. (Note that the ovarian carcinoma cells from this patient are vimentin-negative.) Erythrocytes are indicated by arrowhead in *a*, and presumed mesothelial cells are indicated by arrows in *g* and *j*.

culture from amniotic fluid and concluded on the basis of the keratins they express ($55K_M$, $52K_M$, $44K_M$, and $40K_M$) and their coexpression of high levels of vimentin that they are not derived from amniotic epithelium (see also Virtanen et al. 1981). Our results described above taken together with those of Hoehn and Ochs suggest that the cells that grow in culture from amniotic fluid are kidney epithelial cells, which are expelled in the fetal urine into the amniotic cavity.

Modulation of keratin synthesis and content in mesothelial cells as a function of growth

The behavior of mesothelial cells is particularly interesting as a model for studying extrinsic modulation of keratin gene expression. Our first studies of mesothelial cell keratins (Wu et al. 1982) were done with cells cultured without EGF and in DMEM/F12 medium, a condition in which the cells have a doubling time of about 3 days. The addition of EGF and improvement of the defined component of the medium resulted in doubling times of 26 hours or less and, surprisingly, in the reversible loss of keratin by rapidly growing mesothelial cells (Connell and Rheinwald 1983). Indirect immunofluorescence microscopy using antikeratin antisera discloses a rapid reduction in mesothelial cell keratin content from the normal, in vivo level within the first several days of primary culture, such that most cells in sparse, early passage populations have little or no detectable keratin (Fig. 3a). All cells then accumulate keratin to high levels when EGF or HC is removed from the medium or when their growth slows as saturation density is approached. The greatest keratin content is attained by mesothelial cells in a quiescent, confluent monolayer in the absence of EGF (Fig. 3b). The immunofluorescence estimate of relative keratin content is confirmed by gel electrophoretic analysis of the Triton/high salt–insoluble proteins of rapidly growing vs. quiescent mesothelial populations (Fig. 3c,d). These differences in keratin content appear to be due completely to changes in the rate of keratin synthesis (Fig. 3e,f), because there is no apparent difference in keratin turnover between growing and quiescent mesothelial cells (Connell and

Rheinwald 1983). Thus, mesothelial cell keratin content is determined by rate of synthesis and rate of dilution by cell division. Rapidly growing mesothelial cells have much less poly(A)$^+$ keratin mRNA translatable in a reticulocyte lysate (Fig. 3g,h) and detectable on Northern blots with cross-hybridizing, epidermal keratin cDNA probes (Fig. 3i) (S.M. Rybak et al., unpubl.). Thus, the rate of mesothelial cell keratin synthesis is determined by the level of poly(A)$^+$ keratin mRNA in the cytoplasm. It remains to be determined whether a change in the rate of transcription, in the rate of transcript processing, or in the turnover time of cytoplasmic message is responsible for these dramatic fluctuations in keratin mRNA levels in mesothelial cells.

The mesothelial cell is unique among epithelial cell types studied to date in its wide range of keratin expression. Other cell types that express one or more of the "mesothelial" keratins ($40K_M$, $44K_M$, $52K_M$, and $55K_M$) can grow as rapidly as mesothelial cells in culture without dramatic decreases in their keratin synthesis and content, so most cell types in which these particular keratin genes are open for expression do not regulate them this way. The mesothelial cell may be more motile and able to repair denuded areas of the pleural, pericardial, and peritoneal surfaces more rapidly if it becomes keratin-poor and adopts a fibroblastoid morphology after being stimulated to divide in vivo. Regardless of its natural function, the capacity of the mesothelial cell to modulate its keratin content and morphology explains the considerable morphological and keratin-content heterogeneity among the cells in many mesotheliomas (Klemperer and Rabin 1931; Corson and Pinkus 1982).

Vimentin expression in mesothelial cells in vivo

Our original examination of ascites fluid cells by double-label, vimentin/keratin immunofluorescence during the first hours and days of culture indicated that the progenitors of our cultured mesothelial cell populations with high vimentin and variable keratin content are cells of high keratin and low, but detectable, vimentin content in vivo (Fig. 4d-l) (Connell and Rheinwald 1983). Because normal mesothelial cells in ascites fluid and, presumably, also in pleural effusions contain both keratins and vimentin detectable by immunofluorescence microscopy, such coexpression, as reported to be present in some metastatic carcinomas (Ramaekers et al. 1983), is not a useful diagnostic criterion for tumor cells floating in the body cavities.

The vimentin expressed by mesothelial cells sloughed into ascites fluid does not appear to have been triggered by their detachment from the intact mesothelium. We have found that the cells in intact rat lung mesothelium contain vimentin at a similar level (Fig. 4a-c). (Possibly because the mesothelium is very flat, its cytoskeleton so fragile, and the vimentin content only moderate in comparison with other epithelia, vimentin was not detected in an earlier study of cross-sectioned mesothelium [Schmid et al. 1979].) During the first several days of growth in primary culture, mesothelial cells increase their vimentin content to a very high level, similar to that of fibroblasts. Although vimentin synthesis and content relative to total protein synthesis and content fall severalfold when the mesothelial cells are quiescent (Fig. 3c-f), the vimentin content never returns to the low-to-moderate in vivo levels under the conditions in which we culture them.

Expression of vimentin by other epithelial cell types in culture

Our electrophoretic (Wu et al. 1982) and immunofluorescence (Connell and Rheinwald 1983) analyses of the intermediate filament proteins expressed by various types of normal, human epithelial cells in culture clearly showed that high-level vimentin expression is not an inevitable consequence of serial cultivation. Whereas vimentin rapidly increases in mesothelial cells from rather low to very high levels during the first several days of growth in primary culture, vimentin remains undetectable or at only trace levels in keratinocytes throughout their very long replicative life span in culture with the 3T3 feeder layer (Fig. 5b,c). In cultured urothelial cell populations, vimentin appears at high levels only in large cells that have apparently lost replicative capacity (Fig. 5e,f). Observations of vimentin acquisition in culture by rat hepatocytes (Franke et al. 1981b) and human amniotic fluid epithelial cells (Virtanen et al. 1981) were made in culture systems that did not employ a feeder layer and that, in the hepatocytes, were not optimal for progressive and long-term growth of the cell type. To separate effects of the fibroblast feeder layer and of suboptimal media on vimentin expression, we examined epidermal keratinocytes, urothelial cells, and kidney epithelial cells for their vimentin content during rapid growth in the absence of 3T3 feeder cells, using 3T3-conditioned medium with low Ca^{++} concentration.

We found that keratinocytes rapidly acquire moderate-to-high levels of vimentin during growth in the absence of a feeder layer (Fig. 6a,b). These cells lose vimentin again, however, after they are returned to a 3T3 feeder layer (Fig. 6c,d). In contrast, progressively growing, colony-forming urothelial cells continue to suppress vimentin during growth in the absence of the 3T3 feeder layer, whereas larger, nongrowing urothelial cells have a high vimentin content (Fig. 6e,f), the same behavior as when cultured with a feeder layer (Fig. 5e,f). Kidney epithelial cells are also unaffected by growth in the presence or absence of a 3T3 feeder layer: Type-I cells, which do not express vimentin when cultured with feeder cells, also do not in their absence (Fig. 6g,h), whereas type-II cells express high levels of vimentin under both culture conditions (Fig. 6i,j).

Epidermal keratinocytes do not "dedifferentiate" irreversibly as a result of growth in culture, as evidenced by their reacquisition of normal tissue organization and expression of differentiation-specific proteins upon reinoculation or grafting to animal or human hosts (Banks-Schlegel and Green 1980; Doran et al. 1980; O'Connor et al. 1981). Thus, vimentin expression is probably available to the keratinocyte under appropriate conditions in vivo. It is surprising that a protein as apparently useless

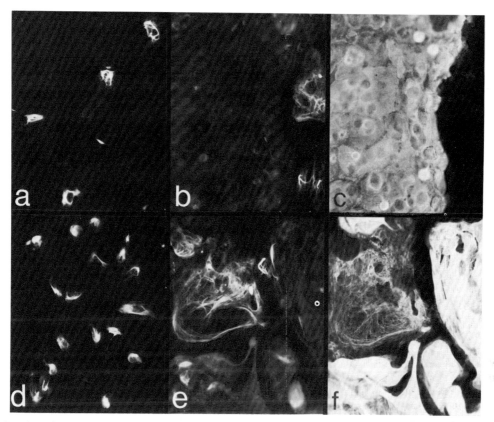

Figure 5 Relative vimentin contents of several normal, human cell types in culture. Cultures were treated with 5×10^{-7} M colcemid 1 day before fixation in order to cause perinuclear aggregation of vimentin filaments and then were stained by fluorescein-coupled indirect immunofluorescence with antivimentin (a,b,d,e). (c,f) Same fields shown in b and e, respectively, counterstained by rhodamine-coupled indirect immunofluorescence with antikeratin. (a) Fibroblast strain A1-F; (b,c) keratinocyte strain N; (d) mesothelial strain LP-9; (e,f) urothelial strain HBI-8.

(for an epidermal keratinocyte) as vimentin is not constitutively suppressed at the level of genomic differentiation in this cell type, as it appears to be in urothelial and type-I kidney epithelial cells. Perhaps vimentin becomes expressed in keratinocytes during wound repair and performs a useful function related to migration or rapid growth. The extrinsic regulatory factor or condition that suppresses vimentin in keratinocytes is as yet unknown but may be sought by experimental manipulation of the culture environment.

At present it can be concluded that some normal epithelial cell types do not express vimentin in culture, whereas some types quickly accumulate high levels during growth in certain conditions of culture. The continued suppression of vimentin in urothelial cells during growth in a dispersed fashion in low Ca^{++}-conditioned medium indicates that, for at least some cell types, loss of intercellular contact is not sufficient to induce vimentin. Such coupling of contact and vimentin suppression has been suggested for parietal endodermal cells in vivo (Lane et al. 1983). Some cell types that do not normally express vimentin might do so as a consequence of malignant transformation or acquisition of replicative immortality (a possible example is HeLa). However, most of our human epidermal and oral squamous cell carcinoma lines do not express vimentin (Wu and Rheinwald 1981), nor do about half of the ovarian carcinoma cell lines (derived from tumors of the ovarian surface epithelium) we have examined (J.G. Rheinwald et al., unpubl.). The best-known, established epithelial cell lines that coexpress keratin and vimentin—PtK_1 and PtK_2—may have arisen from the type-II kidney epithelial cell, described above, which expresses vimentin in its normal, diploid condition in culture. Our data strongly support the conclusion that the maintenance in culture of cell type–specific mechanisms for vimentin regulation is the rule, rather than the exception, for epithelial cells.

Acknowledgments

We thank Dr. Robert Bast for ovarian carcinoma–specific antibody, Dr. Werner Franke for his generous gift of antivimentin, Dr. Elaine Fuchs for epidermal keratin cDNA plasmids, and Dr. Lee Parker for ascites fluid samples. We also wish to acknowledge Mr. Michael Beckett, Ms. Christine Murray, and Ms. Kathleen Finn for excellent technical assistance and Ms. Lynne Dillon for skillful preparation of the manuscript. This research was supported by grants to J.G.R. from the National Cancer Institute and the National Institute on Aging. B.L.A. and P.J.L. were supported by postdoctoral fellowships from the National Institutes of Health.

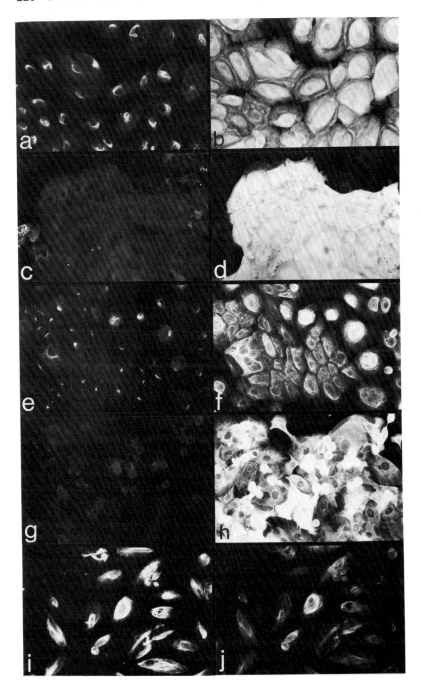

Figure 6 Change of vimentin expression by keratinocytes, but not other epithelial cell types, during growth without 3T3 feeder cells. Cultures were treated with 5×10^{-7} M colcemid 1 day before fixation in order to cause perinuclear aggregation of vimentin filaments and then were stained by fluorescein-coupled indirect immunofluorescence with antivimentin (a,c,e,g,i) and by rhodamine-coupled indirect immunofluorescence with antikeratin (b,d,f,h,j). (a,b) Keratinocyte strain N cultured in 3T3-conditioned medium, in the absence of 3T3 feeder cells. (c,d) Keratinocyte strain N returned to 3T3 feeder layer culture for 10 days after one passage in the absence of 3T3 feeder cells. (e,f) Urothelial strain HBI-6 in its second serial passage in 3T3-conditioned medium, in the absence of 3T3 feeder cells. Note that although some larger urothelial cells have moderate amounts of vimentin, the small, actively dividing cells have very little or no detectable vimentin. (g,h) Kidney epithelial strain HKi-10, type-I colony cultured in the absence of 3T3 feeder cells. (i,j) Kidney epithelial strain HKi-10, type-II colony cultured in the absence of 3T3 feeder cells.

References

Aviv, H. and P. Leder. 1972. Purification of biologically active globin mRNA by chromatography on oligo thymidylic acid-cellulose. *Proc. Natl. Acad. Sci.* **69:** 1408.

Banks-Schlegel, S. and H. Green. 1980. Formation of epidermis by serially cultivated human epidermal cells transplanted as an epithelium to athymic mice. *Transplantation* **29:** 308.

Bast, R.C., Jr., M. Feeney, H. Lazarus, L.M. Nadler, R.B. Colvin, and R.C. Knapp. 1981. Reactivity of a monoclonal antibody with human ovarian carcinoma. *J. Clin. Invest.* **68:** 1331.

Connell, N.D. and J.G. Rheinwald. 1983. Regulation of the cytoskeleton in mesothelial cells: Reversible loss of keratin and increase in vimentin during rapid growth in culture. *Cell* **34:** 245.

Corson, J.M. and G.S. Pinkus. 1982. Mesothelioma: Profile of keratin proteins and carcinoembryonic antigen. An immunoperoxidase study of 20 cases and comparison with pulmonary adenocarcinoma. *Am. J. Pathol.* **108:** 80.

Debus, E., K. Weber, and M. Osborn. 1982. Monoclonal cytokeratin antibodies that distinguish simple from stratified human epithelia: Characterization on human tissues. *EMBO J.* **1:** 1641.

Doran, T.I., A. Vidrich, and T.T. Sun. 1980. Intrinsic and extrinsic regulation of the differentiation of skin, corneal and esophageal epithelial cells. *Cell* **22:** 17.

Felix, J.S. and J.W. Littlefield. 1979. Urinary tract epithelial cells cultured from human urine. *Int. Rev. Cytol.* (suppl.) **10:** 11.

Franke, W.W., D. Mayer, E. Schmid, H. Denk, and E. Borenfreund. 1981b. Differences of expression of cytoskeletal proteins in cultured rat hepatocytes and hepatoma cells. *Exp. Cell Res.* **134:** 345.

Franke, W.W., S. Winter, C. Grund, E. Schmid, D.L. Schiller, and E.D. Jarasch. 1981a. Isolation and characterization of desmosome-associated tonofilaments from rat intestinal brush border. *J. Cell Biol.* **90**: 116.

Fuchs, E. and H. Green. 1978. The expression of keratin genes in epidermis and cultured epidermal cells. *Cell* **15**: 887.

———. 1981. Regulation of terminal differentiation of cultured human keratinocytes by vitamin A. *Cell* **25**: 617.

Fuchs, E., S.M. Coppock, H. Green, and D.W. Cleveland. 1981. Two distinct classes of keratin genes and their evolutionary significance. *Cell* **27**: 75.

Ginzburg, I., S. Rybak, Y. Kimhi, and U.Z. Littauer. 1983. Biphasic regulation by dibutyryl cyclic AMP of tubulin and actin mRNA levels in neuroblastoma cells. *Proc. Natl. Acad. Sci.* **80**: 4243.

Green, H. 1978. Cyclic AMP in relation to proliferation of the epidermal cell: A new view. *Cell* **15**: 801.

Hennings, H., D. Michael, C. Cheng, P. Steinert, K. Holbrook, and S.H. Yuspa. 1980. Calcium regulation of growth and differentiation of mouse epidermal cells in culture. *Cell* **19**: 245.

Hoehn, H., E.M. Bryant, A.G. Fantel, and G.M. Martin. 1975. Cultivated cells from diagnostic amniocentesis in second trimester pregnancies. III. The fetal urine as a potential source of clonable cells. *Humangenetik* **29**: 285.

Kim, K.H., J.G. Rheinwald, and E.V. Fuchs. 1983. Tissue specificity of epithelial keratins: Differential expression of mRNAs from two multigene families. *Mol. Cell. Biol.* **3**: 495.

Klemperer, P. and C.B. Rabin. 1931. Primary neoplasms of the pleura. *Arch. Pathol.* **11**: 385.

Lane, E.B., B.L.M. Hogan, M. Kurkinen, and J.I. Garrels. 1983. Co-expression of vimentin and cytokeratins in parietal endoderm cells of early mouse embryo. *Nature* **303**: 701.

Moll, R., R. Krepler, and W.W. Franke. 1983. Complex cytokeratin polypeptide patterns observed in certain human carcinomas. *Differentiation* **23**: 256.

Moll, R., W.W. Franke, B. Volc-Platzer, and R. Krepler. 1982a. Different keratin polypeptides in epidermis and other epithelia of human skin: A specific cytokeratin of molecular weight 46,000 in epithelia of the pilosebaceous tract and basal cell epitheliomas. *J. Cell Biol.* **95**: 285.

Moll, R., W.W. Franke, D.L. Schiller, B. Geiger, and R. Krepler. 1982b. The catalog of human cytokeratins: Patterns of expression in normal epithelia, tumors and cultured cells. *Cell* **31**: 11.

Ochs, B.A., W.W. Franke, R. Moll, C. Grund, M. Cremer, and T. Cremer. 1983. Epithelial character and morphologic diversity of cell cultures from human amniotic fluids examined by immunofluorescence microscopy and gel electrophoresis of cytoskeletal proteins. *Differentiation* **24**: 153.

O'Connor, N.B., J.B. Mulliken, S. Banks-Schlegel, O. Kehinde, and H. Green. 1981. Grafting of burns with cultured epithelium prepared from autologous epidermal cells. *Lancet* **I**: 75.

O'Farrell, P.H. 1975. High resolution two-dimensional electrophoresis of proteins. *Anal. Biochem.* **250**: 4007.

O'Farrell, P.Z., H.M. Goodman, and P.H. O'Farrell. 1977. High resolution two-dimensional electrophoresis of basic as well as acidic proteins. *Cell* **12**: 1133.

Peehl, D.M. and R.G. Ham. 1980. Clonal growth of human keratinocytes with small amounts of dialyzed serum. *In Vitro* **16**: 526.

Pelham, H.R. and R.J. Jackson. 1976. An efficient mRNA-dependent translation system from reticulocyte lysates. *Eur. J. Biochem.* **67**: 247.

Ramaekers, F.C.S., D. Haag, A. Kant, O. Moesker, P.H.K. Jap, and G.P. Vooijs. 1983. Coexpression of keratin- and vimentin-type intermediate filaments in human metastatic carcinoma cells. *Proc. Natl. Acad. Sci.* **80**: 2618.

Rheinwald, J.G. and H. Green. 1975a. Formation of a keratinizing epithelium in culture by a cloned cell line derived from a teratoma. *Cell* **6**: 317.

———. 1975b. Serial cultivation of strains of human epidermal keratinocytes: The formation of keratinizing colonies from single cells. *Cell* **6**: 331.

———. 1977. Epidermal growth factor and the multiplication of cultured human epidermal keratinocytes. *Nature* **265**: 421.

Rheinwald, J.G., E. Germain, and M.A. Beckett. 1983. Expression of keratins and envelope proteins in normal and malignant human keratinocytes and mesothelial cells. In *Human carcinogenesis* (ed. C.C. Harris and H.N. Autrup), p. 85. Academic Press, New York.

Rigby, W.J., M. Dieckmann, C. Rhodes, and P. Berg. 1977. Labelling deoxyribonucleic acid to high specific activity in vitro by nick translation with DNA polymerase I. *J. Mol. Biol.* **113**: 237.

Schmid, E., S. Tapscott, G.S. Bennett, J. Croop, S.A. Fellini, H. Holtzer, and W.W. Franke. 1979. Differential location of different types of intermediate-sized filaments in various tissues of the chicken embryo. *Differentiation* **15**: 27.

Sun, T.T. and H. Green. 1978. Keratin filaments of cultured human epidermal cells. *J. Biol. Chem.* **253**: 2053.

Sun, T.T., R. Eichner, W.G. Nelson, S.C.G. Tseng, R.A. Weiss, M. Jarvinen, and J. Woodcock-Mitchell. 1983. Keratin classes: Molecular markers for different types of epithelial differentiation. *J. Invest. Dermatol.* **81**: 109.

Thomas, P.S. 1980. Hybridization of denatured RNA and small DNA fragments transferred to nitrocellulose. *Proc. Natl. Acad. Sci.* **77**: 5201.

Todaro, G.J. and H. Green. 1963. Quantitative studies of the growth of mouse embryo cells in culture and their development into established lines. *J. Cell Biol.* **17**: 299.

Tseng, S.C.G., M.J. Jarvinen, W.G. Nelson, J.W. Huang, J. Woodcock-Mitchell, and T.T. Sun. 1982. Correlation of specific keratins with different types of epithelial differentiation: Monoclonal antibody studies. *Cell* **30**: 361.

Virtanen, I., V.P. Lehto, E. Lehtonen, T. Vartio, S. Stenman, P. Kurki, O. Wager, J.V. Small, D. Dahl, and R.A. Badley. 1981. Expression of intermediate filaments in cultured cells. *J. Cell Sci.* **50**: 45.

Woodcock-Mitchell, J., R. Eichner, W.G. Nelson, and T.T. Sun. 1982. Immunolocalization of keratin polypeptides in human epidermis using monoclonal antibodies. *J. Cell Biol.* **95**: 580.

Wu, Y.J. and J.G. Rheinwald. 1981. A new small (40 kd) keratin filament protein made by some human squamous cell carcinomas. *Cell* **25**: 627.

Wu, Y.J., L.M. Parker, N. Binder, M.A. Beckett, J.H. Sinard, C.T. Griffiths, and J.G. Rheinwald. 1982. The mesothelial keratins: A new family of cytoskeletal proteins identified in cultured mesothelial cells and nonkeratinizing epithelia. *Cell* **31**: 693.

Multistage Carcinogenesis Involves Multiple Genes and Multiple Mechanisms

I.B. Weinstein, S. Gattoni-Celli, P. Kirschmeier, M. Lambert, W. Hsiao, J. Backer, and A. Jeffrey

Division of Environmental Sciences and Cancer Center/Institute of Cancer Research, Columbia University, New York, New York 10032

The development of a fully malignant tumor involves complex interactions between several factors, both exogenous (i.e., environmental) and endogenous (i.e., genetic, hormonal, etc.) (Weinstein 1981; Weinstein et al. 1982). In addition, carcinogenesis often proceeds through multiple, discernible stages (i.e., initiation, promotion, progression; see Fig. 1), and the overall process can occupy a major fraction of the life span of the individual. The transitions between successive stages can be enhanced or inhibited by different types of agents. Thus, it appears that the individual stages may involve qualitatively different mechanisms at the cellular and genetic levels. These aspects predict that the establishment and maintenance of a malignant tumor involve multiple cellular genes and multiple types of changes in genomic structure and function. We believe, therefore, that it is a gross oversimplification to speak of "the" cancer gene, even if one is referring to oncogenes, or to assume that multistage carcinogenesis simply involves a successive series of random point mutations (Weinstein et al. 1982). Because of the above considerations, an experimental approach that simply displays the genetic changes present in a fully evolved malignant tumor may not in itself indicate the role such changes play in determining the biologic properties of the particular tumor studied.

In this paper we will briefly review recent studies on the biochemical effects of tumor initiators and promoters and then consider how they might contribute to the phenotype of fully evolved tumors.

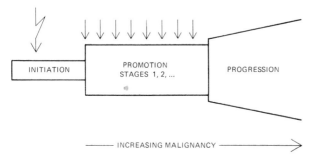

Figure 1 Schematic representation of multistage carcinogenesis based on studies on mouse skin. Initiation requires only a single exposure to a carcinogen and appears to involve DNA damage. Promotion involves multiple exposures to agents that do not damage DNA directly; promotion can also be subdivided into at least two stages. Progression involves the conversion of tumors from benign to malignant and can be considered an open-ended process since tumors may continue to increase in their degree of malignancy and heterogeneity.

Early events in the action of initiating carcinogens

Several types of chemicals that initiate the carcinogenic process yield highly reactive species or metabolites that bind covalently to cellular DNA (Miller 1978; Weinstein 1981). Although there is considerable evidence that damage to DNA can play a critical role in carcinogenesis, the subsequent biochemical events that lead to the conversion of a normal cell to a cancer cell are not known with certainty. Several features of the carcinogenic process, particularly the high efficiency of carcinogen-induced cell transformation, the latent period, and the multistep aspects, suggest that it does not involve simple random point mutations at sites of carcinogen-damaged DNA. Other mechanisms by which carcinogen-induced DNA damage could bring about genomic changes include induction of gene rearrangements or gene amplification (Weinstein 1981; Weinstein et al. 1982). Consistent with this possibility is the increasing evidence that fully established tumors often display rearranged and amplified genes, including the oncogenes (Cooper 1982; Bishop 1983; Weinberg 1983).

We have recently utilized a model system that dramatically demonstrates the ability of certain carcinogens to induce a much more complex change in DNA replication than random point mutation (Lambert et al. 1983). We employed a cell line, ts-A H3, that contains 1.3 viral equivalents of integrated polyoma DNA derived from a virus that carries a temperature-sensitive large T antigen. The exposure of these cells to benzo[a]pyrene (BP) (0.25 μg/ml) at 33°C, the permissive temperature, led to a marked enhancement in the production of extrachromosomal copies of polyoma DNA. This effect was not seen at 39°C, indicating that it was dependent on the function of the T antigen of polyoma virus. Further studies utilizing the highly reactive carcinogenic metabolite of BP, BP-dihydrodiol-9,10-epoxide (BPDE), on a rat cell line transformed by wild-type polyoma virus demonstrated that the enhancement of polyoma DNA synthesis persisted for at least 5 days after a single exposure to BPDE, despite the rapid decay of this compound. In addition, enhanced synthesis of polyoma DNA could be induced by fusion of normal rat fibroblasts previously exposed to BPDE with polyoma-transformed rat fibroblasts not exposed to BPDE. It appears, therefore, that the carcinogen enhancement of polyoma DNA replication is not due to direct damage to the integrated polyoma virus DNA but may instead involve the induction of a cellular factor(s) that can function in *trans* to enhance asynchronous replication of polyoma DNA (Lambert et al. 1983).

Our results are consistent with other evidence that a variety of DNA-damaging agents can enhance the asynchronous replication of integrated polyoma or SV40 virus DNA sequences in mammalian cells (Lavi and Etkin 1981; Lambert et al. 1983). It is not clear, however, to what extent carcinogens might induce the asynchronous replication of cellular genes, since we found that BP enhancement of polyoma virus DNA production is strictly dependent on the function of the polyoma virus T antigen. Studies are in progress to identify the above-mentioned inducible factor and to determine its effects on the replication of various genes and its possible role in the initiation of carcinogenesis. This putative factor may also play a role in the phenomenon by which a number of carcinogens enhance the de novo transformation of cells by certain DNA viruses (Fisher and Weinstein 1980), perhaps by enhancing the replication of the viral DNA prior to integration into target cells.

Mechanism of action of tumor promoters

Tumor promoters can be defined as compounds that have very weak or no carcinogenic activity when tested alone but that markedly enhance tumor yield when applied repeatedly following a low or suboptimal dose of a carcinogen (initiator) (Berenblum 1982; Hecker et al. 1982). Most of our information on the mechanism of action of this class of agents derives from studies on the potent skin-tumor promoter 12-O-tetradecanoylphorbol-13-acetate (TPA) and related phorbol esters. At the biochemical level, it appears that the major difference between initiators and the phorbol ester tumor promoters is that whereas initiators (or their metabolites) bind covalently to cellular DNA, the primary site of action of the phorbol ester tumor promoters appears to be cell membranes. The phorbol ester tumor promoters can exert highly pleiotropic effects on the growth, function, and differentiation of a variety of cell types. We have found it convenient to classify these effects into three categories: (1) mimicry and enhancement of transformation, (2) modulation of differentiation, and (3) membrane effects (for review, see Hecker et al. 1982; Weinstein et al. 1982).

Utilizing ^{3}H-labeled phorbol-12,13-dibutyrate (PDBu), several laboratories have obtained direct evidence for specific high-affinity, saturable "phorboid receptors" in membrane preparations and intact cells from various avian, rodent, and human tissues (Delclos et al. 1980; Horowitz et al. 1981; Hecker et al. 1982). In general, the abilities of a series of phorbol esters to compete with [^{3}H]PDBu for binding to cell-surface receptors correlates with their known potencies in cell culture and with their activities as tumor promoters on mouse skin (Hecker et al. 1982), providing evidence that the phorboid receptors mediate the biologic action of the phorbol esters. We have also found that the antileukemic plant diterpenes gnidilatin and gnilatimacrin are potent inhibitors of [^{3}H]PDBu binding (Horowitz and Weinstein 1982). It appears that the phorboid receptors also mediate the TPA-like effects of two new classes of tumor promoters, teleocidin (and structurally related indole alkaloids) and aplysiatoxin (Umezawa et al. 1981; Horowitz et al. 1983). This could explain why, despite their quite different chemical structures, these compounds are highly potent tumor promoters on mouse skin (Sugimura 1982) and at nanomolar concentrations produce many of the same effects as TPA in cell culture (Fisher et al. 1982; Hecker et al. 1982; Sugimura 1982).

The chemical structures of teleocidin, aplysiatoxin, and TPA are quite different. This prompted us to study the stereochemistries of these compounds to see if they might display structural similarities at the three-dimensional level (Horowitz et al. 1983). All three types of compounds are amphiphilic since they have both hydrophobic and hydrophilic domains. In the case of the phorbol esters there is evidence that all of the biologically active compounds have a highly hydrophobic residue at position 12, although the precise chemical structure of this residue is not critical. Presumably this region of the molecule is required for a relatively nonspecific hydrophobic interaction with a region on the phorboid receptor or the adjacent lipid microenvironment. The saturated, six-membered ring of teleocidin and the side chain attached to the polyacetate ring of aplysiatoxin might play analogous roles. Structure-activity studies of the phorbol esters on mouse skin and in cell culture indicate that the region of the molecule containing the three keto, four OH, and six CH_2OH residues displays marked structural and steric specificity (Hecker et al. 1982). Our model-building studies indicate that the nine-membered lactam region of teleocidin and the macrocyclic polyacetate ring system of aplysiatoxin can assume conformations that are remarkably similar to the corresponding region present in the biologically active phorbol esters. We postulate, therefore, that the respective regions of the phorbol esters, teleocidin, and aplysiatoxin form highly specific chemical bonding and/or steric interactions with the phorboid receptors. These relationships for a phorbol ester and teleocidin are displayed in Figure 2. Data that we have recently obtained with a compound present in tung oil, 12-O-hexadecanoyl-16-hydroxyphorbol-13-acetate (HHPA) (Ito et al. 1983), and 12 of its congeners (provided by Y. Ito) are consistent with this stereochemical model (I.B. Weinstein and E. Okin, unpubl.). Additional compounds are being examined to obtain further data relevant to this model. Information of this type could make it possible to rationally design compounds that would either act as agonists or blockers of the phorboid receptor system.

Until very recently, a major limitation in our knowledge of the action of phorbol ester tumor promoters was the mechanism by which, following receptor binding, they induce their highly pleiotropic effects. Elsewhere we have reviewed the known effects of these compounds on ion flux (Na^+, Rb^+, Ca^{++}), nutrient uptake, phospholipid metabolism, cell-surface receptors, induction of specific enzymes, and modulation of differentiation (Weinstein 1981; Weinstein et al. 1982). In view of the pleiotropic effects of protein kinases and the evidence that the products of certain oncogenes are protein kinases, an attractive mechanism for mediating the cellular

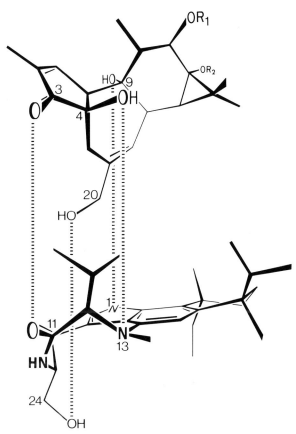

Figure 2 Perspective drawings of TPA (*top*) and dihydroteleocidin B (*bottom*). The dotted lines connect heteroatoms whose spatial positions correspond with one another and represent residues that could form hydrogen bonds with a putative receptor. The hydrophobic R_1 residue on TPA (myristate) and the hydrophobic ring system on the right side of dihydroteleocidin B might interact with a separate hydrophobic domain of the receptor or with its lipid microenvironment. For further details, see text and Horowitz et al. (1983).

effects of tumor promoters would be through specific protein kinases. The recent finding of Castagna et al. (1982) that low concentrations of TPA stimulate the in vitro activity of protein kinase C (PKC), a Ca^{++}- and phospholipid-dependent protein kinase, is of considerable interest. Further evidence that a phospholipid-PKC complex may be the actual receptor, and the effector, for TPA and related phorbol ester tumor promoters has been obtained from several laboratories (for review, see Kikkawa et al. 1983; Weinstein 1983). We have found that teleocidin and aplysiatoxin are also potent activators of PKC (J. Arcoleo et al., unpubl.). These findings can be readily accommodated in the above-described stereochemical model (Fig. 2) if we assume that these amphiphilic agents facilitate the formation of a quaternary complex between the phospholipid, PKC apoenzyme, Ca^{++}, and the tumor promoter (Weinstein 1983). Since the phorbol esters can bind with rather high affinity and also alter the physical properties of simple phospholipid membranes (Deleers et al. 1982; Tran et al. 1983a), it is possible that they act, at least in part, by inducing changes in lipid structure that favor the activation of PKC.

Several findings suggest that there may be heterogeneity (or subclasses) of receptors for phorbol esters and related tumor promoters. Scatchard analyses of [^{3}H]PDBu-receptor binding to intact cells are consistent with at least two classes of binding sites (Horowitz et al. 1981; Horowitz and Weinstein 1982). Although the compound mezerein is equipotent with TPA with respect to certain biologic effects, it competes less well than TPA in inhibiting [^{3}H]PDBu-receptor binding and also is much weaker than TPA as a complete tumor promoter on mouse skin (Delclos et al. 1980; Slaga et al. 1980). Therefore, some cell types may have a subset of receptors that discriminate between TPA and mezerein. Differential effects of aplysiatoxin and debromoaplysiatoxin are also consistent with receptor heterogeneity (Horowitz et al. 1983). A recent study with C3H 10T1/2 cells indicates that the dose response curves for TPA, PDBu, and teleocidin inhibition of the binding of [^{3}H]PDBu to high-affinity receptors were quite different from those obtained when the same compounds were tested for their ability to alter membrane lipid fluidity in the same cells, as measured with fluorescence polarization probes (Tran et al. 1983b). Receptor heterogeneity could contribute to the tissue specificity and pleiotropic effects of these compounds. The basis for this heterogeneity is not known, but it could involve the following mechanisms: (1) heterogeneity of PKCs; (2) variations in lipid domains associated with PKC; and (3) interactions with other lipid-modulated enzymes, in addition to PKC (e.g., see studies with isolated mitochondria and also effects on tyrosine kinase activity, described below).

Obviously, a fundamental area for future research is the identification of critical protein targets phosphorylated by PKC, particularly with respect to the process of tumor promotion. A related question is whether or not the multiple effects of tumor promoters represent a simple linear cascade of events. The circuitry may be quite complex. For example, tumor promoters induce phospholipid turnover, which may be associated with the release of diacylglycerol (Mufson et al. 1981), itself an activator of PKC (Castagna et al. 1982); and alterations in ion flux (i.e., Ca^{++}) could also further modulate protein kinase activities. Recent studies indicate that treatment of cells with TPA can alter the state of phosphorylation of epidermal growth factor (EGF) receptors, with an increase in phosphorylation of serine and threonine residues (Iwashita and Fox 1983; T. Hunter, pers. comm.) and a decrease in the phosphorylation of tyrosine residues (M. Rosner, pers. comm.). TPA also stimulates the phosphorylation of receptors for insulin and somatomedin C (Jacobs et al. 1983). These effects are presumably mediated via activation of PKC and could explain the fact that TPA treatment leads to an indirect inhibition of EGF-receptor (Lee and Weinstein 1980) and insulin-receptor binding (Jacobs et al. 1983). Furthermore, even though PKC does not phosphorylate tyrosine residues, the exposure of cells to TPA can also enhance the phosphorylation of tyrosine residues on a

43K protein, the same protein that is a target for the action of pp60^{src} (Bishop et al. 1983; Gilmore and Martin 1983). It appears, therefore, that in addition to its effect on PKC, TPA can enhance, either directly or indirectly, the activity of a cellular tyrosine-specific protein kinase.

Although nuclear events are usually emphasized in the action of chemical carcinogens, there is evidence that carcinogens can also cause extensive damage to mitochondrial DNA (Backer and Weinstein 1982). In addition, we have obtained evidence that TPA impairs mitochondrial respiration in intact fibroblasts (Backer et al. 1982) and also accelerates the hydrolysis of ATP by isolated rat liver mitochondria (Backer and Weinstein 1983). It is possible, therefore, that disturbances in mitochondrial energy metabolism, and/or related disturbances in intracellular ion homeostasis, may play an important role at specific stages of the carcinogenic process.

Since with repeated applications, BP and certain other polycyclic aromatic hydrocarbon (PAH) carcinogens can act as complete carcinogens on mouse skin, it seemed possible that PAHs might induce certain effects on membranes that are similar to those induced by the phorbol ester tumor promoters. We have indeed found that the exposure of C3H 10T1/2 cells to BP and certain other PAHs led to a loss of EGF-receptor binding (Ivanovic and Weinstein 1982). There is evidence that these compounds can also mimic other effects of TPA, including increases in membrane phospholipid turnover (for review, see Ivanovic and Weinstein 1982). We have obtained indirect evidence that these responses are mediated via the Ah receptor system. Our results have recently been confirmed and extended by D. Nebert et al. (pers. comm.). Our findings may also be relevant to recent studies indicating that 2,3,7,8-tetrachlorodibenzo-p-dioxin (TCDD), which binds to the Ah receptor, acts as a potent mouse skin tumor promoter in an appropriate strain of mice (Poland et al. 1982), and can also lead to an inhibition of EGF-receptor binding in cell culture (L. G. Hudson et al., pers. comm.).

Although our research group has stressed the membrane and cytoplasmic effects of tumor promoters, other investigators have provided evidence that the phorbol ester tumor promoters can also produce chromosomal aberrations and DNA damage, perhaps via activated forms of oxygen (Kinsella and Radman 1978; Nagasawa and Little 1979; Birnboim 1982; Dzarlieva and Fusenig 1982; Troll et al. 1982; Emerit et al. 1983). These effects, however, are most prominent in inflammatory cells and may reflect the specialized responses (i.e., oxidative burst) of these cells in response to activators. We would also stress that since the process of tumor promotion on mouse skin is often reversible (Berenblum 1982), it seems unlikely that the critical events during tumor promotion involve DNA damage and chromosomal aberrations. It is possible, of course, that with prolonged exposure to TPA, chromosomal changes might occur and contribute to the process of tumor progression. The finding that tumor promoters can induce mitotic aneuploidy in yeast (Parry et al. 1981) is of interest but may also be more relevant to tumor progression since chromosomal anomalies are more prominent during the late rather than the early stages of carcinogenesis.

A related question is whether tumor promoters act entirely via inductive or hormonelike effects, thus mediating clonal expansion of previously mutated cells, or whether they can themselves produce stable and heritable changes in cell phenotype. Although most of the pleiotropic effects produced by tumor promoters in cell culture are dependent upon the continuous presence of the promoter (Weinstein 1981), there are a few examples in which the action of tumor promoters is associated with irreversible effects. These include (1) the enhancement of cell transformation induced by certain oncogenic viruses, including adenovirus, Epstein-Barr virus, simian virus 40, and polyoma virus (Fisher et al. 1982; Hecker et al. 1982); (2) the enhancement of anchorage-independent growth of either adenovirally transformed rat fibroblasts (Fisher et al. 1979) or of certain murine epidermal cell lines (Colburn et al. 1979); and (3) the enhancement of outgrowth of cell variants displaying amplified genes for dihydrofolate reductase (Barsoum and Varshavsky 1983) or metallothionein (T. Sugimura, pers. comm.). We have recently found that TPA can enhance the stable transformation of C3H 10T1/2 cells induced by transfection with an activated human oncogene (W. Hsiao et al., unpubl.). The mechanisms by which tumor promoters produce these irreversible effects is not known. These phenomena may be relevant to our understanding of how tumors can eventually evolve to a stage where they become autonomous of tumor promoters, growth factors, and/or specific hormones.

It is of interest that recent studies show that DNA-damaging agents are more effective than tumor promoters in enhancing the progression of papillomas to carcinomas on mouse skin (Hennings et al. 1983). This suggests that the evolution of a fully malignant tumor during multistage carcinogenesis may involve more than one cycle of damage to DNA rather than a simple linear sequence of DNA damage induced by the initiator and subsequent nongenomic effects induced by the promoter. This is, however, a highly speculative area since even less is known about the mechanism of tumor progression than about the earlier stages of initiation and promotion. In future studies it will be important to focus on this question because at a clinical level the major challenges presented by tumors relate to invasion and metastasis, aspects of the tumor phenotype that probably result from the process of tumor progression.

Cellular genes involved in chemical carcinogenesis

In contrast to oncogenic viruses, chemical carcinogens and tumor promoters cannot introduce new genetic information into cells. During the transformation process they must, therefore, call upon genes already present in the target cells to bring about and maintain the transformed state. Until recently, there did not appear to be a direct method available for the identification of these cellular genes. The hypothesis that chemical agents might act by inducing alterations in the state of integration and/or expression of cellular oncogenes (Bishop 1983), or of cellular sequences analogous to the long-terminal-repeat (LTR) sequences of retroviruses (Kirschmeier et al. 1982), is an attractive one. We have explored aspects of these hypotheses in the studies described below.

Studies on the cellular oncogene c-mos
We utilized a cloned DNA fragment representing a portion of the oncogene (*v-mos*) of Moloney murine sarcoma virus (Mo-MSV) as a probe to determine whether or not transformation of rodent cells by chemical carcinogens or radiation is associated with alterations in the state of integration or transcription of the normal cellular sequence (*c-mos*) that is homologous to the *v-mos* gene. Our results with a series of transformed murine embryo fibroblast lines were, however, entirely negative (Gattoni-Celli et al. 1982). More recently (Gattoni-Celli et al. 1983) we have extended our studies on the *c-mos* oncogene to a murine myeloma cell line designated P3-X63-Ag-653 (abbreviated P3), which was derived from the MOPC-21 myeloma. The latter tumor is of interest since it was originally induced by the intraperitoneal injection of a rather simple chemical agent, mineral oil (Potter et al. 1975). We have found that DNA of these cells displays evidence of rearrangement of one of the *c-mos* alleles when analyzed by restriction enzyme digestion and the Southern blot hybridization procedure. Since the P3 line is used routinely to produce hybridomas that synthesize monoclonal antibodies, it is of interest that we have found that this rearranged *c-mos* locus persists in at least three murine hybridoma cell lines, even after prolonged serial passage. We have also obtained evidence that the *c-mos* locus is hypomethylated in the P3 and hybridoma cell lines. Furthermore, analysis of the poly(A)$^+$ RNA fraction of P3 and the hybridoma cells indicates that, in contrast to a variety of other cell types, the *c-mos* locus is transcribed in these cells (Gattoni-Celli et al. 1983).

Rechavi et al. (1982) have found that in a separate mouse myeloma, designated XRPC24, which was originally induced by pristane (2,6,10,14-tetramethylpentadecane), the *c-mos* gene is also rearranged and transcriptionally active. Furthermore, they cloned the rearranged *c-mos* gene present in the XRPC24 myeloma and found that it transformed mouse fibroblasts. Their sequence studies indicate that it differs from the normal *c-mos* by the presence at the 5' end of a 159-bp insertion that is derived from murine intracisternal A-particle DNA, although the insert is oriented in a direction opposite to that of *c-mos*. A comparison of their restriction map with our results suggests that the rearranged *c-mos* allele we have found in P3 cells was derived by a similar mechanism. Yet it is clear that the two myelomas were derived independently, using two different chemicals (M. Potter, pers. comm.). Kuff et al. (1983) have also detected the insertion of an A-particle LTR sequence within a "κ" light-chain intron of a murine hybridoma. Since the genome of *Mus musculus* contains about 1000 copies of these A-particle sequences (Kuff et al. 1983), they may play a rather frequent role in producing gene rearrangements, perhaps following induction of their replication by certain chemical agents.

Studies related to LTR sequences
We have also pursued an alternative approach to identifying retrovirus-related sequences that might be involved in multistage chemical carcinogenesis. Several types of evidence indicate that the LTR regions of the proviral DNA copies of retroviruses play a crucial role in enhancing transcription and that the specific sequences involved (so-called enhancer sequences) are present in the U3 portion of the LTR sequence (Blair et al. 1981; Temin 1982; Weiss et al. 1982). LTR sequences also share structural features with transposable elements of bacteria and *Drosophila* (Temin 1982; Shiba and Saigo 1983). Normal murine cells contain a large number of copies of DNA sequences homologous to the LTR sequence of murine retrovirus proviral DNA (see studies below). This suggests that if damage to cellular DNA caused the rearrangement and/or activation of these endogenous LTR sequences, this could lead to the constitutive expression of host genes whose products might contribute to the multistage process of malignant cell transformation (Kirschmeier et al. 1982).

To test this hypothesis, we examined whether there are differences between normal C3H 10T1/2 cells transformed by chemical carcinogens and cells transformed by radiation in terms of the expression of RNA species containing sequences homologous to a probe prepared to the LTR sequence of Mo-MSV (Kirschmeier et al. 1982). Although the poly(A)$^+$ RNA from normal C3H 10T1/2 cells showed negligible hybridization to the LTR probe, the poly(A)$^+$ RNA from eight independently transformed C3H 10T1/2 cell lines showed appreciable hybridization to this probe, displaying at least five distinct species ranging from about 38S to 18S. Utilizing probes for the *gag*, *pol*, and *env* genes of murine retroviruses, we obtained evidence that the higher-molecular-weight RNAs (35S–24S) present in the transformed cells also contain transcripts of these genes. In the transformed fibroblasts, we detected a prominent 24S RNA transcript homologous to an *env* probe. It is of interest that radiation-induced murine T-cell lymphomas also express endogenous *env* genes in the absence of production of infectious virus (Fischinger et al. 1982; Boccara et al. 1983). It appears that the 20S and 18S transcripts we detected in the transformed C3H 10T1/2 cells may reflect the expression of nonvirally related host sequences that utilize the LTR sequences as promoters. We are currently analyzing the possible role of these mRNAs in initiation or maintenance of the transformed state in rodent systems.

Studies in progress utilizing Southern blot analyses indicate that normal C3H 10T1/2 cells contain more than 30 copies of sequences homologous to our LTR probe. We have also obtained evidence that in C3H 10T1/2 cells the U5 regions of the endogenous LTR sequences are much more highly conserved than the U3 regions (S.

Gattoni-Celli et al., unpubl.). This finding is consistent with studies of endogenous LTR sequences in other strains of mice (Boone et al. 1983; Khan and Martin 1983), in avian cells (Hughes et al. 1981), and in feline cells (Casey et al. 1981). Since enhancer regions are present in the U3 regions, this may have implications with respect to the control of transcription by these endogenous elements during normal development and carcinogenesis. We have been unable, however, to detect any evidence for the rearrangement or amplification of these sequences in the carcinogen or radiation-transformed cells. Thus, the mechanism for the constitutive expression of the above-described RNAs in the transformed cells is not apparent at the present time. We suspect that this may reflect a gross disturbance in mechanisms controlling the transcription of specific families of enhancer sequences (see Discussion).

As part of our interest in LTR sequences, we have recently developed a transfection and expression vector in which specific genes can be conveniently inserted between a 5' and 3' LTR sequence present in the vector (Perkins et al. 1983). This construct can then be transfected into rodent cells, and the inserted gene will then be expressed in the recipient cells, utilizing the transcription signals present in the LTR sequences of the vector. The vector also permits the encapsidation of RNA sequences into viruslike particles that then mediate gene transfer with extremely high efficiency. This should greatly facilitate the further analysis of genes involved in chemical carcinogenesis.

Discussion

As mentioned above, the complexity of multistage carcinogenesis predicts that the evolution and maintenance of fully malignant cancer cells involves changes in multiple types of cellular genes. We have described rearrangement and transcriptional activation of the oncogene c-mos in a cell line derived from a murine myeloma that was induced by a relatively simple chemical agent. There are numerous previous reports of alterations in several other cellular oncogenes (particularly in the ras family of oncogenes) in various types of rodent and human tumors (for review, see Cooper 1982; Reddy et al. 1982; Bishop 1983; Shimizu et al. 1983; Weinberg 1983; and this volume). Most of these studies have been performed with fully malignant tumor cells and, therefore, it is not known at which stage in the carcinogenic process these changes occurred.

In considering the possible involvement of cellular oncogenes in chemical carcinogenesis, it is important to stress that they differ in several respects from their viral homologs. For example, the cellular oncogenes usually contain introns, they may contain coding sequences that differ appreciably from the corresponding viral oncogenes, and they are not usually flanked by LTR sequences (Bishop 1983; Duesberg 1983). The murine c-mos is actually flanked by a sequence that inhibits transcription (G. Vande Woude, pers. comm.). Therefore, it is likely that the products of cellular oncogenes will be qualitatively different and will be expressed at lower levels and under a greater degree of host control in carcinogen-induced tumors than the products of the corresponding oncogenes introduced into cells by the acute transforming viruses. For this reason, cell transformation induced by a chemical carcinogen might often depend on changes in the function of multiple cellular oncogenes, as well as other types of host genes, rather than a single dominant gene. It seems unlikely that during the course of chemical carcinogenesis a given cellular oncogene would undergo the full evolution in structure that occurred in the origin of viral oncogenes.

Consistent with this formulation are studies on the sequences of oncogenes obtained from tumors, as well as increasing evidence that a given tumor may display alterations in more than one cellular oncogene. Thus, in addition to the rearrangement and expression of the c-mos sequence seen in certain murine myelomas (see above), murine myelomas often display chromosomal translocations involving the oncogene c-myc (Klein 1983). Transfection assays are also providing evidence that the conversion of normal primary cell cultures to tumor cells requires multiple transforming genes. These assays have identified at least two classes of genes, one related to immortalization and the other to morphologic and growth properties (Land et al. 1983; Newbold and Overell 1983; Ruley 1983). Probably, additional genes will be found that relate to specific properties of fully malignant tumors (i.e., invasion, metastasis, drug resistance, etc.). The types of changes seen in cellular oncogenes in various tumors include simple point mutations, sequence deletions, sequence insertions, gene amplification, and chromosomal translocations (Bishop 1983; Cooper 1982; Weinberg 1983). At the present time it is difficult to relate these qualitatively different changes to the action of specific initiating agents, tumor promoters, or other causative agents that might be involved in the multistage carcinogenic process. Furthermore, it is already clear that the repertoire of genomic changes that occurs during malignant cell transformation is not confined to the types of oncogenes originally found in retroviruses. Certain human lymphomas display alterations in a family of genes ("B-lym") that appear to code for polypeptide growth factors (Diamond et al. 1983). Specific cellular DNA sequences involved in mammary tumorigenesis also appear to be unrelated to the known oncogenes found in retroviruses (Cooper 1982; Nusse and Varmus 1982). In addition, it appears that the genes involved in the inheritance of certain familial forms of human cancer are not simply activated oncogenes (Needleman et al. 1983), and yet in a sense the inherited predisposing genes may represent a stage in the carcinogenic process. Finally, only about 20% of the human tumors examined yield DNA that is active in the NIH-3T3 transfection assay (Cooper 1982; Weinberg 1983), raising the possibility that in the majority of human tumors, the tumor phenotype is actually recessive or a complex function of multigene interactions (for supporting evidence, see Stanbridge et al. 1982; Cavenee et al. 1983).

The constitutive expression of a series of poly(A)$^+$

RNAs that contain murine leukemia virus LTR–like sequences in cells transformed by carcinogens or radiation (Kirschmeier et al. 1982) suggests that carcinogenesis can also involve a generalized disturbance in mechanisms controlling the transcription of certain families of genes. Consistent with this concept is the evidence that carcinogen-transformed murine cells can also express RNAs homologous to endogenous VL30 sequences. These elements are a dispersed class of moderately repetitive retroviruslike sequences (~100 copies/haploid genome) present in the mouse genome that are 5.2 kb in length and also contain LTR sequences (Courtney et al. 1982). In terms of targets for the action of carcinogens, it is worth noting that, although there are about 18 known cellular oncogenes, the eukaryotic genome contains a vastly greater number of LTR-like elements. In the mouse genome we estimate that, taken together, the LTR sequences associated with endogenous leukemia virus–like sequences, VL30 sequences and intracisternal A-particle sequences amount to more than 1000 copies per haploid genome. The structural resemblance of LTRs to known transposable elements (Temin 1982; Shiba and Saigo 1983), their ability to enhance the transcription of a variety of genes (Blair et al. 1981; Temin 1982), and the phenomena of LTR promoter insertion (Hayward et al. 1981; Bishop 1983) and of insertion mutation by intracisternal A-particle sequences (Rechavi et al. 1982; Gattoni-Celli et al. 1983; Kuff et al. 1983) indicate the capacity of LTRs to act as mobile elements. Thus these elements and other transcriptional enhancer sequences yet to be identified constitute a major repertoire for potential aberrations in the control of gene expression during the course of chemical carcinogenesis. Furthermore, it has long been known that spontaneous and carcinogen-induced rodent tumors often synthesize intracisternal A-particle and C-type retroviruses, presumably reflecting derepression of the corresponding endogenous proviral sequences (for review, see Weinstein et al. 1975). Harel and his colleagues (Hanania et al. 1983) have described the expression of a multigenic family designated ts DNA in a variety of human tumors, and Scott et al. (1983) have identified at least three cellular genes that are often expressed in murine cells transformed by SV40 or a variety of other agents. LTR-like sequences have also been identified in the human genome (Bonner et al. 1982; Noda et al. 1982; Repaske et al. 1983). Disturbances in the function of these and other endogenous enhancer sequences might be related to the frequent aberrations in gene expression (fetal genes, inappropriate isozymes, etc.) and cellular differentiation seen in cancer cells.

Alterations in the state of methylation of cytidine residues in specific DNA sequences may play an important role in some of the above-described alterations in transcription seen in carcinogen-transformed cells. The following findings are consistent with this:

1. There is considerable evidence that in mammalian cells transcriptionally active genes are often hypomethylated (Razin and Riggs 1980).
2. The DNAs of certain tumors show evidence of hypomethylation (Boehm and Drahovsky 1983; Lu et al. 1983).
3. Carcinogens can interfere with DNA methylation, at least in in vitro model systems (Boehm and Drahovsky 1983; Lu et al. 1983; Wilson and Jones 1983).
4. The drug 5-azacytidine, which interferes with DNA methylation, enhances cell transformation (Jones and Taylor 1980; W. Hsiao et al., unpubl.).

Unfortunately our current understanding of the biochemical effects of initiating carcinogens and of tumor promoters is limited largely to describing the initial encounters between these agents and target cells. Entirely new approaches will be required to understand the possible relationship of these early effects to the multifaceted types of genomic changes seen in fully evolved cancer cells. We would emphasize that the various attractive hypotheses described above need not be mutually exclusive, because of the complex and multistage nature of the carcinogenic process, the diversity of causative factors, and the heterogeneity of tumor-cell phenotypes.

Acknowledgments

This research was supported by National Cancer Institute grants CA-021111 and CA-26056, a grant from the National Foundation for Cancer Research, and funds from the Dupont Company. The authors wish to acknowledge the valuable collaboration with Drs. T. Sugimura and H. Fujiki of the National Cancer Center Research Institute (Tokyo), I. Umezawa of the Cancer Institute (Tokyo), and R.E. Moore of the University of Hawaii in the studies on teleocidin and aplysiatoxin. We also thank Dr. Y. Ito for providing the HHPA compounds. We are grateful to Drs. L. Cleveland and B. Erlanger for providing cell lines and to G. Vande Woude for providing c-mos probes. We thank Patricia Kelly for assistance in preparing this manuscript.

References

Backer, J.M. and I.B. Weinstein 1982. Interaction of benzo[a]pyrene and its dihydrodiol epoxide with nuclear and mitochondrial DNA in C3H 10T1/2 cell cultures. *Cancer Res.* **42:** 2764.

———. 1983. A phorbol ester tumor promoter accelerates hydrolysis of ATP by isolated rat liver mitochondria. *Proc. Am. Assoc. Cancer Res.* **24:** 109.

Backer, J.M., M.R. Boersig, and I.B. Weinstein. 1982. Inhibition of respiration by a phorbol ester tumor promoter in murine cultured cells. *Biochem. Biophys. Res. Commun.* **105:** 855.

Barsoum, J. and A. Varshavsky. 1983. Mitogenic hormones and tumor promoters greatly increase the incidence of colony-forming cells bearing amplified dihydrofolate reductase genes. *Proc. Natl. Acad. Sci.* **80:** 5330.

Berenblum, I. 1982. Sequential aspects of chemical carcinogenesis: Skin. In *Cancer: A comprehensive treatise*, 2nd ed. (ed. F.F. Becker), vol. 1, p. 451. Plenum Press, New York.

Birnboim, H.C. 1982. Factors which affect DNA strand breakage in human leukocytes exposed to a tumor promoter, phorbol myristate acetate. *Can. J. Physiol. Pharmacol.* **60:** 1359.

Bishop, J.M. 1983. Cancer genes come of age. *Cell* **32:** 1018.

Bishop, R., R. Martinez, K.D. Nakamura, and M.J. Weber. 1983.

A tumor promoter stimulates phosphorylation on tyrosine. *Biochem. Biophys. Res. Commun.* **115**: 536.

Blair, D.G., M. Oskarsson, T.G. Wood, W.L. McClements, P.J. Fischinger, and G.F. Vande Woude. 1981. Activation of the transforming potential of a normal cell sequence: A molecular model for oncogenesis. *Science* **212**: 941.

Boccara, M., M. Souyri, C. Magarian, E. Stavnezer, and E. Fleissner. 1983. Evidence for a new form of retroviral *env* transcript in leukemic and normal mouse lymphoid cells. *J. Virol.* **48**: 102.

Boehm, T.L.J. and D. Drahovsky. 1983. Alteration of enzymatic methylation of DNA cytosines by chemical carcinogens: A mechanism involved in the initiation of carcinogenesis. *J. Natl. Cancer Inst.* **71**: 429.

Bonner, T.I., C.O. O'Connell, and M. Cohen. 1982. Cloned endogenous retroviral sequences from human DNA. *Proc. Natl. Acad. Sci.* **79**: 4709.

Boone, L.R., F.E. Myer, D.M. Yang, J.O. Kiggans, C. Koh, R.W. Tennant, and W.K. Yang. 1983. Analysis of recombinant DNA clones of the endogenous Balb/c murine leukemia virus WN1802N: Variation in long terminal repeat length. *J. Virol.* **45**: 484.

Casey, J.W., A.R. Roach, J.I. Mullins, K. Bauman Burck, M.O. Nicolson, M.B. Gardner, and N. Davidson. 1981. The U3 portion of feline leukemia virus DNA identifies horizontally acquired proviruses in leukemic cats. *Proc. Natl. Acad. Sci.* **78**: 7778.

Castagna, M., U. Takai, K. Kaibuchi, K. Sano, U. Kikkawa, and Y. Nishizuka. 1982. Direct activation of calcium-activated, phospholipid-dependent protein kinase by tumor promoting phorbol esters. *J. Biol. Chem.* **257**: 7847.

Cavenee, W.K., T.P. Dryja, R.A. Phillips, W.F. Benedict, R. Bodbout, B.L. Gallie, A.L. Murphree, L.C. Strong, and R.L. White. 1983. Expression of recessive alleles by chromosomal mechanisms in retinoblastoma. *Nature* 305: 779.

Colburn, N.H., B.F. Former, K.A. Nelson, and S.H. Yuspa. 1979. Tumor promoter induces anchorage independence irreversibly. *Nature* **281**: 589.

Cooper, G.M. 1982. Cellular transforming genes. *Science* **218**: 801.

Courtney, M.G., L.J. Schmidt, and M.J. Getz. 1982. Organization and expression of endogenous virus-like (VL30) DNA sequences in nontransformed and chemically transformed mouse embryo cells in culture. *Cancer Res.* **42**: 569.

Delclos, K.B., D.S. Nagle, and P.M. Blumberg. 1980. Specific binding of phorbol ester tumor promoters to mouse skin. *Cell* **19**: 1025.

Deleers, M., J.M. Ruysschaert, and W.J. Malaisse. 1982. Interaction between phorbol esters and phospholipid in a monolayer model membrane. *Chem. Biol. Interact.* **42**: 271.

Diamond, A., G.M. Cooper, J. Ritz, and M.A. Lane. 1983. Identification and molecular cloning of the human Blym transforming gene activated in Burkitt's lymphomas. *Nature* **305**: 112.

Duesberg, P.H. 1983. Retroviral transforming genes in normal cells. *Nature* **304**: 219.

Dzarlieva, R.T. and N.E. Fusenig. 1982. Tumor promoter 12-O-tetradecanoyl-phorbol-13-acetate enhances sister chromatid exchanges and numerical and structural chromosome aberrations in primary mouse epidermal cell cultures. *Cancer Lett.* **16**: 7.

Emerit, I., A. Levy, and P. Cerutti. 1983. Suppression of tumor promoter phorbol-myristate acetate-induced chromosome breakage by antioxidants and inhibitors of arachidonic acid metabolism. *Mutat. Res.* **110**: 327.

Fischinger, P.J., H.J. Thiel, M. Lieberman, H.S. Kaplan, N.M. Dunlop, and W.G. Robey. 1982. Presence of a novel recombinant murine leukemia virus-like glycoprotein on the surface of virus-negative C57BL lymphoma cells. *Cancer Res.* **42**: 4650.

Fisher, P.B. and I.B. Weinstein. 1980. Chemical-viral interactions and multistep aspects of cell transformation. *IARC Sci. Publ.* **27**: 113.

Fisher, P.B., J.H. Bozzone, and I.B. Weinstein. 1979. Tumor promoters and epidermal growth factor stimulate anchorage-independent growth of adenovirus transformed rat embryo cells. *Cell* **18**: 695.

Fisher, P.B., R.A. Mufson, A.F. Miranda, H. Fujiki, T. Sugimura, and I.B. Weinstein. 1982. Phorbol ester tumor promoters and teleocidin have similar effects on cell transformation, differentiation and phospholipid metabolism. *Cancer Res.* **42**: 2829.

Gattoni-Celli, S., W.-L. Hsiao, and I.B. Weinstein. 1983. Rearranged c-mos locus in a MOPC-21 murine myeloma cell line and its persistence in hybridomas. *Nature* **306**: 795.

Gattoni-Celli, S., P. Kirschmeier, I.B. Weinstein, J. Escobedo, and D. Dina. 1982. Cellular moloney murine sarcoma ("c-mos") sequences are hypermethylated and transcriptionally silent in normal and transformed rodent cells. *Mol. Cell. Biol.* **2**: 42.

Gilmore, T. and G.S. Martin. 1983. Phorbol ester and diacylglycerol induce phosphorylation at tyrosine. *Nature* **306**: 487.

Hanania, N., D. Shaool, J. Harel, J. Wiels, and T. Tursz. 1983. Common multigenic activation in different human neoplasias. *EMBO J.* **2**: 1621.

Hayward, W.S., B.G. Neel, and S.M. Astrin. 1981. Activation of a cellular onc gene by promoter insertion in ALV-induced lymphoid leukosis. *Nature* **290**: 475.

Hecker, E., N.E. Fusenig, W. Kunz, F. Marks, and H.W. Thielmann, eds. 1982. *Carcinogenesis: A comprehensive survey*, vol. 7. Raven Press, New York.

Hennings, H., R. Shores, M.L. Wenk, E.F. Spangler, R. Tarone, and S. Yuspa. 1983. Malignant conversion of mouse skin tumours is increased by tumour initiators and unaffected by tumour promoters. *Nature* **304**: 67.

Horowitz, A. and I.B. Weinstein. 1982. Receptor binding and cellular effects of tumor promoters. In *Prostaglandins and cancer: First international conference* (ed. T.J. Powles et al.), p. 217. Alan R. Liss, New York.

Horowitz, A., E. Greenebaum, and I.B. Weinstein. 1981. Identification of receptors for phorbol ester tumor promoters in intact mammalian cells and of an inhibitor of receptor binding in biologic fluids. *Proc. Natl. Acad. Sci.* **78**: 2315.

Horowitz, A., H. Fujiki, I.B. Weinstein, A. Jeffrey, E. Okin, R.E. Moore, and T. Sugimura. 1983. Comparative effects of aplysiatoxin, debromoaplysiatoxin and teleocidin on receptor binding and phospholipid metabolism. *Cancer Res.* **43**: 1529.

Hughes, S.H., K. Toyoshima, J.M. Bishop, and H.E. Varmus. 1981. Organization of the endogenous proviruses of chickens: Implications for origin and expression. *Virology* **108**: 189.

Ito, Y., S. Yanase, H. Tokuda, M. Kishishita, H. Ohigashi, M. Hirota, and K. Koshimizu. 1983. Epstein-Barr virus activation by tung oil, extracts of *Aleurites fordii* and its diterpene ester 12-O-hexadecanoyl-16-hydroxyphorbol-13-acetate. *Cancer Lett.* **18**: 87.

Ivanovic, V. and I.B. Weinstein. 1982. Benzo[a]pyrene and other inducers of cytochrome P_1-450 inhibit binding of epidermal growth factor to cell surface receptors. *Carcinogenesis* **3**: 505.

Iwashita, S. and C.F. Fox. 1983. Tumor promoters modulate epidermal growth factor (EGF) receptor phosphorylation in human epidermal carcinoma A431 cells. *Fed. Proc.* **42**: 1902.

Jacobs, S., N.E. Sayyoun, A.R. Saltiel, and P. Cuatracases. 1983. Phorbol esters stimulate the phosphylation of receptors for insulin and somatomedin C. *Proc. Natl. Acad. Sci.* **80**: 6211.

Jones, P.A. and S.M. Taylor. 1980. Cellular differentiation, cytidine analogs and DNA methylation. *Cell* **20**: 85.

Khan, A.S. and M.A. Martin. 1983. Endogenous murine leukemia proviral long terminal repeats contain a unique 190-base-pair insert. *Proc. Natl. Acad. Sci.* **80**: 2699.

Kikkawa, U., Y. Takai, Y. Tanaka, R. Miyake, and Y. Nishizuka. 1983. Protein kinase C as a possible receptor protein of tumor-promoting phorbol esters. *J. Biol. Chem.* **258**: 11442.

Kinsella, A.R. and M. Radman. 1978. Tumor promoters induce sister chromatid exchanges: Relevance to mechanisms of carcinogenesis. *Proc. Natl. Acad. Sci.* **75**: 6149.

Kirschmeier, P., S. Gattoni-Celli, D. Dina, and I.B. Weinstein. 1982. Carcinogen and radiation transformed C3H 10T1/2 cells contain RNAs homologous to the LTR sequence of a murine leukemia virus. *Proc. Natl. Acad. Sci.* **79**: 273.

Klein, G. 1983. Specific chromosomal translocations and the genesis of B-cell-derived tumors in mice and men. *Cell* **32**: 311.

Kuff, E.L., A. Feenstra, K. Lueders, L. Smith, R. Harvey, N. Hozumi, and M. Shulman. 1983. Intracisternal A-particle genes as movable elements in the mouse genome. *Proc. Natl. Acad. Sci.* **80**: 1992.

Lambert, M.E., S. Gattoni-Celli, P. Kirschmeier, and I.B. Weinstein. 1983. Benzo[a]pyrene induction of extrachromosomal viral DNA synthesis in rat cells transformed by polyoma virus. *Carcinogenesis* **4**: 587.

Land, H., L.F. Parada, and R.A. Weinberg. 1983. Tumorigenic conversion of primary embryo fibroblasts requires at least two cooperating oncogenes. *Nature* **304**: 596.

Lavi, S. and S. Etkin. 1981. Carcinogen-mediated induction of SV40 DNA synthesis in SV40 transformed hamster embryo cells. *Carcinogenesis* **2**: 417.

Lee, L.S. and I.B. Weinstein. 1980. Studies on the mechanism by which a tumor promoter inhibits binding of epidermal growth factor to cellular receptors. *Carcinogenesis* **1**: 669.

Lu, L.J.W., E. Randerath, and K. Randerath. 1983. DNA hypomethylation in Morris hepatomas. *Cancer Lett.* **19**: 231.

Miller, E. 1978. Some current perspectives on chemical carcinogenesis in humans and experimental animals: Presidential address. *Cancer Res.* **38**: 1479.

Morris, A.G. 1981. Neoplastic transformation of mouse fibroblasts by murine sarcoma virus; A multi-step process. *J. Gen. Virol.* **53**: 39.

Mufson, R.A., E. Okin, and I.B. Weinstein. 1981. Phorbol esters stimulate the rapid release of choline from cellular phosphatidyl choline. *Carcinogenesis* **2**: 1095.

Nagasawa, H. and J.B. Little. 1979. Effect of tumor promoters, protease inhibitors, and repair processes on X-ray-induced sister chromatid exchanges in mouse cells. *Proc. Natl. Acad. Sci.* **76**: 1943.

Needleman, S.W., Y. Yuasa, S. Srivastava, and S.A. Aaronson. 1983. Normal cells of patients with high cancer risk syndromes lack transforming activity in the NIH/3T3 transfecting assay. *Science* **222**: 173.

Newbold, R.F. and R.W. Overell. 1983. Fibroblast immortality is a prerequisite for transformation by EJ c-Ha-ras oncogene. *Nature* **304**: 648.

Noda, M., M. Kurihara, and T. Takano. 1982. Retrovirus-related sequences in human DNA: Detection and cloning of sequences which hybridize with the long terminal repeat of baboon endogenous virus. *Nucleic Acids Res.* **10**: 2865.

Nusse, R. and H. Varmus. 1982. Many tumors induced by the mouse mammary tumor virus contain a provirus integrated in the same region of the host genome. *Cell* **31**: 99.

Parry, J.M., E.M. Parry, and J.C. Barrett. 1981. Tumor promoters induce mitotic aneuploidy in yeast. *Nature* **294**: 263.

Payne, G.S., J.M. Bishop, and H.E. Varmus. 1982. Multiple arrangements of viral DNA and an activated host oncogene in bursal lymphomas. *Nature* **295**: 209.

Perkins, A., P. Kirschmeier, S. Gattoni-Celli, and I.B. Weinstein. 1983. Development of a new transfection vector containing LTR sequences. *Mol. Cell. Biol.* **3**: 1123.

Poland, A., D. Palen, and E. Glover. 1982. Tumor promotion of TCDD in skin of HRS/J hairless mice. *Nature* **300**: 271.

Potter, M., J.G. Pumphrey, and D.W. Bailey. 1975. Genetics of susceptibility to plasmacytoma induction. Balb/CAuN(c), C57B1/GN(B6), C57BL/Ka(BK), (CXB6)F_1(CXBK)F_1 and CXB recombinant-inbred strains. *J. Natl. Cancer Inst.* **54**: 1413.

Razin, A. and A.D. Riggs. 1980. DNA methylation and gene function. *Science* **210**: 604.

Rechavi, G., D. Givol, and L. Canaani. 1982. Activation of a cellular oncogene by DNA rearrangement: Possible involvement of an IS-like element. *Nature* **300**: 607.

Reddy, E.P., R.K. Reynolds, E. Santos, and M. Barbacid. 1982. A point mutation is responsible for the acquisition of transforming properties by the T24 human bladder carcinoma oncogene. *Nature* **300**: 149.

Repaske, R., R.R. O'Neill, P.E. Steele, and M.A. Martin. 1983. Characterization and partial nucleotide sequence of endogenous type C retrovirus segments in human chromosomal DNA. *Proc. Natl. Acad. Sci.* **80**: 678.

Ruley, H.E. 1983. Adenovirus early region 1A enables viral and cellular transforming genes to transform primary cells in culture. *Nature* **304**: 602.

Scott, M.R.D., K.-H. Westphal, and P.W. Rigby. 1983. Activation of mouse genes in transformed cells. *Cell* **34**: 557.

Shiba, T. and K. Saigo. 1983. Retrovirus-like particles containing RNA homologous to the transposable element *copia* in *Drosophila melanogaster*. *Nature* **302**: 119.

Shimizu, K., D. Birnbaum, M.A. Ruley, O. Fasano, Y. Suard, L. Edlund, E. Taparowsky, M. Goldfarb, and M. Wigler. 1983. Structure of the Ki-*ras* gene of the human lung carcinoma cell line Calu-1. *Nature* **304**: 497.

Slaga, T.J., S.M. Fisher, K. Nelson, and G.L. Gleason. 1980. Studies on the mechanism of skin tumor promotion: Evidence for several stages in promotion. *Proc. Natl. Acad. Sci.* **77**: 3659.

Stanbridge, E.J., C.J. Der, C.J. Doerson, Y. Nishimi, D.M. Peehl, E. Weissman, and J.E. Wilkinson. 1982. Human cell hybrids: Analysis of transformation and tumorigenicity. *Science* **215**: 252.

Sugimura, T. 1982. Potent tumor promoters other than phorbol esters and their significance. *Gann* **73**: 499.

Temin, H.M. 1982. Function of the retrovirus long terminal repeat. *Cell* **28**: 3.

Tran, P.L., L. Ter-Minassian-Saraga, G. Madelmont, and M. Castagna. 1983a. Tumor promoter 12-O-tetradecanoyl phorbol 13-acetate alters state, fluidity and hydration of 1,2-diacyl-sn-glycero-3-phosphocholine bylayers. *Biochim. Biophys. Acta* **727**: 31.

Tran, P.L. M. Castagna, M. Sala, G. Vassent, A.D. Horowitz, D. Schachter, and I.B. Weinstein. 1983b. Differential effects of tumor promoters on phorbol ester receptor binding and membrane fluorescence anisotrophy in C3H 10T1/2 cells. *Eur. J. Biochem.* **130**: 155.

Troll, W., G. Witz, B. Goldstein, D. Stone, and T. Sugimura. 1982. The role of free oxygen radicals in tumor promotion and carcinogenesis. In *Cocarcinogenesis and biological effects* (ed. E. Hecker et al.), p. 593. Raven Press, New York.

Umezawa, K., I.B. Weinstein, A. Horowitz, H. Fujiki, T. Matsushima, and T. Sugimura. 1981. Similarity of teleocidin B and phorbol ester tumor promoters in effects on membrane receptors. *Nature* **290**: 411.

Weinberg, R.A. 1983. A molecular basis of cancer. *Science Am.* **249**: 126.

Weinstein, I.B. 1981. Current concepts and controversies in chemical carcinogenesis. *J. Supramol. Struct.* **17**: 99.

———. 1983. Protein kinase, phospholipid and control of growth. *Nature* **302**: 705.

Weinstein, I.B., N. Yamaguchi, R. Gebert, and M.E. Kaighn. 1975. The use of epithelial cell cultures for studies on the mechanism of transformation by chemical carcinogens. *In Vitro* **2**: 130.

Weinstein, I.B., A.D. Horowitz, P.B. Fisher, V. Ivanovic, S. Gattoni-Celli, and P. Kirschmeier. 1982. Mechanisms of multistage carcinogenesis and their relevance to tumor cell heterogeneity. In *Tumor cell heterogeneity: Origins and implications* (ed. A.H. Owens et al.), p. 261. Academic Press, New York.

Weiss, R., N. Teich, H. Varmus, and J. Coffin, eds. 1982. *Molecular biology of tumor viruses, 2nd edition: RNA tumor viruses*. Cold Spring Harbor Laboratory, Cold Spring Harbor, New York.

Wilson, V.L. and P.A. Jones. 1983. Inhibition of DNA methylation by chemical carcinogens *in vitro*. *Cell* **32**: 329.

Protein Kinase C and the Mechanism of Action of Tumor Promoters

U. Kikkawa,* R. Miyake,* Y. Tanaka,* Y. Takai,*† and Y. Nishizuka*†

*Department of Biochemistry, Kobe University School of Medicine, Kobe 650, Japan;
and †Department of Cell Biology, National Institute for Basic Biology, Okazaki 444, Japan

Pleiotropic actions of tumor-promoting phorbol esters such as 12-O-tetradecanoylphorbol-13-acetate (TPA) appear to be initiated by interaction with a specific cell-surface receptor. However, neither the mechanism of action nor the precise nature of the receptor has been clarified. Preceding reports from our laboratory have provided evidence that protein kinase C (Ca^{++}-activated, phospholipid-dependent protein kinase) is fully activated by tumor-promoting phorbol esters in both in vivo and in vitro systems (Castagna et al. 1982; Yamanishi et al. 1983). Under normal conditions, protein kinase C is activated by reversible association with a membrane phospholipid in the presence of physiologically low concentrations of Ca^{++}, and this activation process is dependent on the presence of diacylglycerol, which may be transiently produced in situ from inositol phospholipids through a signal-mediated reaction (Takai et al. 1979). Tumor-promoting phorbol esters and perhaps some other tumor promoters, such as mezerein, may be intercalated into the membrane phospholipid bilayer, where they activate the protein kinase by substituting for diacylglycerol. It is plausible that most of the pleiotropic actions of the tumor promoter may be mediated through the action of protein kinase C. This protein kinase is distributed widely in various tissues and appears to fulfill the requirements for a receptor that eventually leads to the activation of cellular functions and proliferation (Nishizuka 1983).

Experimental Procedures

Materials and chemicals
Homogeneous protein kinase C was prepared from rat brain soluble fraction as described by Kikkawa et al. (1982). [^{3}H]Phorbol-12,13-dibutyrate (PDBu) (17.5 Ci/mmole) was a product of New England Nuclear. TPA and PDBu were obtained from P-L Biochemicals, and other phorbol derivatives were obtained from P. Borchert (Eden Prairie, Minn.). 1-Oleoyl-2-acetylglycerol was synthesized as described by Kaibuchi et al. (1983). Other chemicals and materials were prepared as described earlier by Takai et al. (1979).

Binding assay
Binding of the phorbol ester to protein kinase C was assayed with [^{3}H]PDBu. The complete assay mixture (0.2 ml) contained 4 µmoles of Tris/malate at pH 6.8, 20 µmoles of KCl, 30 nmoles of $CaCl_2$, 20 µg of a chromatographically pure sample of phosphatidylserine, 10 pmoles of [^{3}H]PDBu (1.51×10^4 cpm/pmole), 100 ng of protein kinase C (this preparation contained 10 nmoles of EGTA), and 0.5% (final concentration) of dimethyl sulfoxide. After 20 minutes, a 0.1-ml aliquot of the mixture was taken and subjected to gel filtration on a column of TSK 3000SW, which was attached to HPLC (Toyo Soda Model HLC-803D). The column was developed with a buffer solution containing 20 mM Tris/malate at pH 6.8, 100 mM KCl, 0.15 mM $CaCl_2$, 0.05 mM EGTA, and 0.5% dimethyl sulfoxide at 23°C. Free Ca^{++} concentration is calculated to be 0.1 mM under these conditions. Other detailed conditions were described elsewhere (Kikkawa et al. 1983). The radioactivity of nonspecific binding was negligible.

Determinations of platelet activation
Human platelets were labeled with ^{32}P or [2-^{14}C]serotonin under the conditions specified earlier (Sano et al. 1983). The radioactive platelets were stimulated as indicated in each experiment. Radioactive platelet proteins were subjected to SDS–slab gel electrophoresis, stained, dried on a filter paper, and then exposed to an X-ray film to prepare the autoradiograph. The relative intensity of each band was quantitated by densitometric tracing at 430 nm. The release of radioactive serotonin from platelets was determined as described by Sano et al. (1983).

Other procedures
Protein kinase C was assayed with H1 histone as phosphate acceptor as described by Kikkawa et al. (1982). Ca^{++}, the phospholipid, and the phorbol ester were added to the reaction mixture as indicated. The radioactivity of ^{3}H-, ^{32}P-, and ^{14}C-labeled samples was determined in a toluene-based scintillator. The amount of protein was quantitated as described previously (Kikkawa et al. 1982).

Results and Discussion

A proposed pathway of signal transduction
A wide variety of hormones, neurotransmitters, growth factors, and many other biologically active substances that activate cellular functions and proliferation provoke rapid breakdown of inositol phospholipids (phosphatidylinositol and/or its polyphosphates) in their target tissues. In general, the stimulation of most of these receptors immediately mobilizes Ca^{++}, often releases arachidonic acid, and increases cGMP but not cAMP. A recent series of studies in this laboratory has clarified

that this inositol phospholipid breakdown may serve as a signal for transmembrane control of cellular functions and proliferation through activation of protein kinase C (Nishizuka 1983). Figure 1 schematically outlines the pathway of the proposed signal transduction.

Protein kinase C absolutely requires Ca^{++} and phospholipids for enzymatic activity. Under normal conditions diacylglycerol, which is transiently produced from the inositol phospholipids in a signal-dependent manner, increases the affinity of this protein kinase for Ca^{++} dramatically to less than the 10^{-6} M range and thereby renders the enzyme fully active without a net increase in the Ca^{++} concentration (Takai et al. 1979). Thus, the signal-induced activation of protein kinase C is biologically independent of Ca^{++}, since its Ca^{++} sensitivity is modulated. Among the various membrane phospholipids, phosphatidylserine is indispensable, but other phospholipids show positive or negative cooperativity in the activation of this enzyme. For instance, in the presence of phosphatidylethanolamine as an additional lipid component, the enzyme is fully active at the 10^{-7} M range of Ca^{++}, whereas phosphatidylcholine and sphingomyelin are inhibitory. Presumably, the asymmetric distribution of various phospholipids in the lipid bilayer seems to favor the activation of protein kinase C.

Recent experiments with intact cell systems have shown that under appropriate conditions a synthetic diacylglycerol such as 1-oleoyl-2-acetylglycerol activates protein kinase C directly without interaction with cell-surface receptors, although this diacylglycerol is rapidly metabolized in situ to its corresponding phosphatidic acid (Kaibuchi et al. 1983). Using the synthetic diacylglycerol and Ca^{++} ionophore, it is possible to demonstrate that protein kinase C activation and Ca^{++} mobilization are both essential and synergistically effective for eliciting full physiological cellular response.

Direct action of phorbol ester on protein kinase C

Preceding reports from our laboratory (Castagna et al. 1982; Yamanishi et al. 1983) have shown that the tumor-promoting phorbol ester TPA activates protein kinase C when tested in vitro as well as in vivo. Kinetic analysis indicates that the tumor promoter is able to substitute for diacylglycerol and greatly increases the affinity of this enzyme for Ca^{++} to the 10^{-7} M range, as well. Experiments with [^{3}H]PDBu, another potent tumor-promoting phorbol ester, indicate that this radioactive tumor promoter may bind to a homogeneous preparation of protein kinase C only in the presence of Ca^{++} and phospholipid, resulting in the concomitant activation of this enzyme. This phorbol ester does not bind simply to protein kinase C nor to phospholipid per se irrespective of the presence and absence of Ca^{++}, and all four components mentioned above are needed for the binding as well as for the enzymatic activity.

Figure 2 shows the effect of PDBu concentration on the activation of the protein kinase and also on the formation of the quaternary complex mentioned above. An apparent activation constant (K_a) of PDBu is calculated to be 8 nM by Lineweaver-Burk double reciprocal plots. The K_a value is identical to the dissociation binding constant (K_d) of the tumor promoter, i.e., 8 nM. This value is remarkably similar to the K_d value of 5.6 nM that has been reported for brain particulate fractions (Dunphy et al. 1981), and also to the K_d values of 8 nM and 10 nM that are estimated for the specific binding sites located on intact cell membranes of rat embryo fibroblasts (Horowitz et al. 1981) and mouse epidermal cells (Solanki and Slaga 1981), respectively. Ashendel et al. (1983) and Sando and Young (1983) have reported recently that both Ca^{++} and a phospholipid are necessary for the binding of phorbol ester to its receptor, which is perhaps protein kinase C.

Scatchard analysis of the binding indicates that approximately one molecule of [^{3}H]PDBu binds to every molecule of protein kinase C in the presence of physiological concentrations of Ca^{++} and apparently an excess of phospholipid. Among membrane phospholipids tested, phosphatidylserine is most effective to support the binding as observed for the activation of protein kinase C, and the relative activities of various phospholipids are nearly equal to those for this protein kinase activation. The amount of phosphatidylserine needed for the binding far exceeds the amount of protein kinase C

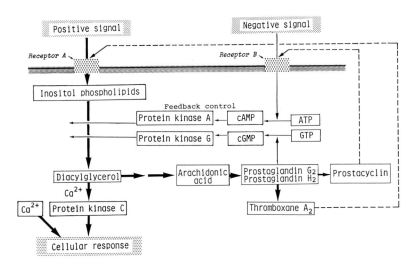

Figure 1 A proposed pathway of signal transduction. (Protein kinase A) cAMP-dependent protein kinase; (Protein kinase G) cGMP-dependent protein kinase. (Reprinted, with permission, from Nishizuka 1983.)

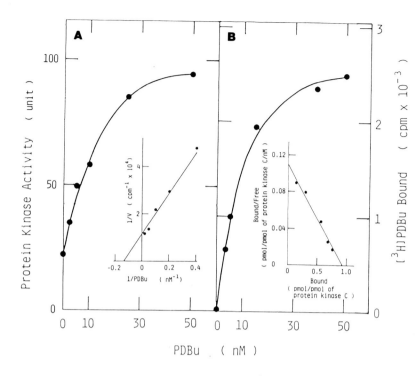

Figure 2 Effect of PDBu concentration on activation of protein kinase C and on its binding to protein kinase C. (*A*) Protein kinase activity. Protein kinase C was assayed at various concentrations of PDBu, as indicated, in the presence of 1×10^{-4} M of $CaCl_2$ and 8 μg/ml of phosphatidylserine. (*Inset*) Lineweaver-Burk double reciprocal plots. (*B*) [^{3}H]PDBu-binding. Binding assay was carried out at various concentrations of [^{3}H]PDBu as indicated. (*Inset*) Scatchard plots. Other detailed conditions are described elsewhere (Kikkawa et al. 1983).

as well as that of PDBu, and phosphatidylserine in the binding assay mixture is saturated at approximately 100 μg/ml. However, since the phospholipid is water-insoluble, the concentration mentioned above does not necessarily indicate the actual physiological picture. Nevertheless, it is suggestive that in intact cell systems one molecule of tumor-promoting phorbol ester, which is intercalated into the lipid bilayer, may attract one molecule of protein kinase C to the membrane, where the enzyme is fully activated to exhibit its apparently multifunctional catalytic activity.

Protein kinase C as a possible receptive protein of phorbol ester

It has been previously described that various phorbol derivatives that show tumor-promoting activity are all capable of activating protein kinase C in in vitro—purified systems and that the structural requirements of phorbol-related diterpenes for tumor promotion appear to be roughly similar to those for protein kinase C activation (Castagna et al. 1982). It has also been shown for intact cell systems that TPA is a potent competitor for [^{3}H]PDBu binding, whereas phorbol and 4-α-phorbol-12,13-didecanoate, which lack tumor-promoting activity, do not interfere with the binding of [^{3}H]PDBu to its receptor (Shoyab and Todaro 1980; Horowitz et al. 1981; Solanki and Slaga 1981). Consistent with these observations, Table 1 shows that TPA, nonradioactive PDBu, and phorbol-12,13-didecanoate, which are all able to promote tumor development in vivo, can activate protein kinase C in vitro and also inhibit the binding of [^{3}H]PDBu in a competitive manner. Inversely, phorbol derivatives having no tumor-promoting activity are all unable to activate the enzyme and cannot compete with [^{3}H]PDBu for the binding. Under similar conditions, diacylglycerols, such as diolein, also prevent the binding of [^{3}H]PDBu to produce the quaternary complex, whereas mono-olein, triolein, and free oleic acid are inactive in this capacity.

The receptor of the tumor-promoting phorbol esters is shown to be distributed widely in many tissues, with brain tissue having the highest concentrations (Shoyab and Todaro 1980; Shoyab et al. 1981). Protein kinase C, first found as a proteolytically activated protein kinase, is also distributed widely in various tissues and organs (Inoue et al. 1977; Kuo et al. 1980; Minakuchi et al. 1981). The distribution patterns of both entities are apparently very similar. For instance, in rat brain homogenates approximately 12 pmoles of [^{3}H]PDBu bind to 1 mg of protein (Shoyab et al. 1981). This value roughly matches the amount of protein kinase C in this tissue, which is estimated to be approximately 16 pmoles/mg protein, assuming the molecular weight as 77,000 with an overall purification of 800-fold (Kikkawa et al. 1982). Under normal conditions, in most tissues protein kinase C is present largely in the soluble fractions but is activated by association with membrane phospholipids as directly by diacylglycerol as by a tumor-promoting phorbol ester, as described above. Thus, Kraft and Anderson (1983) have observed that in the presence of TPA, most of the protein kinase C may be recovered in a form tightly associated with particulate fractions. The results briefly presented above strongly suggest that protein kinase C is the phorbol ester–receptive protein and that the pleiotropic actions, if not all the actions, of the tumor promoter may be mediated through the action of this enzyme. Preliminary experiments with mezerein show that this tumor promoter also activates protein kinase C directly (Table 1). It is possible that protein kinase C may lie on a common pathway

Table 1 Effects of Tumor Promoters on Activation of Protein Kinase C and Inhibition of [³H]PDBu-binding to Protein Kinase C

Tumor promoter added	Activation of protein kinase C (%)	Inhibition of [³H]PDBu-binding (%)	Tumor-promoting activity
TPA	100	100	+++
PDBu	88	100	++
Phorbol-12,13-didecanoate	81	100	++
Phorbol-12-tetradecanoate	0	18	−
Phorbol-13-acetate	0	0	−
4-α-Phorbol-12,13-didecanoate	0	0	−
Phorbol	0	23	−
Mezerein	87[a]	—[b]	+

Activity of protein kinase C was measured in the presence of 1×10^{-5} M of $CaCl_2$, 20 µg/ml of phospholipid, and 10 ng/ml each of phorbol derivatives. Inhibition of [³H]PDBu-binding to protein kinase C was assayed in the presence of a 100-fold excess of each nonradioactive phorbol derivatives. Detailed experimental conditions are described elsewhere (Kikkawa et al. 1983).
[a]Mezerein was added at the concentration of 100 ng/ml.
[b]Not determined.

eventually leading to tumor promotion. It is noted that Niedel et al. (1983) have partially purified a PDBu-binding protein together with protein kinase C from rat brain.

Role of Ca^{++} in the action of phorbol ester

Like the synthetic diacylglycerol described above, TPA is shown to be intercalated into the membrane and to directly activate protein kinase C in intact cell systems without inositol phospholipid breakdown nor with Ca^{++} mobilization (Castagna et al. 1982; Yamanishi et al. 1983). When platelets are stimulated by TPA, an endogenous protein having an approximate molecular weight of 40,000 (40K protein) is as rapidly and heavily phosphorylated as it is by natural extracellular messengers such as thrombin. The radioactive 40K protein has been isolated from platelets stimulated by TPA and then subjected to fingerprint analysis. The mapping pattern of the tryptic phosphopeptides thus obtained is exactly identical with the mapping pattern obtained from the 40K protein preparation that is phosphorylated in vitro by homogeneous protein kinase C. The results provide evidence that TPA is capable of activating protein kinase C in intact cell systems, as well.

It has been well known that when platelets are activated by natural messengers, the myosin light chain having a molecular weight of 20,000 (20K protein) is rapidly phosphorylated (Lyons et al. 1975). This reaction is calmodulin-dependent, and Ca^{++} mobilization is absolutely needed for the phosphorylation of the 20K protein. When platelets are stimulated by thrombin, both 40K and 20K proteins are concurrently phosphorylated, since inositol phospholipid breakdown and Ca^{++} mobilization are induced. On the other hand, it is shown that TPA, like synthetic diacylglycerol (Kaibuchi et al. 1983), induces the phosphorylation of only the 40K protein but not the 20K protein, particularly at lower concentrations. As mentioned above, TPA dramatically increases the affinity of protein kinase C for Ca^{++} and thereby fully activates the enzyme without a net increase of Ca^{++} concentrations. Inversely, the 20K protein is selectively phosphorylated by the addition of low concentrations of A23187 (0.2–0.4 µM), and this reaction rate is not significantly affected by TPA. Thus, under appropriate conditions protein kinase C activation and Ca^{++} mobilization are selectively and independently induced by TPA and the Ca^{++} ionophore, respectively, as shown in Figure 3.

In the experiment shown in Figure 4, platelets are stimulated by various concentrations of TPA. Under the given conditions, the 40K protein is phosphorylated to nearly the same extent in both the presence and absence of A23187. However, serotonin is not released sufficiently by the addition of TPA alone, but the full cellular response may be observed by the simultaneous addition of TPA and A23187. TPA alone at higher concentrations (100 ng/ml) causes release of serotonin in a significant quantity (Castagna et al. 1982), although the reason for this enhanced release is unknown. Perhaps, TPA itself at higher concentrations perturbs the membrane structure and slightly mobilizes Ca^{++} by an unknown mechanism. It is also noted that the Ca^{++} ionophore at higher concentrations (>1.0 µM) induces phosphorylation of both the 40K and 20K proteins and causes release reaction. This is most likely due to activation of various phospholipases as well as of protein kinase C by a large increase in Ca^{++} concentrations. Nevertheless, it is reasonable to suggest that the protein kinase activation and Ca^{++} mobilization are equally essential and synergistically effective for eliciting the full physiological response. TPA, particularly at lower concentrations, appears to induce only the activation of this protein kinase system so that the cell may always be ready to function and proliferate when Ca^{++} is available.

Relation to cyclic nucleotides

In *bidirectional control* systems in many tissues, such as platelets, lymphocytes, neutrophils, and smooth muscles, the receptors that induce inositol phospholipid breakdown and Ca^{++} mobilization generally promote the activation of cellular functions and proliferation, whereas the receptors that produce cAMP normally antagonize such activation by preventing the signal-dependent

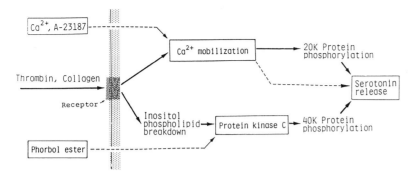

Figure 3 Synergistic roles of phorbol esters and calcium.

breakdown of inositol phospholipids and perhaps mobilization of Ca^{++} concomitantly (Fig. 1) (Nishizuka 1983). In contrast, in *monodirectional control* systems in some tissues, such as hepatocytes and adipocytes, these two classes of receptors appear not to interact with each other but to function independently (Kaibuchi et al. 1982). It has also been described earlier (Takai et al. 1982) that cGMP does not antagonize cAMP but similarly inhibits the signal-dependent breakdown of inositol phospholipids and thereby provides an immediate feedback control that prevents overresponse, although the exact signal pathway to increase cGMP has not been established unequivocally (Fig. 1). In the experiment summarized in Figure 5, it is shown that both dibutyryl cAMP and 8-bromo cGMP are inhibitory for the thrombin-induced activation of protein kinase C, whereas none of these cyclic nucleotides blocks the TPA-induced activation of this enzyme, as judged by phosphorylation of the 40K protein described above. TPA is intercalated into the membrane for prolonged periods of time since this diterpene is hardly metabolizable. Thus, the cell tends to function and proliferate since protein kinase C is always active, despite the feedback control of cyclic nucleotides. In contrast, diacylglycerol is produced only transiently and disappears rapidly when cell-surface receptors are stimulated.

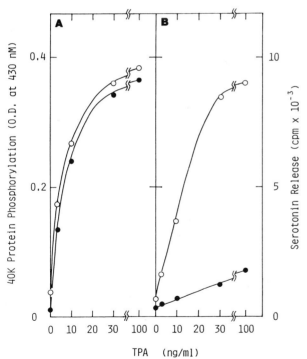

Figure 4 Effects of TPA and the Ca^{++} ionophore on 40K protein phosphorylation and serotonin release. Human platelets were labeled with either ^{32}P or [2-^{14}C]serotonin and then stimulated by various concentrations of TPA in the presence or absence of A23187 (0.2 μM). Phosphorylation of the 40K protein and release of serotonin were estimated as described in Experimental Procedures. Other detailed conditions are described elsewhere (Yamanishi et al. 1983). (*A*) 40K protein phosphorylation; (*B*) serotonin release. (○) In the presence of A23187; (●) in the absence of A23187.

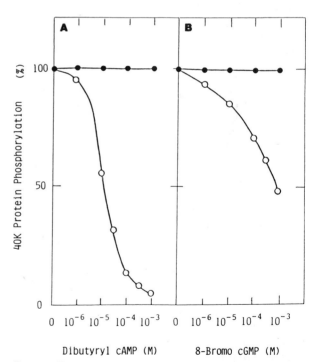

Figure 5 Effects of cyclic nucleotide derivatives on 40K protein phosphorylation induced by thrombin and TPA. Human platelets labeled with ^{32}P were incubated with various concentrations of dibutyryl cAMP or 8-bromo cGMP as indicated and then stimulated by thrombin or TPA. Phosphorylation of the 40K protein was estimated as described in Experimental Procedures; other detailed conditions are described elsewhere (Yamanishi et al. 1983). (*A*) Effect of dibutyryl cAMP; (*B*) effect of 8-bromo cGMP. (○) Stimulated by thrombin; (●) stimulated by TPA.

Summary

Tumor-promoting phorbol esters, and perhaps other tumor promoters such as mezerein, may be intercalated into the membrane phospholipid bilayer and selectively bind to protein kinase C to exhibit its full enzymatic activity. This process is *biologically* Ca^{++}-independent, and it is plausible that when Ca^{++} is increased, this protein kinase may play roles of crucial importance to activate cellular functions and proliferation. Protein kinase C is most likely a receptive protein of tumor promoters, and exploration of the roles of this enzyme seems to provide critical clues for clarifying the mechanism of cell growth and differentiation.

Acknowledgments

The authors express their indebtedness to collaborators who have so efficiently carried out the experiments related to the present article. The authors are particularly grateful to Drs. M. Castagna, J. Yamanishi, K. Kaibuchi, and K. Sano. The skillful secretarial assistance by Mrs. S. Nishiyama and Miss K. Yamasaki are also greatly appreciated. This investigation has been supported in part by research grants from the Research Fund of the Ministry of Education, Science and Culture, the Intractable Diseases Division, Public Health Bureau, the Ministry of Health and Welfare, a Grant-in-Aid of New Drug Development from the Ministry of Health and Welfare, the Science and Technology Agency, Japan, and the Yamanouchi Foundation for Research on Metabolic Disorders.

References

Ashendel, C.L., J.M. Staller, and R.K. Boutwell. 1983. Identification of a calcium- and phospholipid-dependent phorbol ester binding activity in the soluble fraction of mouse tissues. *Biochem. Biophys. Res. Commun* **111:** 340.

Castagna, M., Y. Takai, K. Kaibuchi, K. Sano, U. Kikkawa, and Y. Nishizuka. 1982. Direct activation of calcium-activated, phospholipid-dependent protein kinase by tumor-promoting phorbol esters. *J. Biol. Chem.* **257:** 7847.

Dunphy, W.G., R.J. Kochenburger, M. Castagna, and P.M. Blumberg. 1981. Kinetics and subcellular localization of specific [³H]phorbol 12,13-dibutyrate binding by mouse brain. *Cancer Res.* **41:** 2640.

Horowitz, A.D., E. Greenebaum, and I.B. Weinstein. 1981. Identification of receptors for phorbol ester tumor promoters in intact mammalian cells and of an inhibitor of receptor binding in biologic fluids. *Proc. Natl. Acad. Sci.* **78:** 2315.

Inoue, M., A. Kishimoto, Y. Takai, and Y. Nishizuka. 1977. Studies on a cyclic nucleotide-independent protein kinase and its proenzyme in mammalian tissues. II. Proenzyme and its activation by calcium-dependent protease from rat brain. *J. Biol. Chem.* **252:** 7610.

Kaibuchi, K., Y. Takai, M. Sawamura, M. Hoshijima, T. Fujikura, and Y. Nishizuka. 1983. Synergistic functions of protein phosphorylation and calcium mobilization in platelet activation. *J. Biol. Chem.* **258:** 6701.

Kaibuchi, K., Y. Takai, Y. Ogawa, S. Kimura, Y. Nishizuka, T. Nakamura, A. Tomomura, and A. Ichihara. 1982. Inhibitory action of adenosine 3',5'-monophosphate on phosphatidylinositol turnover: Difference in tissue response. *Biochem. Biophys. Res. Commun.* **104:** 105.

Kikkawa, U., Y. Takai, R. Minakuchi, S. Inohara, and Y. Nishizuka. 1982. Calcium-activated, phospholipid-dependent protein kinase from rat brain. Subcellular distribution, purification, and properties. *J. Biol. Chem.* **257:** 13341.

Kikkawa, U., Y. Takai, Y. Tanaka, R. Miyake, and Y. Nishizuka. 1983. Protein kinase C as a possible receptor protein of tumor-promoting phorbol esters. *J. Biol. Chem.* **258:** 11442.

Kraft, A.S. and W.B. Anderson. 1983. Phorbol esters increase the amount of Ca^{2+}, phospholipid-dependent protein kinase associated with plasma membrane. *Nature* **301:** 621.

Kuo, J.F., R.G.G. Andersson, B.C. Wise, L. Mackerlova, I. Salomonsson, N.L. Brackett, N. Katoh, M. Shoji, and R.W. Wrenn. 1980. Calcium-dependent protein kinase: Widespread occurrence in various tissues and phyla of the animal kingdom and comparison of effects of phospholipids, calmodulin, and trifluoperazine. *Proc. Natl. Acad. Sci.* **77:** 7039.

Lyons, R.M., N. Stanford, and P.W. Majerus. 1975. Thrombin-induced protein phosphorylation in human platelets. *J. Clin. Invest.* **56:** 924.

Minakuchi, R., Y. Takai, B. Yu, and Y. Nishizuka. 1981. Widespread occurrence of calcium-activated, phospholipid-dependent protein kinase in mammalian tissues. *J. Biochem.* **89:** 1651.

Niedel, J.E., L.J. Kuhn, and G.R. Vandenbark. 1983. Phorbol diester receptor copurifies with protein kinase C. *Proc. Natl. Acad. Sci.* **80:** 36.

Nishizuka, Y. 1983. Phospholipid degradation and signal translation for protein phosphorylation. *Trends Biochem. Sci.* **8:** 13.

Sando, J.J. and M.C. Young. 1983. Identification of high-affinity phorbol ester receptor in cytosol of EL4 thymoma cells: Requirement for calcium, magnesium, and phospholipids. *Proc. Natl. Acad. Sci.* **80:** 2642.

Sano, K., Y. Takai, J. Yamanishi, and Y. Nishizuka. 1983. A role of calcium-activated phospholipid-dependent protein kinase in human platelet activation. Comparison of thrombin and collagen actions. *J. Biol. Chem.* **258:** 2010.

Shoyab, M. and G.J. Todaro. 1980. Specific high affinity cell membrane receptors for biologically active phorbol and ingenol esters. *Nature* **288:** 451.

Shoyab, M., T.C. Warren, and G.J. Todaro. 1981. Tissue and species distribution and developmental variation of specific receptors for biologically active phorbol and ingenol esters. *Carcinogenesis* **2:** 1273.

Solanki, V. and T.J. Slaga. 1981. Specific binding of phorbol ester tumor promoters to intact primary epidermal cells from Sencar mice. *Proc. Natl. Acad. Sci.* **78:** 2549.

Takai, Y., K. Kaibuchi, K. Sano, and Y. Nishizuka. 1982. Counteraction of calcium-activated, phospholipid-dependent protein kinase activation by adenosine 3',5'-monophosphate and guanosine 3',5'-monophosphate in platelets. *J. Biochem.* **91:** 403.

Takai, Y., A. Kishimoto, U. Kikkawa, T. Mori, and Y. Nishizuka. 1979. Unsaturated diacylglycerol as a possible messenger for the activation of calcium-activated, phospholipid-dependent protein kinase system. *Biochem. Biophys. Res. Commun.* **91:** 1218.

Yamanishi, J., Y. Takai, K. Kaibuchi, K. Sano, M. Castagna, and Y. Nishizuka. 1983. Synergistic functions of phorbol ester and calcium in serotonin release from human platelets. *Biochem. Biophys. Res. Commun.* **112:** 778.

Membrane and Cytosolic Receptors for the Phorbol Ester Tumor Promoters

P.M. Blumberg, N.A. Sharkey, B. König, S. Jaken, K.L. Leach, and A.Y. Jeng

Molecular Mechanisms of Tumor Promotion Section, Laboratory of Cellular Carcinogenesis and Tumor Promotion, National Cancer Institute, Bethesda, Maryland 20205

Considerable evidence both from human epidemiology and from animal experiments indicates that carcinogenesis is a multistage process. At least in the mouse-skin model system, carcinogenesis can be divided into the mechanistically distinct stages of initiation and tumor promotion (Scribner and Süss 1978; Yuspa 1981, 1982; Hecker et al. 1982). The process of tumor promotion itself can be further subdivided into first- and second-stage promotion, distinguished by different apparent structure-activity relations and profiles of inhibitors for each stage. Among mouse-skin tumor promoters, the phorbol esters are one of the two most potent classes of compounds and the class that has been most extensively studied.

In addition to their activity as tumor promoters, the phorbol esters have profound effects on a variety of biological systems (Blumberg 1980; Diamond et al. 1980; Hecker et al. 1982). These effects include (1) potentiation of other cellular effectors (e.g., epidermal growth factor), (2) inhibition and/or induction of cellular differentiation and differentiated cell functions, (3) modulation of membrane activities, and (4) reversible induction in normal cells of a partially transformed phenotype. A second motivation for understanding the mechanism of action of the phorbol esters is that these studies may shed light on control of these other biological processes as well.

Substantial excitement has been generated over the past year in the tumor promotion field by the fusion of two lines of research. Nishizuka and co-workers have characterized a Ca^{++}-phospholipid-dependent protein kinase (protein kinase C), which they believe partially mediates responses to a large class of cellular effectors whose action is associated with enhanced phosphatidylinositol turnover (Takai et al. 1982; Nishizuka 1983). Our laboratory, and subsequently others, has demonstrated the existence of receptors for the phorbol esters and documented their role in phorbol ester responses (Blumberg et al. 1982, 1983a,b,c, and references therein). The properties of the phorbol ester receptors and of protein kinase C showed marked resemblances (Table 1). Over the past year, Castagna, Nishizuka, and co-workers (Castagna et al. 1982) have reported that the phorbol ester can directly stimulate protein kinase C. We and other laboratories, including those of Niedel, Sando, and Boutwell, have shown that the phorbol ester aporeceptor and protein kinase C copurify (Leach et al. 1983; Niedel et al. 1983; Sando and Young 1983), and Ashendel et al. (1983) have shown copurification with solubilized membrane receptor.[1]

The two approaches, analysis of enzymatic and binding activities, provide complementary insights into the nature of the phorbol ester receptor/C kinase. On the one hand, the enzymatic assays permit dissection of the subsequent steps in the biochemical cascades activated by the phorbol esters. Conversely, the phorbol ester binding provides the best evidence for the functional role of protein kinase C in most systems and for analyzing the membrane-bound form of the enzyme.

Labeling of brain particulate preparations with a photoactivatable phorbol ester derivative

To obtain information about the nature of the phorbol ester binding site, we synthesized the photoactivatable phorbol ester derivative phorbol-12-p-azidobenzoate 13-benzoate (PaBB) (Delclos et al. 1983). PaBB is an analog of phorbol-12,13-dibenzoate, a derivative that possesses moderate tumor-promoting activity and that has similar binding affinity to phorbol-12,13-dibutyrate (PDBu). In the dark, [^{3}H]PaBB bound reversibly to brain particulate preparations with high affinity ($K_d = 0.8$ nM). Following incubation of the [^{3}H]PaBB in the dark with the brain particulate preparation, UV irradiation rendered the specific binding irreversible with good efficiency (35-45%).

The specific irreversible binding was protease-sensitive; pretreatment of membranes with papain reduced the specific reversible and irreversible binding in parallel. Nonetheless, analysis of the specific adducts indicated that they were extractable into chloroform-methanol. Fractionation by thin-layer chromatography (TLC) yielded two peaks of specific label (Fig. 1). Both peaks were degraded by phospholipase A_2, which shifted their

Table 1 Similarities between the Phorbol Ester Receptor and Protein Kinase C

1. Tissue distribution
2. Absolute level in brain
3. Evolutionary conservation
4. High Ca^{++} sensitivity
5. Phospholipid association

---

[1]The term phorbol ester aporeceptor will be used for activity characterized operationally by binding, and protein kinase C will be used for activity characterized by enzymatic assay.

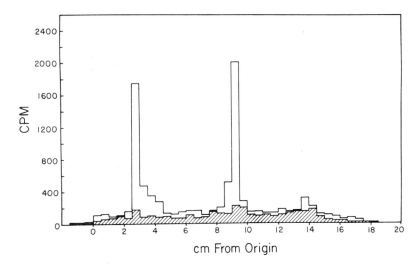

Figure 1 Analysis by TLC of the lipid extract from [³H]PaBB-labeled mouse brain particulate fraction. (□) Sample incubated and irradiated in the absence of nonradioactive PDBu; (▨) sample incubated and irradiated in the presence of 30 μM nonradioactive PDBu (Reprinted, with permission, from Delclos et al. 1983.)

mobilities to a position near the solvent front, consistent with the phorbol ester being attached to the side chains released from position 2 of phospholipids. The lower-mobility peak was identified as a phorbol ester–phosphatidylserine adduct based on its susceptibility to degradation by the specific enzyme phosphatidylserine decarboxylase. The resulting product, a phorbol ester–phosphatidylethanolamine adduct, cochromatographed with the second peak.

Brain contains a significant amount of alk-1-enyl 2-acyl phosphoglycerides (plasmalogens), particularly in the phosphatidylethanolamine fraction. These are not distinguishable from the corresponding diacyl phosphoglycerides by their mobility on TLC but can be identified by their sensitivities to $HgCl_2$. Treatment with $HgCl_2$ degraded the putative phorbol ester–phosphatidylethanolamine adduct by 60%, which corresponds to the proportion of plasmalogen in brain phosphatidylethanolamine.

Of the two phospholipids specifically labeled by [³H]PaBB, phosphatidylserine is most efficient at reconstitution of protein kinase C (Kaibuchi et al. 1981) and of the phorbol ester aporeceptor (see below). Phosphatidylethanolamine further enhances reconstitution of the enzymatic activity, although not of the binding activity.

There are two possible explanations for why PaBB reacts selectively with lipids rather than with the protein portion of the receptor. One explanation is that the reactive nitrene generated from the aryl azide on the PaBB is at the far end of the side chain from the phorbol moiety. If the terminus of the phorbol ester side chain were buried in the lipid bilayer rather than bound in a hydrophobic pocket formed by both protein and phospholipid, it might not be in close enough proximity to react with the protein. Alternatively, the nitrene might be in adequate proximity to both protein and lipid but react selectively. Such behavior has been observed in some (Quay et al. 1981) but not in other (Brunner and Richards 1980) systems with nitrene-generating phospholipid derivatives.

Characterization of the phorbol ester aporeceptor

Although many of the properties of protein kinase C resembled those of the phorbol ester receptors, a striking apparent difference was the subcellular localization. Whereas protein kinase C was largely found in the cytosolic fraction, the phorbol ester binding activity in brain was found to be particulate (Dunphy et al. 1981). A possible explanation for the apparent difference was that the kinase measured in cytosol was an apoenzyme, which was reconstituted into phospholipid to yield active enzyme. We therefore reexamined brain cytosol for phorbol ester binding in the presence of added phospholipid (Leach et al. 1983). Under such conditions, specific [³H]PDBu binding could be detected with a specific activity (20–40 pmoles/mg protein) similar to that for the particulate fraction. Scatchard analysis was consistent with a single class of binding sites having an affinity of 3.1 nM (0.9–4 nM, depending on assay conditions). Since the cytosolic binding activity requires phospholipid for reconstitution, we refer to it as a phorbol ester aporeceptor. The phorbol ester holo-receptor would represent the reconstituted protein-lipid complex.

Different phospholipids vary in their ability to reconstitute specific [³H]PDBu binding activity (Leach et al. 1983 and in prep.). Using partially purified aporeceptor to reduce contamination by lipids retained in the cytosol fraction, we found that phosphatidylserine was most effective at reconstitution, with a half maximally effective dose (ED_{50}) of 7 μg/ml. The negatively charged phospholipids phosphatidylinositol and phosphatidic acid were also active, with similar ED_{50} values. In contrast, the uncharged phospholipids phosphatidylethanolamine and phosphatidylcholine failed to reconstitute activity if added alone, although they did not prevent reconstitution if added as mixtures with phosphatidylserine. The observed phospholipid specificity for reconstitution of binding activity generally resembled that reported by Nishizuka and co-workers for reconstitution of protein kinase C, but there were several exceptions we observed, such as no activity for phosphatidylethanolamine.

Fractionation of the cytosolic aporeceptor from brain (Leach et al. 1983; Niedel et al. 1983) and mouse thymoma cells (Sando and Young 1983) indicated coelution with protein kinase C, consistent with the two activities being the same.

Although the relationship of the cytosolic aporeceptor to the membrane receptor remains to be conclusively determined, the two activities probably represent different states of the same protein. We have found that in rat pituitary cells the distribution of cellular binding recovered in the membrane fraction as compared with the cytosolic fraction depends on the lysis conditions. More is associated with the membranes in the presence of Ca^{++}; more is found in the cytosol if the cells are lysed in the presence of EGTA. Under both conditions, total activity recovered in lysates is similar to that for the intact cells. For rat brain, likewise, a shift of phorbol ester binding and of protein kinase C activity (Kikkawa et al. 1982; Niedel et al. 1983) between cytosol and membranes as a function of free Ca^{++} has been observed. These observations, incidently, help explain the similarity in tissue distribution between cytosolic protein kinase C and membrane phorbol ester binding activity. The former analyses were routinely carried out in the presence of EGTA, the latter in the absence of chelators.

A second treatment leading to a shift in distribution (as determined in cells lysed in the presence of EGTA) is exposure of cells to phorbol esters. Such treatment of intact EL4 mouse thymoma and parietal yolk sac cells causes virtually complete transfer of protein kinase C to the particulate fraction (Kraft et al. 1982; Kraft and Anderson 1983). In rat pituitary cells, transfer, measured for [^{3}H]PDBu binding, still occurs but is less complete.

We have carried out limited structure-activity comparisons between mouse brain aporeceptor reconstituted into phosphatidylserine and the membrane receptor (Leach et al. 1983). Among active derivatives, affinities for the aporeceptor ranged from 0.041 nM for phorbol-12-myristate-13-acetate (PMA, sometimes referred to as TPA) to 440 nM for phorbol-12,13-diacetate (Table 2). These values were all within a factor of 3 of those previously determined for the membrane receptor. Our measured affinity for PMA is approximately 30-fold higher than that reported by Castagna et al. (1982) for activation of protein kinase C. This discrepancy may reflect the conditions under which the enzyme was assayed (e.g., an enzyme concentration greater than the K_d for PMA). We find good agreement between the equilibrium dissociation constant (K_d) for PDBu binding to the aporeceptor and for protein kinase C activation.

The agreement in quantitative structure-activity relations between the brain aporeceptor reconstituted into phosphatidylserine and the brain membrane receptor may reflect a similarity in lipid microenvironments for the protein under the two conditions. In a different system, namely that of GH_4C_1 rat pituitary cell membranes, the affinity of the membrane receptor for [^{3}H]PDBu is markedly less (30–40 nM) than is found for mouse brain membrane binding. Addition of a large excess of phosphatidylserine to the GH_4C_1 membranes increased the affinity ($K_d = \sim 5$ nM) to a value more like that of the reconstituted cytosolic aporeceptor. Conversely, reconstitution of the brain aporeceptor, using a phospholipid mixture resembling that of human red blood cells, yielded an affinity for [^{3}H]PDBu of approximately 20 nM (Fig. 2), a value typical for intact cells and membranes but lower than that for the apo-receptor in phosphatidylserine.

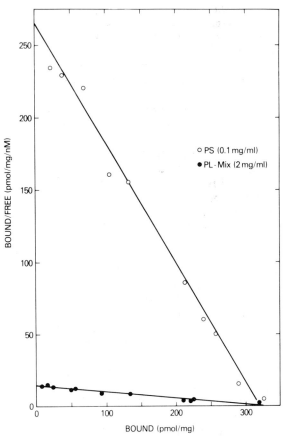

Figure 2 Scatchard plot of specific [^{3}H]PDBu binding to partially purified mouse brain aporeceptor reconstituted into a mixture of phosphatidylcholine, phosphatidylethanolamine, sphingomyelin, phosphatidylserine, cardiolipin, and phosphatidylinositol (30:20:25:13:1:1 molar ratio), 300 μg/ml. Assays were carried out in 0.1 mM $CaCl_2$, 0 $MgCl_2$, 0.05 M Tris·Cl, pH 7.4; precipitation was with 15% polyethylene glycol. Other conditions were as described by Sharkey et al. (1984).

Table 2 Comparison of Binding Affinities for Mouse Brain Membranes and Cytosol

Compound	K_i (nM) membrane	K_i (nM) cytosol
PMA	.066	.041
Phorbol-12,13-didecanoate	<2.8	<7.7
PDBu	7.4	1.3
Phorbol-12,13-diacetate	560	440
4-O-methyl PMA	n.d.	510

Data taken from Leach et al. (1983).

Diacylglycerol as an endogenous phorbol ester analog

The phorbol ester receptor has been highly conserved over evolution. Both the number of binding sites and the [^{3}H]PDBu binding affinity for sea urchins, fruit flies, and nematodes fall within the range observed for vertebrate systems (Blumberg et al. 1983a). In the case of the nematode, moreover, the structure-activity relations for binding have been characterized and found to closely resemble those for mouse skin. In analogy with evolutionarily conserved binding sites for other exogenous ligands such as the opiates, the existence of an endogenous ligand that could exert a constraint on the divergence of the binding site over the course of evolution seemed probable. Although initial efforts to identify endogenous, competitive inhibitors of [^{3}H]PDBu binding in tissue extracts were unsuccessful (for review, see Blumberg et al. 1983c), the screening assays were carried out under conditions that would have failed to detect the activity of transient or highly lipophilic compounds.

The similarity in the activation of protein kinase C by diacylglycerol and by phorbol esters suggested that diacylglycerol was a possible candidate for the predicted endogenous analog of the phorbol esters. To test this hypothesis, we examined the ability of the diacylglycerol derivative diolein to inhibit [^{3}H]PDBu binding to the mouse brain aporeceptor reconstituted into phosphatidylserine (Sharkey et al. 1984). Diolein inhibited binding greater than 90%. Scatchard analysis indicated that the decreased binding reflected an altered affinity for [^{3}H]PDBu rather than a change in maximal binding. The change in binding affinity could reflect either a general perturbation of the phospholipid environment into which the aporeceptor was reconstituted or else specific competition at the binding site itself. For a competitive inhibitor, the change in apparent [^{3}H]PDBu binding affinity as a function of diolein concentration should fit the relationship $K_a = K_d (1 + I/K_i)$, where K_a is the dissociation constant for [^{3}H]PDBu in the presence of diolein, K_d is the dissociation constant for [^{3}H]PDBu in the absence of diolein, I is the concentration of diolein, and K_i is the apparent dissociation constant for diolein. This relationship was in fact found, yielding values of K_d and K_i of 3.69 nM and 3.5 µg/ml, respectively. These values agree well with the K_d for [^{3}H]PDBu of 3.1 nM, determined by Scatchard analysis, and of the K_i of 3.6 µg/ml, determined by competition at a fixed [^{3}H]PDBu concentration.

The K_i for diolein could represent either its concentration in aqueous solution or else its concentration in the phosphatidylserine lipid phase into which the aporeceptor is reconstituted. The latter result is the predicted one, since diolein should be insoluble in aqueous solution and thus be present dissolved in the phosphatidylserine. To distinguish the two possibilities experimentally, the K_i for diolein was determined as a function of the amount of phosphatidylserine used for reconstitution of the aporeceptor. Over a 400-fold range in phosphatidylserine concentration, the K_i for diolein showed little variation expressed relative to phosphatidylserine (0.35–1.1% of the phospholipid) and a corresponding marked variation expressed relative to volume (6.9–0.05 µg/ml). For [^{3}H]PDBu, in contrast, the K_i expressed relative to its aqueous concentration showed little variation with phosphatidylserine concentration, which is once again the expected result since only a few percent of this more hydrophilic ligand partitions into the liposomes under these conditions (the concentration of [^{3}H]PDBu in the liposomes is also essentially independent of the amount of phosphatidylserine under these conditions).

We compared a series of different diacylglycerol derivatives to determine which structural features were required for inhibition of [^{3}H]PDBu binding. In addition to diolein, the symmetrically substituted glycerol derivatives with short, saturated fatty acid side chains (dicaproin, dicaprylin, and dilaurin) were also active, whereas those with longer saturated side chains were not. We interpret these findings to indicate that the side chains may need to be in a disordered or fluid state at 37°C for the compounds to have maximal activity.

Although diolein acts as a competitive inhibitor of [^{3}H]PDBu binding, it is much less potent. Expressing the K_d for [^{3}H]PDBu in terms of its actual concentration in the liposomes, one obtains a difference in affinities of 6×10^4. Despite their lower absolute potency, the effective concentrations of diacylglycerols do not appear to be unphysiological. Evidence at least in the platelet system, as described by Nishizuka and co-workers, suggests the functional role of diacylglycerols as endogenous activators of protein kinase C (Kawahara et al. 1980; Kaibuchi et al. 1983; Sano et al. 1983).

Activity of highly insoluble phorbol esters

An ongoing uncertainty in the analysis of phorbol ester structure-activity relations has been whether the receptor recognizes that proportion of the phorbol ester which is dissolved in the lipid bilayer or that which is free in aqueous solution. The indirect evidence so far available has provided some support for both viewpoints. On the one hand, derivatives more lipophilic than PMA (e.g., phorbol-12,13-didecanoate) are less potent, although they should partition more strongly into the membranes. On the other hand, since diacylglycerols, which are insoluble, compete for phorbol ester binding, the simplest model would be that the phorbol esters also are recognized after they dissolve in the lipid bilayer. Likewise, the thermodynamics of binding indicate that binding is largely driven by hydrophobic interactions, consistent with a major contribution to the binding energy by the lipid bilayer (Dunphy et al. 1981).

An experimental approach to this question was to synthesize a series of highly lipophilic phorbol derivatives —phorbol-12,13-dimyristate, phorbol-12,13-distearate, and phorbol-12,13-dioleate. Based on aqueous solubilities of 2×10^{-6} M and 5×10^{-8} M for PMA and phorbol-12,13-didecanoate, respectively, these derivatives would be expected to show aqueous solubilities too low to achieve an effective inhibitory concentration. Like diolein, however, these derivatives could be mixed in organic solvents with phosphatidylserine and be incorporated directly into the liposomes used to reconstitute the aporeceptor. The results with all three derivatives were similar. Incorporated directly into the liposomes, they displayed apparent K_d values between 5 nM and 60 nM. Added to the aqueous phase of the binding assay, they only inhibited with apparent K_d values of 4 μM to more than 20 μM.

We interpret these results as follows. (1) The receptor recognizes with good efficiency phorbol ester dissolved in the lipid phase. (2) Highly lipophilic phorbol esters show low inhibition of binding or of biological activity because they are unable to transfer efficiently from phorbol ester micelles to the phospholipid bilayer. Detergents facilitate the exchange of lipophilic molecules and would be predicted to permit this transfer to occur. In analogy with phospholipid exchange proteins, the phorbol ester binding protein from serum described by Shoyab and Todaro (1982) may be able to carry out a similar process. (3) The low observed activity of the lipophilic compounds may or may not be real. A low level (0.01%) of phorbol-13-monoester, for example, possibly generated by a phorbol ester hydrolase, could account for the observed residual activity.

The phorbol esters cause an apparent transfer of protein kinase C from the cytosolic to the membrane fraction. Recognition by the receptor of phorbol ester dissolved in the membranes suggests that the "cytosolic receptor" is either loosely or transiently membrane-associated.

Heterogeneity of phorbol ester binding and its relation to the aporeceptor

A large body of literature exists describing the biological and biochemical actions of the phorbol esters. In contrast, for only a very few processes has the direct involvement of protein kinase C been documented. If protein kinase C is the sole phorbol ester receptor, then the information obtained with the phorbol esters should provide a shortcut for identifying the processes in which protein kinase C is important. Unfortunately, considerable evidence indicates heterogeneity of phorbol ester receptors.

Biological evidence for heterogeneity is suggested by different dose-response curves and structure-activity requirements for different biological responses, at least in certain systems. The subdivision of tumor promotion into first- and second-stage promotion has been described above. Other examples are cited in Blumberg et al. (1983b). Direct evidence for heterogeneity of high-affinity phorbol ester binding has been obtained for mouse-skin particulate preparations (Dunn and Blumberg 1983), for NRK cells, and, in unconfirmed studies, for myeloid cells (Colburn et al. 1983). Preliminary reports also suggest the possible existence of distinct nuclear binding activity (Perella et al. 1982).

The origin of the heterogeneity in phorbol ester binding remains to be determined. It could reflect multiple targets for the phorbol esters in addition to protein kinase C. Alternatively, it could reflect the protein kinase C being differentially modified, whether by proteolysis, phosphorylation, or different lipid environments. Modulation of receptor affinity by varying the phospholipid composition or the diacylglycerol content in the phospholipids has been described above. By maintaining heterogeneity of the lipid environment used for reconstitution, curved Scatchard plots can be obtained. For example, Scatchard plots of [³H]PDBu-binding data for receptors reconstituted into liposomes of phosphatidylserine, 1.8% diolein in phosphatidylserine, or a 1:1 mixture of the two types of liposomes were determined. The 1:1 mixture yielded a curved Scatchard plot fitting that predicted for reconstitution into the two distinct liposome populations (Fig. 3). Similarly, addition of small amounts of phosphatidylserine (0.05 μg/ml) to the cytosol fraction from brain (which contains small amounts of phospholipid if not further fractionated) yielded a biphasic Scatchard plot, whereas an excess of phosphatidylserine yielded a linear plot.

A crucial, unresolved issue is whether the heterogeneity observed in structure-activity relations, and not simply the heterogeneity in [³H]PDBu-binding affinity, can be generated from the aporeceptor in the presence of the appropriate lipid environment. A second issue is whether heterogeneity in the lipid environment does in fact occur in particulate preparations or intact cells. The availability of photoaffinity ligands such as PaBB should provide a means of probing this microenvironment. In

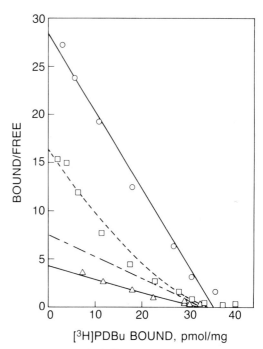

Figure 3 Heterogeneity in [³H]PDBu binding to the mouse brain cytosolic aporeceptor reconstituted into a heterogeneous lipid environment. The cytosolic aporeceptor was reconstituted into liposomes of phosphatidylserine (○), 1.8% diolein in phosphatidylserine (△), or a 1:1 mixture of the above liposomes (□). Specific binding was assayed as described by Sharkey et al. (1984) in the presence of phosphatidylserine at 40 μg/ml, 0.1 mM Ca⁺⁺, and no Mg⁺⁺. Precipitation of receptors was with 12% polyethylene glycol. The binding data were plotted by the Scatchard method. The predicted results for binding to the 1:1 liposome mixture were calculated assuming either no (− − −) or complete (— — —) exchange of diolein between liposomes.

any case, the binding data clearly emphasize the concept that the aporeceptor protein depends critically on its lipid environment for its binding properties and that the holoreceptor is a protein-lipid complex.

Discussion

The emerging understanding of the mechanism of phorbol ester action suggests marked analogies between the phorbol esters and cholera toxin. Both agents penetrate the plasma membrane to chronically activate an intracellular component of a message transduction pathway common to an extensive series of hormones and cellular effectors. A difference, however, is in the step along the pathway at which activation occurs. In the case of cholera toxin the formation of cAMP, the activator for cAMP-dependent protein kinase, is enhanced. The phorbol esters, in contrast, are themselves analogs of an activator for protein kinase C.

The phorbol esters and Rous sarcoma virus induce similar although distinguishable changes in the phenotype of chicken embryo fibroblasts (Weinstein et al. 1979; Blumberg 1980). The finding that protein kinase C is the major phorbol ester receptor and that several oncogenes including that carried by Rous sarcoma virus code for tyrosine kinases further enhances this parallelism. Whether a sequence coding for the catalytically active fragment of the protein kinase C would function as an oncogene remains to be determined.

That people should have receptors for a plant toxin such as the phorbol esters at first sight appears puzzling. A possible rationale is that many of the *Euphorbias*, which produce the phorbol esters, grow in arid regions. The acute toxicity of the phorbol esters, irritation to mucus membranes, would discourage predation by herbivores. The chronic toxicity (viz., tumor promotion) would be incidental.

Although the phorbol esters appear to be diacylglycerol analogs, it does not follow that the two classes of compounds should function identically. First, whereas the phorbol esters are relatively stable, diacylglycerols are rapidly metabolized. The phorbol esters should therefore lead to abnormal, chronic stimulation of protein kinase C. Second, the hormonally activated phosphatidylinositol turnover should generate diacylglycerol at the plasma membrane. The phorbol esters, in contrast, should be able to equilibrate with internal membranes in the cell. They may therefore cause an abnormal distribution of activated kinase. The altered distribution of kinase in turn may result in an altered pattern of phosphorylation. Third, protein kinase C can be cleaved proteolytically to an active fragment (m.w. 51,000) that no longer requires phospholipids, Ca^{++}, or diacylglycerols for activation and that no longer binds to membranes (Kishimoto et al. 1983). The susceptibility to cleavage of protein kinase C is enhanced when the enzyme is in the activated state (Kishimoto et al. 1983). Chronic activation of kinase C by phorbol esters may therefore lead to enhanced conversion to the constitutive, cytosolic form and again lead to an altered pattern of phosphorylation.

The studies on phorbol ester receptors are yielding important insights into the inital steps in the mechanism of tumor promoter action. Since the phorbol esters bind at a modulatory rather than a catalytic site on their receptor, it should be possible to develop antagonists that bind to this site but fail to activate the kinase. With the exception of teleocidin and lyngbyatoxin, tumor promoters structurally unrelated to the phorbol esters fail to compete for binding. Similarly, binding is not inhibited by inhibitors of tumor promotion. Experiments can now be done to determine whether the activities of these agents feed into the biochemical cascade triggered by the phorbol esters at the subsequent steps of protein kinase C activation or of phosphorylation of specific substrates. It is important to emphasize, however, that further understanding of the nature of the heterogeneity of binding and biological response to the phorbol esters will be essential to evaluate whether protein kinase C is the phorbol ester receptor or only the quantitatively most abundant one.

Acknowledgment

K.L.L. was supported by Public Health Service grant CA-07054, awarded by the National Cancer Institute.

References

Ashendel, C.L., J.M. Staller, and R.K. Boutwell. 1983. Protein kinase activity associated with a phorbol ester receptor purified from mouse brain. *Cancer Res.* **43:** 4333.

Blumberg, P.M. 1980. In vitro studies on the mode of action of the phorbol esters, potent tumour promoters. *Crit. Rev. Toxicol.* **8:** 153.

Blumberg, P.M., K.B. Delclos, and S. Jaken. 1983a. Tissue and species specificity for phorbol ester receptors. In *Organ and species specificity in chemical carcinogenesis* (ed. R. Langenbach et al.), p. 201. Plenum Press, New York.

Blumberg, P.M., K.B. Delclos, W.G. Dunphy, and S. Jaken. 1982. Specific binding of phorbol ester tumour promoters to mouse tissues and cultured cells. In *Carcinogenesis — A comprehensive survey. Cocarcinogenesis and biological effects of tumour promoters* (ed. F. Hecker et al.), vol. 7, p. 519. Raven Press, New York.

Blumberg, P.M., K.B. Delclos, J.A. Dunn, S. Jaken, K.L. Leach, and E. Yeh. 1983b. Phorbol ester receptors and the *in vitro* effects of tumour promoters. *Ann. N.Y. Acad. Sci.* **407:** 303.

Blumberg, P.M., J.A. Dunn, S. Jaken, A.Y. Jeng, K.L. Leach, N.A. Sharkey, and E. Yeh. 1983c. Specific receptors for phorbol ester tumour promoters and their involvement in biological responses. In *Mechanisms of tumour promotion: Tumour promotion and cocarcinogenesis in vitro* (ed. T.J. Slaga), vol. 3. CRC Press, Boca Raton. (In press.)

Brunner, J. and F.M. Richards. 1980. Analysis of membranes photolabeled with lipid analogs. *J. Biol. Chem.* **255:** 3319.

Castagna, M., Y. Takai, K. Kaibuchi, K. Sano, U. Kikkawa, and Y. Nishizuka. 1982. Direct activation of calcium-activated, phospholipid-dependent protein kinase by tumour promoting phorbol esters. *J. Biol. Chem.* **257:** 7847.

Colburn, N.H., T.D. Gindhart, B. Dalal, and G.A. Hegamyer. 1983. The role of phorbol ester receptor binding in responses to promoters by mouse and human cells. In *Organ and species specificity in chemical carcinogenesis* (ed. R. Langenbach et al.), p. 189. Plenum Press, New York.

Delclos, K.B., E. Yeh, and P.M. Blumberg. 1983. Specific labeling of mouse brain membrane phospholipids with [20-^{3}H] phorbol 12-p-azidobenzoate 13-benzoate, a photolabile phorbol ester. *Proc. Natl. Acad. Sci.* **80:** 3054.

Diamond, L., T.G. O'Brien, and W.M. Baird. 1980. Tumor promoters and the mechanism of tumor promotion. *Adv. Cancer Res.* **32:** 1.

Dunn, J.A. and P.M. Blumberg. 1983. Specific binding of [20-^{3}H] 12-deoxyphorbol 13-isobutyrate to phorbol ester receptor subclasses in mouse skin particulate preparations. *Cancer Res.* **43:** 4632.

Dunphy, W.G., R.J. Kochenburger, M. Castagna, and P.M. Blumberg. 1981. Kinetics and subcellular localization of specific [^{3}H]phorbol 12,13-dibutyrate binding by mouse brain. *Cancer Res.* **41:** 2640.

Hecker, E., N.E. Fusenig, W. Kunz, F. Marks, and H.W. Thielmann, eds. 1982. *Carcinogenesis — A comprehensive survey. Cocarcinogenesis and biological effects of tumor promoters*, vol. 7. Raven Press, New York.

Kaibuchi, K., Y. Takai, and Y. Nishizuka. 1981. Cooperative roles of various membrane phospholipids in the activation of calcium-activated phospholipid-dependent protein kinase. *J. Biol. Chem.* **256:** 7146.

Kaibuchi, K., Y. Takai, M. Sawamura, M. Hoshijima, T. Fujikara, and Y. Nishizuka. 1983. Synergistic functions of protein phosphorylation and calcium mobilization in platelet activation. *J. Biol. Chem.* **258:** 6701.

Kawahara, Y., Y. Takai, R. Minakuchi, K. Sano, and Y. Nishizuka. 1980. Phospholipid turnover as a possible transmembrane signal for protein phosphorylation during human platelet activation by thrombin. *Biochem. Biophys. Res. Commun.* **97:** 309.

Kikkawa, U., Y. Takai, R. Minakuchi, S. Inohara, and Y. Nishizuka. 1982. Calcium-activated, phospholipid dependent protein kinase from rat brain. Subcellular distribution, purification, and properties. *J. Biol. Chem.* **257:** 13341.

Kishimoto, A., N. Kajikawa, M. Shiota, and Y. Nishizuka. 1983. Proteolytic activation of calcium-activated, phospholipid-dependent protein kinase by calcium-dependent neutral protease. *J. Biol. Chem.* **258:** 1156.

Kraft, A.S. and W.B. Anderson. 1983. Phorbol esters increase the amount of Ca^{++}, phospholipid-dependent protein kinase associated with plasma membrane. *Nature* **301:** 621.

Kraft, A.S., W.B. Anderson, H.L. Cooper, and J.J. Sando. 1982. Decrease in cytosolic calcium/phospholipid-dependent protein kinase activity following phorbol ester treatment of EL4 thymoma cells. *J. Biol. Chem.* **257:** 13193.

Leach, K.L., M.L. James, and P.M. Blumberg. 1983. Characterization of a specific phorbol ester aporeceptor in mouse brain cytosol. *Proc. Natl. Acad. Sci.* **80:** 4208.

Niedel, J.E., L.J. Kuhn, and G.R. Vandenbark. 1983. Phorbol diester receptor copurifies with protein kinase C. *Proc. Natl. Acad. Sci.* **80:** 36.

Nishizuka, Y. 1983. Phospholipid degradation and signal translation for protein phosphorylation. *Trends Biochem. Sci.* **8:** 13.

Perella, F.W., C.L. Ashendel, and R.K. Boutwell. 1982. Specific high affinity binding of the phorbol ester tumor promoter 12-O-tetradecanoyl-phorbol-13-acetate to isolated nuclei and nuclear macromolecules in mouse epidermis. *Cancer Res.* **42:** 3496.

Quay, S.C., R. Radhakrishnan, and H.G. Khorana. 1981. Incorporation of photosensitive fatty acids into phospholipids of *Escherichia coli* and irradiation-dependent cross-linking of phospholipids to membrane proteins. *J. Biol. Chem.* **256:** 4444.

Sando, J.J. and M.C. Young. 1983. Identification of high-affinity phorbol ester receptor in cytosol of EL4 thymoma cells: Requirement for calcium, magnesium, and phospholipids. *Proc. Natl. Acad. Sci.* **80:** 2642.

Sano, K., Y. Takai, J. Yamanishi, and Y. Nishizuka. 1983. A role of calcium-activated phospholipid-dependent protein kinase in human platelet activation. Comparison of thrombin and collagen actions. *J. Biol. Chem.* **258:** 2010.

Scribner, J.D. and R. Süss. 1978. Tumor initiation and promotion. *Int. Rev. Exp. Pathol.* **18:** 137.

Sharkey, N.A., K.L. Leach, and P.M. Blumberg. 1984. Competitive inhibition by diacylglycerol of specific phorbol ester binding. *Proc. Natl. Acad. Sci.* **81:** 607.

Shoyab, M. and G.J. Todaro. 1982. Partial purification and characterization of a binding protein for biologically active phorbol and ingenol esters from murine sera. *J. Biol. Chem.* **257:** 439.

Takai, Y., A. Kishimoto, and Y. Nishizuka. 1982. Calcium and phospholipid turnover as transmembrane signaling for protein phosphorylation. In *Calcium and cell function* (ed. W.Y. Cheung), vol. 2, p. 385. Academic Press, New York.

Weinstein, I.B., L.-S. Lee, P.B Fisher, A. Mufson, and H. Yamasaki. 1979. Action of phorbol esters in cell culture: Mimicry of transformation, altered differentiation, and effects on cell membranes. *J. Supramol. Struct.* **12:** 195.

Yuspa, S.H. 1981. Chemical carcinogenesis related to the skin. Part 1. *Prog. Dermatol.* **15:** 1.

———. 1982. Chemical carcinogenesis related to the skin. Part 2. *Prog. Dermatol.* **16:** 1.

Isolation and Characterization of a Specific Receptor for Biologically Active Phorbol and Ingenol Esters from Murine Brain: The Homogeneous Receptor Is a Ca^{++}-independent Protein Kinase

M. Shoyab
Laboratory of Viral Carcinogenesis, National Cancer Institute, National Institutes of Health, Frederick, Maryland 21701

Tumor promoters are compounds that are themselves noncarcinogenic but that can induce tumors in animals previously treated with a suboptimal dose of certain chemical carcinogens (Van Duuren 1969; Hecker 1971; Boutwell 1974; Slaga et al. 1978; Sivak 1979; Hecker et al. 1982). Most of the experimental work on tumor promotion has been carried out with phorbol esters, especially 12-O-tetradecanoylphorbol-13-acetate (TPA), initially isolated from the seed of the plant *Croton tiglium* (Hecker 1968; Slaga et al. 1978; Hecker et al. 1982). TPA and other biologically active phorbol diesters elicit a variety of biological and biochemical responses in vitro as well as in vivo (Boutwell 1974; Slaga et al. 1978; Sivak 1979; Hecker et al. 1982).

Several biochemical and biological studies provide evidence that the initial site of action of tumor-promoting phorbol esters may be the membranes of target cells (Boutwell 1974; Slaga et al. 1978; Sivak 1979; Hecker et al. 1982). Furthermore, tumor-promoting phorbol esters have been found to modulate the interaction between epidermal growth factor (EGF) and its receptors in a variety of cells in culture (Lee and Weinstein 1978; Shoyab et al. 1979; Shoyab and Todaro 1980a,b 1981). However, the effect, although rapid in modulating the EGF receptors, is indirect and cannot be shown using low temperature and/or fixed cells or in isolated membranes (Shoyab et al. 1979; Shoyab and Todaro 1980a,b, 1981). This suggests that TPA produces its membrane effects in a manner distinct from the EGF-receptor interaction.

Specific high-affinity receptors for biologically active phorbol and ingenol esters as well as mezerein in a variety of tissue-culture cells and in animal tissues using ³H-labeled phorbol-12,13-dibutyrate (PDBu) as a ligand have been reported (Shoyab and Todaro 1980b; Driedger and Blumberg 1980; Shoyab et al. 1981a; Hecker et al. 1982). Although the binding characteristics of these receptors have been extensively studied, very little is known about their physicochemical, biochemical, and functional natures.

The present paper describes the purification of these receptors to homogeneity in an active form and their initial characterization. The purified receptor is a hydrophobic protein and exhibits an absolute requirement of phospholipids for binding. Several divalent cations enhance the ligand binding in the presence of phospholipids. The purified receptor possesses a protein kinase activity. The physicochemical and kinetic properties of the purified receptor kinase are similar but not identical with that of protein kinase C (Takai et al. 1982; Castagna et al. 1982).

Experimental Procedures
Isolation of receptors
All operations were performed at 0–4°C unless otherwise specified.

Step 1: Extraction. Pooled frozen murine brains (58 g) were homogenized in 5 volumes (w/v) of ice-cold extraction buffer (20 mM Tris·HCl, pH 7.4, containing 2 mM EDTA and 10 mM EGTA) for 1–2 minutes. The homogenate was further disrupted with a Polytron P-10 at 21,000 rpm for 90 seconds. Three 30-second pulses with intervals of 30 seconds were used. The final homogenate was centrifuged at 105,000*g* for 1 hour. The supernatant fluid contained almost all of the binding activity with 3,800 mg of protein in 255 ml.

Step 2: Ammonium sulfate fractionation. The supernatant was placed in a beaker with a stirring magnetic bar. Ammonium sulfate (36.7 g) was slowly added to the beaker over a period of about 20 minutes. The contents were stirred for an additional 20 minutes and centrifuged for 15 minutes at 15,000*g*. The supernatant fluids were carefully drained and pooled (263 ml). Additional ammonium sulfate (41.6 g) was added to the supernatant to raise the ammonium sulfate saturation from 25% to 50%. The addition and stirring were performed as in the first ammonium sulfate step. The mixture was centrifuged for 20 minutes at 15,000*g*. The supernatant fluids were drained off and discarded. The pellets were dissolved in 51 ml of cold buffer A (20 mM Tris·HCl, pH 7.4, containing 2 mM EDTA and 50 mM 2-mercaptoethanol) and dialyzed overnight against 2 liters of buffer A. Fifty-seven ml of dialyzed solution contained 1043 mg of protein.

Step 3: DEAE-cellulose chromatography. The dialyzed fraction was applied to a DE-52 column (2.6 × 34 cm) previously equilibrated with buffer A. The column was washed with 1 liter of the same buffer and the receptors were eluted with a linear gradient of NaCl (0–0.3 M; total vol., 2 liters) in buffer A. The flow rate was approximately 80 ml/hr and 260 drop fractions (20.6 ml) were collected. The fractions containing activity (23–40) were pooled and concentrated to approximately 9 ml using an Amicon concentrating system (PM-10 membrane). One ml of 50% sucrose in buffer A was added to the DE-52 concentrate and mixed. The NaCl concentration of the fractions was estimated from their conductivity.

Step 4: Sephadex G-200 chromatography. The resulting concentrate was applied to a Sephadex G-200 column (2.6 × 92 cm) previously equilibrated with buffer A. The flow rate was maintained at 14 ml/hr. One hundred drop fractions (~6.3 ml) were collected into tubes containing 1 ml of 50% sucrose in buffer A. Fractions 45–55 were pooled and concentrated to 7.5 ml using an Amicon concentrating system.

Step 5: Affi-Gel blue chromatography. The concentrated G-200 fraction was applied to a column of Affi-Gel blue (0.9 × 7.8 cm) previously equilibrated with buffer B (5% sucrose in buffer A). The column was washed with 80 ml of 1 M NaCl in buffer B. The receptor was eluted with 3 M NaCl in the same buffer. The flow rate was 25 ml/hr. Fifty drop fractions (~3.1 ml) were collected into tubes, each containing 1 ml of 50% sucrose in buffer A. Fractions 3–16 were pooled and contained 275 μg of protein.

Step 6: Phenyl–Sepharose CL-4B chromatography. The pooled Affi-Gel blue fractions were passed over a column of phenyl–Sepharose CL-4B (0.9 × 3.2 cm) previously equilibrated with 3 M NaCl in buffer B. The column was washed with 20 ml of 0.5 M NaCl in buffer B, and the receptor was then eluted with buffer B. The flow rate was maintained at 20 ml/hr. Thirty-two drop fractions (~2 ml) were collected in tubes, each containing 2 ml of 50% sucrose in buffer A.

PDBu binding assay
The binding of [^{3}H]PDBu to soluble or insoluble receptors was performed in the presence or absence of 20 μg/ml unlabeled PDBu as described previously (Shoyab et al. 1981a,b, 1982; Shoyab and Todaro 1982). The binding mixture, in duplicate, contained in a final volume of 0.25 ml, 5 μmoles of Tris·HCl (pH 7.4), 0.5 μmole of MgCl$_2$, 0.125 μmole of CaCl$_2$, 1–2 μg of phosphatidylserine, 5 ng of [^{3}H]PDBu (~4 × 10^4 cpm), 25 μg of bovine serum albumin (BSA), 0.2% final concentration of dimethyl sulfoxide, and an appropriate amount of receptor protein. After 10 minutes at 23°C, the tubes were chilled, 0.5 ml of cold 4% calf serum (Colorado Serum Co., Denver), and 0.7 ml of cold 25% polyethylene glycol 6000 in 1 mM Tris·HCl (pH 7.4) was added to each tube. The tubes were then processed exactly as described previously (Shoyab et al. 1981a,b, 1982; Shoyab and Todaro 1982).

Gel filtration assay for PDBu binding
The reaction mixture was the same as described above. Following incubation at 23°C for 10 minutes, the bound PDBu was separated from unbound PDBu by Bio-Gel P-10 chromatography as described by Shoyab and Todaro (1982).

Protein estimation
Protein concentration was determined by the method of Lowry et al. (1951) or by the method of Bradford (Bradford 1976), using BSA as the standard.

Sodium dodecyl sulfate–polyacrylamide gel electrophoresis
SDS-PAGE analysis of proteins was performed on a vertical slab gel, using the method of Laemmli (1970). Proteins on the gel were detected by the silver-staining method of Merrill et al. (1981).

Assay for protein kinase
The standard reaction mixture, in duplicate, contained in a final volume of 0.25 ml, 5 μmoles of Tris·HCl, pH 7.4, 0.5 μmole of MgCl$_2$, 0.125 μmole of CaCl$_2$, 1–2 μg of phosphatidylserine, 50 μg of histone, 0.06 nmole of [ν-^{32}P]ATP (3.8 × 10^6 cpm, on the arrival time), and an appropriate amount of enzyme protein. The reaction was carried out for 15 minutes at 23°C. The phosphorylated proteins were precipitated with 5% cold trichloroacetic acid (TCA), using 250 μg of BSA as carrier. The mixture in the tubes was allowed to settle at 4°C for 15–30 minutes. Precipitates were collected on cellulose nitrate filters (0.45 μm) and washed with 75 ml of cold 5% TCA. Filters were dried and counted in Betaflour scintillation fluid (Beckman).

Results

Purification of receptor

A summary of the purification is presented in Table 1. A 4248-fold purification with a 3.1% yield has been achieved, and the method is reproducible; we have purified the receptor on 11 different occasions with similar results. Affi-Gel blue and phenyl-Sepharose chromatographies are the most crucial steps in this purification. A very high salt concentration was needed to elute the receptor from Affi-Gel blue, suggesting a strong hydrophobic interaction between the receptor and the blue dye attached to the gel.

Purity and physical properties of the receptor

Figure 1 shows the analysis of the purified receptor at various concentrations on a 7.5% acrylamide gel under reducing conditions. The purified protein migrated in the gel as a single band. Similarly, a single band was seen in a 10% gel under reducing conditions. The receptor protein has a molecular weight (M_r) of approximately 81,500. Thus, the receptor appears to be a single-chain, highly hydrophobic protein. Linear sucrose density gradient ultracentrifugation of the enzyme preparation of the receptor gave a single peak corresponding to a sedimentation coefficient of 5.2S (data not shown). The receptor had a Stokes radius of 30.3 Å. A corresponding M_r of 52,000 was obtained based upon its elution position from a Sephadex G-200 column, although the receptor appeared as a single, symmetrical peak. Using the above mentioned values of the Stokes radius, sedi-

Table 1 Purification of Receptor from Murine Brain

Step	Volume (ml)	Total protein (mg)	Total units[a]	Sp. act. (unit/mg protein)	Yield	Purification (-fold)
1. Crude extract	255	3800	3315	0.87	100	1
2. Ammonium sulfate (25–50%)	57	1043	3135	3.01	94.6	3.5
3. DE-52	9.8	84.3	1215	14.41	36.7	16.6
4. G-200	7.5	17.3	1005	58.3	30.3	67
5. Affi-Gel blue	55	0.275	248	900	7.5	1034
6. Phenyl Sepharose	12	0.027	102	3696	3.1	4248

[a]One unit is defined as 10^5 cpm of [^{3}H]PDBu bound under our assay conditions.

mentation coefficient, and the M_r of 81,500, the frictional coefficient was calculated to be 0.99, indicating the symmetrical nature of the receptor (Martin and Ames 1961; Siegel and Monty 1966). Chromatofocusing of the receptor from the Sephadex G-200 step revealed the presence of one species of receptor with an isoelectric point of 5.5. It should be noted that there was a tremendous loss of activity during chromatofocusing. The purified receptor was very heat- and acid-labile. A high concentration of sucrose or glycerol was needed in order to stabilize the binding activity. Freezing and thawing also greatly reduced the binding activity.

Binding and kinetic properties

The requirements for PDBu binding to the purified receptor are presented in Tables 2 and 3. A phospholipid was absolutely required for PDBu-receptor interaction. Phosphatidylserine (PS) and phosphatidylinositol (PI) were found to be the most effective phospholipids. Phosphatidylethanolamine (PE) was slightly active, whereas phosphatidylcholine (PC), phosphatidic acid (PA), sphingomyelin (SM), and diolein (DO) were each inactive as cofactors in PDBu binding to its receptor. Recently, we have found that fatty acyl-CoAs (C_{10}–C_{20}) as well as SDS were also effective as cofactors in PDBu binding to its receptor. Furthermore, DO did not significantly affect the PDBu-receptor interaction, even in the presence of PS (Table 3). Membrane-bound receptor did not exhibit a requirement for an exogenous phospholipid (Shoyab and Todaro 1980b, 1982; Shoyab et al. 1981a,b). Other investigators have reported similar phospholipid requirements for binding to crude or partially purified preparations of the receptor (Ashendel et al. 1983; Leach et al. 1983; Niedel et al. 1983; Sando and Young 1983). We did not find an absolute requirement of metal ions for PDBu binding. However, divalent cations such as Mg^{++}, Ca^{++}, Mn^{++}, Zn^{++} Ni^{++}, Co^{++}, Ba^{++}, and Str^{++} stimulated binding in the presence of PS (Table 5). The optimum concentrations of Mg^{++} and Ca^{++} were found to be 5–10 mM and 0.5–1.0 mM, respectively. Metal chelating agents (EDTA and EGTA) at a 10 mM concentration reduced binding to approximately 40% (Table 2). Monovalent cations, such as Na^+, K^+, or NH_4^+, were ineffective in modulating the binding. Optimum binding with the isolated receptor was achieved at pH 6.5–7.5. PDBu bound very rapidly to the purified receptor. Binding was complete within minutes even at 4°C. We have previously reported that PDBu binding to cells or membranous fractions is time- and temperature-dependent (Shoyab and Todaro 1980b, 1982; Shoyab et al. 1981a,b, 1982). The release of receptor molecules

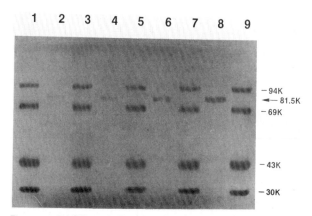

Figure 1 PAGE analysis of the receptor from the phenyl–Sepharose CL-4B step. The running and stacking gels were 7.5% and 5%, respectively, and the gel was run at 30 V. Lanes *1, 3, 5, 7,* and *9* have molecular weight markers: Phosphorylase b, 94,000; BSA, 69,000; ovalbumin (OVA), 43,000; carbonic anhydrase (CA), 30,000. Approximately 13, 26, 52, and 104 ng of receptor protein were used in lanes *2, 4, 6,* and *8,* respectively.

Table 2 Requirement for PDBu Binding to Purified Receptor

	Binding (cpm)	
Addition	–PS	+PS (4 µg/ml)
None	263	2998
2 mM $MgCl_2$	239	5941
0.5 mM $CaCl_2$	486	6978
2 mM $MgCl_2$ + 0.5 mM $CaCl_2$	350	6859
10 mM EGTA	233	1263
10 mM EDTA	239	1271

The basal binding mixture contained 5 moles of Tris·HCl, pH 7.4, 5 ng of [^{3}H]PDBu, 25 µg of BSA, and 50 ng of receptor protein in a final volume of 0.25 ml. The binding assays were performed as described in the text.

Table 3 Effect of Various Lipids on PDBu Binding to Purified Receptor

Lipid (4 µg/ml)	PDBu binding (cpm)
None	390
Phosphatidylserine (PS)	5301
Phosphatidylcholine (PC)	431
Phosphatidylethanolamine (PE)	1383
Phosphatidylinositol (PI)	4677
Phosphatidic acid (PA)	425
Sphingomyelin (SM)	385
Diolein (0.2 µg/ml) (DO)	448
Phosphatidylserine + diolein	5330

The binding assays were performed as described in the text, except 4 µg/ml of various lipid was used; 50 ng of receptor protein were used in the binding assays.

from membranes without the constraints of other neighboring molecules may be responsible for this rapid ligand-receptor interaction.

Saturation of PDBu binding

The effect of PDBu concentration on PDBu binding to its homogeneous receptor is shown in Figure 2. PDBu binding was linear up to a PDBu concentration of 4 nM. As the concentration of PDBu was increased beyond 4 nM, the binding of PDBu to its receptor gradually began to deviate from linearity and leveled off. PDBu binding sites were almost saturated at a PDBu concentration of 25 nM (Fig. 2, inset). Figure 2 also includes a Scatchard plot of PDBu binding. These data produced a linear plot indicating a single type of binding site on the receptor. In contrast, membrane-bound receptors gave a curvilinear plot (Shoyab and Todaro 1982). At the saturating concentration of PDBu, one receptor molecule was calculated to bind 1–2 molecules of PDBu. The apparent K_d value was calculated to be 4.2 nM. The apparent K_d value for PDBu interaction to brain membranous fraction was about 6 nM (Shoyab and Todaro 1982).

Relationship between structure of phorbol and ingenol derivatives and binding to their homogeneous receptors

Several natural and synthetic analogs of phorbol and ingenol derivatives with various degrees of promoting activity in the two-stage tumorigenesis model are now available (Van Duuren 1969; Hecker 1971; Boutwell 1974; Slaga et al. 1978; Sivak 1979; Hecker et al. 1982). We compared their effect on PDBu binding to the purified receptor with that previously reported for the membrane-bound receptor (Shoyab and Todaro 1980b). The doses required for 50% inhibition of PDBu binding to purified receptor for various compounds are summarized in Table 4. TPA was the most potent competitor of PDBu binding among the phorbol, ingenol, and mezerein derivatives tested. The relative potency of these agents was TPA > DPTD > Mz > IHD > PDBu > PDD > PDB > PDA > ITA > Me-TPA = 4α-PDD = phorbol = ingenol. The inhibition of PDBu binding to the purified receptor by different phorbol and ingenol derivatives correlated very well with their tumor-promoting activity. The doses required for 50% inhibition (ID_{50} values) with pure receptor are quite similar to the corresponding values previously obtained with whole cells or membrane-bound receptors (Shoyab and Todaro 1980b, 1982; Shoyab et al. 1981a,b).

Protein kinase activity of homogeneous receptor

A protein kinase activity was found to be associated with the homogeneous receptor protein. This kinase activity was strongly activated by Mg^{++}, Mn^{++}, and Fe^{++} and moderately by Co^{++} among all the divalent cations tested (Tables 5 and 6). Other divalent cations such as

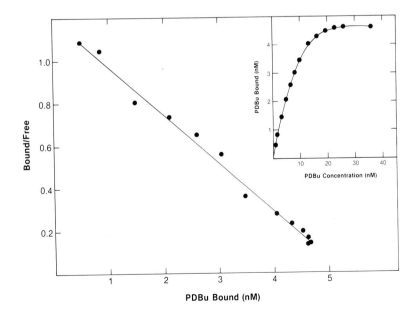

Figure 2 Effect of PDBu concentration on binding to the purified receptor. The indicated concentrations of [^{3}H]PDBu and 50 ng of receptor protein were used. The binding assays were performed as described in the text.

Table 4 Correlation between the Potency of Phorbol and Ingenol Derivatives for Promoting Skin Tumors and Their Ability to Inhibit the Binding of [^{3}H]PDBu to Purified Receptor

Compounds	Dose required for 50% inhibition (ng/ml)	Tumor-promoting activity
12-O-tetradecanoylphorbol-13-acetate (TPA)	2.7	+++
Phorbol-12-13-dibutyrate (PDBu)	35.2	++
Phorbol-12-13-didecanoate (PDD)	49.4	++
Phorbol-12,13-dibenzoate (PDB)	106.9	+
4α-Phorbol-12,13-didecanoate (4α-PDD)	>100,000	−
Phorbol-12,13-diacetate (PDA)	2,900	−
4-O-Methyl-TPA (Me-TPA)	>100,000	−
Phorbol	>100,000	−
12-Deoxyphorbol-13-tetradecanoate (DPTD)	12.5	++
Mezerein	22.7	?
Ingenol-3-hexadecanoate (IHD)	29.3	++
Ingenol-3,5,20-triacetate (ITA)	3,100	−
Ingenol	>100,000	−

The binding of [^{3}H]PDBu to purified receptor (50 ng) in the absence and presence of varying concentrations of different compounds was performed. The competition curves were plotted as previously described (Shoyab and Todaro 1982). ID50 (dose required for 50% inhibition) values were read from these curves.

Ca^{++}, Zn^{++}, Ni^{++}, Ba^{++}, Str^{++}, Cu^{++}, and Hg^{++}, at concentrations of 0.2 mM, did not affect the basal protein kinase activity, although several divalent cations, including Ca^{++}, enhanced PDBu binding to the receptor (Table 2). Interestingly, Ca^{++} at concentrations between 0.05 mM and 5.0 mM did not significantly activate the protein kinase activity (Table 6) either in the presence or absence of PS. Furthermore, the protein kinase activity of the receptor was slightly increased when 10 mM EGTA was included in the reaction mixture with Mg^{++}. In contrast, EGTA completely inhibited the Mn^{++}-primed protein kinase activity (Table 6). This clearly indicates that the protein kinase activity of the receptor, unlike protein kinase C (Takai et al. 1982), is Ca^{++} independent. However, it is possible, although unlikely, that the protein kinase activity of the receptor was converted from a Ca^{++}-dependent form to a Ca^{++}-independent state during purification.

Table 5 Effect of Various Divalent Metal Ions on Protein Kinase Activity and Binding Activity of Purified Receptor

Metal ions	Concentration (mM)	PDBu binding (cpm)	Protein kinase activity ([^{32}P] incorporated cpm × 10^{-5}])
None	—	2683	0.21
Mg^{++}	0.2	2465	0.31
Mg^{++}	4	4746	2.82
Ca^{++}	0.2	5268	0.16
Mn^{++}	0.2	5072	3.24
Zn^{++}	0.2	5218	0.17
Co^{++}	0.2	5495	0.70
Ni^{++}	0.2	5284	0.16
Ba^{++}	0.2	4501	0.19
Str^{++}	0.2	4977	0.20
Cu^{++}	0.2	2384	0.16
Fe^{++}	0.2	3294	2.76
Hg^{++}	0.2	3975	0.19

The PDBu binding assays in the presence of the indicated concentration of cations and 50 ng of receptor proteins were performed as described in the text.

The reaction mixture for protein kinase, contained in a final volume of 0.25 ml, was 5 μmoles Tris·HCl (pH 7.4), 50 μg of histone, 0.06 nmole of [ν-^{32}P]ATP, 25 ng of purified protein, and the indicated concentrations of cations. The mixture was incubated for 15 mM at 23°C. The 5% TCA-insoluble radioactivity was determined as described in the text.

Table 6 Requirement for Protein Kinase Activity of Purified Receptor

Addition	Protein kinase activity ($[^{32}P]$ incorporated [cpm × 10^{-5}])	
	−PS	+PS (4 µg/ml)
None	0.14	0.12
0.01 M EDTA	0.14	0.11
0.01 M EGTA	0.13	0.10
2 mM $MgCl_2$	1.54	1.48
0.01 M EGTA + 2 mM $MgCl_2$	1.73	1.69
4 mM $MgCl_2$	1.91	1.85
0.01 M EGTA + 4 mM $MgCl_2$	2.85	2.41
0.2 mM $MnCl_2$	2.12	2.15
0.01 M EGTA + 0.2 mM $MnCl_2$	0.08	0.07
0.05 mM $CaCl_2$	0.13	0.12
0.25 mM $CaCl_2$	0.26	0.24
0.5 mM $CaCl_2$	0.18	0.17
2 mM $MgCl_2$ + 0.25 mM $CaCl_2$	1.56	3.84
2 mM $MgCl_2$ + 0.5 mM $CaCl_2$	1.52	3.74
2 mM $MgCl_2$ + 0.5 mM $CaCl_2$ - histone	0.06	0.11

The basal reaction mixture in 0.25 ml contained 5 µmoles Tris·HCl (pH 7.4), 50 µg of histone, 0.06 nmole [ν-^{32}P]ATP, and 20 ng of purified protein. The protein kinase assays were performed as in Table 5.

PS did not stimulate the protein kinase activity in the presence of either 2 mM $MgCl_2$ or 0.05–0.5 mM $CaCl_2$, whereas PS enhanced the protein kinase activity approximately twofold when 0.25–0.5 mM $CaCl_2$ was present with 2 mM $MgCl_2$. Other lipids tested did not modulate the protein kinase activity of the receptor. Thus, the protein kinase activity of the receptor exhibited a PS sensitivity but not a PS dependence. It should be emphasized here that the binding of active diterpene esters to the receptor was found to be phospholipid dependent. Interestingly, fatty acyl-CoAs (C_{10}–C_{20}) as well as SDS were found to be very potent stimulators of the protein kinase activity of receptor.

It has been previously reported that active phorbol diesters can directly enhance protein kinase C activity by substituting for diacylglycerol (Castagna et al. 1982). We tested the effects of diolein and various phorbol and ingenol derivatives on the protein kinase activity of the homogeneous receptor (Table 7). Diolein was found to be ineffective in modulating the protein kinase activity either in the presence or absence of PS, irrespective of the divalent cations present. However, TPA stimulated the protein kinase activity approximately twofold only in the presence of PS (Table 7). This stimulation was seen even in the absence of $CaCl_2$. Thus, PS appeared to be required for PDBu-binding activity but not for the protein

Table 7 TPA Stimulation of Protein Kinase Activity of Purified Receptor

$CaCl_2$ (M)	Addition	Protein kinase activity ($[^{32}P]$ incorporated [cpm × 10^{-5}])	
		−PS	+PS (4 µg/ml)
0	none	0.10	0.09
0	2 mM $MgCl_2$	2.01	2.02
0	2 mM $MgCl_2$ + 0.4 µg/ml diolein	1.94	2.01
0	2 mM $MgCl_2$ + 100 ng/ml TPA	2.23	3.94
1×10^{-5}	none	0.13	0.16
1×10^{-5}	2 mM $MgCl_2$	1.98	2.24
1×10^{-5}	2 mM $MgCl_2$ + 0.4 µg/ml diolein	1.94	2.29
1×10^{-5}	2 mM $MgCl_2$ + 100 ng/ml TPA	2.12	3.92
1×10^{-4}	none	0.16	0.23
1×10^{-4}	2 mM $MgCl_2$	2.07	2.79
1×10^{-4}	2 mM $MgCl_2$ + 0.4 µg/ml diolein	1.97	2.18
1×10^{-4}	2 mM $MgCl_2$ + 100 ng/ml TPA	2.29	4.19
0	0.2 mM $MnCl_2$	2.45	2.63
0	0.2 mM $MnCl_2$ + 0.4 µg/ml diolein	2.38	2.57
0	0.2 mM $MnCl_2$ + 100 ng/ml TPA	2.36	4.01

The basal reaction mixture was the same as described in Table 6, except 25 ng of purified protein were used.

Table 8 Effect of Various Phorbol and Ingenol Derivatives on the Protein Kinase Activity of Purified Receptor

Compound	Protein kinase activity (p mole p incorporated)	Stimulation (%)
None	1.89	0
12-O-tetradecanoylphorbol-13-acetate (TPA)	3.55	88
Phorbol-12,13-dibutyrate (PDBu)	2.72	44
Phorbol-12,13-didecanoate (PDD)	3.28	74
Phorbol-12,13-dibenzoate (PDB)	2.59	37
4α-Phorbol-12,13-didecanoate (4α-PDD)	1.86	−2
Phorbol-12,13-diacetate (PDA)	2.17	15
4-O-Methyl-TPA (Me-TPA)	1.88	−1
Phorbol	1.71	−10
12-Deoxyphorbol-13-tetradecanoate (DPTD)	4.19	121
Mezerein	3.25	72
Ingenol-3-hexadecanoate (IHD)	3.84	103
Ingenol-3,5,20-triacetate (ITA)	1.74	−8
Ingenol	1.78	−6

The reaction mixture in 0.25 ml contained 5 μmoles of Tris·HCl (pH 7.4), 0.5 μmole of $MgCl_2$, 0.125 μmole of $CaCl_2$, 50 μg histone, 0.06 nmole of [ν-^{32}P]ATP, 25 ng of purified protein, 1 μg of the indicated lipid, and 25 ng of the indicated compounds.

kinase activity of the receptor. Those phorbol and ingenol derivatives that are biologically active stimulated the protein kinase activity of receptors, whereas inactive congeners failed to do so (Table 8). This is similar to the inhibition of [^{3}H]PDBu binding to the receptor or the affinity modulation of EGF-receptor interaction (Shoyab et al. 1979; Shoyab and Todaro 1980b).

A slight autophosphorylation of the receptor was observed (Table 6) independent of PS. This phosphorylation was greatly increased in the presence of an exogenous protein substrate such as histone. The ability of some histone subfractions and other proteins to serve as substrates for the protein kinase activity of the receptor was investigated (Table 9). Lysine-rich histone (H1) was the best substrate compared with slightly lysine-rich (H2b) or arginine-rich histone (H4) (all from Worthington). Histone H1 has been reported to be the best substrate for brain and heart protein kinase C (Kikkawa et al. 1982; Takai et al. 1982; Wise et al. 1982a,b). The apparent K_m for histone H1 for the receptor protein kinase was approximately 55.5 μg/ml. One enzyme molecule of receptor protein kinase transferred approximately 23 molecules of phosphate from ATP to histone H1 under the experimental conditions. Protamine-free base was a very poor phosphate acceptor, whereas protamine sulfate served as a very good phosphate acceptor. Guinea pig myelin basic protein was also a good substrate for the enzyme. Surprisingly, partially dephosphorylated casein or phosvitin was a very poor substrate, whereas BSA, myosin, phosphorylase kinase, pyruvate kinase, and hemoglobin were even poorer substrates, as expected (Table 9).

The K_m value for ATP was approximately 0.5 μM. The V_{max} for ATP was found to be 83 nM under the experimental conditions used, or about 70 molecules of phosphate ion from ATP were maximally transferred to histone H1. GTP, CTP, or UTP did not serve as phosphate donors for the receptor protein kinase, but dATP could serve as a phosphate donor. Diadenosine oligophosphates competitively inhibited the protein kinase activity of the receptor, whereas tosylamide phenylethyl chloromethyl ketone (TPCK) was found to be an uncompetitive inhibitor of the receptor protein kinase. The optimum pH for the receptor protein kinase was 6.5–7.0 with 20 mM Tris·HCl as a test buffer.

As mentioned previously, the receptor protein kinase autophosphorylated itself. About 1–2 moles of phosphate were incorporated into each mole of the enzyme. Chromatographic analysis following acid hydrolysis revealed that both serine and threonine residues were phosphorylated. The ratio of phosphoserine to phosphothreonine was approximately 2. In contrast, the ratio of phosphoserine to phosphothreonine in histone H1 phosphorylated by receptor protein kinase was about 88. The receptor protein kinase did not phosphorylate any tyrosyl residue of itself or histone H1 (data not shown).

Table 9 Comparative Ability of Various Proteins as Substrates for Protein Kinase Activity of Purified Receptor

Protein	Protein kinase activity (p mole p incorporated)
None	0.09
Histone H2b (slightly lysine rich)	8.11
Histone H1 (lysine rich)	17.8
Histone H4 (arginine rich)	4.33
Bovine serum albumin	0.17
Hemoglobin	0.22
Partially dephosphorylated casein	0.51
Protamine	0.13
Protamine sulfate	9.39
Phosvitin	0.69
Myosin	0.12
Guinea pig myelin basic protein	4.55
Pyruvate kinase	0.08
Phosphorylase kinase	0.09

The assay conditions were the same as described in the text, except 50 μg of various proteins and 25 ng of purified protein were used.

The protein kinase activity of the receptor was not affected by cyclic nucleotides, calmodulin, or antibody to calmodulin either in the presence or absence of PS. Also, the receptor preparation was free from any calmodulin, as judged from SDS-PAGE.

Coinactivation of PDBu binding activity and protein kinase activity of the purified receptor

Both of these activities of the receptor were heat-, acid-, and freezing- and thawing-labile. Both were stabilized by the high concentrations of sucrose and glycerol. These activities followed the same kinetics of inactivation. A typical coinactivation profile at 40°C is shown in Figure 3. Inactivation of 50% of both activities was noted between 9 and 10 minutes at 40°C. Both activities were lost almost completely within 3 minutes at 55°C. The binding activity of the receptor is not separated from the protein kinase activity by any physical or chemical treatments. These data strongly suggest that both activities reside on the same receptor protein.

Discussion

Our results show that we have purified a protein from murine brains to homogeneity that specifically interacts with biologically active phorbol and ingenol derivatives but not with biologically inert congeners of these diterpenes. PDBu binding to homogeneous receptors exhibits specificity and saturability. The purification method described here is simple and reproducible. The receptor is a very hydrophobic protein (M_r 81,000), since it is retained by phenyl-Sepharose as well as by Affi-Gel blue. Conditions that dissociate the protein into subunits do not affect the electrophoretic mobility or gel filtration behavior of the receptor, indicating that it is a single-chain polypeptide. It is an acidic protein with a pI of 5.5.

The purified receptor and membrane-bound receptor exhibit almost similar affinities for the ligand. The membrane-bound receptors do not require a phospholipid for ligand binding, whereas the purified receptors bind negligible amounts of ligand in the absence of a phospholipid or a fatty acyl-CoA. This lack of phospholipid dependence of the membranous receptor might simply reflect the presence of a phospholipid cofactor(s) in the membranes. Recently, it has been reported that phorbol esters bind with high affinity to liposomes even in the absence of the receptor protein (Deleers and Malaisse 1982). We failed to find any specific binding to free PS or PS encapsulated in the lipid bilayers in the absence of receptors.

The receptors appear to be different from the soluble phorbol-binding protein found in serum (Shoyab and Todaro 1982). This conclusion is based on the different physical, chemical, and kinetic properties of the two proteins. Furthermore, PS does not increase the binding of PDBu to serum-binding protein, as is the case with the homogeneous receptors.

It has been known for some time that enzymes (kinases and dehydrogenases) with the "dinucleotide fold" interact specifically with Cibacron blue F3GA, the chromophore of Affi-Gel blue (Thompson et al. 1975; Wilson 1976). The strong and selective binding of the receptors to Affi-Gel blue suggests that the receptor contains the "dinucleotide fold" and may be an enzyme with a kinase or a dehydrogenase activity. However, nucleotides, nucleosides, cyclic nucleotides, or calmodulin do not significantly affect the binding of PDBu to its receptor.

Data presented in this paper clearly indicate that the receptor for biologically active phorbol and ingenol esters is a protein kinase. The physiochemical and biochemical properties of the receptor protein kinase are quite different from other protein kinases (Flockhart and Corbin 1982) except protein kinase C (Castagna et al. 1982; Kikkawa et al. 1982; Takai et al. 1982; Wise et al. 1982a,b). Receptor protein kinase and protein kinase C share certain common features. The molecular weights (determined by SDS-PAGE), sedimentation coefficients, isoelectric points, K_ms for histone H1, and optimum Mg^{++} values for both the receptor protein kinase and protein kinase C are quite similar (Kikkawa et al. 1982; Takai et al. 1982). Both types of protein kinases exhibit a preference for histone H1 among various histone subfractions and other proteins tested. Also, both protein kinase C and the receptor protein kinase phosphorylate seryl and threonyl residues of proteins. Both enzymes are stimulated by active phorbol ester derivatives, although the degrees of enhancement are different for these two kinases.

The optimum pH, Stokes radius, and K_m for the ATP values of the receptor protein kinase and protein kinase C are quite different (Castagna et al. 1982; Takai et al. 1982). The receptor protein kinase has a cofactor requirement distinguishable from that reported for protein

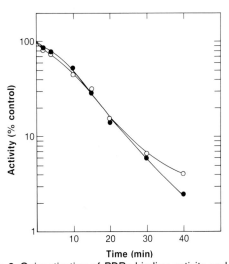

Figure 3 Coinactivation of PDBu-binding activity and protein kinase activity of the purified receptor. Purified receptor (~2.5 μg/ml) in buffer containing 25% sucrose was maintained at 40°C in a siliconized tube. At the indicated time, aliquots of 100 μl were withdrawn into ice-cold tubes. The PDBu binding and kinase activities were determined as described in the text. 100% PDBu binding corresponds to approximately 200 cpm, whereas 100% kinase activity represents 3.42×10^5 cpm. (○) Binding; (●) kinase activity.

kinase C. As previously noted, the receptor protein kinase is a Ca^{++}-independent and PS-sensitive enzyme, whereas protein kinase C is a Ca^{++}- and phospholipid-dependent enzyme. Diacylglycerol specifically activates protein kinase C by increasing the affinity of both Ca^{++} and the phospholipid to the enzyme (Castagna et al. 1982). However, neither the ligand-binding activity nor the protein kinase activity of receptor is significantly affected by diolein.

It has been reported that protein kinase C is converted into hydrophilic ($M \simeq 51{,}000$) and hydrophobic components by a Ca^{++}-dependent, neutral thiol protease. The hydrophilic component exhibits catalytic activity in the absence of Ca^{++}, phospholipids, and diacylglycerol (Kikkawa 1982). The situation is different with the Ca^{++} independence of receptor protein kinase, because phospholipid-dependent and divalent cation–stimulated PDBu-binding activity and Ca^{++}-independent protein kinase activity are present on the same protein with an M_r of 81,000.

It appears from the above discussion that the receptor protein kinase and protein kinase C are closely related but not identical. Functionally, the receptor protein kinase and protein kinase C may be components of a cascade system in which phosphorylation of one enzyme might trigger a series of phosphorylation reactions with important physiological consequences. Purification, identification, and the characterization of the endogenous substrate proteins and endogenous effectors and modulators of the receptor protein kinase from various tissues would uncover new knowledge of the functional role of the receptor as well as provide new clues to the mechanism of tumor promotion and to cell growth and differentiation.

It is gratifying to find that the receptor for tumor-promoting diterpene ester is a protein kinase. The receptors for EGF, platelet-derived growth factor, and insulin as well as for proteins coded by certain oncogenes are tyrosine-protein kinases (Kolata 1983). Thus, the qualitative and/or quantitative alteration in protein phosphorylation and dephosphorylation may be crucial in growth, differentiation, oncogenesis, and tumor promotion.

Acknowledgment

I gratefully acknowledge the expert technical assistance of Mr. Raleigh Boaze, Jr.

References

Ashendel, C.L., J.M. Staller, and R.K. Boutwell. 1983. Identification of a calcium- and phospholipid-dependent phorbol ester binding activity in the soluble fraction of mouse tissues. *Biochem. Biophys. Res. Commun.* **111:** 340.

Boutwell, R.K. 1974. The function and mechanism of promoters of carcinogenesis. *Crit. Rev. Toxicol.* **2:** 419.

Bradford, M. 1976. A rapid and sensitive method for quantitation of microgram quantities of protein utilizing the principle of protein-dye binding. *Anal. Biochem.* **72:** 248.

Castagna, M., Y. Takai, K. Kaibuchi, K. Sano, U. Kikkawa, and Y. Nishizuka. 1982. Direct activation of calcium-activated phospholipid-dependent protein kinase by tumor-promoting phorbol esters. *J. Biol. Chem.* **257:** 7884.

Deleers, M. and W.J. Malaisse. 1982. Binding of tumor-promoting and biologically inactive phorbol esters to artificial membranes. *Cancer Lett.* **17:** 135.

Driedger, P.E. and P.M. Blumberg. 1980. Specific binding of phorbol ester tumor promoters. *Proc. Natl. Acad. Sci.* **77:** 567.

Flockhart, D.A. and J.D. Corbin. 1982. Regulatory mechanisms in the control of protein kinases. *CRC Crit. Rev. Biochem.* **12:** 133.

Hecker, E. 1968. Cocarcinogenic principle from the seed of *Croton tiglium* and from other Euphorbiaceae. *Cancer Res.* **28:** 2338.

———. 1971. Isolation and characterization of the cocarcinogenic principles from croton oil. *Methods Cancer Res.* **6:** 439.

Hecker, E., N.E. Fuesing, W. Kunz, F. Marks, and H.W. Thielman, eds. 1982. *Cocarcinogenesis and biological effects of tumor promoters.* Raven Press, New York.

Kikkawa, U., Y. Takai, R. Minakuchi, S. Inohara, and Y. Nishizuka. 1982. Calcium-activated, phospholipid-dependent protein kinase from rat brain. *J. Biol. Chem.* **257:** 13341.

Kolata, G. 1983. Is tyrosine the key to growth control? *Science* **219:** 377.

Laemmli, U.K. 1970. Cleavage of structural proteins during the assembly of the head of bacteriophage T4. *Nature* **227:** 680.

Leach, K.L., M.L. James, and P.M. Blumberg. 1983. Characterization of a specific phorbol ester aporeceptor in mouse brain cytosol. *Proc. Natl. Acad. Sci.* **80:** 4208.

Lee, L.S. and I.B. Weinstein. 1978. Tumor promoting phorbol esters inhibit binding of epidermal growth factor to cellular receptors. *Science* **202:** 313.

Lowry, O.H., N.J. Rosebrough, A.L. Farr, and R.J. Randall. 1951. Protein measurements with the Folin phenol reagent. *J. Biol. Chem.* **193:** 265.

Martin, R.G. and B.N. Ames. 1961. A method for determining the sedimentation behavior of enzymes: Application to protein mixtures. *J. Biol. Chem.* **236:** 1372.

Merrill, C.R., D. Goldman, S.A. Sedman, and M.H. Ebert. 1981. Ultrasensitive stain for protein in polyacrylamide gels shows regional variation in cerebrospinal variation in cerebrospinal protein. *Science* **211:** 1437.

Niedel, J.E., L.J. Kuhn, and G.R. Vandenbark. 1983. Phorbol diester receptor copurifies with protein kinase C. *Proc. Natl. Acad. Sci.* **80:** 36.

Sando, J.J. and M.C. Young. 1983. Identification of high-affinity phorbol diester receptor in cytosol of EL4 thymoma cells: Requirements for calcium, magnesium and phospholipids. *Proc. Natl. Acad. Sci.* **80:** 2642.

Shoyab, M. and G.J. Todaro. 1980a. Vitamin K3 and related quinones, like tumor-promoting phorbol esters, alter the affinity of epidermal growth factor for its membrane receptors. *J. Biol. Chem.* **255:** 8735.

———. 1980b. Specific high affinity cell membrane receptors for biologically active phorbol and ingenol esters. *Nature* **288:** 451.

———. 1981. Perturbation of membrane phospholipids alters the interaction between epidermal growth factor and its membrane receptors. *Arch. Biochem. Biophys.* **206:** 222.

———. 1982. Partial purification and characterization of a binding protein for biologically active phorbol and ingenol esters from murine sera. *J. Biol. Chem.* **257:** 439.

Shoyab, M., J.E. De Larco, and G.J. Todaro. 1979. Biologically active phorbol esters specifically alter the affinity of epidermal growth factor membrane receptors. *Nature* **279:** 387.

Shoyab, M., G.J. Todaro, and J.F. Tallman. 1982. Chlorpromazine and related antipsychotic tricyclic compounds competitively inhibit the interaction between tumor-promoting phorbol esters and their specific receptors. *Cancer Lett.* **16:** 171.

Shoyab, M., T.C. Warren, and G.J. Todaro. 1981a. Tissue and species distribution and developmental variation of specific receptors for biologically active phorbol and ingenol esters. *Carcinogenesis* **2:** 1273.

———. 1981b. Isolation and characterization of an ester hydrolase active on phorbol diesters from murine liver. *J. Biol. Chem.* **256:** 12529.

Siegel, L.M. and K.G. Monty. 1966. Determination of molecular weight and fractional ratio by use of gel filtration and density gradient centrifugation. *Biochim. Biophys. Acta* **112:** 346.

Sivak, A. 1979. Cocarcinogenesis. *Biochim. Biophys. Acta* **560:** 67.

Slaga, T.J., A. Sivak, and R.K. Boutwell, eds. 1978. *Mechanisms of tumor promotion and cocarcinogenesis.* Raven Press, New York.

Takai, Y., A. Kishimoto, and Y. Nishizuka. 1982. Calcium and phospholipid turnover as transmembrane signalling for protein phosphorylation. In *Calcium and cell function* (ed. W.Y. Cheung), vol. 2, p. 385. Academic Press, New York.

Thompson, S.T., K.H. Cass, and E. Stellwagen. 1975. Blue dextran-sepharose: An affinity column for the dinucleotide fold in proteins. *Proc. Natl. Acad. Sci.* **72:** 669.

Van Duuren, B.L. 1969. Tumor-promoting agents in two-stage carcinogenesis. *Prog. Exp. Tumor Res.* **11:** 31.

Wilson, J.E. 1976. Application of blue dextran and Cibacron blue F3GA in purification and structural studies of nucleotide requiring enzymes. *Biochem. Biophys. Res. Commun.* **72:** 816.

Wise, B.C., R.L. Raynor, and J.F. Kuo. 1982a. Phospholipid-sensitive Ca^{2+}-dependent protein kinase from heart. I. Purification and general properties. *J. Biol. Chem.* **257:** 8481.

Wise, B.C., D.B. Glass, C.-H. JenChou, R.L. Raynor, N. Katch, R.C. Schatzman, R.S. Turner, R.F. Kilber, and J.F. Kuo. 1982b. Phospholipid-sensitive Ca^{2+}-dependent protein kinase from heart. II. Substrate specificity and inhibition by various agents. *J. Biol. Chem.* **257:** 8489.

Antibody and Cellular Detection of SV40 T-antigenic Determinants on the Surfaces of Transformed Cells

L.R. Gooding, R.W. Geib, K.A. O'Connell, and E. Harlow*

Department of Microbiology and Immunology, Emory University School of Medicine, Atlanta, Georgia 30322; and *Cold Spring Harbor Laboratory, Cold Spring Harbor, New York 11724

Cells transformed by simian virus 40 (SV40) possess a strong tumor-specific transplantation antigen (TSTA) in that prior immunization with SV40 tumor cells or virus can afford protection against challenge with large doses of otherwise lethal tumor cells. The finding that immunization with purified SV40 tumor (T) antigen can also induce SV40-specific transplantation resistance (Chang et al. 1979; Flyer and Tevethia 1982) indicates that the T antigen and TSTA are at least partially overlapping, i.e., they share antigenic determinants. In the mouse, cytotoxic T lymphocytes (CTL) specific for SV40 transformants are readily detected following tumor challenge (Gooding 1977; Pfizenmaier et al. 1978) and provide the best in vitro correlate of the TSTA phenomenon (Gooding 1982). Tevethia et al. (1980) have shown that immunization with purified T antigen is sufficient for generation of SV40-specific CTL. Therefore at least a portion of the cellular immune response leading to destruction of SV40-transformed cells is directed at determinants of large T antigen. Given the assumption that immune destruction of intact tumor cells requires recognition of cell-surface determinants, it follows that a molecule with at least partial cross-reactivity with T antigen must be exposed on the external surface of the cell.

Several laboratories have reported detectable antibody binding to the surface of SV40-transformed cells by using some hyperimmune anti-T-antigen antisera (Deppert et al. 1980; Soule et al. 1980; Henning et al. 1981); however, the genesis of the surface-associated protein and its relationship to nuclear T antigen remain in doubt. For example, the molecule may arise from posttranslational modification of T antigen, followed by insertion into the plasma membrane (Anderson et al. 1977). Alternatively, T antigen released from dead cells may be passively adsorbed to live cells in the culture (Lange-Mutschler and Henning 1982). A third possibility is that a subpopulation of viral mRNA produces a protein distinct from but serologically cross-reactive with T antigen. We have employed a panel of monoclonal anti-T-antigen antibodies to investigate the expression of individual T-antigenic determinants on the cell surface. These results are then compared with information available on the specificity of SV40-immune cytotoxic T cells.

Experimental Procedures

Mice
C3H/HeJ mice were obtained from the Jackson Laboratories, Bar Harbor, Maine.

Cell lines and viruses
PS-C3H and SV-C3H are SV40-transformed C3H/HeJ lymphoblasts and embryo fibroblasts, respectively (Gooding 1977). SV-3T3 are SV40-transformed BALB/c 3T3 cells. SV-C3H-V1, SV-C3H-V2, and SV-C3H-V3 were derived as described previously (Gooding 1982). MM5 was derived from a C3H mouse mammary tumor. C3HA is a spontaneously transformed C3H/HeJ embryo fibroblast (Gooding 1979a). pVBtlTK-1-clone 10 (termed "clone 10" in the text) is an LMTK$^-$ cell transfected with the pVBtlTK-1 plasmid and produces a 33K aminoterminal fragment of T antigen (Reddy et al. 1982). Wild-type SV40, strain 776, was obtained from J. Pipas. The adeno-SV40 hybrid viruses Ad2$^+$ ND$_1$ and Ad2$^+$ ND$_2$ were obtained from A. Lewis.

Antisera and monoclonal antibodies
Mouse anti-SV40 sera were prepared by biweekly injection of cultured SV40-transformed cells as described previously (Gooding 1979b). Sera from hamsters bearing SV40-induced tumors were obtained from the National Institutes of Health serum bank. Cells producing monoclonal anti-H-$2K^k$ antibody 11-4.1 (Oi et al. 1978) were obtained from the Salk Distribution Center. Cells producing monoclonal anti-T antibodies were obtained from L. Crawford and E. Harlow (PAb400 series; Harlow et al. 1981) or were produced by us (PAb600 series; L. Gooding et al., in prep.). Antibodies were purified from culture fluid by absorption to and elution from protein A–Sepharose columns (Ey et al. 1978).

Indirect immunofluorescence
Indirect immunofluorescent staining of live cells, harvested from monolayers by incubation with 0.02% EDTA in phosphate-buffered saline (PBS), was performed as described previously (Gooding 1979b). Briefly 1 × 10^6 cells were incubated on ice for 45 minutes with 2.5 μg of purified antibody in a final volume of 50 μl. Cells were washed in cold PBS containing 0.1% sodium azide and then incubated on ice for 45 minutes with 50 μl of fluorescein-conjugated rabbit anti-mouse IgG (Miles-Yeda, Ltd., lot no. S 168) diluted 1:20. After additional washes, stained cells were fixed for 1 hour in 1% formaldehyde in PBS. In control experiments, fixation was found to have no immediate effect on levels of fluorescence of stained or control cells, but fixation was necessary to permit storage of stained cells prior to examination (e.g., 24 hr). Just before reading, cells were resuspended in a solution of 90% glycerol, 10% PBS, 0.1% p-phen-

ylenediamine to prevent fading of the fluorescene dye during illumination (Johnson and Nogueira Aravjo 1981). Fluorescence was observed on a Leitz Orthoplan microscope equipped with a 100-W mercury-vapor lamp for incident light fluorescence. Fluorescence was quantitated using a Leitz 63× objective (1.3 N.A.) on the Orthoplan microscope and a Leitz MPV-2 microphotometer. Individual cells (a minimum of 40 per sample) were measured using a circular reading diaphragm superimposed on a minimum-diameter illuminated field. The MPV-2 output, in arbitrary units, is linearly related to the amount of fluorescent light detected in the field. Empty fields were also read and this background value was subtracted. In each experiment a control sample of cells, stained as above using the IgG_1 myeloma protein MPC 21, was included. For individual experiments, results are expressed in arbitrary units of fluorescence. Where results from several experiments are compared, data are expressed as the fluorescence index (F.I.), where:

$$F.I. = \frac{\text{units of fluorescence on stained sample}}{\text{units of fluorescence on control sample}}$$

Radioimmunoassay for T antigen and anti-T-antigen antibody
A double-antibody, solid-phase radioimmunoassay (RIA) for T antigen was performed as described previously (Benchimol et al. 1982; Gooding and O'Connell 1983). Monoclonal antibodies used for the experiments shown in Figures 4 and 5 were PAb416 and ^{125}I-labeled PAb423. T-antigen-containing extracts were prepared by incubating 1×10^7 test cells in 1 ml of 1% NP-40 in 0.15 M NaCl and 0.02 M Tris·Cl, pH 8.0. Insoluble material was removed by centrifugation. Relative quantitation was performed by comparing ^{125}I cpm bound with test extracts with ^{125}I cpm bound by an extract of TC7 monkey cells infected 48 hours earlier with 10 pfu/cell of SV40 wild-type strain 776. The latter is assigned the arbitrary value of 100 "T-antigen units" per cell.

Quantitation of antibody to T antigen was performed essentially as described by Benchimol et al. (1982) for antibodies to p53 and is based on the premise that antibodies that bind to the same or closely adjacent sites on T antigen will compete with one another for binding whereas those binding to distant sites will not. T antigen, from TC7 cells infected with wild-type SV40 as above, is bound to microtiter plates with one monoclonal antibody. Test antisera or unlabeled monclonal antibodies are then added along with the ^{125}I-labeled second antibody (~10,000 cpm per sample). The ability of test antisera to inhibit binding of the ^{125}I-labeled antibody is compared with inhibition produced when the unlabeled monoclonal antibody itself is used, and the results are expressed as micrograms of antibody per milliliter of serum.

Production and assay of cytotoxic T-cell clones
Cytotoxic T-cell lines, specific for SV40-infected or transformed cells expressing the $H-2K^k$ allele, were produced by limiting-dilution cloning in T-cell growth factors as described previously (O'Connell and Gooding 1984). Cytolytic activity was assayed by a 4-hour ^{51}Cr-release assay (Gooding 1977).

Results

Purified anti-T-antigen monoclonal antibodies were tested for binding to the surface of an SV40-transformed mouse cell line, PS-C3H, by indirect immunofluorescence. Figure 1 illustrates the surface reaction of one anti-T-antigen antibody, PAb601, relative to staining with antiglobulin alone or the anti-$H-2K^k$ antibody 11-4.1. Visual screening using a panel of 22 anti-T-antigen antibodies revealed a spectrum of reactivities against the surface of PS-C3H, with some antibodies giving strong surface staining and others appearing completely unreactive. Visual observation was then confirmed using a micro-

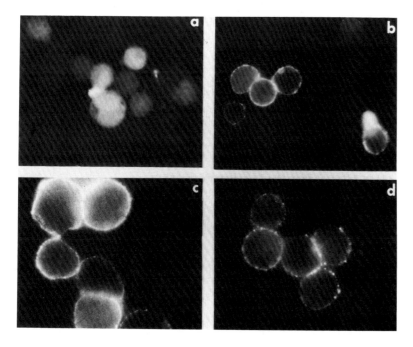

Figure 1 Indirect immunofluorescent staining of live PS-C3H cells. Objectives used were an Olympus 45× (1.0 N.A.) (*a,b*) or a Leitz 63× (1.3 N.A.) (*c,d*). Cells were stained with FITC-rabbit anti-mouse Ig alone (deliberately overexposed to illustrate autofluorescence) (*a*) PAb601 (*b,d*) and 11-4.1 (*c*).

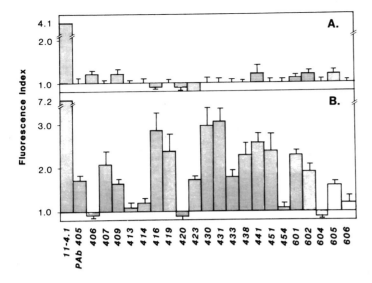

Figure 2 MPV-2 quantitation of monoclonal anti-T-antigen antibody reaction with the surface of living MM5 (A) or PS-C3H (B) cells by indirect immunofluorescence.

photometer system capable of measuring the light emitted from individual stained cells.

Figure 2 summarizes a series of experiments in which the anti-T-antigen antibodies were tested for binding to PS-C3H (Fig. 2B) and to a T-antigen-negative mammary tumor, MM5 (Fig. 2A). The F.I. given for each antibody is the mean value ($\pm$ S.E.) derived from a minimum of six separate experiments. For comparison, the anti-H-$2K^k$ antibody 11-4.1 was also tested. The strong surface staining produced by 11-4.1 on PS-C3H is reflected in MPV-2 readings averaging about sevenfold higher than background. Although none of the anti-T-antigen antibodies achieved this high level of binding, most averaged between 1.5-fold and 3-fold over background. Statistical analysis of the data in Figure 2 indicates that 15 of the 22 antibodies show significant ($P < .005$) reaction with the surface of PS-C3H cells. The seven exceptions (PAb406, 413, 414, 420, 454, 604, and 606) fail to show significant binding ($P > .25$), and none of the antibodies bind significantly to MM5 ($P > .25$). Since all of the antibodies immunoprecipitate nuclear T antigen from PS-C3H (not shown), the differential reactions observed with the cell surface suggest some alteration in the molecule relative to nuclear T antigen.

Harlow et al. (1981) have used T-antigen fragments produced by early-region deletion mutants to determine the site within the T antigen to which monoclonal antibodies bind. We have recently extended this study by using a larger panel of deletion mutants (Peden et al. 1980; Pipas et al. 1980) and the cross-reacting papovaviruses BK and SA12 (L. R. Gooding et al., in prep.). A summary of the information available on binding sites for the monoclonals used in this study is shown in Figure 3. The length of the box within which the monoclonal name appears represents the maximum region of the T antigen within which the antibody binds. Where more than one antibody appears within the same box, these could not be distinguished from one another in our analysis of binding sites. The open background denotes antibodies that do not bind to the surface of PS-C3H; the striped background denotes those that do. The pattern that emerges from this analysis is that all determinants in the aminoterminal third to half of T antigen are readily detected on the cell surface. Likewise, some but not all determinants within the carboxyterminal 185 amino acids are exposed on the surface. Six of the seven antibodies that fail to react with the surface are found in two mapping groups that require the "midregion" of T antigen for binding; i.e., they bind to sites found on the Ad2$^+$ ND$_2$ proteins but not on Ad2$^+$ ND$_1$.

The finding that the majority of monoclonal anti-T-antigen antibodies bind to the surface of SV40-transformed cells was unexpected because we had been repeatedly unsuccessful in using hyperimmune anti-T-

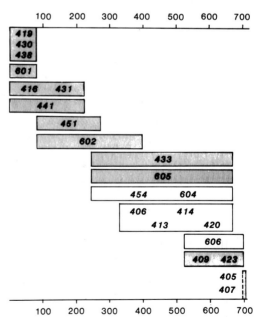

Figure 3 Cell-surface binding of monoclonal anti-T-antigen antibodies as a function of the site within T antigen to which the antibodies bind. (▨) Antibodies that bind to the surface of PS-C3H.

antigen sera to detect surface T antigen (L.R. Gooding et al., unpubl.). It seemed possible that antibodies in immune serum are preferentially produced to T-antigenic determinants that are not represented on the cell surface. The approach used to test this hypothesis is based on studies of E. Harlow and L. Crawford (unpubl.). These workers used competitive inhibition of binding to T antigen to define different binding sites among anti-T-antigen monoclonal antibodies. The 37 antibodies tested could be assigned to 15 separate binding groups. Those within a group inhibited each other's binding to T antigen, whereas those from different groups in general did not. Four ^{125}I-labeled monoclonal antibodies that bind to different regions of T antigen were chosen, and several hyperimmune anti-T-antigen sera were tested for the ability to inhibit monoclonal binding to T antigen. The amount of a given antibody specificity in the serum was calculated from a standard curve generated by using the unlabeled monoclonal antibody. As shown in Table 1, all five sera tested, three from hyperimmune mice and two from tumor-bearing hamsters, contain high levels of antibody activity for determinants requiring the midregion of T antigen (PAb454 and PAb413), while very little antibody binds to regions near the termini (PAb416 and PAb407). Although more-exhaustive studies are required to evaluate all possible antibody binding sites on T antigen, the tentative conclusion from these experiments is that in hyperimmune serum the preponderance of antibodies recognize sites not represented on the cell surface.

Although surface T-antigenic determinants can be detected on PS-C3H and a number of other SV40 cell lines available to us, screening with representative antibodies has failed to detect T antigen on the surface of many SV40 cell lines, particularly those with prolonged culture histories. This suggests that there may be factors regulating the quantity of T antigen on the surface. One possibility, that T antigen on the surface is derived from the nuclear T-antigen pool, predicts that the amounts of surface-associated and nuclear T antigen should vary proportionately. This was tested with a series of cells derived from SV40-transformed mouse embryo fibroblasts. The uncloned cell line, SV-C3H, fails to grow progressively in normal, syngeneic C3H mice. Growth is observed, however, in mice whose cellular immune response has been suppressed by sublethal whole-body irradiation. SV-C3H cells were selected for the ability to grow in vivo by serial passage through irradiated mice. Cells from each passage were explanted to culture (termed SV-C3H-V1, SV-C3H-V2, and SV-C3H-V3). With each successive passage, the selected cells became more capable of transient growth in normal (i.e., unirradiated) mice; however, all are ultimately rejected by immunocompetent recipients (Gooding 1982). The level of T antigen in these cells was quantitated by radioimmunoassay as shown in Figure 4. The cells recovered from in vivo selection of SV-C3H showed only a slight decrease in T-antigen content compared with the SV-C3H parent. This is in marked contrast to the changes in the surface T-antigen level of these cells. Figure 5 summarizes three separate experiments where the levels of *H-2* and surface T antigen assayed by immunofluorescence are compared with total cellular (primarily nuclear) T antigen measured by RIA. In vivo passage caused a sharp decrease in surface exposure of T antigen measured either at the amino terminus (PAb416) or the carboxyl terminus (PAb423, not shown). The 2.5-fold decrease in surface binding of PAb416 is clearly disproportionate to the mere 25% decrease in total cellular T antigen.

Another approach to the relationship between surface-associated and nuclear T antigens is to monitor the dependence of production of the surface molecule on de novo protein synthesis. One might predict that if the surface-associated molecule is derived directly from the nuclear pool by some form of passive intracellular equilibrium, then continued protein synthesis should not be required for continued production of surface T antigen, at least over a period of a few hours. To test this prediction, the following experiment was designed. PS-C3H cells were treated with 2 mg/ml trypsin for 30 minutes at 37°C, a protocol found to remove the determinants to which the anti-T-antigen antibodies bind, and then plated into culture medium with or without 25 μg/ml cyclo-

Table 1 Quantitation of Specificities in Hyperimmune Anti-T-antigen Sera by Competition Radioimmunoassay

	^{125}I-labeled monoclonal PAb			
	416	454	413	407
T-antigen binding site[a]	1–223	~245–670	~523–670	698–708
Binding to SDS-denatured T antigen	+	–	–	+
Self inhibition[b]	14.4	19.2	9.6	11.7
Inhibiting antisera[c]				
(C3H × B10.D2) anti-PS-C3H	6 ± 0	224 ± 13	34 ± 4	15 ± 3
C3H/HeJ anti-PS-C3H	7 ± 1	139 ± 12	53 ± 10	16 ± 2
BALB/c anti-SV-3T3	24 ± 3	160 ± 10	132 ± 13	33 ± 3
Hamster, tumor bearer (7X0004)	5 ± 1	300 ± 24	101 ± 21	14 ± 4
Hamster, tumor bearer (7X0001)	4 ± 1	322 ± 48	137 ± 35	15 ± 4

[a]Amino acid number.
[b]Nanograms for 50% inhibition.
[c]Micrograms per milliliter (± S.E.M.).

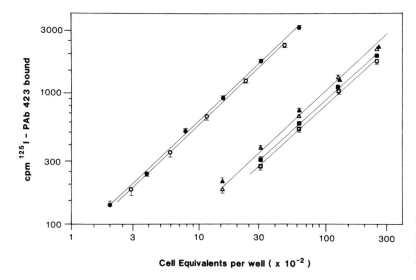

Figure 4 Radioimmunoassay for total cellular T antigen. Lysates were produced from SV40-infected TC7 (○), PS-C3H (●), SVC3H (△), SV-C3H-V1 (▲), SV-C3H-V2 (■), and SV-C3H-V3 (□).

heximide to prevent de novo protein synthesis. The experiment shown in Table 2 demonstrates that although cells that are allowed to synthesize protein reexpress surface T antigen at about half the level of untreated cells after 4.5 hours, reexpression is completely inhibited by cycloheximide. In contrast, total cellular T-antigen content, as monitored by RIA, did not decrease measurably during incubation with cycloheximide (data not shown). This suggests that surface T antigen does not arise from an intracellular pool, whether from its own nucleus or from the nucleus of dead cells in the culture, but rather that expression of T antigen on the surface is directly dependent on the active protein-synthetis machinery of the cell.

Of particular interest is the relationship between the T-antigen-like molecule detected on the cell surface by monoclonal antibodies and determinants recognized in the cellular immune response to SV40-induced tumors. Tevethia et al. (1983) recently reported that cells containing only a 33K aminoterminal fragment of T antigen

are lysed by SV40-immune cytotoxic T cells. We have confirmed this observation and find as well that cells infected with Ad2$^+$ ND$_1$, which produce a 28K protein homologous with the carboxyl terminus of T antigen, are also lysed by SV40-immune T cells (Gooding and O'Connell 1983). These results suggest that if in fact structural determinants of T antigen are involved in cellular recognition, then both amino- and carboxyterminal determinants are recognized. To assess more accurately the dependence of cellular recognition on different regions of T antigen, we derived cloned lines of C3H cytotoxic T lymphocytes responding to syngeneic PS-C3H cells. These "monoclonal" cytotoxic cells were then tested for their target specificity as shown in Figure 6. Of the 13 lines produced, 10 lyse cells infected with Ad2$^+$ ND$_1$, and 3 lyse cells expressing a 33K aminoterminal fragment of T antigen (clone 10). None of the clones lysed cells infected with adenovirus 2 or the LMTK$^-$ cell used to produce clone 10 (not shown). Thus, among the cytotoxic lines tested to date, none requires expression of the midregion of T antigen (i.e., between amino acids 245 and 523) for recognition and cytolysis.

Discussion

Several theories have been proposed to explain how SV40 T antigen, which resides primarily within the cell nucleus, can behave as, or at least cross-react with, the SV40-induced TSTA. This latter term is somewhat ambiguous since the cellular mechanisms underlying overt manifestations of immune resistance almost certainly vary with the genotype of the responding animal, and not all of these mechanisms would require surface expression of T antigen for cellular destruction to occur. Nonetheless, at least one in vitro correlate of the TSTA, lysis by CTL, does require surface exposure of antigenic determinants, and SV40-immune CTL can be elicited by injection of purified T antigen (Tevethia et al. 1980).

One theory, proposed by Lange-Mutschler and Henning (1982), suggests that T antigen is released from

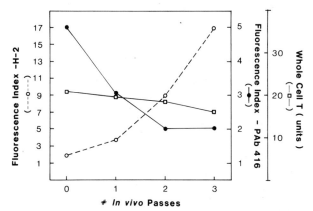

Figure 5 Changes in the antigenic profile of SV-C3H cells with passage through irradiated mice. Whole-cell T antigen (□) was measured by RIA. Surface expression of H-2 (○) and T antigen (●) was measured by indirect immunofluorescence with 11-4.1 and PAb416, respectively.

Table 2 Recovery of Surface T-antigen Expression Following Removal by Trypsin Digestion

Monoclonal antibody	Untreated	Trypsin	Trypsin recovery	
			$-CHI^a$	$+CHI$
—	32 ± 3	18 ± 2	25 ± 2	20 ± 2
11-4.1	130 ± 17	172 ± 25	141 ± 17	160 ± 28
PAb601	81 ± 6	17 ± 2	42 ± 4	23 ± 2
PAb431	108 ± 11	17 ± 1	53 ± 5	19 ± 2
PAb433	64 ± 5	18 ± 2	40 ± 4	17 ± 1

Results are given in arbitrary units of fluorescence.
[a]Cycloheximide.

cells, either living or dead, and then is adsorbed to the cell surface. Three lines of evidence argue against this mechanism for the origin of surface T-antigenic determinants described here. First, expression of "surface T antigen" requires active protein synthesis, which would not be predicted for passive adsorption of T antigen from the medium. Second, cells with similar amounts of nuclear T antigen (the in vivo–selected cells) express very different levels of surface T antigen. Finally, not all T-antigenic determinants could be detected on the cell surface, suggesting either a modification or sequestering of the molecule in a defined, rather than a random, manner.

One simple explanation for the "loss" of some T-antigenic determinants on surface T antigen is that the molecule is denatured, since with one exception (PAb606) all of the antibodies that fail to bind to the cell surface also fail to bind to SDS-denatured T antigen (Harlow et al. 1981; L. R. Gooding et al., in prep.). However, of the 15 antibodies that do bind to the cell surface, three (PAb451, PAb433, and PAb605) bind to denaturation-sensitive sites on T antigen, which argues for a "native" configuration for the surface-associated molecule. Nonetheless, our observation that the majority of T-antigenic determinants exposed on the cell surface are those that survive denaturation of the molecule is in agreement with the finding of Deppert et al. (1980) that the surface-associated molecule is most readily detected using antisera prepared to SDS-denatured T antigen. The failure of high-titered antisera elicited with the native T antigen (e.g., tumor-bearer sera) to bind to the cell surface (Lange-Mutschler et al. 1981) may reflect the immunodominance (vis-à-vis the antibody response) of configurational determinants, most of which are not exposed on the cell surface.

Another possibility for the relationship of nuclear and surface-associated T antigens is that both are synthesized from the same mRNA, but that modifications during or after translation direct most molecules to the nucleus and only a subpopulation to the cell surface. The data reported here are consistent with this hypothesis. The dependence of surface-T-antigenic expression on active protein synthesis suggests that molecules may be transported to the surface only during or shortly after their synthesis, and that surface T antigen is not derived from the nuclear pool. Also, the finding of both amino- and carboxyterminal determinants on the surface indicates that, at least for most of the molecule, the mRNA and reading frame for nuclear and surface T antigen are the same. From monoclonal antibody binding, it appears that one or more regions in the carboxyterminal half of T antigen are not exposed on the cell surface and may be either removed or sequestered from the external environment. Deppert and Walter (1982) reached a similar conclusion by using cells infected with nondefective adeno-SV40 hybrid viruses. Finally, the recent report that

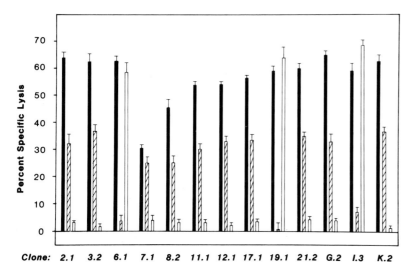

Figure 6 Lysis by cloned CTL of PS-C3H (closed bars), Ad2$^+$ ND$_1$–infected C3HA (hatched bars), and clone 10 (open bars). Effector-to-target ratio was 6.

plasma membrane–associated T antigen is acylated while that in the nucleus is not (Klockmann and Deppert 1983) lends strength to the idea that surface localization of T antigen occurs via a specific pathway. It should be noted, however, that we cannot eliminate the possibility that surface-associated T antigen is translated from a very low frequency early-region mRNA, the splicing of which removes the midregion of T antigen while preserving the carboxyterminal reading frame.

At present, the relationship between the surface-associated T antigen defined serologically and the structures recognized by SV40-immune CTL remains to be determined. There is no direct evidence that the CTL lines used here recognize structural determinants on T antigen, but the correlation between the serologic findings and the domains of T antigen required for CTL recognition is indeed striking. One is tempted to postulate that CTL recognition does in fact require surface exposure of T-antigenic sites and that sequestered regions of surface T antigen (i.e., those in the midregion of the molecule) are accessible neither to antibody nor T cells.

Acknowledgment

This work was supported by U.S. Public Health Service grants CA-30266 and CA-00672 to L.R.G.

References

Anderson, J., R. Martin, C. Chang, P. Mora, and D. Livingston. 1977. Nuclear preparations of SV40-transformed cells contain tumor-specific transplantation antigen activity. *Virology* **76:** 420.

Benchimol, S., D. Pim, and L. Crawford. 1982. Radioimmunoassay of the cellular protein p53 in mouse and human cell lines. *EMBO J.* **1:** 1055.

Chang, C., R. Martin, D. Livingston, S. Luborsky, C. Hu, and P. Mora. 1979. Relationship between T antigen and tumor-specific transplantation antigen in simian virus 40–transformed cells. *J. Virol.* **29:** 69.

Deppert, W. and G. Walter. 1982. Domains of simian virus 40 large T antigen exposed on the cell surface. *Virology* **122:** 56.

Deppert, W., K. Hanke, and R. Henning. 1980. Simian virus 40 T-antigen-related cell surface antigen: Serological demonstration on simian virus 40–transformed monolayer cells *in situ*. *J. Virol.* **35:** 505.

Ey, P., S. Prowse, and C. Jenkin. 1978. Isolation of pure IgG$_1$, IgG$_{2a}$, and IgG$_{2b}$ immunoglobulins from mouse serum using protein A–Sepharose. *Immunochemistry* **15:** 429.

Flyer, D. and S. Tevethia. 1982. Biology of simian virus 40 (SV40) transplantation antigen (TrAg). VIII. Retention of SV40 TrAg sites on purified SV40 large T antigen following denaturation with sodium dodecyl sulfate. *Virology* **117:** 267.

Gooding, L. 1977. Specificities of killing by cytotoxic lymphocytes generated *in vivo* and *in vitro* to syngeneic SV40-transformed cells. *J. Immunol.* **118:** 920.

———. 1979a. Specificities of killing by T lymphocytes generated against syngeneic SV40 transformants: Studies employing recombinants within the *H-2* complex. *J. Immunol.* **122:** 1002.

———. 1979b. Antibody blockade of lysis by T lymphocyte effectors generated against syngeneic SV40-transformed cells. *J. Immunol.* **122:** 2328.

———. 1982. Characterization of a progressive tumor from C3H fibroblasts transformed *in vitro* with SV40 virus. Immunoresistance *in vivo* correlates with phenotypic loss of H-2K^k. *J. Immunol.* **129:** 1306.

Gooding, L. and K. O'Connell. 1983. Recognition by cytotoxic T lymphocytes of cells expressing fragments of the SV40 tumor antigen. *J. Immunol.* **131:** 2580.

Harlow, E., L. Crawford, D. Pim, and N. Williamson. 1981. Monoclonal antibodies specific for simian virus 40 tumor antigens. *J. Virol.* **39:** 861.

Henning, R., J. Lange-Mutschler, and W. Deppert. 1981. SV40-transformed cells express SV40 T-antigen-related antigens on the cell surface. *Virology* **108:** 325.

Johnson, G. and G. Nogueiro Aravjo. 1981. A simple method of reducing the fading of immunofluorescence during microscopy. *J. Immunol. Methods* **43:** 349.

Klockmann, U. and W. Deppert. 1983. Acylated simian virus 40 large T antigen: A new subclass associated with a detergent-resistant lamina of the plasma membrane. *EMBO J.* **2:** 1151.

Lange-Mutschler, J. and R. Henning. 1982. Cell surface binding affinity of simian virus 40 T antigen. *Virology* **117:** 173.

Lange-Mutschler, J., W. Deppert, K. Hanke, and R. Henning. 1981. Detection of simian virus 40 T-antigen-related antigens by a ^{125}I-Protein A–binding assay and by immunofluorescence microscopy on the surface of SV40-transformed monolayer cells. *J. Gen. Virol.* **52:** 301.

O'Connell, K.A. and L. Gooding. 1984. Cloned cytotoxic T lymphocytes recognize cells expressing discrete fragments of the SV40 tumor antigen. *J. Immunol.* **132:** 953.

Oi, V., P. Jones, J. Goding, L. Herzenberg, and L. Herzenberg. 1978. Properties of monoclonal antibodies to mouse Ig allotypes, H-2 and Ia antigens. *Curr. Top. Microbiol. Immunol.* **81:** 115.

Peden, K., J. Pipas, S. Pearson-White, and D. Nathans. 1980. Isolation of mutants of an animal virus in bacteria. *Science* **209:** 1392.

Pfizenmaier, K., G. Trichièri, D. Solter, and B. Knowles. 1978. Mapping of *H-2* genes associated with T cell-mediated cytotoxic responses to SV40 tumor–associated specific antigens. *Nature* **274:** 691.

Pipas, J., S. Adler, K. Peden, and D. Nathans. 1980. Deletion mutants of SV40 that affect the structure of viral tumor antigens. *Cold Spring Harbor Symp. Quant. Biol.* **44:** 285.

Reddy, V., S. Tevethia, M. Tevethia, and S. Weissman. 1982. Nonselective expression of simian virus 40 large tumor antigen fragments in mouse cells. *Proc. Natl. Acad. Sci.* **79:** 2064.

Soule, H., R. Lanford, and J. Butel. 1980. Antigenic and immunogenic characteristics of nuclear and membrane-associated simian virus 40 tumor antigen. *J. Virol.* **33:** 887.

Tevethia, S., D. Flyer, and R. Tjian. 1980. Biology of simian virus 40 (SV40) transplantation antigen (TrAg). VI. Mechanism of induction of SV40 transplantation immunity in mice by purified SV40 T antigen (D2 protein). *Virology* **107:** 13.

Tevethia, S., M. Tevethia, A. Lewis, V. Reddy, and S. Weissman. 1983. Biology of the simian virus 40 (SV40) transplantation antigen (TrAg). IX. Analysis of TrAg in mouse cells synthesizing truncated SV40 large T antigen. *Virology* **128:** 319.

Localization of Antigenic Sites Reactive with Cytotoxic Lymphocytes on the Proximal Half of SV40 T Antigen

S.S. Tevethia and M.J. Tevethia
Department of Microbiology and Cancer Research Center, The Pennsylvania State University, College of Medicine, Hershey, Pennsylvania 17033

The presence of specific surface antigens on simian virus 40 (SV40)–transformed or tumor cells was demonstrated 20 years ago by the in vivo transplantation rejection test (for review, see Tevethia 1980, 1983). The observations that these antigens, referred to as tumor-specific transplantation antigen (TSTA), cross-reacted among SV40-transformed cells of various species and that TSTA was expressed during the viral lytic cycle (Girardi and Defendi 1970; Tevethia and Tevethia 1976) and during the abortive cycle (Pretell et al. 1979) led to the speculation that the SV40 genome directs the synthesis or expression of TSTA. The additional finding that the early region of the SV40 genome was sufficient to induce the synthesis of TSTA during the viral infection of permissive and nonpermissive cells led to the tentative conclusion that one or more gene products of this region is responsible for inducing the expression of TSTA (for review, see Tooze 1980; Tevethia 1980). Only two proteins are coded for by the early region of the SV40 genome—a 94,000 dalton (94K) and a 17K protein, the large T and small T antigen, respectively (for review, see Tooze 1980; Rigby and Lane 1983). T antigen is mainly localized in the nucleus, as shown by the immunofluorescence test (Rapp et al. 1964) and by cell fractionation and subsequent immunoprecipitation of T antigen and analysis using polyacrylamide gel electrophoresis (Tegtmeyer et al. 1975).

Anderson et al. (1977) were the first to demonstrate that the T antigen from the nucleus of SV40-transformed human cells can in fact immunize BALB/c mice against syngeneic tumor cells. This observation was followed by the demonstration that T antigen purified to homogeneity was able to induce in vivo protection (Chang et al. 1979; Tevethia et al. 1980). In addition, the denatured T antigen extracted from SDS gels immunized BALB/c mice against tumor-cell challenge (Flyer and Tevethia 1982). These findings lay the foundation for the prediction that the T antigen, the product of SV40 A gene, possesses antigenic sites to which the host can respond immunologically, leading to the rejection of tumor cells. Since immunological events involved in tumor rejection take place at the cell surface (Cerottini and Brunner 1974), it was concluded that the SV40 T antigen must be expressed at the cell surface. This view was further strengthened by the observation (Tevethia et al. 1980; Chang et al. 1982) that T antigen is involved in the generation of thymus-derived cytotoxic lymphocytes (CTL) that specifically killed syngeneic SV40-transformed cells, further defining the specificity of immunological events occurring during tumor rejection. With the availability of a cytotoxic assay involving CTL that follows the rules of H-2-restricted killing (Zinkernagel and Doherty 1979), it was possible to demonstrate convincingly that T antigen is processed in vivo and is presented in the context of the host H-2 antigen during the generation of CTL (Gooding and Edwards 1980). The evidence cited above and reviewed by Tevethia (1980, 1983) is sufficient to conclude that T antigen actually carries TSTA sites.

If this conclusion is valid, and there is no reason to doubt that it is, based on the overwhelming evidence cited above, then the questions that must be answered are, (1) what portions of the SV40 early region specify TSTA, and (2) are there distinct antigenic sites that act as TSTA on T antigen or are antigenic sites duplicated on the T antigen molecule? We have explored these questions, using cellular immunology and molecular biological approaches.

Role of the aminoterminal region of T antigen in specifying TSTA

Previous attempts to determine the regions of the SV40 genome that specify the TSTA sites have utilized the adenovirus 2 (Ad2)–SV40 hybrid viruses (Lewis and Rowe 1973; Jay et al. 1978, 1979). Jay et al. (1978) demonstrated that immunization of BALB/c mice with a 30K protein representing the carboxyterminal end of T antigen and specified on the SV40 genome between 0.28–0.17 map units (m.u.) in Ad2$^+$ND$_1$ hybrid virus was sufficient to immunize BALB/c mice against tumor-cell challenge. Jay et al. (1979) also reported that the TSTA sites were present on the 42K and 56K proteins coded in the SV40 genome from 0.38–0.17 m.u. in Ad2$^+$ND$_2$ virus. Since both the 42K and the 56K proteins include the entire protein complement coded for by the SV40 DNA (0.28–0.17 m.u.) in AD2$^+$ND$_1$ virus, it was not possible from these results to determine whether the 42K and 56K proteins carry additional TSTA sites that are not present on the 30K protein coded by Ad2$^+$ND$_1$ virus.

We have examined the question of whether the SV40 DNA that codes for the aminoterminal region of T antigen (0.65–0.42 m.u.) also carries TSTA sites. For this purpose, we have generated cell lines by transfecting Ltk$^-$

cells with plasmid pBR322 carrying either the entire early region of the SV40 genome (0.72–0.14 m.u.) or only a portion of the SV40 early region (0.72–0.42 m.u.). The cloning of SV40 DNAs of varying lengths and their expression in mouse Ltk⁻ cells has been described (Reddy et al. 1982). The plasmids carrying SV40 DNA were transfected into Ltk⁻ cells, using the method described previously by Wigler et al. (1977), and the TK⁺ transformants were selected in HAT medium. Cell lines were established from individual clones and maintained in HAT medium. The transformed cells were analyzed for the expression of large and small T antigens by the immunoprecipitation assay (Tevethia et al. 1983). The results shown in Figure 1 demonstrate that Ltk⁺ cells transformed with pVBETK-1 plasmid, which carries the entire SV40 early region, synthesize normal-sized large and small T antigens. In addition, the 53K cellular protein (NVT) (Lane and Crawford 1979; Linzer and Levine 1979) was also found to be complexed with large T antigen. We identified one clone that was transformed with pVBETK-1 plasmid but did not synthesize any of the SV40 tumor antigens. The cells transformed with pVBt2TK-1 plasmid, which carries SV40 DNA extending from 0.72 m.u. to 0.52 m.u., synthesized a normal-sized small T antigen and a truncated large T antigen of 12,300 daltons. A truncated large T antigen of 30,000–33,000 daltons was synthesized by cells transformed with plasmid pVBt1TK-1 containing SV40 DNA of 0.72–0.42 m.u.

in addition to small T antigen. An additional plasmid, pVBdlATK-1, which carries SV40 DNA of 0.72–0.42 m.u. and has a 300-bp deletion in the 0.59–0.54-m.u. region (Sleigh et al. 1978), was used to generate transformed cell lines that synthesized a 33K, truncated large T antigen but no small T antigen (Fig. 2). The truncated products did not stably associate with NVT.

An in vitro ^{51}Cr release assay was used to examine the expression of SV40 TSTA at the cell surface. This assay involved cytolytic thymus-derived lymphocytes (CTL), which have been shown to be involved in the rejection of SV40 tumor cells in vivo (Tevethia et al. 1974; Gooding 1982). Since Ltk cells are of the H-2^k haplotype, the lymphocyte-mediated cytotoxicity assay using C3H (H-2^k) mice was employed according to procedures described by Gooding (1977). C3H/HeJ mice were immunized with the PSC3H, an SV40-transformed cell line of H-2^k origin, and the spleen cells were stimulated in vitro with γ-irradiated PSC3H cells for 5 days. The lymphocytes were then reacted with ^{51}Cr-labeled target cells (Tevethia et al. 1983). The results in Table 1 show that SV40-immune CTL lysed not only syngeneic cells that synthesized the full-length large T antigen (pVBETK-1 cl 15), but also cells synthesizing the 33K, truncated large T antigen (pVB1TK-1 cl 10). The results also show that the pVBdlA2-3-cl-1 cells, which synthesize a truncated large T antigen of 33,000 daltons but no small T antigen, were susceptible to lysis by SV40-immune CTL, indicating that the 33K truncated large T antigen is a target for CTL and that small T antigen may not be involved. It should be noted that Ltk⁺ cells transformed with plasmid pVBdlA2-4-cl-2, which did not synthesize the truncated large T antigen because a portion of SV40 DNA is inserted in the wrong orientation in the

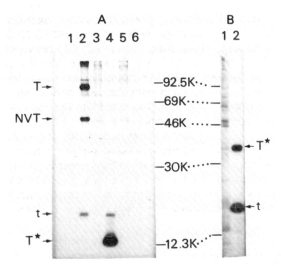

Figure 1 (A) Synthesis of large T, NVT, and small T antigens in pVBETK-1-transformed cells, and synthesis of small T and truncated large T antigens in cells transformed by pVBt2TK-1. ^{35}S-labeled extracts from pBVETK-1-cl-15 cells (lanes 1,2), pBVt2TK-1 cells (lanes 3,4), and the large-T-antigen-negative cell line pBVETK-1 cl 1 (lanes 5,6) were immunoprecipitated with anti–large T antigen T sera (lanes 2,4,6) or normal hamster sera (lanes 1,3,5) and analyzed by SDS-polyacrylamide gel electrophoresis (PAGE). (T*) Truncated large T antigen. (B) Synthesis of small T and truncated large T antigens in cells transformed by pVBt1TK-1. ^{35}S-labeled extracts from pBVt1TK-1-cl-10 cells were immunoprecipitated with normal hamster sera (lane 1) or anti–large T antigen sera (lane 2) and analyzed by SDS-PAGE. (Reprinted, with permission, from Tevethia et al. 1983.)

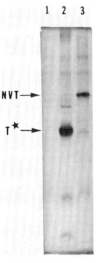

Figure 2 Synthesis of truncated large T antigen in pVBdlATK-1-transformed cells. ^{35}S-labeled extracts from pVBdlA2-3-cl-1 cells were immunoprecipitated with normal hamster sera (lane 1), anti–large T antigen T sera (lane 2), or monoclonal antibody to NVT (lane 3) and analyzed by SDS-PAGE (Harlow et al. 1981). (T*) Truncated large T antigen. (Reprinted, with permission, from Tevethia et al. 1983.)

Table 1 Expression of SV40 TSTA in Transfected Ltk+ ($H\text{-}2^k$) Cells Synthesizing Truncated Large T Antigen

Target cells	Proteins			% Specific ^{51}Cr release
	large T	small T	NVT	
PSC3H	+	+	+	55
pVBETK-1 cl 15	+	+	+	58
pVBt1TK-1 cl 10	+ (30K–35K)	+	–	53
pVBdlA2-3 cl 1	+ (30K–35K)	–	–	51
pVBdlA2-4 cl 2	–	–	–	2
pVBETK-1 cl 1	–	–	–	2
B6/WT-19 ($H\text{-}2^b$)	+	+	+	1.2

Effector cells were generated in C3H ($H\text{-}2^k$) mice by immunization with SV40-transformed syngeneic cells (PSC3H), and spleen cells were stimulated in vitro with γ-irradiated PSC3H cells for 5 days. The effector lymphocytes were used at an effector/target cell ratio of 10:1. Numbers in parentheses represent approximate m.w. of proteins. (+) Positive; (−) negative. (Reprinted, with permission, from Tevethia et al. 1983.)

plasmid, also did not express SV40 TSTA. The dependence of TSTA expression on the synthesis of large T antigen is further shown by the fact that pVBETK-1-cl-1 cells, which do not synthesize any of the SV40 tumor antigens, also do not express TSTA. It should be pointed out that SV40-transformed $H\text{-}2^b$ cells (B6/WT-19) were not susceptible to lysis by SV40-immune CTL from C3H mice.

The cells transformed by plasmid pVBt2TK-1, which synthesize normal-sized small T antigen and a truncated large T antigen of approximately 12,300 daltons, did not express SV40 TSTA at the cell surface (Table 2). None of the five separate clones of pVBt2TK-1 cells expressed TSTA at a level higher than the SV40-transformed allogeneic ($H\text{-}2^b$) cell line (B6/WT-19), which is not expected to react with SV40-immune CTL from $H\text{-}2^k$ mice. The PSC3H cells were highly susceptible to lysis by SV40-immune CTL. One possibility that remains to be explored is that the 12.3K truncated large T antigen does possess some portion of the unique large T antigen amino acid residues that might provide a target for CTL if such CTL could be expanded and cloned.

Attempts to determine whether the aminoterminal fragments of SV40 large T antigen immunize mice against tumor transplantation in vivo have indicated that only immunization of mice with cells that contain the entire large T antigen provided protection against SV40 tumor-cell challenge (Tevethia et al. 1983). The lack of immunogenicity of pVBt1TK-1-cl-10 cells could be due to the fact that the 33K large T antigen synthesized by these cells is highly unstable, having a half-life of approximately 40 minutes (Reddy et al. 1982). We have recently been able to transform primary mouse embryo fibroblasts (MEF) of C57BL/6 origin with plasmid pSV3T3-20-GV (Clayton et al. 1982), which contains SV40 early-region sequences of 0.65–0.37 m.u. These cells, B6/pSV-20-GV, synthesize a 48K large T antigen, are susceptible to lysis by CTL, and can immunize mice against tumor-cell challenge (S.S. Tevethia, unpubl.). The truncated large T antigen in these cells is, however, stable.

The results discussed above have clearly shown that SV40 early-region sequences that code for the proximal half of large T antigen can induce the expression of large T antigen and that the TSTA sites may be distributed throughout the large-T-antigen molecule and may not be limited to the carboxyterminal region of large T antigen.

Table 2 Expression of SV40 TSTA in pVBt2TK-1 Transfected Ltk+ ($H\text{-}2^k$) Clones

Target cells	Proteins			% Specific ^{51}Cr release
	large T	small T	NVT	
PSC3H	+	+	+	72
pVBt2TK-1 cl 1	–	–	–	16
pVBt2TK-1 cl 2	+ (12.3K)	+	–	16
pVBt2TK-1 cl 3	+ (12.3K)	+	–	14
pVBt2TK-1 cl 6	+ (12.3K)	U	–	8
pVBt2TK-1 cl 7	+ (12.3K)	U	–	13
pVBETK-1 cl 1	–	–	–	14
B6/WT-19 ($H\text{-}2^b$)	+	+	+	22

Effector cells were generated in C3H ($H\text{-}2^k$) mice as described in Table 1 and were used at an effector/target-cell ratio of 20:1. Numbers in parentheses represent approximate m.w. of proteins. (+) Positive; (−) negative; (U) undetectable. (Reprinted, with permission, from Tevethia et al. 1983.)

Existence of distinct TSTA sites

The observation that TSTA sites are located on widely separated portions of the large-T-antigen molecule that are exposed at the cell surface raises the question of whether the TSTA sites on different parts of the molecule are distinct from each other. This question is important, since a multiplicity of TSTA sites differing in antigenicity would enhance immunosurveillance mechanisms, since the development of immunoresistance to one site would not lower the immune response of the host to other sites.

Since CTL generated in bulk cultures to SV40 TSTA are directed against many antigenic sites, clones of CTL specific for SV40 TSTA that would recognize a single antigenic determinant were isolated. For this purpose, SV40-immune lymphocytes were generated in vivo by immunization of C57B1/6 mice with B6/WT-19 cells ($H-2^b$), and splenic lymphocytes were secondarily stimulated in vitro by culturing 1×10^7 lymphocytes with 2×10^5 γ-irradiated B6/WT-19 cells in RPMI medium supplemented with 10% fetal calf serum and 1×10^{-5} M of β-mercaptoethanol. After 4 days in culture, the lymphocytes were passaged at 5×10^6 cells/ml with 5×10^3 γ-irradiated B6/WT-19 cells in Iscove's modified Dulbecco's medium supplemented as the RPMI medium and containing 5 units/ml of purified rat T-cell growth factor (TCGF; Collaborative Research, Waltham, Mass.). Cloning of CTL was performed by limiting dilution in 96-well tissue-culture plates in the presence of 1×10^6 γ-irradiated syngeneic spleen cells, 3×10^5 γ-irradiated B6/WT-19 cells, and 5 units/ml TCGF in Iscove's medium. After 10 days, lymphocytes were harvested from cultures containing single clones. The clones were expanded in 24-well tissue-culture plates and were passaged in continuous culture in 12-well tissue-culture plates. Two of the independently derived clones were further subcloned and designated as K11 and K19.

The results of cytotoxicity assays using clones K11 and K19 indicated (Table 3) that both clones lysed only the syngeneic SV40-transformed cells. The CTL clones lysed cells that were transformed by SV40 virus, viral DNA, or cellular DNA from SV40-transformed cells and demonstrated $H-2$-restricted cytotoxicity since they failed to lyse either SV40-transformed $H-2^d$ (mKSA) or $H-2^k$ (PSC3H) cells. The clonal nature of K11 and K19 was shown by the fact that both the CTL clones recognized SV40 TSTA in association with $H-2D^b$ antigen. Both clones lysed SV40-transformed KHTGSV ($H-2^{d/b}$) cells but not K5RSV ($H-2^{b/d}$) target cells (Table 3). This result was further confirmed by blocking assays using monoclonal antibodies against $H-2K^b$ and $H-2D^b$. Only the monoclonal antibody against $H-2D^b$ inhibited cell lysis (Campbell et al. 1983).

The demonstration that K11 and K19 recognize distinct antigenic sites on SV40 TSTA was accomplished by using target cells transformed by SV40 and by a related human papovavirus, BKV. The large T antigens of SV40 and BKV possess extensive amino acid homology, but sequences unique to each virus do exist (Seif et al. 1979). The cytotoxicity experiments shown in Table 4 indicated that the two CTL clones recognize a different determinant on SV40 TSTA. The B6/BKVD-1 cell line derived by transfection of B6/MEF with BKV DNA was not lysed by CTL clone K11 but was lysed by K19. Both clones lysed B6/WT-19 cells, indicating that TSTA sites recognized by K11 and K19 are distinct and may reflect differences in the domains of large T antigen expressed at the surface of SV40- and BKV-transformed cells. Our results suggest that the immune response to SV40-transformed cells comprises populations of CTL recognizing distinct antigenic determinants of SV40 TSTA. By the use of target cells synthesizing truncated large T antigens and additional CTL clones, it should be pos-

Table 3 Viral Specificity and $H-2$ Restriction of the Cytotoxicity of SV40-immune CTL Clones

Target cells	Transforming agent	$H-2$ haplotype K	$H-2$ haplotype D	% Specific ^{51}Cr release by effectors[a] SV40-immune[b]	clones K11	clones K19
B6/WT-19	SV40	b	b	24.5	52.9	45.7
B6/WT-3D	DNA from SV40-transformed mouse cells	b	b	42.1	63.0	55.8
B6/SVD	SV40 DNA	b	b	N.T.[c]	25.6	17.1
B6/HCMVD	human cytomegalovirus DNA	b	b	N.T.	2.7	1.8
B6/293D	DNA from adenovirus-transformed human cells	b	b	1.6	−1.5	−2.2
B6/Py	polyomavirus	b	b	−0.7	−4.5	−4.2
KHTGSV	SV40	d	b	35.6	60.1	60.3
K5RSV	SV40	b	d	16.2	0	−0.7
mKSA	SV40	d	d	N.T.	1.2	2.0
PSC3H	SV40	k	k	−0.1	−1.3	0.5

[a]Effector/target-cell ratio = 5:1.
[b]SV40-immune lymphocytes were generated by a secondary in vitro stimulation of B6/WT-19-primed spleen cells as described in the text.
[c](N.T.) Not tested.
(Reprinted, with permission, from Campbell et al. 1983.)

Table 4 Antigenic Specificity of SV40-immune CTL Clones

Effectors	Experiment no.	% Specific ^{51}Cr release	
		B6/WT-19	B6/BKVD-1
SV40-immune[a]	1[b]	42.5	36.1
	2[b]	60.0	42.9
K11	1[b]	31.7	−0.7
	2[b]	52.9	0.8
	3[c]	52.3	0.8
K19	1[c]	26.8	29.1
	2[c]	45.7	29.3
	3[c]	55.4	47.7

[a]SV40-immune lymphocytes were generated by a secondary in vitro stimulation of B6/WT-19-primed spleen cells as described in the text.
[b]Effector/target-cell ratio = 10:1.
[c]Effector/target cell ratio = 5:1.
(Reprinted, with permission, from Campbell et al. 1983.)

sible to map more precisely the antigenic sites that play a dominant role in tumor rejection. Toward this end, we have recently shown that the antigenic sites that react with K11 and K19 reside in the proximal half of large T antigen (data not shown).

Cell-surface large T antigen and TSTA expression

The demonstration that large T antigen purified to homogeneity can immunize mice against SV40 tumor-cell challenge (Chang et al. 1979; Tevethia et al. 1980) and can induce a cellular immune response has led to the conclusion that large T antigen must be present at the cell surface and accessible to immunological components participating in tumor rejection. This conclusion is based on the fact that all immunological reactions involved in tumor rejection take place at the cell surface (Cerottini and Brunner 1974). A number of investigators have attempted to demonstrate cell-surface large T antigen by a variety of techniques (for review, see Rigby and Lane 1983) that utilize anti–large T antibody. It is now agreed that large T antigen is expressed at the cell surface; however, the nature of surface large T antigen is still unsettled. The following issues need to be resolved.

1. What domains of large T antigen are exposed at the cell surface (Deppert and Walter 1982)?
2. How stable is large T antigen at the cell surface? Lanford and Butel (1982) have presented evidence that the surface large T antigen constantly turns over in the membranes of SV40-transformed and infected cells (J.S. Butel, pers. comm.).
3. Is surface large T antigen selectively modified? A recent report (Klockmann and Deppert 1983) that surface large T antigen is acylated is encouraging and will provide a rationale for the presence of large T antigen at the cell membrane. The modification of surface large T antigen may also include glycosylation (Schmidt-Ullrich et al. 1982; J.S. Butel, pers. comm.).
4. How does the antigenicity of surface large T antigen compare with nuclear large T antigen? It has been reported that antisera prepared in heterologous hosts against denatured large T antigen recognize surface large T antigen more efficiently than the sera from animals bearing tumors induced by SV40-transformed cells (Deppert and Walter 1982). In addition, some anti–large T antigen sera detect surface large T antigen more efficiently than others (Soule et al. 1980).
5. Does free large T antigen have affinity for membranes, as reported by Lange-Mutschler and Henning (1983). It would be interesting to determine if the free large T antigen associates with Ia$^+$ macrophages.
6. Under what conditions can surface large T antigen be demonstrated without disturbing cell permeability?

As discussed extensively by Rigby and Lane (1983), most of the approaches utilized for the demonstration of surface large T antigen have drawbacks due to the fact that most large T antigen is present in the nucleus. The most convincing evidence relating the surface large T antigen with TSTA is provided by Pan and Knowles (1983), who have isolated a single CTL clone specific for SV40 TSTA, the activity of which could be blocked by a monoclonal antibody to large T antigen.

By using recombinant DNA technology to generate target cells synthesizing fragments of large T antigen and large T antigens deleted in certain segments in combination with CTL clones, it should be possible to seek more direct evidence for the role of large T antigen in specifying TSTA at the surface of SV40-transformed cells. The evidence so far available has shown that multiple, antigenically distinct TSTAs on SV40-transformed cells play a role in suppressing tumor potential. The tumors that do arise from the SV40-transformed cells in immunocompetent hosts may develop because of the reduction in or loss of the expression of *H-2* gene products (Gooding 1982; Flyer et al. 1983), which are required for CTL recognition (Zinkernagel and Doherty 1979). A systematic elucidation of the biochemical nature of surface large T antigen and the conditions required for its association with the *H-2* gene products will lead to a better understanding of how the tumorigenic potential of viruses like SV40 is suppressed by the immune response of the host.

Acknowledgments

This study was supported by research grants CA-25000 to S.S.T., CA-24694 to M.J.T., and in part by CA-18450 awarded by the National Cancer Institute. The invaluable assistance of Ms. Carol Buck and Ms. Elaine Neidigh in the preparation of this manuscript is greatly appreciated.

References

Anderson, J.L., R.G. Martin, C. Chang, P.T. Mora, and D.M. Livingston. 1977. Nuclear preparations of SV40 transformed cells contain tumor specific transplantation activity. *Virology* **76:** 420.

Campbell, A.E., L.F. Foley, and S.S. Tevethia. 1983. Demonstration of multiple antigenic sites of the SV40 transplantation rejection antigen by using cytotoxic lymphocyte clones. *J. Immunol.* **130:** 490.

Cerottini, J.C., and K.T. Brunner. 1974. Cell mediated cytotoxicity: Allograft rejection and tumor immunity. *Adv. Immunol.* **18:** 67.

Chang, C., R. Chang, P.T. Mora, and C.-P. Hu. 1982. Generation of cytotoxic lymphocytes by SV40-induced antigens. *J. Immunol.* **128:** 2160.

Chang, C., R.G. Martin, D.M. Livingston, S.W. Luborsky, C.-P. Hu, and P.T. Mora. 1979. Relationship between T-antigen and tumor specific transplantation antigen in simian virus 40-transformed cells. *J. Virol.* **29:** 69.

Clayton, C.R., D. Murphy, M. Lovett, and P.W.J. Riggy. 1982. A fragment of the SV40 large T-antigen gene transforms. *Nature* **299:** 59.

Deppert, W. and G. Walter. 1982. Domains of simian virus 40 large T-antigen exposed on the cell surface. *Virology* **122:** 56.

Flyer, D.C. and S.S. Tevethia. 1982. Biology of simian virus 40 (SV40) transplantation antigen (TrAg). VIII. Retention of SV40 TrAg sites on purified SV40 large T antigen following denaturation with sodium dodecyl sulfate. *Virology* **117:** 267.

Flyer, D.C., J. Pretell, A.E. Campbell, W.S.L. Liao, M.J. Tevethia, J.M. Taylor, and S.S. Tevethia. 1983. Biology of simian virus 40 (SV40) transplantation in antigen (TrAg). X. Tumorigenic potential of mouse cells transformed by SV40 in high responder C57Bl/6 mice and correlation with the persistence of SV40 TrAg, early proteins and viral sequences. *Virology* **131:** 207.

Girardi, A.J. and V. Defendi. 1970. Induction of SV40 transplantation antigen (TrAg) during the lytic cycle. *Virology* **42:** 688.

Gooding, L.R. 1977. Specificities of killing by cytotoxic lymphocytes generated in vivo and in vitro to syngeneic SV40-transformed cells. *J. Immunol.* **118:** 920.

———. 1982. Characterization of a progressive tumor from C3H fibroblasts transformed *in vitro* with SV40 virus. Immunoresistance *in vivo* correlates with phenotypic loss of H-$2K^k$. *J. Immunol.* **129:** 1306.

Gooding, L.R. and C.B. Edwards. 1980. H-2 antigen requirements in the in vitro induction of SV40- specific cytotoxic T lymphocytes. *J. Immunol.* **124:** 1258.

Harlow, E., L.V. Crawford, D.C. Pim, and N.M. Williamson. 1981. Monoclonal antibodies specific for simian virus 40 tumor antigens. *J. Virol.* **39:** 861.

Jay, G., F.T. Jay, C. Chang, R.M. Friedman, and A.S. Levine. 1978. Tumor specific transplantation antigen: Use of the $Ad2^+ND_1$ hybrid virus to identify the proteins responsible for simian virus 40 tumor rejection and its genetic origin. *Proc. Natl. Acad. Sci* **75:** 3055.

Jay, G., F.T. Jay, C. Chang, A.S. Levine, and R.M. Friedman. 1979. Induction and simian virus 40–specific tumor rejection by the $Ad2^+ND2$ hybrid virus. *J. Gen. Virol.* **44:** 287.

Klockman, H. and W. Deppert. 1983. Acylated simian virus 40–specific proteins in the plasma membrane of HeLa cells infected with adenovirus 2–simian virus 40 hybrid virus $Ad2^+ND2$. *Virology* **126:** 717.

Lane, D.P. and L.W. Crawford. 1979. T antigen is bound to a host cell protein in SV40 transformed cells. *Nature* **278:** 261.

Lanford, R.E. and J.S. Butel. 1982. Intracellular transport of SV40 large tumor antigen. A mutation which abolishes migration to the nucleus does not prevent association with the cell surface. *Virology* **119:** 169.

Lange-Mutschler, J. and R. Henning. 1983. A subclass of simian virus 40 T antigen with a high cell surface binding affinity. *Virology* **127:** 333.

Lewis, A.M., Jr. and W.P. Rowe. 1973. Studies of nondefective adenovirus 2–simian virus 40 hybrid viruses. VIII. Association of simian virus 40 transplantation antigen with a specific region of the early viral genome. *J. Virol.* **12:** 836.

Linzer, D.I.H. and A. Levine. 1979. Characterization of a 54K dalton cellular SV40 tumor antigen present in SV40-transformed cells and uninfected embryonal carcinoma cells. *Cell* **17:** 43.

Pan, S., and B.B. Knowles. 1983. Monoclonal antibody to SV40 T antigen blocks lysis of cloned cytotoxic T cell line specific for SV40 TASA. *Virology* **125:** 1.

Pretell, J., R.S. Greenfield, and S.S. Tevethia. 1979. Biology of simian virus (SV40) transplantation rejection antigen (TrAg). V. *In vitro* demonstration of SV40 TrAg in SV40 infected nonpermissive mouse cells by the lymphocyte mediated cytotoxicity assay. *Virology* **97:** 32.

Rapp, F., J.S. Butel, and J.L. Melnick. 1964. Virus induced intracellular antigen in cells transformed by papovavirus SV40. *Proc. Soc. Exp. Biol. Med.* **116:** 1131.

Reddy, V.B., S.S. Tevethia, M.J. Tevethia and S.M. Weissman. 1982. Nonselective expression of simian virus 40 large tumor antigen fragments in mouse cells. *Proc. Natl. Acad. Sci.* **79:** 2064.

Rigby, P.W.J. and D.P. Lane. 1983. The structure and function of the simian virus 40 large T antigen. *Adv. Viral Oncol.* **3:** 31.

Schmidt-Ullrich, R., W.S. Thompson, S.J. Kahn, M.J. Monroe, and D.F.H. Wallach. 1982. Simian virus 40 (SV40)–specific isoelectric point 4.7-94,000 M_r membrane glycoprotein: Major peptide homology exhibited with the nuclear and membrane associated 94,000-M_r SV40 T antigen in hamsters. *J. Natl. Cancer Inst.* **69:** 839.

Seif, I., G. Khoury, and R. Dhar. 1979. The genome of human papovavirus BKV. *Cell* **18:** 963.

Sleigh, M.J., W.C. Topp, R. Hunich, and J.F. Sambrook. 1978. Mutants of SV40 with an altered small t protein are reduced in their ability to transform cells. *Cell* **14:** 79.

Soule, H.R., R.E. Lanford, and J.S. Butel. 1980. Antigenic and immunogenic characteristics of nuclear and membrane associated SV40 tumor antigen. *J. Virol.* **33:** 887.

Tegtmeyer, P., M. Schwartz, J.K. Collins, and K. Rundell. 1975. Regulation of tumor antigen synthesis by simian virus 40 gene A. *J. Virol.* **16:** 168.

Tevethia, M.J. and S.S. Tevethia. 1976. Biology of SV40 transplantation antigen (TrAg). I. Demonstration of SV40 TrAg on glutaraldehyde fixed SV40-infected African green monkey kidney cells. *Virology* **69:** 474.

Tevethia, S.S. 1980. Immunology of simian virus 40. In *Viral oncology* (ed. G. Klein), p. 581. Raven Press, New York.

———. 1983. Cytolytic T lymphocyte response to SV40. *Surv. Immunol. Rev.* **2:** 312.

Tevethia, S.S., D.C. Flyer, and R. Tjian. 1980. Biology of simian virus 40 (SV40) transplantation antigen (TrAg). VI. Mechanism of induction of SV40 transplantation immunity in mice by purified SV40 T antigen (D2 protein). *Virology* **107:** 13.

Tevethia, S.S., J.W. Blasecki, G. Waneck, and A.L. Goldstein. 1974. Requirement of thymus-derived θ-positive lymphocytes for rejection of DNA virus (SV40) tumors in mice. *J. Immunol.* **113:** 1417.

Tevethia, S.S., M.J. Tevethia, A.J. Lewis, V.B. Reddy, and S. M. Weissman. 1983. Biology of simian virus 40 (SV40) transplantation antigen (TrAg). IX. Analysis of TrAg in mouse cells synthesizing truncated SV40 large T antigen. *Virology* **128:** 319.

Tooze, J., ed. 1980. *Molecular biology of tumor viruses*, 2nd edition: *DNA tumor viruses*. Cold Spring Harbor Laboratory, Cold Spring Harbor, New York.

Wigler, M., S. Silverstein, L.-S. Lee, A. Pellicer, Y.-C. Chang, and R. Axel. 1977. Transfer of purified herpesvirus thymidine kinase gene to cultured mouse cells. *Cell* **11:** 223.

Zinkernagel, R.M. and P.C. Doherty. 1979. MHC restricted cytotoxic T cells: Studies on the biological role of polymorphic major transplantation antigens detemining T-cell restriction-specificity, function and responsiveness. *Adv. Immunol.* **27:** 51.

Unique Molecular Structure of the Leukemogenic Cell Membrane Glycoprotein (gp55) Encoded by the Friend Spleen Focus-forming Virus

Y. Ikawa, M. Obata, N. Sagata, and H. Amanuma*

Laboratory of Molecular Oncology, Institute of Physical and Chemical Research (RIKEN), Wako, Saitama 351, Japan; *Department of Viral Oncology, Cancer Institute, Toshima-ku, Tokyo 170, Japan

Friend leukemia virus complex (FrLV) was isolated from the leukemic spleens of Swiss mice that had been inoculated with the cell-free extracts of Ehrlich ascites tumor cells (Friend 1957). FrLV can induce an acute erythroleukemia in susceptible adult mice by arresting the maturation of the early erythroid precursor cells (Ikawa et al. 1972).

A polycythemia-inducing strain of FrLV was given by C. Friend to the Cancer Institute, Tokyo, in 1960 and thereafter has been passaged through DDD mice or the cells originated from DDD mice. This FrLV strain induced focal leukemic lesions in the spleens of mice when the diluted viral material was applied (Ikawa et al. 1967). These splenic lesions consisted of cells with proerythroblastoid morphology and with erythrocyte-membrane antigens (Ikawa et al. 1967, 1973). Spleen foci induced in DDD mice 9–12 days after the FrLV injection were transplanted subcutaneously to syngeneic mice in order to establish solid and ascitic forms of Friend erythroleukemia cells (Ikawa and Sugano 1966). Some of them were converted to culture cell lines, and clonal sublines were used for molecular genetical analyses of the inducer-mediated erythrodifferentiation (Furusawa et al. 1971; Ross et al. 1972; Ikawa et al. 1976, 1978b). A certain subline has been shown to produce fetal hemoglobin upon induction of erythrodifferentiation (Brown et al. 1982).

One of the differentiation-resistant Friend erythroleukemia cell lines (T3-K-1 or K-1) was found to produce an excess amount of the replication-defective spleen focus-forming virus (SFFV), which is one of the viral components of FrLV and is primarily responsible for induction of an acute erythroleukemia (Steeves 1975), over the replication-competent, helper Friend murine leukemia virus (Fr-MLV) (Ikawa and Yoshida 1977). The use of this cell line facilitated the molecular analysis of the specific sequences (i.e., sequences not homologous with the helper Fr-MLV genome) in the SFFV genome 32S RNA (Ikawa and Yoshida 1979; Ikawa et al. 1980).

Two proteins have been identified as the SFFV-specific products. One of them, designated gp55 (or gp52), is a cell-membrane glycoprotein with a molecular weight of 52,000–55,000 (Racevskis and Koch 1977; Ikawa et al. 1978a; Dresler et al. 1979; Ruscetti et al. 1979). gp55 is immunoprecipitated from the Friend erythroleukemia cells or the FrLV-induced leukemic spleen cells by the monospecific antiserum against gp70, a major viral envelope glycoprotein of the ecotropic MLV. Analysis of the SFFV genome RNA by molecular hybridization (Troxler et al. 1977), oligonucleotide fingerprinting (Evans et al. 1979; Yoshida and Yoshikura 1980), or heteroduplex formation (Bosselman et al. 1980) has indicated that the envelope (env)-gene region of the SFFV genome that codes for gp55 contained the sequence homologous to the env-gene sequence of Fr-MLV and also the specific sequence that was partially homologous to the env-gene sequences of xenotropic MLVs, suggesting that gp55 may be an env recombinant protein.

Recently, from studies on the pathogenic properties of SFFV mutants either constructed in vitro by using recombinant DNA technology from the molecularly cloned SFFV proviral DNA (Linemeyer et al. 1982) or isolated as spontaneous mutants (Ruta et al. 1983), the essential role of gp55 in acute erythroleukemogenesis has been well established. For the elucidation of the molecular mechanism by which this env-related cell-membrane glycoprotein of SFFV elicits the abnormal proliferation of a certain erythroid precursor cell to result in acute erythroleukemia, the knowledge of its structure in detail is indispensable. In this paper we describe the molecular cloning of biologically active proviral DNAs of SFFV and Fr-MLV, nucleotide sequence analysis of the env-gene regions of SFFV and Fr-MLV, and considerations on the derivation of the gp55 gene.

Experimental Procedures

Molecular cloning

The FrLV material produced by the K-1 Friend erythroleukemia cell line was the virus source for molecular cloning of both SFFV and Fr-MLV proviral DNAs.

SFFV

NIH-3T3 cells were freshly infected with viruses prepared from the culture supernatant of the K-1 cells. An unintegrated, closed circular proviral DNA of SFFV was isolated from the infected NIH-3T3 cells together with that of Fr-MLV, which existed in a very minor amount, by extraction of DNA by the Hirt procedure (Hirt 1967). Preliminary restriction enzyme analysis revealed that EcoRI cleaved SFFV proviral DNA once. The circular

SFFV proviral DNA was thus linearized with EcoRI, isolated by agarose gel electrophoresis, and ligated with the plasmid vector pBR322 at the site of EcoRI. The ligated DNA was used for transformation of E. coli, and several colonies that harbored the SFFV recombinant plasmid were selected for further analyses.

Fr-MLV

Fr-MLV produced by K-1 cells was first cloned biologically by limiting dilution on SC-1 cells (H. Yoshikura, unpubl.). This Fr-MLV isolate, when injected into newborn DDD mice, induced erythroblastosis in association with severe anemia (hematocrit, 22–25). Since the circular Fr-MLV proviral DNA recovered from the freshly infected NIH-3T3 cells had a single EcoRI site like SFFV DNA, molecular cloning of the Fr-MLV proviral DNA was accomplished by the same method as used for the cloning of the SFFV proviral DNA.

DNA sequence analysis

An appropriate DNA fragment generated by restriction enzymes and isolated by agarose gel electrophoresis was labeled at its 5′ ends by using [γ-^{32}P]ATP (Amersham) and T4 polynucleotide kinase (Boehringer-Mannheim). End-labeled fragments were digested with restriction enzymes and separated by polyacrylamide gel electrophoresis. The nucleotide sequence was determined by the chemical cleavage method as described by Maxam and Gilbert (1980). The sequences were confirmed by sequence analysis of the complementary strand or the fragments generated by digestion with different enzymes.

Results and Discussion

Molecularly cloned SFFV and Fr-MLV proviral DNAs

The purpose of the present study was to determine the total primary structure of gp55 in expectation of obtaining a framework for further studies on its function. For this purpose, we isolated a molecular clone of SFFV proviral DNA and determined the nucleotide sequence of its env-gene region. The representative SFFV clone (pMSF4) had a size of 6.0 kb. The restriction enzyme cleavage map (Fig. 1a) indicated that the pMSF4 represented the permuted form of the entire SFFV(K-1) proviral genome, having two long-terminal-repeat (LTR) sequences (Sagata and Ikawa 1982). As for the restriction map, our SFFV clone closely resembled the other SFFV proviral DNA molecularly cloned from cells nonproductively infected with the polycythemia-inducing strain of SFFV (Mak-Bernstein strain) (Yamamoto et al. 1981), but it was significantly dissimilar to another polycythemia-inducing SFFV proviral clone (Lilly-Steeves strain) (Linemeyer et al. 1980).

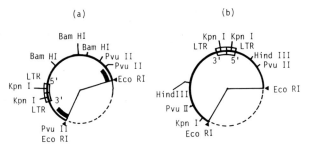

Figure 1 Restriction enzyme maps of the cloned proviral DNAs of SFFV(K-1) (a) and Fr-MLV(K-1) (b). The closed box in a shows the approximate location of the gp55 gene.

Before using the cloned proviral genome for the sequence analysis, it is important to confirm that it has its biological activity. The pMSF4 plasmid DNA was cleaved with EcoRI to release the SFFV genome and then transfected into NIH-3T3 cells by the calcium phosphate coprecipitation method (Graham and Van der Eb 1973). The transfected NIH-3T3 cells were superinfected with Fr-MLV cl 57 (Oliff et al. 1980), and the resulting cell-culture supernatant was injected into adult DDD mice. The mice developed spleen foci, splenomegaly, and erythroleukemia in association with polycythemia, as expected for the polycythemia-inducing strain of SFFV.

The restriction map of the molecularly cloned Fr-MLV(K-1) proviral DNA (9.0 kb in size) is shown in Fig. 1b. The map was nearly identical with that of the other Fr-MLV proviral clone (Fr-MLV cl 57) (Oliff et al. 1980). The Fr-MLV proviral clone was biologically active, since the XC-positive Fr-MLV was produced upon transfection into NIH-3T3 cells. The recovered Fr-MLV could induce erythroblastosis with anemia when injected into newborn DDD mice.

Sequence of gp55

The complete nucleotide sequence of the gp55 gene was determined by using the pMSF4 plasmid DNA (Amanuma et al. 1983). The sequence is presented in Figure 2 together with the deduced amino acid sequence. There is a long, open reading frame composed of 1227 bp that codes for gp55. The amino acid composition (total, 409 amino acids) indicates that the primary translation product of gp55 has a molecular weight of 44,752. About 290 bp upstream from the initiation ATG codon, there is a 3′ consensus sequence for mRNA splicing (Sharp 1981). This splice sequence is presumably used for the generation of the subgenomic 18S SFFV RNA that codes for gp55 (Ruscetti et al. 1980).

The amino acid sequence of gp55 shows that near the carboxyl terminus there is a stretch of sequence (residues 380–407; residue number starts at methionine, encoded by the initiation ATG codon) that is completely

Figure 2 (see facing page) Nucleotide sequences of the env-gene regions of SFFV(K-1), Fr-MLV(K-1) (F-MuLV), and Fr-MCF (F-MCF) and the deduced amino acid sequence of gp55. Standard one-letter amino acid codes are used. (*) Nucleotide identical with that of SFFV; (-) gap; (CHO) potential glycosylation site. Underlined portions in the Fr-MLV and Fr-MCF sequences have the identical amino acid sequence as SFFV. Arrows 1–5 are explained in the text.

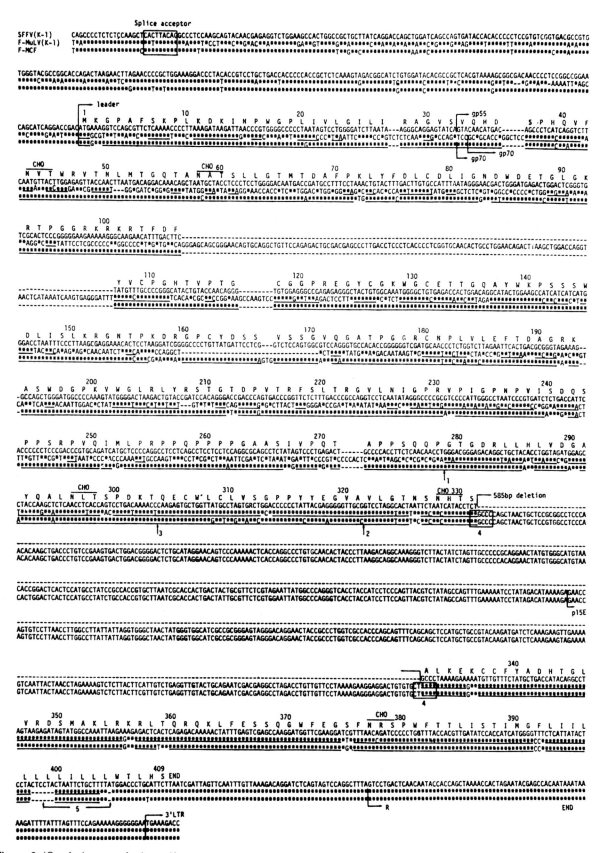

Figure 2 (See facing page for legend.)

uncharged and mostly hydrophobic; 16 amino acids of this stretch are either leucine or isoleucine. This portion of gp55 is probably embedded in the lipid bilayer of the membranes. gp55 is known to contain four N-asparagine-linked, high-mannose-type oligosaccharide chains (Polonoff et al. 1982). In the sequence of gp55, there are five canonical sequences, Asn-X-Thr/Ser, that can serve as asparagine-linked glycosylation sites (Marshall 1974). For reasons described below, the site located at the carboxyl terminus (residues 377–379) may not actually be glycosylated. The schematic representation of the peptide structure of gp55 is shown in Figure 3.

Structural features of gp55 in relation to those of the viral envelope proteins

Since the previous studies have indicated that the gp55 gene is closely related to the env genes of MLVs, comparisons of the gp55 sequence with the env-gene sequences of MLVs would reveal structural features of gp55. We determined the complete nucleotide sequence of the env gene of Fr-MLV(K-1) by using the molecularly cloned proviral DNA (M. Obata et al., in prep.), and this sequence was used for the sequence comparison with the gp55 gene (Fig. 2). We also compared the gp55 sequence with the env-gene sequence of the molecularly cloned proviral DNA of the Friend mink-cell focus-inducing virus (Fr-MCF) (A. Adachi et al., pers. comm.) (Fig. 2). Fr-MCF is a replication-competent, dualtropic MLV and is an env-gene recombinant virus arising from Fr-MLV and an endogenous xenotropic MLV env-like sequence (Ishimoto et al. 1981).

For the sake of the following argument, it is appropriate first to describe the nonecotropic sequence in the Fr-MCF genome. It is clear from Figure 2 that the 5' limit of the nonecotropic sequence in the Fr-MCF genome is located at least 300 bp upstream of the env ATG initiation codon and that the nonecotropic sequence extends downstream in the gp70 coding region to amino acid residue 279 (arrow 1 in Fig. 2). The Fr-MCF nonecotropic sequence is, as expected, highly homologous with the env-gene sequence of the xenotropic MLV (Repaske et al. 1983). The sequence of Fr-MCF 3' to the position indicated by arrow 1 (i.e., the carboxyterminal 39% of gp70, Prp15E, and a portion of the 3' LTR) is quite homologous with the Fr-MLV sequence. In this region both sequences use the same reading frame. The sequence of Fr-MCF between arrows 1 and 2 of Figure 2 (129 bp) may be nonecotropic, although it is difficult to come to this conclusion by comparing the amino acid sequences.

The sequence of gp55 between amino acid residues 1 and 278 is highly homologous with the Fr-MCF env-protein sequence between residues 1 and 279. This region of Fr-MCF env is nonecotropic, as described above. Between residues 166 and 167, gp55 lacks one amino acid, valine, in the Fr-MCF gp70 sequence. Amino acids are different at 16 positions between the two sequences (i.e., 94% homology). Recently, the sequence of the aminoterminal 25 amino acids of the Rauscher MCF virus gp69 has been determined by direct protein sequencing (Schultz et al. 1983). This sequence almost exactly corresponds to the gp55 sequence starting from residue 33 (24/25 match). This suggests that, although the aminoterminal amino acid of gp55 has not yet been identified, it may be valine at residue 33 and that the preceding sequence (between residues 1 and 32) may serve as a leader peptide for membrane protein synthesis (Engelman and Steitz 1981). The SFFV sequence (~300 bp long) 5' to the gp55 ATG initiation codon is also highly homologous with the corresponding Fr-MCF sequence. This indicates that the 5' limit of the nonecotropic sequence in the SFFV genome is located at least 300 bp upstream of the gp55 initiation codon.

Extensive sequence homology between gp55 and Fr-MCF env-gene product extends further downstream beyond the amino acid residue 278 of gp55 (i.e., amino acid 279 of Fr-MCF env-gene product). Since the Fr-MCF env-protein sequence 3' to the amino acid 279 (or 3' to the position indicated by arrow 2) is highly homologous with the Fr-MLV env-protein sequence, the gp55 sequence 3' to amino acid 278 is also highly homologous with the Fr-MLV env sequence at both nucleotide and amino acid levels. Actually, the sequence of SFFV between the position indicated by arrow 1 and the 3' end of the U3 portion of the 3' LTR is almost identical with the Fr-MLV sequence except for one minor and two major differences. The minor difference is evident at the nucleotide level but not at the amino acid level. The sequence of gp55 between arrows 1 and 3 (75 bp) does not have so high a sequence homology with the Fr-MLV env sequence as the sequence of SFFV 3' to arrow 3. This 75-bp sequence of gp55 is more homologous with the Fr-MCF sequence than with the Fr-MLV sequence, indicating that it may be nonecotropic and that gp55 may have a slightly shorter nonecotropic sequence than the Fr-MCF env gene.

One major difference is a large deletion (585 bp) in the gp55 sequence. This 585-bp portion of the Fr-MLV env sequence encodes 195 amino acids that correspond to the carboxyterminal 108 amino acids of gp70 and the aminoterminal 87 amino acids of p15E and consequently contains the site of proteolytic cleavage to produce gp70 and p15E. The deletion in gp55 is in frame, and gp55 uses the same reading frame as the Fr-MLV env gene on both sides of the deletion. The deletion causes gp55 to be a fusion protein consisting of the aminoterminal portion of gp70 and the carboxyterminal portion of p15E. This configuration is similar to that of the precursor pro-

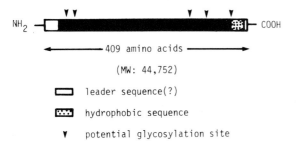

Figure 3 Primary peptide structure of gp55.

tein for gp70 and p15E, but, unlike the latter, gp55 would not be proteolytically processed, because of the absence of the site for such processing. It has been shown by a pulse-chase experiment that gp55 did not undergo major proteolytic processing (Racevskis and Koch 1977). It should be noted that there are 6-bp repeated sequences (CTGCCC) that flank the deletion in the Fr-MLV and Fr-MCF env sequences (arrow 4 in Fig. 2). By the deletion, one of the two repeated sequences is lost. These repeated sequences are probably involved in the genetic event that generates the deletion.

The other major difference between gp55 and Fr-MLV env sequences is a 7-bp insertion in the gp55 sequence near its carboxyl terminus and within the sequence that encodes the hydrophobic stretch (arrow 5 in Fig. 2). The 5′ 6 bp (CTCCTA) of this insertion is the duplication of the 5′ adjacent sequence. This 6-bp insertion encodes Leu-Leu and thus adds two more hydrophobic amino acids to the hydrophobic stretch of gp55. The insertion of a single base (A) causes a frameshift that leads gp55 to an earlier encounter with the termination codon (TAA) than the Fr-MLV or Fr-MCF env sequence. After the single-base insertion, there are 6 amino acids in gp55, three of which are hydrophobic, while the corresponding Fr-MLV (or Fr-MCF) env sequence encodes 40 amino acids that constitute the carboxyterminal 23 amino acids of p15E and the R peptide (17 amino acids).

The total primary structure of the gp55 peptide in relation to those of the MLV env products is schematically represented in Figure 4. The whole sequence of gp55 can be accounted for by the murine retroviral env sequences. To put it briefly, gp55 is the Fr-MCF env protein that has lost the carboxyterminal 26% of gp70 and the aminoterminal 48% and carboxyterminal 17% of Prp15E. This unique molecular structure of gp55 has been confirmed independently in two other laboratories by the nucleotide sequence analysis of the different polycythemia-inducing SFFV strains (Clark and Mak 1983; Wolff et al. 1983). Moreover, the same unique structural features could be found in the homologous protein (gp54) encoded by the SFFV component contained in the Rauscher leukemia virus preparation, another acute erythroleukemia-inducing virus complex isolated independently from FrLV (R. Bestwick, pers. comm.).

Fr-MCF virus causes erythroleukemia relatively rapidly when injected into newborn mice (Ishimoto et al. 1981). It does not induce spleen foci, and the erythroleukemia is accompanied by anemia. There is an idea that the leukemogenic potential of Fr-MCF may reside in the env gene. Both the Fr-MCF env protein and SFFV gp55 have an endogenous, nonecotropic MLV env-like sequence at their aminoterminal ends. This may be a prerequisite condition for both proteins to be leukemogenic. The particular structure of gp55 caused by the deletion and the insertion must be related to its distinguished biochemical and pathogenic properties. Unlike the Fr-MCF env protein, gp55 is not incorporated into the viral envelope but remains primarily in the intracellular membranous organelles (Lyles and McConnell 1981). Correspondingly, gp55 was shown to contain only high-mannose-type oligosaccharides, and in a certain Friend erythroleukemia cell line another, underglycosylated form of gp55 (designated gp51) could be detected (Matsugi et al. 1983). Only a small fraction (<5%) of gp55 is fully glycosylated and transported into the plasma membrane (Srinivas and Compans 1983). The loss of 195 amino acids caused by the deletion may induce a conformation change of the other part of the gp55 molecule, which then may interfere with the recognition by the cellular machinery for processing, or it is possible that the lost 195-amino-acid sequence contains a signal for processing. The absence in gp55 of the carboxyterminal portion of p15E and the R peptide also may be responsible for the inefficient processing and the failure to be incorporated into the viral envelope, since it was reported that the cleavage of the R peptide from p15E must take place before the gp70-p15E complex is incorporated into the viral envelope (Green et al. 1981). One may argue that the inefficient processing of gp55 is somehow related to its leukemogenic function. Recently, however, it has been suggested that the plasma membrane form of gp55 may be functionally active (Ruta et al. 1983). If this is the case, it is tempting to speculate that gp55 would perturb the process of cellular recognition at the cell surface of stimuli effected by humoral factor(s) and/or by cell-cell contact. This perturbation would lead to unrestricted proliferation of certain erythroid precursor cells.

Acknowledgments

We thank Drs. N. Ishimoto and A. Adachi for their kind approval of our using their nucleotide sequence data before publication, and Mrs. M. Kimura for typing this manuscript. The work performed in the authors' labo-

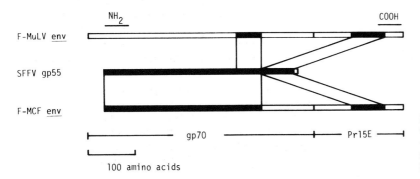

Figure 4 Structural features of gp55 in relation to those of the Fr-MLV (F-MuLV) and Fr-MCF (F-MCF) env proteins. The closed boxes indicate regions with a high degree of sequence homology.

ratories was supported by grants from the Ministry of Education, Science and Culture, the Ministry of Health and Welfare, Japan, and the Tokushima Research Institute of Otsuka Pharmaceutical Co., Ltd.

References

Amanuma, H., A. Katori, M. Obata, N. Sagata, and Y. Ikawa. 1983. Complete nucleotide sequence of the gene for the specific glycoprotein (gp55) of Friend spleen focus-forming virus. *Proc. Natl. Acad. Sci.* **80:** 3913.

Bosselman, R.A., L.J.L.D. van Griensven, M. Vogt, and I.M. Verma. 1980. Genome organization of retroviruses IX. Analysis of the genomes of Friend spleen focus-forming (F-SFFV) and helper murine leukemia viruses by heteroduplex formation. *Virology* **102:** 234.

Brown, B.A., R.W. Padgett, S.C. Hardies, C.A. Hutchison III, and M.H. Edgell. 1982. β-Globin transcript found in induced murine erythroleukemia cells is homologous to the βh0 and βh1 genes. *Proc. Natl. Acad. Sci.* **79:** 2753.

Clark, S.P. and T.W. Mak. 1983. Complete nucleotide sequence of an infectious clone of Friend spleen focus-forming virus: gp55 is an envelope fusion glycoprotein. *Proc. Natl. Acad. Sci.* **80:** 5037.

Dresler, S., M. Ruta, M.J. Murray, and D. Kabat. 1979. Glycoprotein encoded by the Friend spleen focus-forming virus. *J. Virol.* **30:** 564.

Engelman, D.M. and T.A. Steitz. 1981. The spontaneous insertion of proteins into and across membranes: The helical hairpin hypothesis. *Cell* **23:** 411.

Evans, L.H., P.H. Duesberg, D.H. Troxler, and E.M. Scolnick. 1979. Spleen focus-forming Friend virus: Identification of genomic RNA and its relationship to helper virus RNA. *J. Virol.* **31:** 133.

Friend, C. 1957. Cell-free transmission in adult Swiss mice of a disease having the character of a leukemia. *J. Exp. Med.* **105:** 307.

Furusawa, M., Y. Ikawa, and H. Sugano. 1971. Development of erythrocyte membrane-specific antigen(s) in clonal cultured cells of Friend virus-induced tumor. *Proc. Jpn. Acad.* **47:** 220.

Graham, F.L. and A.J. Van der Eb. 1973. A new technique for the assay of infectivity of human adenovirus 5 DNA. *Virology* **52:** 456.

Green, N., T.M. Shinnick, O. Witte, A. Ponticelli, J.G. Sutcliffe, and R.A. Lerner. 1981. Sequence-specific antibodies show that maturation of Moloney leukemia virus envelope polyprotein involves removal of a COOH-terminal peptide. *Proc. Natl. Acad. Sci.* **78:** 6023.

Hirt, B. 1967. Selective extraction of polyoma DNA from infected mouse cell cultures. *J. Mol. Biol.* **26:** 365.

Ikawa, Y. and H. Sugano. 1966. An ascitic tumor derived from early splenic lesion of Friend's disease; a preliminary report. *Gann* **57:** 641.

Ikawa, Y. and M. Yoshida. 1977. Friend leukemia cell line (K-1) propagating mainly spleen focus-forming virus (SFFV). *Proc. Am. Assoc. Cancer Res.* **18:** 215.

———. 1979. Erythroblastosis-inducing RNA sequences in Friend leukemia virus (FLV) of various sources. In *Oncogenic viruses and host cell genes* (ed. Y. Ikawa and T. Odaka), p. 163. Academic Press, New York.

Ikawa, Y., M. Furusawa, and H. Sugano. 1973. Erythrocytic membrane-specific antigens in Friend virus–induced leukemia cells. *Bibl. Haematol.* **39:** 955.

Ikawa, Y., H. Sugano, and M. Furusawa. 1972. Pathogenesis of Friend virus-induced leukemia in mice. *Gann Monogr.* **12:** 33.

Ikawa, Y., H. Sugano, and K. Oota. 1967. Spleen focus in Friend's disease; a histological study. *Gann* **58:** 61.

Ikawa, Y., M. Yoshida, and H. Yoshikura. 1978a. Identification of proteins specific to Friend strain of spleen focus-forming virus (SFFV). *Proc. Jpn. Acad.* **54:** 651.

Ikawa, Y., Y. Inoue, M. Aida, R. Kameji, C. Shibata, and H. Sugano. 1976. Phenotypic variants of differentiation-inducible Friend leukemia lines; isolation and correlation between inducibility and virus release. *Bibl. Haematol.* **43:** 37.

Ikawa, Y., R. Kameji, Y. Uchiyama, Y. Inoue, M. Aida, and M. Obinata. 1978b. The Friend leukemia cell system as a model for the molecular genetics of erythrodifferentiation. In *Mechanisms of cell change* (ed. J.D. Ebert and T.S. Okada), p. 52. Wiley, New York.

Ikawa, Y., Y. Kobayashi, M. Obinata, F. Harada, S. Hino, and Y. Yoshikura. 1980. RNA sequences and proteins specific to Friend strain of spleen focus-forming virus. *Cold Spring Harbor Symp. Quant. Biol.* **44:** 875.

Ishimoto, A., A. Adachi, K. Sakai, T. Yorifuji, and S. Tsuruta. 1981. Rapid emergence of mink cell focus-forming (MCF) virus in various mice infected with NB-tropic Friend virus. *Virology* **113:** 644.

Linemeyer, D.L., S.K. Ruscetti, J.G. Menke, and E.M. Scolnick. 1980. Recovery of biologically active spleen focus-forming virus from molecularly cloned spleen focus-forming virus–pBR322 circular DNA by cotransfection with infectious type C retroviral DNA. *J. Virol.* **35:** 710.

Linemeyer, D.L., J.G. Menke, S.K. Ruscetti, L.H. Evans, and E.M. Scolnick. 1982. Envelope gene sequences which encode the gp52 protein of spleen focus-forming virus are required for the induction of erythroid cell proliferation. *J. Virol.* **43:** 223.

Lyles, D.S. and K.A. McConnell. 1981. Subcellular localization of the env-erythroleukemia cells. *J. Virol.* **39:** 263.

Marshall, R.D. 1974. The nature and metabolism of the carbohydrate-protein linkages of glycoproteins. *Biochem. Soc. Symp.* **40:** 17.

Matsugi, T., H. Amanuma, and Y. Ikawa. 1983. Presence of a novel spleen focus-forming virus-specific glycoprotein (gp51) in a Friend leukemia cell line and its decrease during erythrodifferentiation. *Gann* **74:** 509.

Maxam, A.M. and W. Gilbert. 1980. Sequencing end-labeled DNA with base-specific chemical cleavages. *Methods Enzymol.* **65:** 499.

Oliff, A.I., G.L. Hager, E.H. Chang, E.M. Scolnick, H.W. Chan, and D.R. Lowy. 1980. Transfection of molecularly cloned Friend murine leukemia virus DNA yields a highly leukemogenic helper-independent type C virus. *J. Virol.* **33:** 475.

Polonoff, E., C.A. Machida, and D. Kabat. 1982. Glycosylation and intracellular transport of membrane glycoproteins encoded by murine leukemia viruses. *J. Biol. Chem.* **257:** 14023.

Racevskis, J. and G. Koch. 1977. Viral protein synthesis in Friend erythroleukemia cell lines. *J. Virol.* **21:** 328.

Repaske, R., R.R. O'Neill, A.S. Khan, and M.A. Martin. 1983. Nucelotide sequence of the env-specific segment of NFS-Th-1 xenotropic murine leukemia virus. *J. Virol.* **46:** 204.

Ross, J., Y. Ikawa, and P. Leder. 1972. Globin messenger-RNA induction during erythroid differentiation of cultured leukemia cells. *Proc. Natl. Acad. Sci.* **69:** 3620.

Ruscetti, S., D. Troxler, D. Linemeyer, and E. Scolnick. 1980. Three laboratory strains of spleen focus-forming virus: Comparison of their genomes and translational products. *J. Virol.* **33:** 140.

Ruscetti, S.K., D. Linemeyer, J. Field, D. Troxler, and E.M. Scolnick. 1979. Characterization of a protein found in cells infected with the spleen focus-forming virus that shares immunological cross-reactivity with the gp70 found in mink cell focus-inducing virus particles. *J. Virol.* **30:** 787.

Ruta, M., R. Bestwick, C. Machida, and D. Kabat. 1983. Loss of leukemogenicity caused by mutations in the membrane glycoprotein structural gene of Friend spleen focus-forming virus. *Proc. Natl. Acad. Sci.* **80:** 4704.

Sagata, N. and Y. Ikawa. 1982. Molecular cloning and identification of oncogenic subgenome of Friend virus. (suppl. 1) *Jpn. J. Cancer Chemother.* **9:** 290.

Schultz, A., A. Rein, L. Henderson, and S. Oroszlan. 1983. Biological, chemical, and immunological studies of Rauscher ecotropic and mink cell focus-forming viruses from JLS-V9 cells. *J. Virol.* **45:** 995.

Sharp, P.A. 1981. Speculations on RNA splicing. *Cell* **23:** 643.

Srinivas, R.V. and R.W. Compans. 1983. Glycosylation and intracellular transport of spleen focus-forming virus glycoproteins. *Virology* **125:** 274.

Steeves, R.A. 1975. Spleen focus-forming virus in Friend and Rauscher leukemia virus preparations. *J. Natl. Cancer Inst.* **54:** 289.

Troxler, D.H., D. Lowy, R. Howk, H. Young, and E.M. Scolnick. 1977. Friend strain of spleen focus-forming virus is a recombinant between ecotropic murine type C virus and the *env* gene region of xenotropic type C virus. *Proc. Natl. Acad. Sci.* **74:** 4671.

Wolff, L., E. Scolnick, and S. Ruscetti. 1983. Envelope gene of the Friend spleen focus-forming virus: Deletion and insertions in 3′ gp70/p15E-encoding region have resulted in unique features in the primary structure of its protein product. *Proc. Natl. Acad. Sci.* **80:** 4718.

Yamamoto, Y., C.L. Gamble, S.P. Clark, A. Joyner, T. Shibuya, M.E. MacDonald, D. Mager, A. Bernstein, and T.W. Mak. 1981. Clonal analysis of early and late stages of erythroleukemia induced by molecular clones of integrated spleen focus-forming virus. *Proc. Natl. Acad. Sci.* **78:** 6893.

Yoshida, M. and H. Yoshikura. 1980. Analysis of spleen focus-forming virus-specific RNA sequences coding for spleen focus-forming virus-specific glycoprotein with a molecular weight of 55,000 (gp55). *J. Virol.* **33:** 587.

Detection and Enhancement (by Recombinant Interferon) of Carcinoma Cell Surface Antigens, Using Monoclonal Antibodies

J. Greiner,* P. Horan Hand,* S. Pestka,† P. Noguchi,‡ P. Fisher,§ D. Colcher,* and J. Schlom*

*Laboratory of Tumor Immunology and Biology, National Cancer Institute, National Institutes of Health, Bethesda, Maryland 20205; †Roche Institute of Molecular Biology, Nutley, New Jersey 07110; ‡Office of Biologics, National Center for Drugs and Biologics, Food and Drug Administration, Bethesda, Maryland 20205; §Department of Microbiology and Cancer Center/Institute of Cancer Research, Columbia University, College of Physicians and Surgeons, New York, New York 10032

The rationale for the studies summarized here was to utilize membrane-enriched extracts of human metastatic mammary tumor cells as immunogens in an attempt to generate and characterize monoclonal antibodies reactive with determinants that would be maintained on primary as well as metastatic human mammary carcinoma cells. Multiple assays (Nuti et al. 1981) have been employed to reveal the range of reactivities and diversity of the various antibodies.

Some of the monoclonal antibodies generated showed a strong selective reactivity for some carcinomas (not including those of the breast), especially colon carcinoma. In an attempt to overcome the antigenic heterogeneity observed among cells within carcinoma masses, recombinant interferon was used and was shown to enhance the expression of specific cell-surface tumor antigens.

Generation of monoclonal antibodies

Mice were immunized with membrane-enriched fractions of human metastatic mammary carcinoma cells from either of two involved livers of two different patients. Spleens of immunized mice were fused with nonimmunoglobulin (non-Ig)-secreting NS-1 murine myeloma cells to generate 4250 primary hybridoma cultures. Supernatant fluids from hybridoma cultures were first screened in solid-phase radioimmunoassays (RIAs) for the presence of Ig that is reactive with extracts of metastatic mammary tumor cells from involved liver versus normal liver; 11 double-cloned hybridoma cell lines were chosen for further study. The isotypes of all 11 antibodies were determined; 10 were IgG of various subclasses, and one was an IgM.

The 11 monoclonal antibodies could be divided into three major groups based on their differential reactivity to tumor extracts in solid-phase RIA (Colcher et al. 1981). All 11 antibodies were negative when tested against similar extracts from normal human liver, a rhabdomyosarcoma cell line, the HBL-100 cell line derived from cultures of human milk cells, mouse mammary tumor and fibroblast cell lines, disrupted mouse mammary tumor virus and mouse leukemia virus, purified carcinoembryonic antigen (CEA), and ferritin. Two monoclonals were used as positive controls in all these studies: (1) W6/32, a commercially available anti–human histocompatibility antigen (Barnstable et al. 1978) and (2) B139, which was generated in our laboratory against a human breast tumor metastasis and which demonstrates reactivity to all human cells tested.

To determine if the monoclonal antibodies bind cell-surface determinants, each was tested for binding to live cells in culture. The monoclonals grouped together on the basis of their binding to both metastatic cell extracts could be further separated into three different groups on the basis of their differential binding to surface determinants of a breast cancer cell line. Many of the monoclonals also bound to the surface of selected non-breast carcinoma cell lines. None of the 11 antibodies, however, bound to the surface of sarcoma or melanoma cell lines, nor to the surface of more than 24 cell lines derived from apparently normal human tissues (Colcher et al. 1981; Nuti et al. 1981; Horan Hand et al. 1983).

To further define the specificity and range of reactivity of each of the monoclonal antibodies, the immunoperoxidase technique was employed on formalin-fixed tissue sections. All the monoclonals reacted with carcinoma cells of both infiltrating ductal and lobular primary mammary carcinomas. The percentage of primary tumors that were reactive varied for the different monoclonals, and in many of the positive primary and metastatic mammary carcinomas, not all tumor cells stained. A high degree of selective reactivity with mammary tumor cells, but not with apparently normal mammary epithelium, stroma, blood vessels, or lymphocytes of the breast, was also observed.

Experiments were then carried out to determine if the 11 monoclonal antibodies could detect mammary carcinoma cell populations in regional nodes and at distal sites. Since the antibodies were all generated using metastatic mammary carcinoma cells as antigens, it was not unexpected that they all reacted, but with different degrees, to various metastases. None of the monoclonals reacted with normal lymphocytes or stroma from any involved or uninvolved nodes. The monoclonal antibodies were then tested for reactivity to normal and neoplastic nonmammary tissues. Most of the monoclonals showed reactivity with selected nonbreast carcinomas such as adenocarcinoma of the colon, but showed no staining of sarcomas and lymphomas. Monoclonal antibodies (mAbs) B72.3, B6.2, and B38.1 were chosen for further study since they recognized noncoordinately expressed antigens in different breast tumor cells and lesions; these monoclonals reacted with approximately 45%, 75%, and 56%, respectively, of 39 infiltrating duct carcinomas tested via the immunoperoxidase technique. mAb B38.1 also reacted to several normal epithelial tissues such as sweat glands. The major reactivity to normal tissues of mAb B6.2 was subsets of circulating polymorphonuclear leukocytes. Thus far, mAb B72.3 has demonstrated the most selective degree of reactivity for normal versus tumor tissues. However, fetal tissue has not yet been carefully examined nor has every tissue type in the body. It would be naive to assume that any antigenic determinant consisting of a few amino acids would be expressed only on carcinoma cells and at no time during development in the embryo or at various stages of cell differentiation within the spectrum of adult tissues.

Since mAb B72.3 (an IgG1) displayed such a restricted range of reactivity for human mammary tumor versus normal cells, this antibody was used for further studies in immunoperoxidase assays. Using 4 μg of antibody per slide, we found the percentage of positive primary breast tumors was 46% (19/41); 62% (13/21) of the metastatic lesions scored positive. Several histologic types of primary mammary tumors scored positive (Nuti et al. 1982). Metastatic breast carcinoma lesions that were positive were in axillary lymph nodes, and at the distal sites of skin, liver, lung, pleura, and mesentery. A variety of nonbreast cells and tissues were tested and were negative; these included those of the uterus, liver, spleen, lung, bone marrow, colon, stomach, salivary gland, lymph node, and kidney (Nuti et al. 1982).

Mammary carcinoma tissue as immunogen for the preparation of monoclonal antibodies to CEA

Monoclonal antibodies were also generated to membrane-enriched fractions of human mammary carcinoma metastases and screened for reactivity with purified CEA. The differential binding properties of two of these antibodies (B1.1 and F5.5) to CEA and to breast and nonbreast tumors was investigated (Colcher et al. 1983a). Both mAb B1.1 (IgG2a) and mAb F5.5 (IgG1) precipitated iodinated CEA, resulting in a radiolabeled peak at approximately 180,000 daltons. No precipitation of purified CEA was obtained using mAbs B6.2 and B72.3 nor using any of the other monoclonals described above.

Extracts of breast tumor metastases were assayed for reactivity with mAbs B1.1, B6.2, and B72.3 by solid-phase RIA. The appropriate immunoreactive fractions were then pooled and labeled with ^{125}I. mAb B72.3 immunoprecipitated a complex of four bands with molecular weights of approximately 220,000, 250,000, 285,000, and 400,000; mAb B1.1 immunoprecipitated a heterogenous component with an average estimated molecular weight of 180,000; and mAb B6.2 immunoprecipitated a 90K component.

Antigenic heterogeneity, modulation, and evolution within human mammary carcinoma cell populations

Phenotypic variation was usually observed in the expression of tumor-associated antigens (TAAs) within a given mammary tumor (Horan Hand et al. 1983). One pattern sometimes observed was that one area of a mammary tumor contained cells with TAAs reactive with a particular monoclonal antibody, while another area of the same tumor contained unreactive cells. A more common type of antigenic heterogeneity was observed among cells in a given area of a tumor mass. Tumor cells expressing a specific TAA were seen directly adjacent to tumor cells negative for the same antigen.

In an attempt to elucidate the phenomenon of antigenic heterogeneity in human mammary tumors, model systems were examined. The MCF-7 human mammary tumor cell line was tested for the presence of TAAs, using the cytospin/immunoperoxidase method. The cell line was shown to contain various subpopulations of cells as defined by variability in expression of TAA reactive with mAb B6.2; i.e., positive MCF-7 cells are seen adjacent to cells that scored negative (Horan Hand et al. 1983).

Studies were then conducted to determine if the antigenic heterogeneity observed in MCF-7 cells was (1) the result of at least two stable genotypes or phenotypes, (2) the reflection of a modulation of cell-surface antigen expression of a single phenotype, or (3) the result of both phenomena. In experiments designed to monitor cell-surface antigen expression with different phases of cell growth, it was observed that MCF-7 cells at contact inhibition expressed less antigen on their surfaces, as detected by mAb B6.2, than cells in active proliferation. Using fluorescent activated-cell-sorter analyses, it has been shown that at least two of the monoclonal antibodies developed (B6.2 and B38.1) are most reactive with the surface of MCF-7 cells during S phase of the cell cycle (Kufe et al. 1983).

To further understand the nature of antigenic heterogeneity of human mammary tumor cell populations, MCF-7 cells were cloned by endpoint dilution, and 10 different clones were obtained and assayed for cell-surface TAAs (Horan Hand et al. 1983). The parent MCF-7 culture reacts most strongly with mAbs B1.1 and B6.2 and least with mAb B72.3. One clone exhibited a phenotype similar to that of the parent, but at least three additional major phenotypes were observed among the other clones. To determine the stability of the cell-surface phenotype of the MCF-7 clones, each line was monitored through a 4-month period and assayed during log phase at approximately every other passage. Although some of the MCF-7 clones maintained a stable antigenic phenotype throughout the observation period, a dramatic change in antigenic phenotype, i.e., antigenic evolution, was observed in some of the clones (Horan Hand et al. 1983).

Radiolocalization of human mammary tumors in athymic mice by a monoclonal antibody

With the development of hybridoma technology, homogeneous populations of monoclonal antibodies to TAAs can now be utilized in either lymphangiography, to detect mammary tumor lesions in nodes of the axilla and internal mammary chain, or to detect distal metastases. In the studies described here, mAb B6.2, which may be useful in lymphangiography procedures, was utilized. mAb B6.2 IgG was purified and $F(ab')_2$ and Fab' fragments were generated by pepsin digestion. All three forms of the antibody were radiolabeled and assayed to determine their utility in the radioimmunolocalization of transplanted human mammary tumor masses (Colcher et al. 1983b).

The IgG and its fragments were labeled with ^{125}I, using the iodogen method to specific activities of 15–50 µCi/µg. The labeled antibody was shown to bind to the surface of live MCF-7 cells and retained the same specificity as the unlabeled antibody. Better than 70% of the antibody remained immunoreactive in sequential-saturation, solid-phase RIAs after labeling.

Athymic mice bearing the Clouser transplantable human mammary tumor were injected with 0.1 µg of ^{125}I-labeled mAb B6.2. The ratio of radioactivity per milligram of tissue in the tumor compared with that of various tissues rose over a 4-day period and then fell at 7 days. The tumor-to-tissue ratios were 10:1 or greater in the liver, spleen, and kidney at day 4. Ratios of the counts in the tumor to those found in the brain and muscle were greater than 50:1 and as high as 110:1. When the mammary tumor–bearing mice were injected with ^{125}I-labeled $F(ab')_2$ fragments of mAb B6.2, higher tumor-to-tissue ratios were obtained. The tumor-to-tissue ratios in the liver and spleen were 15:1 to 20:1 at 96 hours. This is probably due to the faster clearance of the $F(ab')_2$ fragments as compared with the IgG. Athymic mice bearing a human melanoma (A375), a tumor that shows no surface reactivity with mAb B6.2 in live-cell RIAs, were used as controls and were negative for nonspecific binding of the labeled antibody or antibody fragments to tumor tissue. Similarly, no localization was observed when either normal murine IgG or MOPC-21 IgG1 (the same isotype as mAb B6.2 from a murine myeloma) or their $F(ab')_2$ fragments were radiolabeled and inoculated into athymic mice bearing human mammary tumor or melanoma transplants.

Studies were then undertaken to determine whether the localization of the ^{125}I-labeled antibody and fragments in the tumors was sufficient to detect by using a gamma camera. Mice were injected with ^{125}I-labeled mAb B6.2 $F(ab')_2$ fragments. Due to the rapid clearance of the fragments, a significant amount of radioactivity was observed in the two kidneys and bladder at 24 hours; tumors were also clearly positive (Fig. 1A). The activity was cleared from the kidneys and bladder by 48 hours, and the tumor-to-background ratio increased over the 4-day period of scanning, with little background, and good tumor localization was observed at 96 hours (Fig. 1B). No localization was observed with the radiolabeled mAb B6.2 $F(ab')_2$ fragments in the athymic mice bearing the control A375 melanoma (Fig. 1C).

Monoclonal antibodies to human colon, lung, and breast carcinomas have been generated in a number of laboratories. A surprising finding in these studies has been the large number of antigenic cross-reactivities observed among these three major carcinoma groups. The studies reported here are designed to define the range of reactivities of mAbs B72.3, B6.2, and B1.1 to colon adenocarcinomas versus benign colon lesions and normal colon epithelium.

Reactivity of monoclonal antibodies with malignant and benign colon lesions

Purified IgG of mAb B1.1 was reacted with fixed, 5-µm tissue sections of malignant and benign colon lesions at

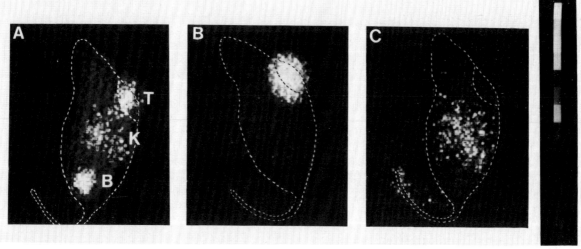

Figure 1 Gamma camera scanning with ^{125}I-labeled mAb B6.2 $F(ab')_2$ of athymic mice bearing transplanted human tumors. Mice bearing a transplantable human mammary tumor (Clouser, A and B) or a human melanoma (A375, C) were inoculated with ~30 µCi of ^{125}I-labeled mAb B6.2 $F(ab')_2$. The mice were scanned after various time intervals (A, 24 hr; B and C, 96 hr) until an equal number of counts were detected in each field.

concentrations of 10, 1, and 0.1 μg/ml of purified IgG per slide, using the avidin-biotin complex (ABC) method of immunoperoxidase staining (Stramignoni et al. 1983). The characteristic dark reddish-brown positive reaction was contrasted with that of the blue hematoxylin counterstain. Scoring of carcinomas was based on the percentage of carcinoma cells in samples that were positive, as well as the intensity of the stain.

Using 10 μg/ml of mAb B1.1, 15 of 16 (94%) colon carcinomas from different patients scored positive (Fig. 2A). All the positive carcinomas contained greater than 25% reactive cells, with three of the tumors having 100% of their cells positive. A heterogeneity of antigenic expression in terms of luminal versus cytoplasmic reactivity in a given tumor mass was usually observed. Approximately two thirds of coexistent adenoma lesions and normal colon epithelium were also positive. The majority of adenomas from noncarcinoma patients scored positive. Using lower dilutions of mAb B1.1 (1 μg/ml and 0.1 μg/ml), the majority of colon carcinoma specimens remained positive. Some of the adenomas in noncarcinoma patients also remained positive, scoring from 71% positive at 1 μg/ml to 44% positive at 0.1 μg/ml.

Using 10 μg/ml of mAb B72.3, 14 of 17 (82%) of colon carcinomas demonstrated a variable number of positive-staining cells (Fig. 2B). The majority of the carcinomas that were positive had more than 50% of tumor cells present reacting. The cellular location of the reactivity also varied within a given tumor mass; luminal reactivity was often observed in glandular structures, with cytoplasmic staining in less-differentiated structures. As a negative control for the specificity of the staining observed, parallel slides with phosphate-buffered saline (instead of primary antibody B72.3), isotype-identical IgG (MOPC-21), as well as normal murine IgG were used; all of these gave negative results.

Adenomas in patients with colon carcinoma were also examined for expression of antigen reactive with mAb B72.3; only 1 of 10 such lesions showed a positive reaction. This lesion was immediately adjacent to a carcinoma and had less than 5% cells positive. All other adenomas and adjacent normal epithelium in carcinoma patients were negative, with one exception in which less than 1% of the cells scored positive. Adenomas from patients without apparent carcinoma were examined using 10 μg/ml of mAb B72.3. None of the 18 lesions showed reactivity with greater than a few percent of adenoma cells positive. Five of the 18 samples, however, did show staining in a few percent of adenoma cells.

Malignant and benign colon lesions were then examined using 10-fold less (1 μg/ml) mAb B72.3. At this antibody dilution, 8 of 16 carcinomas showed a positive reaction; 1 of 10 adjacent adenomas showed a focal reactivity of less than 1% of adenomatous cells. None of 16 adenomas from noncarcinoma patients and none of normal epithelium from 19 carcinoma or adenoma patients reacted at this antibody dilution. Thus, a positive reactivity with mAb B72.3 at this dilution appears to be an even stronger marker for malignancy.

As mentioned above, monoclonals B1.1 and B6.2 can be distinguished on the basis of the molecular weights of the proteins precipitated from tumor cells. From reactivity patterns on tumor and normal tissues, it has been difficult thus far to distinguish the reactivities of these two monoclonals. Used at 10 μg/ml, mAb B6.2 reacted with all 15 colon carcinomas tested, with 3 of the tumors showing 100% reactivity. Three of 10 adenomas, as well as 2 of 11 (18%) normal epithelia, adjacent to carcinoma lesions were positive. Using 10 μg/ml of mAb B6.2 per slide, 11 of 17 (65%) adenomas from noncarcinoma patients also scored positive.

Using 1 μg/ml of mAb B6.2 per slide, 14 of 15 (93%) carcinoma lesions scored positive. Two of 10 adenomas in carcinoma patients and 5 of 16 (30%) adenomas from noncarcinoma patients also were positive. However, with the exception of 1 lesion, all adenomas had less than 10% tumor cells positive. Using 0.1 μg/ml of mAb B6.2, 8 of 15 (53%) carcinomas were positive, whereas 3 of 18 (17%) adenomas were positive with only a few percent of cells reacting. Thus, mAb B6.2 could clearly be distinguished from mAb B1.1 on the basis of its more selective reactivity with colon carcinoma versus adenomas or normal colon epithelium. The reactivity of mAb B6.2, however, was not as selective for carcinomas as was that observed for mAb B72.3 (Stramignoni et al. 1983).

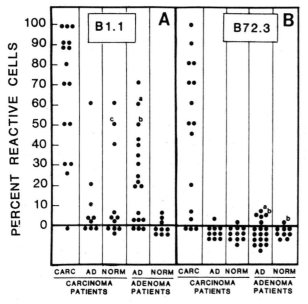

Figure 2 Reactivities of mAbs B1.1 and B72.3 with colon carcinomas. The scoring of immunoperoxidase staining with mAb B1.1 (A) and mAb B72.3 (B) used at 10 μg/ml is based on a semiquantitative rating by two pathologists (Stramignoni et al. 1983). The first three columns of each panel show the reactivity of formalin-fixed carcinomas (CARC), adenomas (AD), and normal epithelium (NORM) in carcinoma-bearing patients; the next two columns show the reactivity of formalin-fixed adenomas (AD) and normal epithelium (NORM) of adenoma-bearing patients. The lesions from the 18 adenoma patients studied could be divided into two groups: 10 were classified as tubular adenomas and 8 as tubulovillous adenomas. (a) Tubular adenoma in a patient with "probable recurrence" of colon carcinoma; (b) tubulovillous adenoma with atypia and diagnosis of in situ carcinoma; (c) colon cancer patient with ulcerative colitis.

Colon carcinomas and adenomas were also antigenically phenotyped into several distinct groups based on their reactivity with the three monoclonal antibodies employed in this study. Large retrospective studies using fixed tissue sections can now be conducted to determine if a given antigenic phenotype correlates with a specific biologic property, such as response to a specific therapeutic modality or prognosis.

Recombinant interferon enhances tumor antigen detection

Previous studies with monoclonal antibodies have shown a great deal of antigenic heterogeneity among cells within both primary and metastatic human breast and colon carcinoma lesions. If monoclonals are to be used successfully for the in situ detection or therapy of human carcinoma lesions, this phenomenon of antigenic heterogeneity must be modified so that most or all cells of a tumor mass express a given TAA. One approach to enhance the expression of TAAs on the surface of cancer cells would be the use of those substances that (1) have been shown to alter states of cell differentiation and (2) have potential clinical applicability. Native interferon (Imai et al. 1981) and recombinant human leukocyte clone-A interferon (IFN) (Pestka 1983) meet the above two criteria. Because of its homogeneity, stability, and extensive degree of characterization (Pestka 1983), IFN was thus evaluated for its ability to enhance the detection of TAAs on the surface of human carcinoma cells by monoclonal antibodies (J. Greiner et al., in prep.).

IFN was first titrated on the human mammary carcinoma cell line MCF-7 for its ability to enhance the detection of TAAs as well as normal cell-surface antigens. A solid-phase RIA with live cells (Horan Hand et al. 1983) and various concentrations of purified monoclonal Ig were used to detect antigen expression. As seen in Figure 3, 10, 100, or 1000 units/ml IFN enhanced the detection of the 90K (Fig. 3C), the 180K CEA (Fig. 3D), and the 220K–400K (Fig. 3E) TAAs by mAbs B6.2, B1.1, and B72.3, respectively, in a dose-dependent manner. mAbs W6/32 and B139, directed against two antigens found on the surface of normal and neoplastic human cells, were used as controls. As seen in Figure 3A, IFN enhanced the cell-surface binding of mAb W6/32 but had no effect on the surface binding of mAb B139 (Fig. 3B). Addition of 5,000 and 10,000 units/ml IFN resulted in less-effective binding of all the monoclonals than with 1000 units/ml. The effect of time of exposure of cells to IFN was also examined. The enhanced expression of TAA was first detected within 4 to 12 hours after the

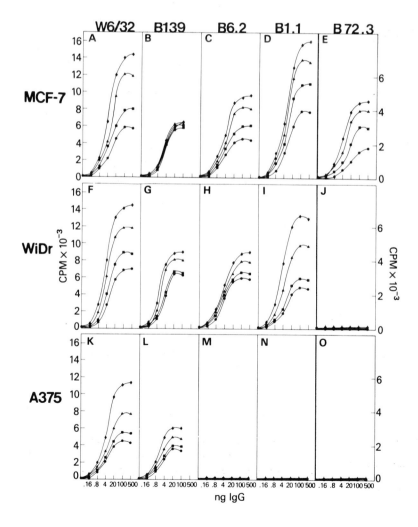

Figure 3 The effect of molecularly cloned interferon (IFN) on the expression of human tumor-associated and normal cell-surface antigens. The live-cell, solid-phase RIA used has been described previously (Horan Hand et al. 1983). Cells (5×10^4) were added to 96-well microtiter plates. After 24 hr at 37°C, in medium with or without IFN, medium was removed and monoclonal antibody was added. After 1 hr at 37°C, cells were washed twice and 75,000 cpm of ^{125}I-labeled goat anti-mouse IgG was added. After 1 hr at 37°C, cells were washed and the cpm bound was determined. (A–E) MCF-7 human mammary carcinoma cells; (F–J) WiDr colon carcinoma cells; (K–O) A375 melanoma cells (Flow 4000 human embryonic kidney and WI-38 human embryonic lung cell gave similar results). Monoclonal IgGs were purified and used at the amounts indicated. mAb W6/32 (A,F,K), mAb B139 (B,G,L), mAb B6.2 (C,H,M), mAb B1.1 (D,I,N), and mAb B72.3 (E,J,O) have been previously described (Colcher et al. 1981). IFN was obtained and purified as previously described (Pestka 1983). The specific activity of the preparation used was 2×10^8 units/mg protein when tested on MDBK (bovine kidney) cells. Amounts of IFN per milliliter used were 10 units (■), 100 units (▲), 1000 units (♦). (●) Control buffer RPMI 1640 containing 1% BSA.

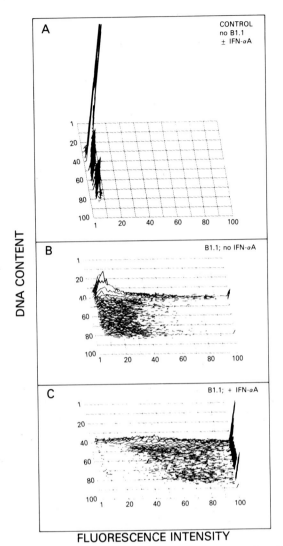

Figure 4 Flow cytometric analyses of mAb B1.1 binding the surface of WiDr carcinoma colon cells following treatment with IFN. Each section is a three-dimensional isometric display of DNA content (Y axis), fluorescence intensity (i.e., cell-surface 180K antigen expression) (X axis), number of cells (Z axis). WiDr cells (10^7) were incubated for 24 hr at 37°C with or without 1000 units/ml IFN. The cells were then harvested, washed, and incubated for 2 hr at 4°C in medium containing 2 μg of mAb B1.1 per 10^6 cells. The cells were then washed with PBS and incubated with fluoresceinated sheep anti-mouse antibody for 1 hr at 4°C. The cells were washed, centrifuged at 500g, and resuspended at a concentration of 10^6 cells/0.3 ml. The cells were then fixed by adding 0.7 ml of ice-cold 100% ethanol while vortexing. After fixing in ethanol for 24 hr, the cells were pelleted, resuspended at 10^6 cells/ml, and stained for 4 hr at room temperature in PBS with propidium iodide (PI, 18 μg/ml) and ribonuclease A (2000 units/ml). The stained cells were analyzed on an Ortho Cytofluorograf System 50H with blue laser excitation of 200 mW at 488 nm. Under these conditions PI bound to nuclear DNA fluoresces red, whereas surface immunofluorescence bound to the 180K antigen fluoresces green. Data from 25,000 cells were stored on an Ortho Model 2150 computer system and used to generate the figures shown here. (A) WiDr cells stained for DNA content but with no mAb B1.1; (B) WiDr cells stained for both nuclear DNA and mAb B1.1 binding; (C) WiDr cells treated with 1000 units of IFN for 24 hr and stained as in B.

addition of 1000 units/ml IFN; optimal enhancement was observed 16 to 24 hours after IFN addition. IFN was then analyzed for its ability to enhance the detection of TAAs on the surface of the human colon carcinoma cell line WiDr. IFN at a concentration of 10–1000 units/ml enhanced the binding of mAbs B6.2 (Fig. 3H) and B1.1 (Fig. 3I) in a dose-dependent manner. The 220K–400K TAA is not expressed on the surface of WiDr cells, and various levels of IFN up to 1000 units/ml failed to elicit any mAb B72.3 binding (Fig. 3J). Although IFN enhanced the binding of mAb W6/32 to the surface of WiDr (Fig. 3F) and MCF-7 cells (Fig. 3A), it enhanced the binding of mAb B139 only to WiDr (Fig. 3G) and not to MCF-7 cells (Fig. 3B).

The effect of 10, 100, and 1000 units/ml IFN on the growth of MCF-7 and WiDr cells was monitored. Incubation of cells in growth medium containing 10 or 100 units/ml IFN for 6 days had no appreciable effect on cell growth; incubation with 1000 units/ml IFN for 6 days reduced the total cell number by 46%. However, no effect on cell number was observed 1, 2, or 4 days following IFN addition at any dose level. Since all determinations of surface antigen expression were made 24 hours after the addition of IFN and since in many cases similar results were observed using 100 or 1000 units/ml, the enhanced TAA expression appears to be independent of changes in cell proliferation.

Studies were carried out to determine if various concentrations of IFN would induce the expression of the 90K, 180K, and 220K–400K TAAs on normal as well as noncarcinoma neoplastic cells not normally expressing these surface antigens. Three cell lines were chosen for these studies: WI-38 (normal human embryonic lung), Flow 4000 (normal human embryonic kidney), and A375 (human melanoma). All three cell lines were previously shown to be negative for the expression of the three TAAs (Colcher et al. 1981; Horan Hand et al. 1983) and remained so following exposure to various concentrations (10–1000 units/ml) of IFN (see Fig. 3 M–O). IFN did, however, increase the expression of the normal cell-surface antigens that bind mAbs B139 (Fig. 3L) and W6/32 (Fig. 3K).

The enhanced binding of monoclonal antibodies to TAAs on the surface of human breast and colon carcinoma cells mediated by IFN could be due to (1) the increased expression of TAAs on a subpopulation of cells already expressing the TAAs, (2) the induction of the expression of a given TAA on a population of cells not previously expressing the antigen, or (3) a combination of both phenomena. To explore these possibilities, mAb B1.1 was reacted with WiDr human colon carcinoma cells in the presence or absence of 1000 units/ml IFN, and the cells were analyzed via fluorescent activated cell sorting. The background analysis of WiDr cells in the absence of mAb B1.1, with or without IFN, is shown in Figure 4A. Figure 4B shows a heterogenous population of cells, depicted as a spectrum of fluorescence intensities (X axis), expressing various levels of the cell-surface 180K antigen. Following a 24-hour incubation with 1000 units/ml IFN, a dramatic shift is observed (Fig. 4C) in both the percentage of cells expressing the 180K antigen (vertical Z axis) and the fluorescence intensity per cell (X axis). No difference in DNA content of cells (Fig. 4, Y axis) was observed after IFN treatment. Thus, following the addition of IFN, computer analysis revealed more than 98% of tumor cells now bound mAb B1.1

One of the potential clinical applications of monoclonal antibodies is the use of radiolabeled immunoglobulins to detect micrometastases in regional nodes and at distal sites. mAbs B6.2 and B72.3 have already been used to detect human carcinoma transplants in athymic mice (Colcher et al. 1983b). One of the major considerations of radiolocalization studies is reducing the amount of radiolabeled antibody required to bind and detect a given tumor mass in situ; the use of lower levels of labeled monoclonal antibody would increase "signal-to-noise" ratios and thus make detection of smaller lesions more efficient. Accordingly, we investigated the effect of IFN on the amount of mAb B72.3 required to bind a reference number of counts per minute to the surface of MCF-7 mammary tumor cells. Ninety-four ng of mAb 72.3 were required to bind 1000 cpm to 5×10^4 MCF-7 cells, whereas this could be reduced to 4 ng of mAb B72.3 following exposure of the same number of cells to 1000 units/ml of IFN. This 24-fold reduction in the amount of antibody required to give an equally efficient signal to surface binding of a monoclonal antibody may have potential clinical application for the in situ detection of carcinoma lesions with radiolabeled monoclonals or in the use of monoclonal antibodies for immunotherapy. In addition, the ability of recombinant interferon to selectively enhance the expression of monoclonal-defined TAAs in carcinoma cell lines may prove useful in defining the role of specific TAAs in the expression of the transformed phenotype.

Acknowledgments

We wish to thank R. Riley, D. Poole, D. Simpson, J. Howell, J. Collins, and A. Sloan for expert technical assistance in these studies. We also thank Dr. M. Weeks for many helpful suggestions in the preparation of this manuscript.

References

Barnstable, C., W. Bodmer, G. Brown, G. Galfre, C. Milstein, A. Williams, and A. Ziegler. 1978. Production of monoclonal antibodies to group A erythrocyles, HLA and other human cell surface antigens—new tools for genetic analysis. *Cell* **14**: 9.

Colcher, D., P. Horan Hand, M. Nuti, and J. Schlom. 1981. A spectrum of monoclonal antibodies reactive with human mammary tumor cells. *Proc. Natl. Acad. Sci.* **73**: 3199.

———. 1983a. Differential binding to human mammary and non-mammary tumors of monoclonal antibodies reactive with carcinoembryonic antigen. *J. Cancer Invest.* **1**: 127.

Colcher, D., M. Zalutsky, W. Kaplan, D. Kufe, F. Austin, and J. Schlom. 1983b. Radiolocalization of human mammary tumors in athymic mice by a monoclonal antibody. *Cancer Res.* **43**: 736.

Horan Hand, P., M. Nuti, D. Colcher, and J. Schlom. 1983. Definition of antigenic heterogeneity and modulation among human mammary carcinoma cell populations using monoclonal antibodies to tumor-associated antigens. *Cancer Res.* **43**: 728.

Imai, K., A. Ng, M. Glassy, and S. Ferrone. 1981. Differential effect of interferon on the expression of tumor-associated antigens and histocompatibility antigens on human melanoma cells: Relationship to susceptibility to immune lysis mediated by monoclonal antibodies. *J. Immunol.* **127**: 505.

Kufe, D., L. Nadler, L. Sargent, H. Shapiro, P. Horan Hand, D. Colcher, and J. Schlom. 1983. Biological behavior of human breast carcinoma-associated antigens expressed during cellular proliferation. *Cancer Res.* **43**: 851.

Nuti, M., D. Colcher, P. Horan Hand, F. Austin, and J. Schlom. 1981. Generation and characterization of monoclonal antibodies reactive with human primary and metastatic mammary tumor cells. In *Monoclonal antibodies and development in immunoassay* (ed. A. Albertini and R. Ekins), p. 87. Elsevier/North-Holland, Amsterdam.

Nuti, M., Y. Teramoto, R. Mariani-Costantini, P. Horan Hand, D. Colcher, and J. Schlom. 1982. A monoclonal antibody (B72.3) defines patterns of distribution of a novel tumor-associated antigen in human mammary carcinoma cell populations. *Int. J. Cancer* **29**: 539.

Pestka, S. 1983. The human interferons—From protein purification and sequence to cloning and expression in bacteria: Before, between, and beyond. *Arch. Biochem. Biophys.* **221**: 1.

Stramignoni, D., R. Bowen, B.F. Atkinson, and J. Schlom. 1983. Differential reactivity of monoclonal antibodies with human colon adenocarcinomas and adenomas. *Int. J. Cancer* **31**: 543.

Human Tumor Antigens

H. Koprowski
The Wistar Institute of Anatomy and Biology, Philadelphia, Pennsylvania 19104

The study of human tumor antigens has been greatly facilitated by the availability of monoclonal antibodies that bind to human tumor cells and that are used for the isolation and partial characterization of the antigens. Hybridomas secreting monoclonal antibodies are generated by the fusion of mouse myeloma cells with splenocytes of mice immunized either with intact human cancer cells or cell extracts or with cell-free tissue-culture medium containing "soluble" antigenic fraction(s). As shown in Table 1, deriving a hybridoma that secretes antibodies binding to tumor cells is a major task. For instance, out of 8,000 colonies of hybridomas generated by the immunization of mice with gastrointestinal adenocarcinoma (GIC) cells, only 35 clones showed certain specificities for the tumor cells. Of 10,000 hybridoma colonies generated after immunization of mice with melanoma cells, only 61 clones bound to melanoma cells.

Human tumor cells quite often express either transplantation or blood group antigens that are strong immunogens. Therefore, a monoclonal antibody derived from mice immunized with human tumor preparations is often directed against antigens such as DR after immunization with melanoma and against various blood groups after immunization with GIC. In addition, once a large number of monoclonals is derived against the same tumor species, antibodies with the same specificities are repeatedly produced. For instance, a set of melanoma-derived monoclonal antibodies with similar binding specificities was produced in different laboratories (Morgan et al. 1981; Herlyn et al. 1982; Kantor et al. 1982; Ross et al. 1982). In our hands this set of antibodies binds to melanoma cells, to cells of benign pigment lesions, such as nevi, to cells of astrocytomas, and to fetal epithelioid cells. The antigen extracted either from melanoma cells or from cell-culture supernatants and immunoprecipitated with a monoclonal antibody shows two components (Ross et al. 1983): a chondroitin sulfate proteoglycan of $M_r > 500,000$ (500K) and a glycoprotein of M_r 260,000 (260K). The 500K component can be converted to the gp 260K component by chondroitinase treatment, suggesting that the protein portion of the two components is the same (Ross et al. 1983).

Another interesting protein antigen is one defined by a monoclonal antibody that binds to GIC, breast cancer, melanoma, lymphoma and leukemia cells, and macrophages, but not to human fibroblasts. Two components common to the different tumor species were immunoprecipitated by this antibody (Pak et al. 1983): one polypeptide of M_r 37,000, which predominates in the GIC extract, and another of M_r 65,000, which predominates in the melanoma cell extract. The monoclonal antibody was found to inhibit growth of GIC, breast cancer, and lung cancer xenografts in nude mice, but was ineffective against melanoma cells (Herlyn and Koprowski 1982). The question of whether "vulnerability" of the tumors was related to the predominance of the 37K polypeptide is unresolved at present and awaits a detailed peptide mapping of the two antigenic components.

One of the major problems encountered in studying protein antigens of human tumors is the difficulty in obtaining aminoterminal sequences of the monoclonal antibody-defined antigens. This has been attributed to blocking of the amino terminus. A notable exception is a 97K glycoprotein derived from melanoma cells (Brown et al. 1982), which was isolated by immunoprecipitation with a monoclonal. The aminoterminal sequence of the amino acids of this glycoprotein revealed homology with the aminoterminal sequence of transferrin and lactotransferrin (Brown et al. 1982); since the p97 antigen binds iron, it was also functionally related to transferrin.

Glycolipid antigen

In contrast to the difficulties that have hampered characterization of the human tumor-associated protein antigen, knowledge concerning glycolipid antigens expressed by malignant cells is more advanced. Table 2 lists the number of monoclonal antibodies, derived from the immunization of mice with GIC cells, that bind to carbohydrate determinants. Most of the monoclonals define blood group antigens, and two bind to antigens expressed by GIC cells. This gastrointestinal cancer antigen (GICA) was identified as a sialylated form of the Lewis A antigen. Individuals who are Le^{a-b-} cannot syn-

Table 1 Production of Monoclonal Antibodies by Hybridomas Obtained after Immunization of Mice with Human Tumors

Tumor type[a]	Fusion	Number of colonies screened[b]	Number of clones defining specificities
Colorectal Ca	55	8000	35
Gastric Ca	8	900	8
Pancreatic Ca	4	500	4
Lung Ca	10	600	3
Mammary Ca	25	2000	30
Bladder Ca	5	900	17
Prostate Ca	15	3000	6
Ovarian	4	1000	3
Cervical	3	400	3
Melanoma	75	10,000	61
Astrocytoma	6	400	3

[a]For immunization, intact cells, crude or solubilized membranes, cytoplasmic fractions, or shed antigens were used either from established tissue-culture cell lines or from freshly obtained tumors. (Ca) Carcinoma.
[b]Estimated number. Generally, 40–80% of screened colonies are reactive with the immunizing preparation.

Table 2 Monoclonal Antibodies That Bind to Cells of Gastrointestinal Cancers and Define Carbohydrate Determinants of Glycolipid Antigens

Antigen	Number of monoclonals reacting
Lewis A	1
Lewis B	4
Lewis B and H type 1	2
Lewis A and Lewis B	2
Sialylated Lewis A[a]	2
Lewis X (SSEA-1)[b]	2
Blood group A	1
Blood group B	5

[a]GICA-sialyl lacto-N-fucopentaose II is the carbohydrate determinant of this glycolipid.
[b](SSEA-1) Stage-specific embryonic antigen. Lactofucopentaose III is the carbohydrate determinant of this glycolipid.

Table 4 Detection of GICA in Sera of Patients

Disease	Total no. of sera examined	Percentage sera showing presence of GICA
Pancreatic and gastric Ca	12	75
Colorectal Ca		
primary	39	36
recurrent (early)	16	44
advanced	30	67
Healthy controls	107	3

(GICA) Gastrointestinal cancer antigen; (Ca) carcinoma.

thesize GICA (Table 3), probably because they lack an as yet unidentified sialyltransferase. As shown in Table 3, GICA is present in the saliva of normal control and cancer patients, but only in serum of patients with GIC. The presence of this antigen in serum has permitted the development of a diagnostic procedure based on the binding activity of the patient's serum to polystyrene beads coated with one monoclonal antibody and the reactivity to these adsorbents with another radiolabeled monoclonal of similar specificity (M. Herlyn et al. 1984). By this procedure, serum detection of the antigen has been possible in the majority of patients with GIC. As shown in Table 4 (M. Herlyn et al. 1984), patients with pancreatic and gastric cancer show the highest percentages of elevated GICA in their serum, followed by patients with colorectal cancer. Quite significantly, sera of only 2 out of 107 healthy donors showed elevated GICA levels. GICA has been recovered in lipid extracts of colon cancer, either surgically excised or grown in tissue culture, and in extracts of meconium, but not of other tumors or normal tissue (Magnani et al. 1982). Based on the immunoperoxidase staining reaction with anti-GICA monoclonal antibody (Atkinson et al. 1984), it was possible to demonstrate the presence of GICA in a single layer of epithelial cells lining the large secretory ducts of pancreas, liver, and bronchi.

Lewis antigens in human tumors

Of all the blood group antigens detected by monoclonal antibodies generated in mice immunized with GIC cells, the Lewis antigen shows the most interesting association with human tumors (Ernst et al. 1984). Using the immunoperoxidase staining method, it was possible to show again that Le^{a-b-} individuals do not express Lewis antigens either in normal or tumor tissue. Likewise, Le^{a+b-} subjects do not show the presence of Lewis B antigens. In Le^{a-b+} individuals (secretors), Lewis B and A antigens are expressed by cells of proximal colon and by tumors derived from these cells. In distal colon, however, only Le^a antigen is expressed by normal cells, though Le^b antigen is abundantly expressed in tumors derived from normal distal colon cells (Ernst et al. 1984). Similarly, Le^b antigen is not detected in renal tubular epithelium but is readily detectable in renal carcinoma cells. The absence of H-substance, a substrate for Le^b antigen, in normal tissue and the presence of this substrate in tumor tissue probably account for the differences in the expression of Lewis B antigen.

Localization of human tumor antigens in experimental hosts and in humans

Monoclonal antibodies labeled with radioisotopes such as ^{125}I, ^{131}I, or ^{123}I and injected into nude mice bearing human tumor xenografts can be specifically localized by gammascintigraphy to the tumor, which is bound by the respective monoclonal antibody in an in vitro assay (D. Herlyn et al. 1983). Biodistribution studies also demonstrate the preferential localization of the antibody to the tumor as compared with other organs and blood of the mouse.

Table 3 Lewis Phenotype and GICA in Patients with Colon Carcinoma, Pancreatic Cancer, and Other Conditions

Le phenotype	Colon Ca GICA in		Pancreatic Ca GICA in		Controls GICA in	
	saliva	serum	saliva	serum	saliva	serum
Le^{a+b-}	5/5	4/5	1/2	2/2	23/29	0/7
Le^{a-b+}	5/17	5/17	6/7	7/7	22/52	0/18
Le^{a-b-}	0/5	0/5	0/3	0/3	0/3	0/3
Le^{a+b+}	—	—	1/1	1/1	—	—

Both sera and saliva were tested in these patients. (GICA) Gastrointestinal cancer antigen; (Ca) carcinoma.

Table 5 Detection of Colorectal Cancer Antigen in Human Tumors In Vivo by Radioimmunodetection and by Immunoperoxidase Staining

Antigen	Ratio of tumors detected by radioimmunodetection and immunoperoxidase staining	
	positive	negative
GICA	10/13	4/4
CEA	8/14	2/3

(GICA) Gastrointestinal cancer antigen; (CEA) carcinoembryonic antigen.

These findings prompted the application of radiolabeled monoclonals in the localization of tumors in humans (Mach et al. 1983). These studies were quite successful and occasionally permitted localization of tumors that were undetectable by any other means. As shown in Table 5, there is a good correlation between the radioimmunodetection results and the presence of antigen detected on colorectal cancer cells by the immunoperoxidase staining method (Douillard et al. 1983).

Suppression of human tumor growth in vitro and in vivo

Monoclonal antibodies may lyse human target cells by interaction with complement. Most of these antibodies are of the IgM isotype, which has little if any lytic effect in vivo. On the other hand, there is a set of monoclonal antibodies that cannot interact with complement but that destroys human tumor cells in the presence of macrophages (Herlyn and Koprowski 1982). Most of these antibodies are of the IgG2a isotype, and they interact with macrophages by means of the Fc receptors, since $F(ab')_2$ fragments of the monoclonals are inactive. The reactions are highly specific since only tumor cells that bind the respective monoclonal antibody are destroyed by the interaction of the antibody with the macrophages. Freshly obtained human monocytes show low concentrations of Fc receptors for mouse IgG2a monoclonal antibody. After 10–14 days in culture, however, a sufficient number of human macrophages express Fc receptors to be effective in tumor destruction in the presence of mouse IgG2a monoclonal (Steplewski et al. 1983). The tumoricidal effect is directly related to the concentration of antigenic sites per tumor cell reactive with a given IgG2a (D. Herlyn and J. Powe, pers. comm.). As shown in Table 6, the IgG2a monoclonals that effectively mediate destruction of tumor cells bind to tumor cells at a concentration of higher than 10^6 molecules. Those monoclonals that do not destroy tumor cells bind to a target cell at a concentration of 3.1×10^5 or less molecules. Affinity of the monoclonal antibodies does not seem to play a role.

The tumoricidal effect of monoclonal antibodies can be reproduced in vivo in nude mice bearing human tumor xenografts. Again, only the IgG2a monoclonals that are tumoricidal in vitro effectively inhibit the growth of tumors in nude mice. The tumoricidal effect is mediated by macrophages since injection of mice with a silica preparation that blocks macrophage action interferes with the tumoricidal effect of the antibody (Herlyn and Koprowski 1982). Thus, studies of human tumors and their antigens have advanced sufficiently in the laboratory and in experimental animals to begin developing rational approaches to the control of tumor growth in humans.

Studies on the localization of gastrointestinal tract cancers in humans by radiolabeled monoclonal antibodies have shown that one IgG2a monoclonal, 17-1A, binds to the majority of human tumors in vivo. This antibody was selected for injection into patients with metastatic cancer of the gastrointestinal tract (Sears et al. 1982). The results of the treatment of 55 patients indicated that humans tolerate injection of mouse monoclonal antibody in doses up to 1 g without untoward reactions. Small doses of antibody less than 190 mg induce formation of antimouse antibodies (Koprowski 1984), but it is nevertheless possible to reinject such subjects with the same monoclonal months after the first administration without any ill effects. Several treated individuals, particularly those who showed remission from the disease (Sears et al. 1984), developed anti-idiotype antibody reactive with the monoclonal antibody idiotype (Koprowski et al. 1984). Extracts of cells from cultures of colorectal cancer inhibit this reaction. This "hapten inhibition" suggests that the anti-idiotype antibody may represent the "internal image" of the 17-1A–defined cancer antigen, and patients may develop antibodies that bind to their own tumor as a result of a response to the antigen internal image presented on an immunoglobulin molecule of the anti-idiotype.

Table 6 Affinity of Anti-melanoma Monoclonal Antibody for Tumor Cells and Binding Sites of Tumor Cells Determined by Monoclonal Antibodies

Monoclonal antibody function	code	Sites (max. molecules/cell[a])	Affinity (liters/mole)
Tumoricidal	37-7	3.7×10^6	2.6×10^8
	B_2 28-8	5.4×10^6	1.1×10^7
	O_2 8214	1.2×10^6	1.0×10^9
Nontumoricidal	Nu4-B	5.0×10^4	1.9×10^7
	JC4C1	1.6×10^5	1.5×10^8
	JD6C9	3.1×10^5	1.4×10^9

[a]WM9 melanoma cells.

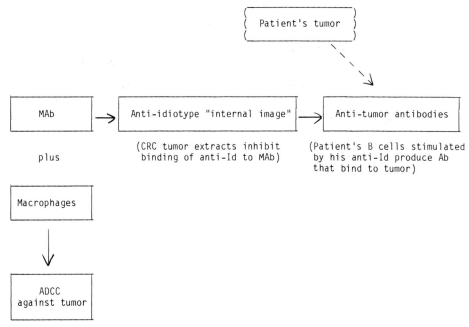

Figure 1 Chain of immunologic reaction of cancer patients treated with a monoclonal antibody (MAb). (CRC) Colorectal cancer; (ADCC) antibody-dependent cell cytotoxicity; (anti-Id) anti-idiotype.

Thus, as shown in Figure 1, there are two pathways for the monoclonal antibody to exert an antitumor effect in the human organism. One pathway involves the interaction of the IgG2a monoclonal (see above) with Fc receptors of human macrophages (Steplewski et al. 1983), which in turn destroys the tumor cells. The second pathway may involve a network of interacting human immunocytes that respond to the antibody binding site by producing an "internal image" of the tumor antigen which, in turn, may immunize the cancer patient against his or her own tumor.

Acknowledgments

The research was supported in part by U.S. Public Health Service grants CA-10815, CA-25874, CA-21124, and CA-33491 awarded by the National Cancer Institute of the National Institutes of Health.

References

Atkinson, B.F., C.S. Ernst, B.F.D. Ghrist, M. Herlyn, M. Blaszczyk, A. Ross, D. Herlyn, Z. Steplewski, and H. Koprowski. 1984. Identification of melanoma-associated antigens using fixed tissue screening of antibodies. *Cancer Res.* (in press).

Brown, J.P., R.M. Hewick, I. Hellstrom, K.E. Hellstrom, R.F. Doolittle, and W.J. Dreyer. 1982. Human melanoma-associated antigen p97 is structurally and functionally related to transferrin. *Nature* **296:** 171.

Douillard, J.-Y., J.-F. Chatal, J.-C. Saccavini, C. Curtet, M. Kremer, P. Peuvrel, and H. Koprowski. 1983. Pharmacokinetic study of radiolabeled anti-colorectal carcinoma monoclonal antibodies in tumor-bearing nude mice. *J. Nucl. Med.* (in press).

Ernst, C.S., B.F. Atkinson, M. Wysocka, M. Blaszczyk, M. Herlyn, H. Sears, Z. Steplewski, and H. Koprowski. 1984. Monoclonal antibody localization of Lewis antigens in fixed tissue. *Lab. Invest.* (in press).

Herlyn, D. and H. Koprowski. 1982. IgG2a monoclonal antibodies inhibit human tumor growth through interaction with effector cells. *Proc. Natl. Acad. Sci.* **79:** 4761.

Herlyn, D., J. Powe, A. Alavi, J.A. Mattis, M. Herlyn, C. Ernst, R. Vaum, and H. Koprowski. 1984. Radioimmunodetection of human tumor xenografts by monoclonal antibodies. *Cancer Res.* **43:** 2731.

Herlyn, M., M. Blaszczyk, and H. Koprowski. 1984. Immunodiagnosis of human solid tumors. Genes and antigens in cancer cells: The monoclonal antibody approach. In *Proceedings of "Experten-Treffen"*, Bonn, Federal Republic of Germany, June 1983, p. 160. Karger, Basel.

Herlyn, M., Z. Steplewski, B.F. Atkinson, C.S. Ernst, and H. Koprowski. 1982. Comparative study of the binding characteristics of monoclonal antimelanoma antibodies. *Hybridoma* **1:** 403.

Kantor, R.R.S., A.K. Ng, P. Giacomini, and S. Ferrone. 1982. Analysis of the NIH workshop monoclonal antibodies to human melanoma antigens. *Hybridoma* **1:** 473.

Koprowski, H. 1984. Mouse monoclonal antibodies in vivo. In Monoclonal antibodies and cancer. In *Proceedings of 4th Armand Hammer Cancer Symposium* (ed. B.D. Boss et al.), p. 17. Academic Press, New York.

Koprowski, H., D. Herlyn, M. Lubeck, E. DeFreitas, and H.F. Sears. 1983. Human anti-idiotype antibodies in cancer patients: Is the modulation of the immune response beneficial for the patient? *Proc. Natl. Acad. Sci.* **81:** 216.

Mach, J.P., J.F. Chatal, J.D. Lumbroso, F. Buchegger, M. Forni, J. Ritschard, C. Berche, J.Y. Douillard, S. Carrel, M. Herlyn, Z. Steplewski, and H. Koprowski. 1983. Tumor localization in patients by radiolabeled antibodies against colon carcinoma. *Cancer Res.* **43:** 5593.

Magnani, J.L., M. Brockhaus, D.F. Smith, V. Ginsburg, M. Blaszczyk, K.F. Mitchell, Z. Steplewski, and H. Koprowski. 1981. A monosialoganglioside is a monoclonal antibody-defined antigen of colon carcinoma. *Science* **212:** 55.

Morgan, A.C., D.R. Galloway, and R.A. Reisfeld. 1981. Production and characterization of monoclonal antibody to a melanoma-specific glycoprotein. *Hybridoma* **1:** 27.

Pak, Y.P., M. Blaszczyk, Z. Steplewski, and H. Koprowski. 1983. Identification and isolation of a common tumor-associated molecule using monoclonal antibody. *Mol. Immunol.* **20:** 1369.

Ross, A.H., K.F. Mitchell, Z. Steplewski, and H. Koprowski. 1982. Identification and isolation of melanoma-associated antigens with monoclonal antibodies. *Hybridoma* **1:** 413.

Ross, A.H., G. Cossu, M. Herlyn, J.R. Bell, Z. Steplewski, and H. Koprowski. 1983. Isolation and chemical characterization of a melanoma-associated proteoglycan antigen. *Arch. Biochem. Biophys.* **225:** 370.

Sears, H.F., D. Herlyn, Z. Steplewski, and H. Koprowski. 1984. Effects of monoclonal antibody immunotherapy on patients with gastrointestinal adenocarcinoma. *J. Biol. Modifiers* (in press).

Sears, H.F., J. Mattis, D. Herlyn, P. Hayry, B. Atkinson, C. Ernst, Z. Steplewski, and H. Koprowski. 1982. Phase-I clinical trial of monoclonal antibody in treatment of gastrointestinal tumours. *Lancet* **1:** 762.

Steplewski, Z., M. Lubeck, and H. Koprowski. 1983. Human macrophages armed with murine immunoglobulin G2a antibodies to tumors destroy human cancer cells. *Science* **221:** 865.

Author Index

Aaronson, S. A., 35
Albrecht-Buehler, G., 87
Allen-Hoffmann, B. L., 217
Altmannsberger, M., 191
Amanuma, H., 277
Antoniades, H. N., 35
Anzano, M. A., 1
Assoian, R. K., 1

Backer, J., 229
Bernal, S. D., 75
Blumberg, P. M., 245
Brady, J., 105
Bravo, R., 147
Brickell, P. M., 97

Callahan, M. A., 51
Celis, A., 123
Celis, J. E., 123
Chen, L. B., 75
Cochran, B. H., 51
Cohlberg, J. A., 177
Colcher, D., 285
Connell, N. D., 217
Cooper, D., 169
Cowley, G., 5

Debus, E., 191
De Larco, J. E., 1
Dermody, M., 17
Deuel, T. F., 25, 43
Devare, S. G., 35

Eichner, R., 169

Feramisco, J. R., 11
Fey, S. J., 123
Fisher, P., 285
Franke, W. W., 177, 201
Franza, B. R., Jr., 137
Frolik, C. A., 1
Fuchs, E., 161

Garrels, J. I., 137
Gattoni-Celli, S., 229
Geib, R. W., 263
Geiger, B., 201
Geisler, N., 153
Gigi, O., 201
Gilmer, T. M., 117
Gooding, L., 263
Grace, M. P., 161
Greiner, J., 285
Gusterson, B., 5

Harlow, E., 263
Hatzfeld, M., 177
Heldin, C.-H., 25
Hendler, F., 5
Horan Hand, P., 285
Hsiao, W., 229
Huang, J. S., 25, 43
Huang, S. S., 43
Hunkapiller, M. W., 35

Ikawa, Y., 277

Jaken, S., 245
Jeffrey, A., 229
Jeng, A. Y., 245
Johnsson, A., 25
Jorcano, J. L., 177, 201

Kakunaga, T., 67
Kamata, T., 11
Khoury, G., 105
Kikkawa, U., 239
Kim, K. H., 161
Kirschmeier, P., 229
König, B., 245
Koprowski, H., 293
Kreis, T. E., 201

Laimins, L. A., 105
Lambert, M., 229
Lampidis, T. J., 75

Lane, D. P., 97
LaRocca, P. J., 217
Larsen, P. M., 123
Latchman, D. S., 97
Leach, K. L., 245
Leavitt, J., 67
Levy, B. T., 117
Lin, J. J.-C., 57
Linzer, D.I.H., 111

Magin, T. M., 177
Maness, P. F., 117
Marchuk, D., 161
Matsumura, F., 57
McClure, D. B., 17
Meyers, C. A., 1
Mittnacht, S., 177, 201
Miyake, R., 239
Murphy, D., 97

Nadakavukaren, K. K., 75
Nathans, D., 111
Nelson, W. G., 169
Nishizuka, Y., 239
Noguchi, P., 285

Obata, M., 277
O'Connell, K. A., 263
O'Connell, T. M., 217
Osborn, M., 191
Ozanne, B., 5

Pestka, S., 285

Quinlan, R. A., 177

Robbins, K. C., 35
Roberts, A. B., 1
Reffel, A. C., 51
Rehwoldt, S. M., 217
Rheinwald, J. G., 217
Rigby, P.W.J., 97
Rybak, S. M., 217

Sagata, N., 277
Schermer, A., 169
Schiller, D. L., 177
Schlom, J., 285
Schmid, E., 177, 201
Scott, M.R.D., 97
Scrace, G. T., 25
Sharkey, N. A., 245
Shepherd, E. L., 75
Shoyab, M., 253
Simanis, V., 97
Smith, J. A., 5
Sorge, L. K., 117
Sporn, M. B., 1
Stiles, C. D., 51
Stockwell, P.A., 25
Stroobant, P., 25
Summerhayes, I. C., 75
Sun, T.-T., 169

Takai, Y., 239
Tanaka, Y., 239
Taniguchi, S., 67
Tevethia, M. J., 271
Tevethia, S. S., 271
Topp, W. C., 17

von Bassewitz, D. B., 201

Wasteson, A., 25
Waterfield, M. D., 25
Weber, K., 153, 191
Weinstein, I. B., 229
Weiss, R. A., 169
Westermark, B., 25
Westphal, K.-H., 97
Whittle, N., 25
Willison, K., 97
Wu, Y.-J., 217

Yamashiro-Matsumura, S., 57

Zullo, J. N., 51

Subject Index

Actin, mutant type β
  characterization, 71–72
  in transformed cells, 68–73
Adenovirus type-5 E1A, effect on insulin expression, 108–110

Carcinogenesis
  cellular genes, 232–233
  chemical. See Tumor promoters
  LTR sequences, 233–234
  multistage, 229–235
Cell-surface tumor antigens
  interferon, 289–291
  mammary tumors, 286–287
  monoclonal antibodies, 285–286
  See also SV40
Centrosphere, 90–91
Cyclin, 128–133, 140–141
Cytokeratins. See Intermediate filaments
Cytoskeleton. See Actin; Centrosphere; Intermediate filaments; Mitochondria; Nucleus; Tropomyosin
Cytotoxic T-cell clone, 264–268

Desmosomes. See Intermediate filaments

Epidermal growth factor
  effects of A431 cells, 6, 148–151
  effects on phosphorylation of v-Ha-ras protein, 12–14
Epidermal growth factor receptor
  effects of anti-ras antibodies, 12
  in human tumors, 7–8
  nucleotide effects on EGF binding, 11–12
  in squamous cell carcinoma, 5–9
Embryonic development, 98

Friend leukemia virus, 277
  spleen focus forming virus (SFFV), 277–278
    gp55 glycoprotein, 278–281

Growth-associated mRNAs, 111
Growth factors. See Platelet-derived growth factor; Transforming growth factors

Human cellular oncogenes. See v-Ha-ras
Human epidermal basal cells
  immunofluorescence, 125
  polypeptides, 125–128
Human tumor antigens
  localization by gammascintigraphy, 294
  monoclonal antibodies, 293–294

Intermediate filaments
  astrocyte-specific, glial fibrillary acidic protein, 153–157
  biochemical structure, 153–158
  cytokeratins. See keratins
  cytology, 197–198
  desmin, 153–157
  epithelial (cyto)keratins, 153–157
  histologic typing, 192–197
  keratins, 201, 219–220
    biochemical structure, 157–158, 173
    classes, 163–165, 169–171
    divergence, 157–158
    expression in normal epithelia, 161–166, 169–172, 177–180
    expression in malignant epithelia, 161–166, 172
    evolutionary heterogeneity, 173–174
    functional significance, 165–166
    interaction with desmosomes, 183–187
    markers for tumor diagnosis, 174–175, 192-199
    messenger RNA, 163–166, 177
    microinjection, 209
    in mitosis, 205–209
    monoclonal antibodies, 191–192, 205
    neuronal neurofilaments, 153–157
    tissue distribution, 171–172
    vimentin, 153–157, 223–224
Insulin gene
  effects of adenovirus, 108–110
  effects of SV40 T antigen, 108–110

Major histocompatibility complex, 98
Mitochondria
  distribution with cytoskeleton, 76–80
  in tumor cells, 76
  See also rhodamine, 123
Mitosis, 201
Monensin
  effects on cells, 89–90
  nuclear movement, 91

Near-infrared microscopy, 88–89
Nucleus, rotational movement, 91–96

Onco-fetal antigens, 97

p53, 141–143
Phorbol esters. See Tumor promoters
Platelet-derived growth factor
  amino acid sequence, 27–28, 35–39
  purification, 26–27
  receptor
    protein kinase activity, 44–48
    purification, 46
    regulated genes, 51–53
  relationship to simian sarcoma virus p28^{sis}, 28–31, 38–40
Polycyclic aromatic hydrocarbons, 232
pp60$^{c\text{-}src}$,
  in chick embryos, 118–119
  in developing retina, 119–121
  in human embryos, 118–119
Protein kinase C, 239–244. See also Tumor promoters
Protein kinases. See pp60$^{c\text{-}src}$; Platelet-derived growth factor receptor; Protein kinase C; Tumor promoters

REF52 cells, 17–22
  growth factor requirements, 19–22
  SV40 transformants, 18–21
  transformants, 137–145
  tumorigenicity, 18–19
Repetitive element, 97
Rhodamine 123
  mitochondria, 76–80
  selective toxicity, 81–85

Simian sarcoma virus
  biological properties, 36
  genomic alalysis, 36–38
  See also Platelet-derived growth factor
SV40
  immunofluorescence, 264–265
  monoclonal antibodies, 263–264
  nuclear T antigen, 266–268
  in REF52 transformants, 141–143
  surface T antigen, 266–268, 274–275
  tumor-specific transplantation antigens, 263–269, 271–275

T antigen
  insulin expression, 108–110
  mutant, 107
  SV40 gene expression, 107–108
Transformation-sensitive proteins, 123–133
  in REF52 cells, 140–143
Transforming growth factors
  type α. See Epidermal growth factor
  type β,
    amino acid composition of bovine kidney, human placenta, human platelet, 3
    biochemical properties, 2–3
    biological properties, 3
    purification, 1–2
  See also Platelet-derived growth factor
Tropomyosin
  effects of transformation, 58–65, 140
  multiple forms, 57–65
Tumor-associated antigens. See Cell-surface tumor antigen; Human tumor antigens; SV40

301

Tumor promoters
  mechanism, 230–235
  phorbol esters, 230–235
    protein kinase C, 240–244
  phorbol ester receptor, 235–250
    phosphorylation, 257–260
    properties, 243–257
    purification, 253–254

Two-dimensional gel electrophoresis
  A431 cells, 147–151
  computer analysis, 138
  human epidermal cells, 123–133
  REF52 cells, 137–145

v-Ha-*ras* protein
  guanine nucleotide binding, 14
  phosphorylation, 12–14
  relationship to EGF receptor, 11–15
  See also EGF receptor

Volume eleven

Symposium on malignancies of the head and neck

Editors

ROBIN ANDERSON, M.D.

*Professor and Head, Department of Plastic Surgery,
Cleveland Clinic Educational Foundation,
Cleveland, Ohio*

JOHN E. HOOPES, M.D.

*Professor of Surgery, Division of Plastic Surgery,
The Johns Hopkins University
School of Medicine, Baltimore, Maryland*

Proceedings of the Symposium of the Educational
Foundation of the American Society of Plastic and
Reconstructive Surgeons, Inc., held at
Baltimore, Maryland, September 12-14, 1973.

With 249 illustrations

The C. V. Mosby Company

Saint Louis   1975

Volume eleven

Copyright © 1975 by The C. V. Mosby Company

All rights reserved. No part of this book may be reproduced
in any manner without written permission of the publisher.

Volume one copyrighted 1969, Volume two copyrighted 1969,
Volume three copyrighted 1971, Volume four copyrighted 1972,
Volume five copyrighted 1973, Volume six copyrighted 1973,
Volume seven copyrighted 1973, Volume eight copyrighted 1974,
Volume nine copyrighted 1974, Volume ten copyrighted 1974

Printed in the United States of America

Distributed in Great Britain by Henry Kimpton, London

Library of Congress Cataloging in Publication Data

Symposium on Malignancies of the Head and Neck, 3d,
    Baltimore, 1973.
    Symposium on Malignancies of the Head and Neck.

    (Proceedings of the symposium of the Education
Foundation of the American Society of Plastic and
Reconstructive Surgeons; v. 11)
    Includes index.
    1. Head—Cancer—Congresses. 2. Neck—Cancer—
Congresses. I. Anderson, Robin, 1919-   ed.
II. Hoopes, John E., ed. III. Series: American
Society of Plastic and Reconstructive Surgeons. Ed-
ucational Foundation. Proceedings of the symposium;
v. 11. [DNLM: 1. Head—Congresses. 2. Neck neo-
plasms—Congresses. 3. Neoplasms—Congresses. W3
SY506C 1973s/WE705 S989 1973s]
RC280.H4S93   1973   616.9′94′91   74-22108
ISBN 0-8016-0183-5

TS/NK/B 9 8 7 6 5 4 3 2 1